TECHNIQUES OF CHEMISTRY

ARNOLD WEISSBERGER, *Editor*

VOLUME I

PHYSICAL METHODS OF CHEMISTRY

PART IIIC
Polarimetry

TECHNIQUES OF CHEMISTRY

ARNOLD WEISSBERGER, *Editor*

TECHNIQUES OF CHEMISTRY

VOLUME I

PHYSICAL METHODS
OF CHEMISTRY

INCORPORATING FOURTH COMPLETELY REVISED AND AUGMENTED
EDITION OF TECHNIQUE OF ORGANIC CHEMISTRY,
VOLUME I, PHYSICAL METHODS OF ORGANIC CHEMISTRY

Edited by
ARNOLD WEISSBERGER
AND
BRYANT W. ROSSITER

Research Laboratories
Eastman Kodak Company
Rochester, New York

PART III
Optical, Spectroscopic, and Radioactivity Methods

PART IIIC
Polarimetry

WILEY-INTERSCIENCE

A DIVISION OF JOHN WILEY & SONS, INC.

New York · London · Sydney · Toronto

Library of Congress Catalog Card Number: 49   48584

ISBN 0   471   92732   5

Printed in the United States of America.

10 9 8 7 6 5 4 3 2 1

PLAN FOR

PHYSICAL METHODS OF CHEMISTRY

PART I

Components of Scientific Instruments, Automatic Recording and Control, Computers in Chemical Research

PART II

Electrochemical Methods

PART III

Optical, Spectroscopic, and Radioactivity Methods

PART IV

Determination of Mass, Transport, and Electrical-Magnetic Properties

PART V

Determination of Thermodynamic and Surface Properties

AUTHORS OF PART IIIC

NICHOLAS M. BASHARA
Department of Electrical Engineering, University of Nebraska, Lincoln, Nebraska

BRUCE BUCKMAN
Department of Electrical Engineering, University of Nebraska, Lincoln, Nebraska

DENNIS J. CALDWELL
Department of Chemistry, University of Utah, Salt Lake City, Utah

PIERRE CRABBÉ
Director of Chemical Research, Research Laboratories Syntex, S.A., Mexico City, Mexico

HENRY G. CURMÈ
Research Laboratories, Eastman Kodak Company, Rochester, New York

HENRY EYRING
Department of Chemistry, University of Utah, Salt Lake City, Utah

ARTHUR C. HALL
Sacony Mobil Oil Company, Dallas, Texas

v

WILFRIED HELLER

Department of Chemistry, Wayne State University, Detroit, Michigan

CATHERINE G. LE FEVRÉ

School of Chemistry, University of Sydney, Sydney, New South Wales, Australia

R. J. W. LE FEVRÉ

School of Chemistry, University of Sydney, Sydney, New South Wales, Australia

PETR MUNK

Department of Chemistry, University of Texas, Austin, Texas

ARTHUR C. PARKER

Glaxco Research Ltd., Greenford, Middlesex, England

ANTON PETERLIN

Director of Research, Research Triangle Institute, Research Triangle Park, Durham, North Carolina

JAMES M. THORNE

Department of Chemistry, Brigham Young University, Provo, Utah

NEW BOOKS AND NEW EDITIONS OF BOOKS OF THE TECHNIQUE OF ORGANIC CHEMISTRY WILL NOW APPEAR IN TECHNIQUES OF CHEMISTRY. A LIST OF PRESENTLY PUBLISHED VOLUMES IS GIVEN BELOW.

TECHNIQUE OF ORGANIC CHEMISTRY
ARNOLD WEISSBERGER, *Editor*

INTRODUCTION TO THE SERIES

Techniques of Chemistry is the successor to the Technique of Organic Chemistry Series and its companion—Technique of Inorganic Chemistry. Because many of the methods are employed in all branches of chemical science, the division into techniques for organic and inorganic chemistry has become increasingly artificial. Accordingly, the new series reflects the wider application of techniques, and the component volumes for the most part provide complete treatments of the methods covered. Volumes in which limited areas of application are discussed can be easily recognized by their titles.

Like its predecessors, the series is devoted to a comprehensive presentation of the respective techniques. The authors give the theoretical background for an understanding of the various methods and operations and describe the techniques and tools, their modifications, their merits and limitations, and their handling. It is hoped that the series will contribute to a better understanding and a more rational and effective application of the respective techniques.

Authors and editors hope that readers will find the volumes in this series useful and will communicate to them any criticisms and suggestions for improvements.

Research Laboratories   ARNOLD WEISSBERGER
Eastman Kodak Company
Rochester, New York

PREFACE

Physical Methods of Chemistry succeeds, and incorporates the material of, three editions of *Physical Methods of Organic Chemistry* (1945, 1949, and 1959). It has been broadened in scope to include physical methods important in the study of all varieties of chemical compounds. Accordingly, it is published as Volume I of the new Techniques of Chemistry series.

Some of the methods described in *Physical Methods of Chemistry* are relatively simple laboratory procedures, such as weighing and the measurement of temperature or refractive index and determination of melting and boiling points. Other techniques require very sophisticated apparatus and specialists to make the measurements and to interpret the data; x-ray diffraction, mass spectrometry, and nuclear magnetic resonance are examples of this class. Authors of chapters describing the first class of methods aim to provide all information that is necessary for the successful handling of the respective techniques. Alternatively, the aim of authors treating the more sophisticated methods is to provide the reader with a clear understanding of the basic theory and apparatus involved, together with an appreciation for the value, potential, and limitations of the respective techniques. Representative applications are included to illustrate these points, and liberal references to monographs and other scientific literature providing greater detail are given for readers who want to apply the techniques. Still other methods that are successfully used to solve chemical problems range between these examples in complexity and sophistication and are treated accordingly. All chapters are written by specialists. In many cases authors have acquired a profound knowledge of the respective methods by their own pioneering work in the use of these techniques.

In the earlier editions of *Physical Methods* an attempt was made to arrange the chapters in a logical sequence. In order to make the organization of the treatise lucid and helpful to the reader, a further step has been taken in the new edition—the treatise has been subdivided into technical families and parts:

Part I   Components of Scientific Instruments, Automatic Recording and Control, Computers in Chemical Research
Part II   Electrochemical Methods
Part III Optical, Spectroscopic, and Radioactivity Methods

Part IV   Determination of Mass, Transport, and Electrical-Magnetic Properties

Part V   Determination of Thermodynamic and Surface Properties

This organization into technical families provides more consistent volumes and should make it easier for the reader to obtain from a library or purchase at minimum cost those parts of the treatise in which he is most interested.

The more systematic organization has caused additional labors for the editors and the publishers. We hope that it is worth the effort. We thank the many authors who made it possible by adhering closely to the agreed dates of delivery of their manuscripts and who promptly returned their proofs. To those authors who were meticulous in meeting deadlines we offer our apologies for delays caused by late arrival of other manuscripts, in some cases necessitating rewriting and additions.

The changes in subject matter from the Third Edition are too numerous to list in detail. We thank previous authors for their continuing cooperation and welcome new authors to the Series. New authors of Part IIIC are Dr. Nicholas M. Bashara, Dr. Bruce Buckman, Dr. Dennis J. Caldwell, Dr. Pierre Crabbé, Dr. Henry G. Curmè, Dr. Henry Eyring, Dr. Arthur C. Hall, Dr. Petr Munk, Dr. Anton Peterlin, and Dr. James M. Thorne.

We are grateful to the many colleagues who advised us in the selection of authors and helped in the evaluation of the manuscripts. These are, for Part IIIC, Dr. Henry G. Curmè, Dr. Charles F. Farran, Dr. Richard T. Klingbiel, Mrs. Ardelle Kocher, Mr. Arthur C. Parker, Mrs. Donna S. Roets, and Dr. Melvin D. Sterman.

The senior editor expresses his gratitude to Bryant W. Rossiter for joining him in the work and taking on the very heavy burden with exceptional devotion and ability.

ARNOLD WEISSBERGER
BRYANT W. ROSSITER

September 1971
Research Laboratories
Eastman Kodak Company
Rochester, New York

CONTENTS

Chapter **1**

THEORY OF OPTICAL ROTATION

Dennis J. Caldwell and Henry Eyring

Part 1 NATURAL OPTICAL ACTIVITY

I. INTRODUCTION

The interaction of light with matter is the basis of the most subtle methods for exploring molecular structure. The information obtained is classed under two broad categories, dispersion and absorption. The first measures the rate of phase propagation in the medium as a function of frequency and the second the amount of energy absorbed by the medium.

Only one independent parameter is being measured by the two experiments. The absorption is completely determined by the dispersion and vice versa; however, it must be noted that this is a purely mathematical observation that

1

will be tempered by the relative sensitivity of the instrumentation and the availability of the necessary regions of the spectrum.

The most precise characterization of a light wave is be means of its polarization. We may expect that the most detailed information on molecular structure is to be obtained from experiments with polarized light, while average, less sensitive properties will be determined from investigations with random polarization.

All material media exhibit the phenomena of dispersion and absorption. Over the years it has been possible to separate the absorption spectra of many compounds into regions controlled by individual groups. In the area of dispersion molar refractivities are found to be nearly additive. This phase of spectroscopy has become a means for identifying the existence of certain groups in a molecule along with any strong interactions among them.

Dissymmetric molecules, which have no improper rotations among their symmetry operations, will exhibit the phenomena of optical rotatory dispersion and circular dichroism. The properties measured are the differences between the indices of refraction and between extinction coefficients for left and right circularly polarized light. Ordinary dispersion and absorption are measured by the average values of these quantities.

It should be borne in mind that despite the traditional emphasis on configuration and asymmetric centers, dissymmetry is more a property of molecular conformation. The difference between the environment of a group in an optically active molecule and that of a group in an inactive one often involves very small energies. Normally ordinary dispersion and absorption are insensitive to conformation changes. On the other hand, rotatory dispersion and circular dichroism depend on molecular dissymmetry.

Any optically active molecule will have conformations of zero rotation. If the conformation is measured by a set of coordinates, Q_i, and $E(Q_i)$ is the associated energy, the observed parameter of optical activity is

$$\bar{A} = \int dQ_1 \cdots \int dQ_n A(Q_1 \cdots Q_n) e^{-E(Q_i)/kT}. \tag{1.1}$$

This gets us used to the idea that conformation is essentially a temperature-dependent property as is also optical activity. In some cases molecules are frozen in nearly rigid conformations with high-energy barriers. Then (Q_i) is sharply peaked at some particular set of Q_i values and the integration may be dropped. In (1.1) we must include only that set of conformations which may be continuously deformed into one another without crossing any barriers. Otherwise, a racemic mixture would always be described.

Every molecule with more than three atoms will have dissymmetric conformations or configurations. Normally only those which form stable mirror image molecules are considered optically active. For example, two of the three staggered forms of *n*-butane (Fig. 1.1) are dissymmetric and are mirror images. The third has both a plane of symmetry and a center of inversion. The molecule is not considered optically active because the barrier separating the first two forms is too small at temperatures above the melting point. In fact this may be taken as a rough criterion for an optically active molecule: that there exists at least one conformation separated from its mirror image by a barrier with $E \gg kT_{\text{fusion}}$.

In earlier days it would have been customary to explain the fact by saying that *n*-butane does not have an asymmetric center. An asymmetric center has been considered important because it means that all so-called rotamers of the molecule would be dissymmetric. There are, of course, many optically active molecules without asymmetric centers such as the α, α' substituted biphenyls (Fig. 1.2). Here a rotational barrier exists similar to the ethane barrier except that it is sufficiently greater than the thermal energy factor kT that the criterion for optical activity is satisfied.

There is evidence for believing that in the series of optically active hydrocarbons the rotations would be an order of magnitude lower than observed if all rotamers were equally probable. It is apparent that optical activity has the potential for being a very sensitive probe into the conformation of molecules. Unfortunately, only the surface has been scratched in the art of interpreting dispersion and absorption spectra.

In any discussion of optical rotatory power it must be remembered that the optical rotation for a given molecule is the composite effect of many conformations. This tool will be most useful when one conformation is believed to predominate greatly over the others. At present there is still considerable difficulty in organizing and interpreting data on molecules with reasonably well-known conformations.

2 ELECTROMAGNETIC THEORY

The theoretical discussion will proceed along two lines: a qualitative presentation of the mechanisms, along with an outline of the quantum mechanical derivations and resulting formulas.

If precise tractable methods for calculating wave functions of complex molecules existed, spectroscopy would largely reduce to an analytical tool for identifying new compounds. As it is, many techniques are geared for deciding the positions of the nuclei. This has been the primary goal in studying optical activity, particularly from the standpoint of the biochemist.

If theoretical and empirical rules for calculating optical activity as a function of arbitrary conformation could be developed, the way would be open for using this as a universal tool in the determination of conformation. From the theoretical standpoint, spectroscopy is used as a guide in finding approximate eigenstates of the Hamiltonian.

In optical activity one is faced with the situation where most investigators are satisfied with the electronic information furnished by ordinary absorption and delve into further refinements of the wave function only as much as is necessary to construct a theory of conformations.

An electromagnetic wave is not a particularly easy phenomenon to characterize. It is, of course, a periodic disturbance with a sinusoidal dependence on space and time:

$$F = F_o \sin 2\pi(vt - x/\lambda), \tag{1.2}$$

where F is a measure of the intensity of the disturbance at x at time t. A velocity is seen to be associated with the wave, since the points of constant phase, the nodes, for example, satisfy the equation, $x/t = c = v\lambda$.

The constant c is the phase velocity of the wave. The simplest of all situations is the monochromatic wave in a vacuum. Here c is a universal constant, the so-called speed of light. In a region pervaded by matter three distinctions must be made for a monochromatic wave. The wavefront velocity is defined as the rate at which the first disturbance is propagated through the medium. It is always equal to c, the vacuum constant. The signal velocity is the rate at which the bulk of the energy is transferred to undisturbed parts of the medium. This velocity is difficult to define precisely, but it is always equal to or less than c. Once a steady state has been achieved throughout the medium, the parameter of interest is the phase velocity described above. Since all parts of the medium are under the influence of the disturbance in the steady state, the propagation of a mathematically defined quantity known as the phase, at velocities greater than c, in no way conflicts with the ideas of relativity.

In optical rotation we are concerned with the phase velocity. Two vector quantities, $\mathbf{E}$ and $\mathbf{B}$, are associated with any electromagnetic disturbance. The constitutive equations of electromagnetic theory reduce their total of six components to four, and in an electromagnetic wave the number is reduced effectively to three. The electric vector $\mathbf{E}$ determines the accelerations of charge at rest in the observer's system, and the magnetic induction vector $\mathbf{B}$ introduces a component of force perpendicular to the instantaneous velocity for a moving charge.

The principles of electromagnetic theory are best discussed in terms of a four-dimensional space-time continuum. It is possible to construct a system of charges with arbitrary values of $\mathbf{B}$ and $\mathbf{E}$ at a given point at some time but their values over all space cannot be specified independently. This fact is implicit in the Maxwell equations:

$$\mathbf{V} \times \mathbf{E} = \frac{-1}{c} \frac{\partial \mathbf{B}}{\partial t} \tag{1.3a}$$

$$\mathbf{V} \times \mathbf{B} = \frac{1}{c} \frac{\partial \mathbf{E}}{\partial t}. \tag{1.3b}$$

If we have detailed knowledge of the charge and current distribution for a given problem, the computed values of **E** and **B** will agree entirely with (1.3a,b).

In material media it is convenient to take account of polarization effects separately. Two auxiliary vectors are defined as **M**, the magnetic moment per unit volume, and **P**, the electric dipole moment per unit volume. In cgs units these vectors have the same dimensions as **B** and **E**. The properties of the medium are manifested in the quantities, **M** and **P**. In general they will be functions of **B** and **E**.

There are several points to remember about a material medium. The force on a macroscopic charge embedded therein is equal to $q\mathbf{E}$, since **E** is defined as the space average over a region large compared with molecular dimensions. The magnetic component of the force on a moving charge is not so easy to characterize and it is preferable to define a new vector, $\mathbf{H} = \mathbf{B} - 4\pi\mathbf{M}$. In a vacuum, (1.3a,b) imply that $\mathbf{V}\cdot\mathbf{E}$ and $\mathbf{V}\cdot\mathbf{B}$ are constants, which may be set equal to zero. A simple argument shows that in the presence of a charge distribution $\mathbf{V}\cdot\mathbf{E}$ is no longer zero, but rather $\mathbf{V}\cdot\mathbf{D} = 4\pi\rho$, where $\mathbf{D} = \mathbf{E} + 4\pi\mathbf{P}$ and ρ is the charge density.

A similar argument could be applied to the magnetic field if a magnetic charge existed. The difference between **B** and **H** may be resolved in the following way: If a test charge is completely free to move through all parts of a small but still macroscopic element of the medium including the electron orbits, it will experience a force derived solely from currents governed by the vector **B** according to the general equation

$$\mathbf{V} \times \mathbf{B} = \frac{4\pi}{c}\,\mathbf{J}_{\text{total}},\tag{1.4}$$

where $\mathbf{J}_{\text{total}}$ is the sum of all types of currents in the medium, including migrating charge (the conventional current), the polarization current (moving bound charge), the atomic magnetization currents, and Maxwell's so-called convection current, $(1/\pi)/(\partial\mathbf{E}/\partial t)$.

It is this last current which made the early theory of radiation possible. Equation (1.4) may also be written

$$\oint \mathbf{B} \cdot d\mathbf{s} = \frac{4\pi}{c} \int_{\mathbf{S}} \mathbf{J}_{\text{total}} \cdot d\mathbf{S},\tag{1.4a}$$

stating that the line integral of **B** over an arbitrary path in the medium is proportional to the total current flux through any surface bounded by the closed line. When the total current is resolved into its components, (1.4) becomes

$$\mathbf{V} \times \mathbf{B} = \frac{1}{c}\frac{\partial\mathbf{E}}{\partial t} + \frac{4\pi}{c}\frac{\partial\mathbf{P}}{\partial t} + \frac{4\pi}{c}(c\mathbf{V} \times \mathbf{M}) + \frac{4\pi}{c}\,\mathbf{J},\tag{1.5}$$

where $\mathbf{J}$ is now the conventional macroscopic current.

It can be shown that a test charge free to move in all regions of the medium without restriction will experience an electric force $\mathbf{E}$ and a magnetic force $\mathbf{B}$. If certain regions are to be excluded in a systematic way, other results follow. For example, it is usually assumed that the regions occupied by molecules are impenetrable. In particular, molecules may be regarded as electric and magnetic dipoles. The spatially averaged fields subject to this restriction may then be shown to be $\mathbf{D} = \mathbf{E} + 4\pi\mathbf{P}$ and $\mathbf{H} = \mathbf{B} - 4\pi\mathbf{M}$ with

$$\mathbf{V} \cdot \mathbf{D} = 4\pi\rho \tag{1.6a}$$

$$\mathbf{V} \cdot \mathbf{B} = 0 \tag{1.6b}$$

These equations supplement (1.3a,b) and are assumed to hold under all conditions. The differences arise in the interpretation of the four vectors, $\mathbf{E}$, $\mathbf{B}$, $\mathbf{D}$, and $\mathbf{H}$.

The customary exposition of Maxwell's equations is

$$\mathbf{V} \times \mathbf{E} = -\frac{1}{c}\frac{\partial \mathbf{B}}{\partial t} \tag{1.7a}$$

$$\mathbf{V} \times \mathbf{H} = \frac{1}{c}\frac{\partial \mathbf{D}}{\partial t} + \frac{4\pi}{c}\mathbf{J} \tag{1.7b}$$

$$\mathbf{V} \cdot \mathbf{D} = 4\pi\rho \tag{1.7c}$$

$$\mathbf{V} \cdot \mathbf{B} = 0 \tag{1.7d}$$

In order to solve these equations to determine the types of waves propagated in materials, constitutive relations connecting $\mathbf{B}$ and $\mathbf{H}$ and $\mathbf{E}$ and $\mathbf{D}$ are required. These will in turn depend on the geometry and composition of the medium.

If both $\mathbf{P}$ and $\mathbf{M}$ are zero, we are dealing in effect with a distribution of charge ρ and a current $\mathbf{J}$. The vectors $\mathbf{B}$ and $\mathbf{E}$ may be found by a special type of integration over the charge and current densities. The solution to (1.7a–d) with $\mathbf{E} = \mathbf{D}$ and $\mathbf{H} = \mathbf{B}$ may be shown to be

$$\mathbf{B} = \mathbf{V} \times \mathbf{A} \tag{1.8a}$$

$$\mathbf{E} = -\mathbf{V}\phi - \frac{1}{c}\frac{\partial \mathbf{A}}{\partial t} \tag{1.8b}$$

$$\mathbf{A} = \int \frac{\mathbf{J}(\text{ret})}{r}\,d\tau \tag{1.8c}$$

$$\phi = \int \frac{\rho(\text{ret})}{r}\,d\tau \tag{1.8d}$$

where r is the distance from the variable volume element of integration to the field point and $\mathbf{J}(\text{ret})$ and $\rho(\text{ret})$ are the retarded potentials to be evaluated at the time $t - r/c$. It is worth noting that $\mathbf{J}$ and ρ are not entirely independent since the equation of continuity must be satisfied:

$$\mathbf{V} \cdot \mathbf{J} + \frac{\partial \rho}{\partial t} = 0. \tag{1.9}$$

If $\mathbf{J} = 0$, then ρ must be independent of time and ϕ becomes the ordinary scalar potential of electrostatics, the sum of the potentials furnished by each volume element of stationary charges. Unfortunately, $(1.8a–d)$ do not provide a practical means to solve many electromagnetic problems, since the final charge and current densities are quantities to be determined in the course of the solution.

The problem is neatly reformulated in material media by absorbing the various currents involved into the vectors $\mathbf{H}$ and $\mathbf{D}$. In the problems of interest there are no macroscopic currents ($\mathbf{J} = 0$) and no time-dependent charge distributions. The electrostatic part of the field, $-\mathbf{V}\phi$, may be discarded in the treatment of the radiation problem; the equations to be solved are

$$\mathbf{V} \times \mathbf{E} = -\frac{1}{c} \frac{\partial \mathbf{B}}{\partial t} \tag{1.10a}$$

$$\mathbf{V} \times \mathbf{H} = \frac{1}{c} \frac{\partial \mathbf{D}}{\partial t}. \tag{1.10b}$$

The relations $\mathbf{V} \cdot \mathbf{D} = 0$, $\mathbf{V} \cdot \mathbf{B} = 0$ are effectively implied in these equations.
Equation $(1.10a)$ is satisfied by the relations

$$\mathbf{B} = \mathbf{V} \times \mathbf{A} \tag{1.11a}$$

$$\mathbf{B} = -\frac{1}{c} \frac{\partial \mathbf{A}}{\partial t}. \tag{1.11b}$$

In a vacuum where $\mathbf{H} = \mathbf{B}$ and $\mathbf{E} = \mathbf{D}$ the equation, $\mathbf{V} \times \mathbf{B} = (1/c)/(\partial \mathbf{E}/\partial t)$, would lead to $\mathbf{V} \times (\mathbf{V} \times \mathbf{A}) + (1/c^2)/(\partial^2 \mathbf{A}/\partial t^2) = 0$ and the result

$$\mathbf{V}^2 \mathbf{A} = \frac{1}{c^2} \frac{\partial^2 \mathbf{A}}{\partial t^2}, \tag{1.12}$$

where use has been made of the vector relation, $\mathbf{V} \times (\mathbf{V} \times \mathbf{A}) = \mathbf{V}(\mathbf{V} \cdot \mathbf{A}) - \mathbf{V}^2 \mathbf{A}$, and the fact that $\mathbf{V} \cdot \mathbf{A}$ may be set equal to zero in most radiation problems with no loss of generality.

One form of general solution to (1.12) is $\mathbf{A} = f\{\omega[t - (\mathbf{k} \cdot \mathbf{r}/c)]\}$, where f is any function, ω an arbitrary constant, and $\mathbf{k}$ a unit vector. In particular, if $\mathbf{A} = \mathbf{A}_o e^{i\omega(t - \mathbf{k} \cdot \mathbf{r}/c)}$, then

$$\mathbf{B} = \nabla \times \mathbf{A} = -\frac{i\omega}{c} \mathbf{k} \times \mathbf{A} \qquad (1.13a)$$

$$\mathbf{E} = -\frac{1}{c}\frac{\partial \mathbf{A}}{\partial t} = -\frac{i\omega}{c}\mathbf{A}. \qquad (1.13b)$$

In this solution, phase is propagated along the direction $\mathbf{k}$ with a velocity c and the vectors, $\mathbf{B}$ and $\mathbf{E}$, are transverse to the direction of travel and vary sinusoidally in phase with one another.

This particular choice of vector potential $\mathbf{A}$ represents a plane polarized wave with the plane of electric polarization determined by the vectors $\mathbf{k}$ and $\mathbf{A}$ and the plane of magnetic polarization by the vectors $\mathbf{k}$ and $\mathbf{k} \times \mathbf{A}$. A third plane, sometimes referred to as a plane of polarization, is the one perpendicular to $\mathbf{k}$ containing the vectors, $\mathbf{E}$ and $\mathbf{B}$. The instantaneous values of these vectors are constant throughout this plane.

In the foregoing we have tacitly assumed that $\mathbf{A}_o$ is real; so that by taking the real parts of $\mathbf{B}$ and $\mathbf{E}$ we obtain

$$\mathbf{B} = \frac{\omega}{c} \mathbf{k} \times \mathbf{A}_o \sin\left[\omega\left(t - \frac{\mathbf{k} \cdot \mathbf{r}}{c}\right)\right],$$

$$\mathbf{E} = \frac{\omega}{c} \mathbf{A}_o \sin\left[\omega\left(t - \frac{\mathbf{k} \cdot \mathbf{r}}{c}\right)\right].$$

If $\mathbf{A}_o = A_o(\mathbf{i} \pm i\mathbf{j})$ the real part of $\mathbf{A}$ is

$$\text{Re } \mathbf{A} = A_o[\mathbf{i} \cos \Delta \mp \mathbf{j} \sin \Delta], \qquad (1.14)$$

where $\Delta = \omega[t - (\mathbf{k} \cdot \mathbf{r}/c)]$. The vector $\mathbf{A}$ is seen to be constant in magnitude and variable in direction. If the vectors $\mathbf{i}$, $\mathbf{j}$, and $\mathbf{k}$ form a right-handed coordinate system, the upper sign will describe a clockwise motion of the vector $\mathbf{A}$, and the lower a counterclockwise motion, when the oncoming wave is viewed along $\mathbf{k}$. These two forms of polarization are conventionally called right and left circular polarization.

If beams of right and left circularly polarized light are superimposed, the result is a plane polarized wave, as can be seen by adding both components of (1.14).

The most general solution to (1.12) may be written in the form

$$\mathbf{A} = \iiint \mathbf{g}(\mathbf{k})e^{i(\mathbf{k} \cdot \mathbf{r} - kct)}\mathbf{k}^2 d\mathbf{k}\, d\Omega,$$

where $\mathbf{k}$ is the wave vector equal in magnitude to the reciprocal of the wavelength. The function $\mathbf{g(k)}$ may be obtained in terms of the initial condition $\mathbf{A}(O)$ by a Fourier inversion. In this form we have a method for completely specifying a radiation field in terms of its behavior at some arbitrary time.

In general the behavior of the electromagnetic field for all times may be determined by specifying the charge and current densities, ρ, $\mathbf{J}$, as functions of space and time or by specifying the scalar and vector potentials, ϕ, $\mathbf{A}$, as functions of space alone at some particular time subject to the gauge condition, $\mathbf{V} \cdot \mathbf{A} + (1/c)(\partial\phi/\partial t) = 0$.

Once again this is not a universally useful solution, since a knowledge of $\mathbf{A}$ even at one particular instant is not generally available. The difficulty in electromagnetic problems is based on insufficient information. It is not possible to describe in detail the individual electronic motion at the molecular level. Instead we are forced to define macroscopic quantities such as $\mathbf{P}$ and $\mathbf{M}$, which summarize the net observable effects of this motion. Equations (1.10a,b) will be solvable only if constitutive relations connecting $\mathbf{H}$ with $\mathbf{B}$ and $\mathbf{D}$ with $\mathbf{E}$ can be found. If it is possible to write

$$\mathbf{H} = \mathbf{H}(\mathbf{B}, \dot{\mathbf{B}}, \mathbf{E}, \dot{\mathbf{E}}),$$

$$\mathbf{D} = \mathbf{D}(\mathbf{B}, \dot{\mathbf{B}}, \mathbf{E}, \dot{\mathbf{E}}),$$

then in view of (1.11a,b) a single equation in $\mathbf{A}$ will result, which may be solved by conventional analytical methods.

For example, in an isotropic medium the induced components of both $\mathbf{M}$ and $\mathbf{P}$ are assumed to be proportional to $\mathbf{B}$ and $\mathbf{E}$. These relations are conventionally written

$$\mathbf{D} = \epsilon\mathbf{E} \tag{1.15a}$$

$$\mathbf{H} = \frac{1}{\mu}\mathbf{B} \tag{1.15b}$$

This leads to the equation

$$\mathbf{V}^2\mathbf{A} = \frac{\mu\epsilon}{c^2}\frac{\partial^2\mathbf{A}}{\partial t^2}. \tag{1.16}$$

The solution proceeds in exactly the same way as in (1.12) with a new phase velocity given by

$$c' = \frac{c}{\sqrt{\mu\epsilon}} \tag{1.17}$$

For diamagnetic materials μ is nearly equal to 1. When the natural frequency ν_o of the oscillating electric dipoles is greater than the driving frequency ν, $\mathbf{P}$ will be in phase with $\mathbf{E}$ and $|\mathbf{D}| = |\mathbf{E} + 4\pi\mathbf{P}| > |\mathbf{E}|$; hence, $\epsilon > 1$

and $c' < c$. When $v_o < v$, the reverse is true. In the neighborhood of an absorption band the ratio $(c'/c) = n$ will first reach a minimum less than 1, become unity near the center of the band, and finally reach a maximum greater than unity on the short-wavelength side. This is the familiar phenomenon of ordinary dispersion. In anisotropic media (1.15a,b) are replaced by tensor relations and more complex phenomena result.

In the ensuing discussions of optical activity we shall confine ourselves to isotropic media. We shall assume that the individual atomic electric and magnetic dipoles will be proportional to the instantaneous values of the effective electromagnetic field. One should remember that $\mathbf{E}$ and $\mathbf{B}$ are not independent and are in effect specified by a single vector function $\mathbf{A}$ for a radiation field. Before dealing with the atomic level we should remember that the vectors $\mathbf{E}$ and $\mathbf{B}$ are the macroscopic average values over all regions of space including those densely occupied by atomic charges. The equation, $\nabla \times \mathbf{E} = (-1/c)(\partial \mathbf{B}/\partial t)$, is satisfied in the presence of matter as well as in a vacuum. The equation, $\nabla \times \mathbf{B} = (1/c)(\partial \mathbf{E}/\partial t)$ is satisfied only in a vacuum. In a material medium it must be replaced by $\nabla \times \mathbf{H} = (1/c)(\partial \mathbf{D}/\partial t)$.

We shall suppose that each molecule in a solution or a gas is subject to an effective field described by the vector potential $\mathbf{A}_{\text{eff}}$. The effective vectors in the force equation governing the motion of a charge q will be

$$\mathbf{F} = q\left[\mathbf{E}_{\text{eff}} + \frac{1}{c}(\mathbf{v} \times \mathbf{B}_{\text{eff}})\right] \tag{1.18a}$$

$$\mathbf{B}_{\text{eff}} = \nabla \times \mathbf{A}_{\text{eff}} \tag{1.18b}$$

$$\mathbf{B}_{\text{eff}} = -\frac{1}{c}\frac{\partial \mathbf{A}_{\text{eff}}}{\partial t}. \tag{1.18c}$$

In an isotropic medium the simplest assumption that can be made about $\mathbf{A}_{\text{eff}}$ is that it is proportional to the macroscopic average value $\mathbf{A}$: $\mathbf{A}_{\text{eff}} = S\mathbf{A}$. If quasi-crystalline structure and specific molecular geometry are ignored, the Lorentz argument gives

$$S_L = \frac{1}{1 - (4\pi N\alpha/3)}. \tag{1.19}$$

This essentially corrects for the effect of a spherical cavity in which the molecule is presumed to be located. From the elementary theory of electrostatics it is evident that the effective field inside a cavity in a dielectric depends on both the geometry of the cavity and its orientation with respect to the applied field. The value given by (1.19) is seen to be at best an estimate of a quantity difficult to obtain in a solution. In crystals the estimate is only good for lattices of the highest symmetry, such as the simple cubic.

The Hamiltonian for the electromagnetic forces described by (1.18a,b,c) is

$$H = \frac{1}{2m}\left[\mathbf{p} - \frac{e\mathbf{A}_{\text{eff}}}{c}\right]^2 + eV \tag{1.20}$$

where V is the potential of all the conservative forces acting on the electron, that is, the forces of the nuclei and other electrons.

3 THE GENERAL EQUATIONS OF OPTICAL ACTIVITY

Space does not permit a detailed derivation of all the quantum mechanical phases of radiation theory. We shall content ourselves with a summary of the standard derivations and concentrate in more detail on those aspects peculiar to optical activity. The results of time-dependent perturbation theory on the above Hamiltonian gives for the induced electric and magnetic dipole moments

$$\boldsymbol{\mu}(\Omega) = \frac{2Re}{h}\left\{\sum_a \frac{v_{oa}}{(v_{oa}^2 - v^2)}\,\boldsymbol{\mu}_{oa}\boldsymbol{\mu}_{ao} \cdot \left(-\frac{1}{c}\frac{\partial \mathbf{A}_{\text{eff}}}{\partial t}\right)\right.$$
$$+ \sum_a \frac{i}{2\pi(v_{oa}^2 - v^2)}\,\boldsymbol{\mu}_{oa}\boldsymbol{\mu}_{ao} \cdot \left(-\frac{1}{c}\frac{\partial^2 \mathbf{A}_{\text{eff}}}{\partial t^2}\right)$$
$$+ \sum_a \frac{v_{oa}}{(v_{oa}^2 - v^2)}\,\boldsymbol{\mu}_{oa}\mathbf{m}_{ao} \cdot (\nabla \times \mathbf{A}_{\text{eff}})$$
$$\left. + \sum_a \frac{i}{2\pi(v_{oa}^2 - v^2)}\,\boldsymbol{\mu}_{oa}\mathbf{m}_{ao} \cdot \left(\nabla \times \frac{\partial \mathbf{A}_{\text{eff}}}{\partial t}\right)\right\} \tag{1.21a}$$

$$\mathbf{m}(\Omega) = \frac{2Re}{h}\left\{\sum_a \frac{v_{oa}}{(v_{oa}^2 - v^2)}\,\mathbf{m}_{oa}\boldsymbol{\mu}_{ao} \cdot \left(-\frac{1}{c}\frac{\partial \mathbf{A}_{\text{eff}}}{\partial t}\right)\right.$$
$$+ \sum_a \frac{i}{2\pi(v_{oa}^2 - v^2)}\,\mathbf{m}_{oa}\boldsymbol{\mu}_{ao} \cdot \left(-\frac{1}{c}\frac{\partial^2 \mathbf{A}_{\text{eff}}}{\partial t^2}\right)$$
$$+ \sum_a \frac{v_{oa}}{(v_{oa}^2 - v^2)}\,\mathbf{m}_{oa}\mathbf{m}_{ao} \cdot (\nabla \times \mathbf{A}_{\text{eff}})$$
$$\left. + \sum_a \frac{i}{2\pi(v_{oa}^2 - v^2)}\,\mathbf{m}_{oa}\mathbf{m}_{ao} \cdot \left(\nabla \times \frac{\partial \mathbf{A}_{\text{eff}}}{\partial t}\right)\right\}. \tag{1.21b}$$

The induced moments, $\boldsymbol{\mu}(\Omega)$ and $\mathbf{m}(\Omega)$, are functions of the orientation of the molecule of the electromagnetic field. In the absence of strong fields the wave functions in the molecule are independent of orientation. An averaging of (1.21a,b) may be performed over all orientations of the vectors. This will entail the averaging of such quantities as $\mu_{x'oa}m_{x'ao}$, $\mu_{x'oa}m_{y'ao}$. We obtain

$\langle \mu_{x'oa} m_{x'ao} \rangle = \frac{1}{3}\mu_{xoa}\mu_{xao}$, $\langle \mu_{x'oa} m_{y'ao} \rangle = 0$, where the primed letters refer to the axes moving with the molecule and the unprimed ones to the fixed frame of reference. The average values of the moments are

$$\langle \boldsymbol{\mu} \rangle = \alpha\left(-\frac{1}{c}\frac{\partial \mathbf{A}_{\text{eff}}}{\partial_t}\right) + \gamma(\mathbf{V} \times \mathbf{A}_{\text{eff}}) - \frac{1}{c}\beta\left(\mathbf{V} \times \frac{\partial \mathbf{A}_{\text{eff}}}{\partial_t}\right) \qquad (1.22a)$$

$$\langle \mathbf{m} \rangle = \kappa(\mathbf{V} \times \mathbf{A}_{\text{eff}}) + \gamma\left(-\frac{1}{c}\frac{\partial \mathbf{A}_{\text{eff}}}{\partial t}\right) + \frac{1}{c}\beta\left(-\frac{1}{c}\frac{\partial^2 \mathbf{A}_{\text{eff}}}{\partial_t^2}\right) \qquad (1.22b)$$

where

$$\alpha = \frac{2}{3h}\sum_a \frac{v_{oa}\,|\boldsymbol{\mu}_{oa}|^2}{v_{oa}^2 - v^2} \qquad (1.23a)$$

$$\kappa = \frac{2}{3h}\sum_a \frac{v_{oa}\,|\mathbf{m}_{oa}|^2}{v_{oa}^2 - v^2} \qquad (1.23b)$$

$$\gamma = \frac{2}{3h}\sum_a \frac{v_{oa}\,Re[\boldsymbol{\mu}_{oa}\cdot\mathbf{m}_{ao}]}{v_{oa}^2 - v^2} \qquad (1.23c)$$

$$\beta = \frac{c}{3\pi h}\sum_a \frac{Im[\boldsymbol{\mu}_{oa}\cdot\mathbf{m}_{ao}]}{v_{oa}^2 - v^2} \qquad (1.23d)$$

The oscillating field in a crystal will be described by a vector potential, $\mathbf{A}(x, y, z, t)$. In determining the steady-state solutions to an electromagnetic problem there will be two values of interest,

$$\mathbf{A}_{\text{cell}} = \frac{1}{V}\int_{\substack{\text{unit}\\\text{cell}}} \mathbf{A}\,d\tau \qquad \text{and} \qquad \mathbf{A}(o, t),$$

the lattice point value giving the effective field acting on the molecule. It is, of course, possible to solve for the forced oscillations at the lattice points and add the individual contributions to obtain the total wave; but it is simpler to take the average of the equation

$$\mathbf{V} \times \mathbf{B} = \frac{1}{c}\frac{\partial \mathbf{E}}{\partial t} + \frac{4\pi}{c}\mathbf{J}$$

over a unit cell:

$$\mathbf{V} \times \mathbf{B}_{\text{cell}} = \frac{1}{c}\frac{\partial \mathbf{E}_{\text{cell}}}{\partial t} + \frac{4\pi}{c}\mathbf{J}_{\text{cell}},$$

where

$$\mathbf{J}_{\text{cell}} = N\left[c\mathbf{V} \times \langle \mathbf{m} \rangle + \frac{\partial\langle \boldsymbol{\mu} \rangle}{\partial t}\right].$$

If, as in (1.22a,b), $\langle \mathbf{m} \rangle$ and $\langle \boldsymbol{\mu} \rangle$ are linear functions of the space and time derivatives of $\mathbf{A}(o, t)$, the problem is solved when $\mathbf{A}_{\text{cell}}$ is related to $\mathbf{A}(o, t)$. The division of a crystal into unit cells is somewhat arbitrary and it is convenient to think of the molecule at the center of such a cell. Then the average phase of $\mathbf{A}(o, t)$ at the center is roughly equal to that of $\mathbf{A}_{\text{cell}}$. In general the two will be related by a tensor relation,

$$\mathbf{A}(o, t) = \mathbf{S}\mathbf{A}_{\text{cell}}. \tag{1.24}$$

We shall concentrate on the isotropic case such as the cubic crystal. Then $\mathbf{S} = S\mathbf{I}$. If magnetic dipole radiation is ignored and the medium is microscopically isotropic, then and only then may the Lorentz value,

$$S_L = 1/[1 - (4\pi N\alpha/3)],$$

be used. It is also worth pointing out that this value has been derived for an electrostatic field. In the region of an absorption band the discrepancy could become quite serious even for an isotropic medium. For this reason it is preferable to use the parameter S to emphasize the importance of further investigation into this matter.

In (1.22a,b) $\mathbf{A}_{\text{eff}}$ will be set equal to $S\mathbf{A}$, where $\mathbf{A}$ is now interpreted as a suitable average value determining the macroscopic behavior of the radiation. The substitution of (1.22a,b) into the equation,

$$\mathbf{V} \times \mathbf{B} = \frac{1}{c}\frac{\partial \mathbf{E}}{\partial t} + \frac{4\pi}{c}\left[\frac{\partial \mathbf{P}}{\partial t} + c\mathbf{V} \times \mathbf{M}\right],$$

where $\mathbf{P} = N \langle \mu \rangle$ and $\mathbf{M} = N \langle m \rangle$, gives

$$\mathbf{V} \times \mathbf{V} \times \mathbf{A} = -\frac{1}{c^2}\frac{\partial^2 \mathbf{A}}{\partial t^2} + \frac{4\pi NS}{c}\left\{\left[\kappa c\mathbf{V} \times \mathbf{V} \times \mathbf{A} - \gamma\mathbf{V} \times \frac{\partial \mathbf{A}}{\partial t} - \frac{\beta}{c}\mathbf{V} \times \frac{\partial^2 \mathbf{A}}{\partial t^2}\right]\right.$$
$$\left. + \left[-\frac{\alpha}{c}\frac{\partial^2 \mathbf{A}}{\partial t^2} + \gamma\mathbf{V} \times \frac{\partial \mathbf{A}}{\partial t} - \frac{\beta}{c}\mathbf{V} \times \frac{\partial^2 \mathbf{A}}{\partial t^2}\right]\right\}. \tag{1.25}$$

Since $\mathbf{V} \cdot \mathbf{A} = 0$ we may write $\mathbf{V} \times \mathbf{V} \times \mathbf{A} = -\mathbf{V}^2\mathbf{A}$, and the equation becomes

$$\mathbf{V}^2\mathbf{A}[1 - 4\pi NS\kappa] - \frac{1}{c^2}\frac{\partial^2 \mathbf{A}}{\partial t^2}[1 + 4\pi NS\alpha] = \frac{8\pi NS\beta}{c^2}\mathbf{V} \times \frac{\partial^2 \mathbf{A}}{\partial t^2}. \tag{1.26}$$

When $N \to 0$ this equation becomes the free space equation $\mathbf{V}^2\mathbf{A} = (1/c^2)(\partial^2\mathbf{A}/\partial t^2)$ with the solution $\mathbf{A} = \mathbf{A}_0 e^{i\omega(t-\mathbf{k}\cdot\mathbf{r}/c)}$. In an inactive medium $\beta = 0$ and the solution to (21.6) is $\mathbf{A} = \mathbf{A}_o e^{i\omega(t-\bar{n}\mathbf{k}\cdot\mathbf{r}/c)}$, where

$$\bar{n} = \sqrt{\frac{1 + 4\pi NS\alpha}{1 - 4\pi NS\kappa}}. \tag{1.2}$$

It will always be true that $\kappa \ll \alpha$, and the denominator may be set equal to 1. In a condensed medium $\alpha \sim (1/N)$ and $S \sim 1$, so it will not be permissible to set $\alpha = 0$.

The solution when $\beta \neq 0$ may be found by solving the algebraic equations that result from substituting $\mathbf{A} = \mathbf{A}_o\, e^{i\omega(t - n\mathbf{k}\cdot\mathbf{r}/c)}$ into (1.26). If κ is set equal to zero and $1 + 4\pi N S = \bar{n}^2$, the result is

$$\left| \frac{\omega^2 n^2}{c^2} - \frac{\omega^2 \bar{n}^2}{c^2} \right| \mathbf{A} = \frac{i 8 \pi N S \beta \omega^3 n}{c^3}\, \mathbf{k} \times \mathbf{A}. \tag{1.28}$$

This expresses the difference between the square of n and the square of its average value in terms of the optical rotation parameter β. Since $\mathbf{k} \cdot \mathbf{A} = 0$, we may write $\mathbf{A}_o = \mathbf{i}A_1 + \mathbf{j}A_2$, and (1.28) becomes

$$(n^2 - \bar{n}^2)A_1 + ihn\, A_2 = 0 \tag{1.29a}$$

$$-ihn\, A_1 + (n^2 - \bar{n}^2)A_2 = 0, \tag{1.29b}$$

where $h = 8\pi N S \beta \omega / c$.

This system of homogeneous equations has solutions for two positive values of n for which the determinant vanishes. The solutions to the equation, $(n^2 - \bar{n}^2)^2 - h^2 n^2 = 0$, are

$$n_{\pm} = \bar{n} \mp \tfrac{1}{2}h \tag{1.30}$$

to first order in the small term h. The corresponding solutions to the linear equations are $(A_1/A_2) = \mp i$ for n_+ and n_- giving

$$\mathbf{A}_+ = A_o(\mathbf{i} + i\mathbf{j})e^{i\omega(t - n_+ + \mathbf{k}\cdot\mathbf{r}/c)} \tag{1.31a}$$

$$\mathbf{A}_- = A_o(\mathbf{i} - i\mathbf{j})e^{i\omega(t - n - \mathbf{k}\cdot\mathbf{r}/c)}. \tag{1.31b}$$

From (1.14) and the preceding discussion it follows that $\mathbf{A}_+$ and $\mathbf{A}_-$ will describe right and left circularly polarized waves, respectively.

The above result indicates that plane polarized waves are not transmitted as such by an optically active medium, but rather only waves which can be resolved into opposite-handed circular components transmitted with different phase velocities. When a plane polarized wave impinges upon an optically active medium, the final steady state will consist of the propagation of these opposite-handed, circularly polarized waves with different phase velocities.

After traversing a distance z in the medium the phase difference is $(\omega z/c)(n_- - n_+)$. If $n_- > n_+$ the phase velocity of the right circularly polarized component is greater, which leads to dextrorotation. If two equal vectors have azimuthal angles, ϕ_1 and ϕ_2, their sum will have the angle $\tfrac{1}{2}(\phi_1 - \phi_2)$. Accordingly the angle of rotation in radians per unit path length will be written

$$\phi = \frac{1}{2}\frac{\omega}{c}(n_- - n_+) = 4\pi N S \frac{\omega^2}{c^2}\beta. \tag{1.32}$$

Upon substitution of the quantum mechanical expression for β, (1.23d), this becomes

$$\phi = \frac{16\pi^2}{3ch} NS \sum_a v^2 \frac{Im[\boldsymbol{\mu}_{oa} \cdot \mathbf{m}_{ao}]}{v_{oa}^2 - v^2}. \tag{1.33}$$

If the Lorentz value, $S_L = 1/[1 - (4\pi N\alpha)/3)]$, is used along with the relation, $\bar{n}^2 = 1 + 4\pi NS\alpha$, S becomes $(\bar{n}^2 + 2)/3$. This factor generally appears in the expressions for optical rotatory dispersion, but its validity is questionable.

Equation (1.33) has been derived for regions outside the absorption band, which would appear to render it useless for investigating the most interesting features of optical activity. Since there are many undesirable instructional features about the elementary quantum theory of absorption, we shall at this point proceed by classical analogy.

An electron with a harmonic restoring force, $-kx$, obeys the differential equation,

$$m\ddot{x} = -kx, \tag{1.34}$$

which has the solution

$$x = Re(x_o e^{-i\omega_o t}), \tag{1.35}$$

where $\omega_o = 2\pi v_o = \sqrt{k/m}$.

If this electron is subject to an oscillating field, $E_o e^{-i\omega t}$, the equation of motion is

$$\ddot{x} + \omega_o^2 x = \frac{e}{m} E_o e^{-i\omega t}. \tag{1.36}$$

The solution is

$$x = \frac{e}{m} \frac{1}{\omega_o^2 - \omega^2} E_o e^{-i\omega t}. \tag{1.37}$$

As with (1.33) this result is valid only in regions well removed from the absorption band. At $\omega = \omega_o$ the above solution is not valid; instead we obtain

$$x = \frac{ieE_o}{2m\omega} te^{-i\omega t} \tag{1.38}$$

as a particular solution. This represents an oscillation of linearly increasing amplitude. Since $E = \frac{1}{2} m\dot{x}^2 + \frac{1}{2}kx^2$, the energy of this system will increase quadratically with time.

There are, of course, no situations in nature where this can occur. The resonance catastrophe is prevented by the dissipation of this energy in the form of heat or reradiation. In the elementary classical theory we assume a damping factor such that the force opposing the direction of motion is $-\gamma\dot{x}$. Equation (1.36) becomes

$$\ddot{x} + \frac{\gamma}{m}\dot{x} + \omega_o^2 x = \frac{eE_o e^{-i\omega t}}{m}. \tag{1.39}$$

The solution obtained by the substitution, $x = x_o e^{-i\omega t}$, gives

$$\mu = ex = \frac{e^2}{m}\frac{1}{\omega_o^2 - \omega^2 - i\Gamma}E_o e^{-i\omega t}. \tag{1.40}$$

From (1.27) with $\kappa = 0$ the complex index of refraction is obtained in the form

$$n^{\neq} = \sqrt{1 + 4\pi NS\frac{e^2}{m}\frac{1}{\omega_o^2 - \omega^2 - i\Gamma}}$$

$$\simeq 1 + 2\pi NS\frac{e^2}{m}\left[\frac{\omega_o^2 - \omega^2}{(\omega_o^2 - \omega^2)^2 + \Gamma^2} + \frac{i\Gamma}{(\omega_o^2 - \omega^2)^2 + \Gamma^2}\right]. \tag{1.41}$$

Under these conditions a monochromatic plane wave propagated in the medium must have the form

$$\mathbf{A} = \mathbf{A}_o e^{-i\omega(t - n^{\pm}\mathbf{k}\cdot\mathbf{r}/c)} = \mathbf{A}_o e^{-(\epsilon/2)\mathbf{k}\cdot\mathbf{r}}e^{-i\omega(t - n\,\mathbf{k}\cdot\mathbf{r}/c)},$$

where

$$n = 1 + 2\pi NS\frac{e^2}{m}\frac{\omega_o^2 - \omega^2}{(\omega_o^2 - \omega^2)^2 + \Gamma^2} \tag{1.42a}$$

$$\epsilon = 4\pi NS\frac{e^2}{m}\frac{\omega}{c}\frac{1}{(\omega_o^2 - \omega^2)^2 + \Gamma^2}. \tag{1.42b}$$

We wish now to use this simplified model to indicate how the area under the observed absorption curve may be related to the index of refraction outside the absorption region.

It will be seen that the intensity of the wave, which varies as the square of $\mathbf{A}$, decays exponentially according to the factor, $e^{-\epsilon\mathbf{k}\cdot\mathbf{r}}$. In this treatment nothing is said about the fate of the lost energy. The result is a consequence of the partial differential equations of electromagnetic theory. These are the conditions under which a monochromatic plane wave may be propagated in the medium. It is, however, tacitly assumed that energy is not reradiated at the frequency ω, but rather is converted into kinetic energy or radiant energy of other frequencies.

Consider the integral of ϵ over all frequencies,

$$\int_0^\infty \epsilon(\omega)\,d\omega = 4\pi NS\frac{e^2}{mc}\int_0^\infty \frac{\omega\Gamma\,d\omega}{(\omega_o^2 - \omega^2)^2 + \Gamma^2}. \tag{1.43}$$

The damping factor Γ will be quite small compared with ω_o^2 and the principal contribution will come from the region near $\omega = \omega_o$. The integral may be approximated by

$$\int_0^\infty \frac{\omega_o \Gamma \, d\omega}{[2\omega_o(\omega - \omega_o)]^2 + \Gamma^2} = \int_{-\omega_0}^\infty \frac{\omega_o \Gamma \, dy}{4\omega_o^2 y^2 + \Gamma^2}$$

and the region of integration extended to $-\infty$. The final result is

$$\int_0^\infty \epsilon(\omega) \, d\omega = 2\pi^2 \frac{NSe^2}{mc}, \qquad (1.44)$$

which is independent of Γ and of line shape. From (1.42a) outside the absorption band the frequency factor becomes $1/(\omega_o^2 - \omega^2)$, and the numerator in this expression is $2\pi NS(e^2/m)$, which differs from the above result by the factor (π/c).

From (1.30) and (1.23) the contribution to the difference between n_- and n_+ from a single transition outside the absorption band is

$$n_- - n_+ = \frac{8}{3h} NS\omega \frac{Im\mu_{oa} \cdot \mathbf{m}_{ao}}{v_{oa}^2 - v^2}. \qquad (1.45)$$

By analogy the area under the corresponding band in the circular dichroism curve, which measures the difference in the absorption of left- and right-hand circularly polarized light, should be related to the numerator of this expression by a similar expression,

$$\Omega \int_{\substack{\text{adsorption} \\ \text{band}}} (\epsilon_l - \epsilon_r) \, d\omega = Im(\boldsymbol{\mu}_{oa} \cdot \mathbf{m}_{ao}). \qquad (1.46)$$

This result is sufficiently important to warrant a digression into the theory of generalized susceptibility. If a periodic force acts on a polarizable medium, oscillations will occur which in time will reach a steady state at the driving frequency. If there are no velocity-dependent forces opposing the motion, it will be exactly in phase or $180°$ out of phase with the driving force, depending on whether ω is greater or less than the appropriate resonance frequency ω_n. A force, $F_o \cos \omega t$, produces a response $\pm x_o \cos \omega t$. The average energy absorbed is zero, since $\int_0^{1/\nu} \sin \omega t \cos \omega t \, dt = 0$.

If there is damping the induced motion has the form, $x_1 \cos \omega t + x_2 \sin \omega t$. In general $x_2 \ll x_1$; the energy dissipated per cycle is

$$E_{\text{cycle}} = \int_0^{1/\nu} F \frac{dx}{dt} \, dt = \int_0^{1/\nu} x_2 F_o \cos^2 \omega t \, dt = \frac{x_2 F_o}{2\omega}. \qquad (1.47)$$

The advantage of the complex formulation is that the nondissipative portion in phase motion is described by the real part of a complex function and the

In the integral of (1.53) $\theta(\omega')$ will differ from zero for values of ω' near the frequencies ω_n. If ω lies outside these regions no singularities result and

$$\phi(\omega) = \frac{\omega^2}{2\pi} \sum_n \frac{\theta_n}{\omega_n(\omega_n^2 - \omega^2)}. \tag{1.54}$$

Comparison of this result with (1.33) gives

$$R_{on} = Im[\boldsymbol{\mu}_{on} \cdot \mathbf{m}_{no}] = \frac{3ch}{32\pi^3} \frac{\theta_n}{NS\omega_n}. \tag{1.55}$$

This now gives us an important relation between a molecular parameter, R_{on}, the rotational strength, and an experimental quantity, θ_n, the area under the dichroism band. The ratio θ_n/N is simply absorption per molecule. For sufficiently narrow bands (1.55) is very accurate, but an appreciable error will occur for very broad bands even if ω_n is taken as the center of the band. The presence of the effective field parameter S reminds us that the oscillations of a molecule are governed by the effective field. If this differs from the macroscopic average, the absorption may be enhanced or diminished. If $S > 1$ the absorption per molecule is greater than it would be if the effective field were completely uniform. The parameter S corrects the situation by providing a quantity that expresses some individual property of the molecule apart from the medium in which it is situated. Almost all optical activity theory to date has centered around the evaluation of $R_n = Im\, \boldsymbol{\mu}_{on} \cdot \mathbf{m}_{no}$.

4 SYMMETRY CONSIDERATIONS

We must not lull ourselves into a false sense of security by regarding $Im[\boldsymbol{\mu}_{on} \cdot m_{no}]$ as the answer to all questions on optical activity. The role of higher order terms in the expansion of $e^{i\mathbf{k} \cdot \mathbf{r}/\lambda}$ has not been settled conclusively, nor has the question of the magnitude of the quantum relativistic corrections to radiation theory. In addition, there are many problems concerned with the effective field. In the preceding derivations it has been assumed that the only reason for a variation in the effective field from one part of the molecule to another is found in the phase factor, $e^{in\mathbf{k} \cdot \mathbf{r}/\lambda}$. If we grant that this whole question arises from the microscopic inhomogeneity of the medium, it would seem reasonable to begin our considerations at the molecule itself. We must refer to the differential and integral forms of Maxwell's equations along with the quantum mechanical current density vector.

In defense of $Im[\boldsymbol{\mu}_{on} \cdot m_{no}]$ it may be said that it represents the lowest order contribution to optical activity under the assumption that the effective field acting on all parts of a molecule varies by only a fractional amount of the same magnitude as the ratio of the molecule's dimension to the wavelength. It has the advantage of simplicity and may be said to prevail by default, since if two or more terms contribute to a phenomenon and neither of them may be evaluated accurately, the wisest course is to concentrate one's mathematical efforts on the lowest order term. It is in this spirit that a résumé of optical activity theory is given.

Certain general properties of the product $\mathbf{\mu}_{on} \cdot \mathbf{m}_{no}$ may be found by symmetry arguments. In only nondegenerate wave functions are considered, the result of a coordinate transformation, $\Omega\mathbf{r}$—where Ω is a linear operator describing the result of one of the molecule's group operations—is $\psi_n(\Omega\mathbf{r}) = \pm\psi_n(\mathbf{r})$. If $T(\mathbf{r}, \mathbf{V})$ is a quantum mechanical operator, the matrix element T_{on}

$$T_{on} = \int \psi_o(\mathbf{r})T(\mathbf{r}, \mathbf{V})\psi_n(\mathbf{r}) \, d\tau(\mathbf{r}) = \int \psi_o(\Omega\mathbf{r})T(\Omega\mathbf{r}, \Omega\mathbf{V})\psi_n(\Omega\mathbf{r}) \, d\tau(\Omega\mathbf{r}). \quad (1.56)$$

For proper rotations through an angle θ about the z-axis, the components of $\mathbf{\mu} = e \Sigma_a \mathbf{r}_a$ and $\mathbf{m} = (eh/2mc)i \Sigma_a \{\mathbf{i}[z_a(\partial/\partial y_a) - y_a(\partial/\partial z_a)] + \mathbf{j}[x_a(\partial/\partial z_a) - z_a(\partial/\partial z_a)] + \mathbf{k}[y_a(\partial/\partial x_a) - x_a(\partial/\partial y_a)]\}$ have the transformation properties

$$\begin{Bmatrix} \mu'_x \\ m'_x \end{Bmatrix} = \begin{Bmatrix} \mu_x \\ m_x \end{Bmatrix}\cos\theta - \begin{Bmatrix} \mu_y \\ m_y \end{Bmatrix}\sin\theta \quad (1.57a)$$

$$\begin{Bmatrix} \mu'_y \\ m'_y \end{Bmatrix} = \begin{Bmatrix} \mu_x \\ m_x \end{Bmatrix}\sin\theta + \begin{Bmatrix} \mu_y \\ m_y \end{Bmatrix}\cos\theta \quad (1.57b)$$

$$\begin{Bmatrix} \mu'_z \\ m'_z \end{Bmatrix} = \begin{Bmatrix} \mu_z \\ m_z \end{Bmatrix} \quad (1.57c)$$

For improper rotations

$$\begin{Bmatrix} \mu'_x \\ m'_x \end{Bmatrix} = \begin{Bmatrix} \mu_x \\ -m_x \end{Bmatrix}\cos\theta - \begin{Bmatrix} \mu_y \\ -m_y \end{Bmatrix}\sin\theta \quad (1.58a)$$

$$\begin{Bmatrix} \mu'_y \\ m'_y \end{Bmatrix} = \begin{Bmatrix} \mu_x \\ -m_x \end{Bmatrix}\sin\theta + \begin{Bmatrix} \mu_y \\ -m_y \end{Bmatrix}\cos\theta \quad (1.58b)$$

$$\begin{Bmatrix} \mu'_z \\ m'_z \end{Bmatrix} = \begin{Bmatrix} -\mu_z \\ m_z \end{Bmatrix}. \quad (1.58c)$$

For proper rotations $d\tau(\Omega\mathbf{r}) = d\tau(\mathbf{r})$, and $d\tau(\Omega\mathbf{r}) = -d\Omega(\mathbf{r})$ for improper rotations. In addition, each variable change of the form $x \rightarrow -x$ will introduce an additional change of sign, since $\int_{-\infty}^{\infty} f(x) \, dx = -\int_{\infty}^{-\infty} f(-x) \, dx$.

If the symmetry group of the molecule contains an improper rotation, (1.56) and (1.58) may be used to obtain the result

$$\mathbf{\mu}_{on} \cdot \mathbf{m}_{no} = (\mu_x)_{on}(m_x)_{no} + (\mu_y)_{on}(m_y)_{no} + (\mu_z)_{on}(m_z)_{no} = -\mathbf{\mu}_{on} \cdot \mathbf{m}_{no}, \quad (1.59)$$

which requires that $\mathbf{\mu}_{on} \cdot \mathbf{m}_{no}$ vanish. The special case of a center of inversion leads to the same result, since $\mathbf{\mu}' = -\mathbf{\mu}$ and $\mathbf{m}' = \mathbf{m}$. If the state n is degenerate, it will be a member of a set of states with equal energy, $\psi_1, \psi_2, \ldots, \psi_s$. The most symmetrical molecules studied so far belong to the octahedral or tetrahedral groups; and s is effectively limited to 3 for all cases of practical interest, although there is no physical reason why a molecule with icosahedral symmetry with $s = 5$ could not be synthesized.

For degenerate states we must write

$$\psi_s(\Omega \mathbf{r}) = \sum_t c_t^{(s)} \psi_t(\mathbf{r}). \tag{1.60}$$

If use is made of the fundamental group property that $\sum_s c_t^{(s)} c_{t'}^{(s)} = \delta_{tt'}$, it follows that $\sum_s \boldsymbol{\mu}_{os} \cdot \mathbf{m}_{so} = -\sum_s \boldsymbol{\mu}_{os} \cdot \mathbf{m}_{so} = 0$; that is, the total contribution of the degenerate band is zero. The above arguments serve to establish the important theorem: In order for a molecule to be optically active it is necessary that its symmetry group contain no improper rotations (a reflection plane, center of inversion, or a rotation-reflection).

In most practical cases attention will be focused on the first two; however, it is interesting that the symmetry groups $S_4, S_8, S_{12}, \ldots$ are based solely on the operations of reflecting and rotating through angles of $90°$, $45°$, $22.5°$, $\ldots$. They contain no pure reflection planes or centers, of inversion, yet they are inactive.

A second theorem on the general behavior of the rotational strength states that the sum of the rotational strengths over all transitions in a molecule is zero. The proof readily follows from the matrix multiplication rule:

$$(PQ)_{oo} = \sum_n P_{on} Q_{no}, \tag{1.61}$$

where P and Q are two quantum mechanical operators, and the summation is taken over all states of the system. This gives

$$\sum_{n \neq 0} \boldsymbol{\mu}_{on} \cdot \mathbf{m}_{no} + \boldsymbol{\mu}_{oo} \cdot \mathbf{m}_{oo} = (\boldsymbol{\mu} \cdot \mathbf{m})_{oo}. \tag{1.62}$$

The term, $\boldsymbol{\mu}_{oo} \cdot \mathbf{m}_{oo}$ arises because the sum in (1.61) is over all states; whereas the sum of the rotational strengths is over all transitions $0 \to n$, which would exclude $n = 0$. Since $\boldsymbol{\mu}_{oo}$, $\mathbf{m}_{oo}$, and $(\boldsymbol{\mu} \cdot \mathbf{m})_{oo}$ are quantum expectation values for the ground states, they will be real; and the taking of imaginary parts in (1.63) gives

$$\sum_n R_{on} = Im \sum_{n \neq 0} \boldsymbol{\mu}_{on} \cdot \mathbf{m}_{no} = 0. \tag{1.63}$$

This strongly suggests that the total area under the dichroism curve is zero. Whether or not this is actually true will depend on a more exact formulation of the theory. Since the summation is over all transitions, the inaccessibility of the far-ultraviolet states seriously hinders experimental verification. The assertion would be true if it could be shown that the area under the individual dichroism band is KR_{on}, where K is the same for all transitions. This is not immediately evident even for the gas phase, although most workers in the field believe that this is the case.

An important question relating to the behavior of rotational strengths arises in connection with the problem of degeneracy. If $\psi_1, \ldots, \psi_N$ are a linearly independent set of N degenerate functions, the total contribution of the band will be

$$R = Im \sum_{i=1}^{N} \boldsymbol{\mu}_{oi} \cdot \mathbf{m}_{io}. \tag{1.64}$$

An arbitrary linearly independent set, ψ_i, may be constructed from the ψ_i such that $\psi_i = \sum_j c_j^{(i)} \psi_j$. In particular, the $c_j^{(i)}$ may be chosen ~ 1 for $j = 1$ and ~ 0 for $j \neq 1$. The total rotational strength will be given by $R = Im[N\boldsymbol{\mu}_{oi} \cdot \mathbf{m}_{io} + \text{vanishingly small terms}]$. Since R must be independent of the basis set used in its computation, we are forced to conclude that the individual terms in (1.64) are all equal to $(1/N)R$; that is, for a degenerate transition

$$R = N \, Im \, \boldsymbol{\mu}_{on} \cdot \mathbf{m}_{no}, \tag{1.65}$$

where ψ_n is an arbitrary function with the degenerate eigenvalue, E_n.

It must be emphasized that this result applies to exact degeneracy, which has not been split by any intramolecular perturbation. This will include the space groups, C_n and D_n $(n > 2)$, along with various three-dimensional groups such as T, O, and I. The most far-reaching consequence of the above result will apply to the question of near degeneracy. For example, consider an octahedral complex such as $Co(NH_3)_6^{+3}$. The molecule will belong to the inactive group O_h. If a perturbing field resulting in O symmetry is superimposed on the complex, the wave functions will be perturbed in such a way as to give equal rotational strengths to the individual components of the triply degenerate T_1 transitions. There will be two sets of such transitions, which are derived from the T_{1u} electric dipole transitions and from the T_{1g} magnetic dipole transitions in the original O_h symmetry complex.

If, in addition to the octahedral perturbation, a low symmetry component is introduced, the degeneracy will be removed; and the individual contributions to the rotational strength of the band are no longer required to be equal. It appears that when an inactive molecule of high symmetry is dissymmetrically substituted, two extremes in the behavior of the dichroism curve may be observed. When a high degree of symmetry remains with a small amount of splitting of the level, we observe a dichroism curve quite similar to what would be expected for a completely degenerate, optically active molecule, a single large extremum adjacent to a very small one of opposite sign. With greater departures from high symmetry we often observe two nearly equal adjacent extremes of opposite sign. This is characteristic of electric dipole transitions in molecules containing identical chromophores such as polypeptides.

5 APPLICATION TO MODELS

We can now discuss these particular features of $Im\,\mu_{on}\cdot m_{no}$ applicable to individual molecules. Allowed electric dipole transitions have moments of the order of a debye unit (10^{-18} cgs units), and allowed magnetic dipole transitions have moments of the order of one Bohr magneton [$(e\hbar/2mc)\sim 10^{-20}$ cgs units]. This sets 10^{-38} erg cc^3 as an upper limit to the rotational strength. The visible and ultraviolet region is to a large degree dominated by the electronic dipole motion, that of the nuclei being neglected to a first approximation. In the infrared, ordinary absorption proceeds on an equal footing with the ultraviolet, since both are governed by $|\mu_{on}|^2$. On the other hand, a factor of $\sim 1/2000$ will appear in a comparison of infrared rotational strengths with visible and ultraviolet, since m in $(e\hbar/2mc)$ must be taken as a nuclear mass. This fact, until recently, has discouraged attempts to study optical activity in the infrared region; however, certain compensating features may exist which may allow the construction of intrumentation to observe the effect.

The figure of 10^{-38} applies to electronic transitions and will be approached for a special class of molecules containing what is termed inherently dissymmetric chromophores. To a good first approximation any transition may be considered localized to some particular group known as the chromophore. In most optically active molecules the chromophore is situated in an environment that differs only slightly from the most closely related inactive derivative. This will lead to rotational strengths one or two orders of magnitude lower than the upper limit in the regions of 10^{-40} erg cc^3. Table 1.1 contains data on typical chromophores.

Table 1.1

Type	R	Examples
Inherently dissymmetric	10^{-38}	Hexahelicene
Strongly dissymmetric	10^{-39}	Transition metal complexes
Conventional, primarily electric or magnetic dipole allowed	10^{-40}–10^{-41}	Phenyl, carbonyl, peptide derivatives
Transitions forbidden both electrically and magnetically	$<10^{-42}$	Most optically active molecules contain such transitions

Optical activity theory has taken three basic directions:

1. Exact treatment of simple quantum mechanical models.
2. Attempts at first-principles calculations on particular molecules.
3. Semiempirical theories designed to correlate optical activity with structure.

It may be said at the outset that no reliable fundamental expression exists relating the rotational strength to molecular geometry and group properties, although partial success has been achieved with certain models such as the polarizability theory.

An instructive example is furnished by a harmonic oscillator in a dissymmetric field. The simplest potential with proper symmetry for optical activity is

$$V = \tfrac{1}{2}(k_1 x_1^2 + k_2 x_2^2 + k_3 x_3^2) + A x_1 x_2 x_3. \tag{1.66}$$

The basic motion of the electron will be governed by the quadratic term with the relatively small perturbing field described by the term in A. The unperturbed chromophore has D_{2h} symmetry with operations consisting of three perpendicular reflection planes, three perpendicular two-fold rotation axes, and a center of inversion. The qualitative features of this model will apply strictly only to chromophores of this symmetry.

The unperturbed wave functions are

$$\psi(n_1, n_2, n_3) = \phi_{n_1}\left(\frac{x_1}{a_1}\right)\phi_{n_2}\left(\frac{x_2}{a_2}\right)\phi_{n_3}\left(\frac{x_3}{a_3}\right), \tag{1.67}$$

where

$$\phi_n\left(\frac{x_i}{a_i}\right) = (2^n n! \sqrt{\pi})^{-1/2} H_n\left(\frac{x_i}{a_i}\right) e^{-x_i^2/2a^{i2}} \quad \text{and} \quad a_i = \tfrac{1}{2}\pi\sqrt{\frac{h}{mv_i}}.$$

The H_n are the Hermite polynomials and v_1, v_2, v_3 are the three frequencies of oscillation.

Next, one of the standard perturbation techniques is used for evaluating $\mu_{on} \cdot \mathbf{m}_{no}$. In general the oscillator has three electric dipole transitions along the x_1, x_2, and x_3 axes with frequencies v_1, v_2, and v_3; three magnetic dipole transitions with frequencies $v_2 - v_3$, $v_1 - v_3$, and $v_1 - v_2$; and three weak magnetic dipole transitions with frequencies $v_2 + v_3$, $v_1 + v_3$, and $v_1 + v_2$. We assume that $v_1 > v_2 > v_3$. In addition there are nine emission transitions with identical frequencies. We not do often deal with emission in actual problems, since the molecules are generally in the ground state.

The coefficients for induced emission and absorption have the identical mathematical form. For the oscillator in an excited state the transitions occur in pairs of emission and absorption types with equal frequencies and probabilities. This means that if spontaneous emission is neglected, the average absorption coefficient will be zero and the average index of refraction will be unity, since the probability of the system's absorbing radiation equals the probability of emission. This fact does not prevent calculation of the differences between these quantities for left and right circularly polarized light. It will merely happen that one of the absorption coefficients is positive and the other negative of equal magnitude. This means that one form of circularly polarized light will lead to a net absorption and the other to a net emission. Such a situation would be unlikely in nature, because a partial equilibrium between molecular and radiation energies is always present; however, the basic features of the model are still instructive.

In the unperturbed form each of the oscillator transitions is allowed either electrically or magnetically, but not both. Consider a transition from a state o to a state n in the presence of a static perturbation V. First-order perturbation theory gives

$$\psi'_n = \psi_n - \sum_s \frac{V_{sn}\psi_s}{E_s - E_n} \tag{1.68a}$$

$$\psi'_o = \psi_o - \sum_t \frac{V_{to}\psi_t}{E_t - E_o}. \tag{1.68b}$$

The electric dipole moment becomes

$$\boldsymbol{\mu}'_{on} = \boldsymbol{\mu}_{on} - \frac{V_{on}}{h\nu_{on}}[\boldsymbol{\mu}_{nn} - \boldsymbol{\mu}_{oo}]$$

$$- \sum_{s \neq o,n} \frac{V_{sn}\boldsymbol{\mu}_{os}}{E_s - E_n}$$

$$- \sum_{t \neq o,n} \frac{V_{to}\boldsymbol{\mu}_{nt}}{E_t - E_o}, \tag{1.69}$$

where the mixing of the ground- and excited-state functions has been explicitly displayed. A similar expression for $\mathbf{m}'_{on}$ is obtained. To first order the rotational strength is the imaginary part of

$$\boldsymbol{\mu}'_{on} \cdot \mathbf{m}'_{no} = \boldsymbol{\mu}_{on} \cdot \mathbf{m}_{no} - \frac{V_{on}}{h\nu_{on}}[\mathbf{m}_{nn} - \mathbf{m}_{oo}] \cdot \mathbf{m}_{no}$$

$$- \frac{V_{on}}{h\nu_{on}}[\mathbf{m}_{nn} - \mathbf{m}_{oo}] \cdot \boldsymbol{\mu}_{on} - \sum_{s \neq o,n} \frac{V_{sn}}{E_s - E_n}[\boldsymbol{\mu}_{os} \cdot \mathbf{m}_{no} + \boldsymbol{\mu}_{on} \cdot \mathbf{m}_{so}]$$

$$- \sum_{t \neq o,n} \frac{V_{to}}{E_t - E_o}[\boldsymbol{\mu}_{tn} \cdot \mathbf{m}_{no} + \boldsymbol{\mu}_{on} \cdot \mathbf{m}_{nt}]. \tag{1.70}$$

The first term is zero for the unperturbed chromophore; the second is often called a charge transfer term, but it must not be confused with those effects caused by a transfer of charge between the chromophore and some other group. In chromophores with D_{2h} symmetry the second term is also zero. In other instances it can be important when a magnetic dipole allowed transition involves a shift of the center of charge; the third term is zero for nondegenerate transitions, because permanent magnetic moments can only occur with degenerate states. The fourth term arises from perturbations of the excited state, and the fifth from perturbations of the ground state. In crude calculations this last term is often omitted because of its larger energy denominator compared with that of the fourth term.

For the harmonic oscillator there are a finite number of terms in (1.70), because the matrix elements $(x)_{ns}$ are zero unless $s = n \pm 1$. The final result for the optical rotatory parameter is

$$\beta_{n_1 n_2 n_3} = \frac{A\hbar e^2}{12(2\pi)^5 m^3} \left\{ \left(\frac{1}{v_2} + \frac{1}{v_3} \right) \frac{(n_2 - n_3)}{(v_2 - v_3)^2 - v_1^2} \left[-\frac{1}{v_1^2 - v^2} + \frac{1}{(v_2 - v_3)^2 - v^2} \right] \right.$$

$$+ \left(\frac{1}{v_3} + \frac{1}{v_1} \right) \frac{(n_3 - n_1)}{(v_3 - v_1)^2 - v_2^2} \left[-\frac{1}{v_2^2 - v^2} + \frac{1}{(v_3 - v_1)^2 - v^2} \right]$$

$$+ \left(\frac{1}{v_1} + \frac{1}{v_2} \right) \frac{(n_1 - n_2)}{(v_1 - v_2)^2 - v_3^2} \left[-\frac{1}{v_3^2 - v^2} + \frac{1}{(v_1 - v_2)^2 - v^2} \right] \right\}, \quad (1.71)$$

where smaller terms in $(1/v_2) - (1/v_3)$, including the contributions from the weak magnetic dipole transitions, have been neglected.

The essential qualitative features of this model may now be explored. The simplest assumption is that the perturbing coefficient A arises from a collection of point charges Q_i with position vectors $\mathbf{X}_i$, $\mathbf{Y}_i$, $\mathbf{Z}_i$ with respect to the center of the chromophore. A multipole expansion on the resulting potential gives

$$V = -\sum_i \frac{Q_i}{\sqrt{(x_1 - X_i)^2 + (x_2 - Y_i)^2 + (x_3 - Z_i)^2}}$$

$$= -\left\{ \cdots + \tfrac{5}{2} x_1 x_2 x_3 \sum_i \frac{X_i Y_i Z_i}{R_i^7} Q_i + \cdots \right\}. \quad (1.72)$$

For a single charge

$$A = -\tfrac{5}{2} \gamma_x \gamma_y \gamma_z Q / R^4, \quad (1.73)$$

where $\gamma_x, \gamma_y, \gamma_z$ are the direction cosines of the charge in the coordinate system centered on the chromophore. The labeling of the axes is arbitrary so long as the system is right-handed. If it is assumed that $v_1 > v_2 > v_3$, there will be two choices of interest: $v_1 - v_2 < v_3$ and $v_1 - v_2 > v_3$. In the first case the lowest transition is magnetic dipole allowed with frequency $v_1 - v_2$. Actually, a pair of transitions, $(n_1 n_2 n_3) \to (n_1 + 1, n_2 - 1, n_3)$ are involved corresponding to emission and absorption. The model can be made more realistic by assuming that $n_1 = 0, n_2 = 1, n_3 = 0$. Then there will be only absorbing transitions along the x_3-axis: $(010) \to (100)$ magnetic, and $(010) \to (011)$ electric.

The value of $n_1 - n_2$ is now -1; if $(v_1 - v_2) < v_3$, the sign of this transition's contribution to β will be the same as that of A in (1.73). For a positive charge in a positive octant, A is negative and the rotation is negative. The coordinate system will have the correspondence $(\mathbf{i}, \mathbf{j}, \mathbf{k}) \to$ (final orientation, initial orientation, direction of polarization). One can see the relation of this model to $p \to p'$-type transitions by noting that

$$(010) \sim y \exp\left[-\frac{1}{2}\left(\frac{x^2}{a_1^2} + \frac{y^2}{b_1^2} + \frac{z^2}{c_1^2} \right) \right] \tag{1.74a}$$

$$(100) \sim x \exp\left[-\frac{1}{2}\left(\frac{x^2}{a_1^2} + \frac{y^2}{b_1^2} + \frac{z^2}{c_1^2} \right) \right]. \tag{1.74b}$$

The rule may now be stated: the sign of rotation of a p-p' magnetic dipole transition in a chromophore of D_{2h} symmetry is determined by establishing at the center of the group an $\mathbf{i}$-axis along the axis of the final p-orbital in either direction, a $\mathbf{j}$-axis along the axis of the initial p-orbital, also in either direction, and the $\mathbf{k}$-axis such that $\mathbf{i} \times \mathbf{j} = \mathbf{k}$. Then if a single positive charge is placed in a positive octant, the rotation will be negative.

When there are several charges in different octants, a more detailed calculation is necessary to determine even the sign. The model may also be used to suggest rules for electric dipole transitions, but the results are not so well defined. One major reason is that there is an important mechanism for the optical activity of electric dipole transitions which has been neglected in the model, the coupled oscillator effect. This subject will be treated later; for now it suffices to say that the approximation of regarding a chromophore as acted upon by a static field is best suited to magnetic dipole transitions.

Unfortunately, the most widely disseminated application of the above result has been made in a somewhat misleading fashion to the carbonyl chromophore. This group has C_{2v} symmetry; but the argument has been made that the $n \to \pi^*$ transition is centered primarily on the oxygen atom in a D_{2h} environment. The application, of the above harmonic oscillator result or a simple calculation using hydrogenlike p-orbitals centered on the oxygen atom led to the so-called octant rule for ketones. Originally, only molecules with substituents in "back octants" on the carbon side of the carbonyl group were studied and the rule was found to work rather well. Subsequently, compounds were prepared with substituents in the "front octants" on the oxygen side with signs opposite to those predicted by the octant rule. It became evident that a quadrant rule was actually in effect, which is just what would be expected on symmetry grounds when one recognizes that, even if the nonbonding orbital could be considered to be in a D_{2h} environment, the π^* certainly is not.

This illustrates a general principle of symmetry. If the electron cloud of a chromophore is effectively dominated by one type of symmetry, the substituents located in planes of symmetry will make no contribution to the optical activity. Furthermore, the sign of the contribution must change as a perturbing group crosses such a plane from one region into another, since reflection in the plane results in the mirror-image molecule with opposite-signed rotation.

If this is true, then the planes of symmetry of a chromophore will divide it into sections in such a way that neighboring ones will be of opposite sign. For example, D_{2h} chromophores like the ethylenic double bond will follow an octant rule, since this group has three mutually perpendicular planes of symmetry; C_{2v} chromophores like carbonyl will follow a quadrant rule, because this group has two perpendicular planes of symmetry; D_{6h} chromophores like benzene will follow a 24-section rule, because the group has one horizontal and six vertical planes of symmetry; chromophores with octahedral symmetry will be divided into 64 sections by a total of 9 planes of symmetry.

Two words of caution must be given at this time. First, these arguments apply only to mechanisms in which the current density in other parts of the molecule may be neglected. This may be a reasonable approximation for magnetic dipole transitions, but a dangerous one for the electric dipole type. Second, it should again be emphasized that the governing symmetry of the chromophores must be of the type specified and not some lower group. For example, benzene may be substituted in such a way as to effectively impart a lower symmetry such as D_{2h} or C_{2v} to the π-electrons.

The complexity of the situation increases as we pass on to a second important mechanism for optical activity—the coupled oscillator. The perturbation calculation that led to (1.70) assumed that the chromophore was perturbed by a constant field provided by the rest of the molecule. If the only change in the charge density of the molecule from the ground to an excited state occurred at the chromophore (1.70) would describe all the effects of optical activity to the degree of approximation used in its derivation. In actuality the charge distribution of one part of a molecule cannot be changed without affecting other parts. It may be assumed that there is a continuous variation in charge density over all parts of the molecule from one state to another. The only questions are how large is this effect and what role does it play in optical activity?

A partial answer to these questions may be found by considering two groups, A and B. Let us first consider two electric dipole transitions, $o \to a$ in group A with transition moment μ_{oa}, and $o \to b$ in group B with transition moment μ_{ob}. The vector distance from group A to group B will be labeled $\mathbf{R}_{AB}$. We shall further assume that each group contains only one electron. If electrons 1 and 2 are subject to fixed force fields with potentials $V_A(\mathbf{r}_1)$ and $V_B(\mathbf{r}_2)$, the potential energy of interaction is

$$V = eV_A(\mathbf{r}_2) + eV_B(\mathbf{r}_1) + \frac{e^2}{r_{12}}. \tag{1.75}$$

The first two terms depend only on the separate invididual electronic coordinates. The third term may be expanded

$$\frac{1}{r_{12}} = \frac{1}{R_{AB}} \left| 1 + \frac{r_1^2 + r_2^2 - 2\mathbf{r}_1 \cdot \mathbf{R}_{AB} + 2\mathbf{r}_2 \cdot \mathbf{R}_{AB} - 2\mathbf{r}_1 \cdot \mathbf{r}_2}{R_{AB}^2} \right|^{1/2}$$

$$= \cdots + \frac{1}{R_{AB}^3} \left| \mathbf{r}_1 \cdot \mathbf{r}_2 - \frac{3(\mathbf{r}_1 \cdot \mathbf{R}_{AB})(\mathbf{r}_2 \cdot \mathbf{R}_{AB})}{R_{AB}^2} \right| + \cdots, \tag{1.76}$$

where the relation, $r_{12}^2 = (\mathbf{r}_2 + \mathbf{R}_{AB} - \mathbf{r}_1)^2$, has been used and only the lowest order terms involving products of the components of $\mathbf{r}_1$ and $\mathbf{r}_2$ have been retained. The reason for this will become evident in the ensuing discussion.

If the groups are sufficiently far apart that exchange can be neglected, the zeroth-order functions may be written as a product

$$\Psi_{oo} = \phi_o^{(A)} \phi_o^{(B)} \tag{1.77a}$$

$$\Psi_{ao} = \phi_a^{(A)} \phi_o^{(B)} \tag{1.77b}$$

$$\Psi_{ob} = \phi_o^{(A)} \phi_b^{(B)} \tag{1.77c}$$

$$\Psi_{ab} = \phi_a^{(A)} \phi_b^{(B)}. \tag{1.77d}$$

Relative to an arbitrary origin in which the position vectors of the two groups are $\mathbf{R}_A$ and $\mathbf{R}_B$, the electric and magnetic moment operators are

$$\mu = e(\mathbf{R}_A + \mathbf{R}_B) + \mu_1 + \mu_2 \tag{1.78a}$$

$$\mathbf{m} = \frac{e}{2mc} |(\mathbf{R}_A + \mathbf{r}_1) \times \mathbf{p}_1 + (\mathbf{R}_B + \mathbf{r}_2) \times \mathbf{p}_2|$$

$$= \frac{e}{2mc} \mathbf{R}_A \times \mathbf{p}_1 + \frac{e}{2mc} \mathbf{R}_B \times \mathbf{p}_2 + \mathbf{m}_1 + \mathbf{m}_2, \tag{1.78b}$$

where μ_1, μ_2 and $\mathbf{m}_1$, $\mathbf{m}_2$ are the local operators referred to the centers of the respective groups, and $\mathbf{p}_1$, $\mathbf{p}_2$ are the momentum operators.

The evaulation of the scalar product, $\langle\psi'_{oo}|\boldsymbol{\mu}|\psi'_{ao}\rangle \cdot \langle\psi'_{ao}|\mathbf{m}|\psi'_{oo}\rangle$, will be required for the contribution of the $o \rightarrow a$ transition. This will be zero to zeroth order, since $(\mathbf{p}_1)_{ao}$ and $(\boldsymbol{\mu}_1)_{oa}$ are collinear. A partial perturbation expansion involving only states a and b can be made. The contributions of other states in group B will be assessed by summing the final result over b. We obtain

$$\Psi'_{oo} = \Psi_{oo} - \frac{V_{oa;\,ob}}{h(v_a + v_b)}\,\Psi_{ab} \tag{1.79a}$$

$$\Psi'_{ao} = \Psi_{ao} - \frac{V_{oa;\,ob}}{h(v_b - v_a)}\,\Psi_{ob}, \tag{1.79b}$$

where

$$V_{oa;\,ob} = \iint \phi_o^{(A)}\phi_a^{(A)}V\phi_o^{(B)}\phi_b^{(B)}\,d\tau_1\,d\tau_2$$

and only terms that give a nonvanishing result have been retained, as will be verified at the end of the discussion. The required rotational strength may be obtained by combining (1.77), (1.78), and (1.79):

$$Im\langle\Psi'_{oo}|\boldsymbol{\mu}|\Psi'_{ao}\rangle \cdot \langle\Psi'_{ao}|\mathbf{m}|\Psi'_{oo}\rangle$$

$$= \frac{2\pi}{ch}\,V_{oa;\,ob}\,\frac{v_a v_b}{v_a^2 - v_b^2}\,\boldsymbol{\mu}_{oa}\cdot\mathbf{R}_{AB}\times\boldsymbol{\mu}_{ob}$$

$$= \frac{2\pi}{ch}\,\frac{v_a v_b}{v_a^2 - v_b^2}\,\boldsymbol{\mu}_{oa}\times\mathbf{R}_{AB}\cdot\boldsymbol{\mu}_{ob}\boldsymbol{\mu}_{ob}\cdot\mathbf{T}\cdot\boldsymbol{\mu}_{oa}, \tag{1.80}$$

where

$$\mathbf{T} = \frac{1}{R_{AB}^3}\left[\mathbf{I} - \frac{3\mathbf{R}_{AB}\mathbf{R}_{AB}}{R_{AB}^2}\right],$$

and the relation, $e\mathbf{p}_{oa} = -2\pi i m v_{oa}\boldsymbol{\mu}_{oa}$, has been used.

The summation over b will give for the rotational strength

$$R_{oa} = \frac{\pi}{c}\,v_a\mathbf{R}_{AB}\times\boldsymbol{\mu}_{oa}\cdot\left[\frac{2}{h}\sum_b\frac{v_b\boldsymbol{\mu}_{ob}\boldsymbol{\mu}_{ob}}{v_b^2 - v_a^2}\right]\cdot\mathbf{T}\cdot\boldsymbol{\mu}_{oa}$$

$$= \frac{\pi}{c}\,[\mathbf{R}_{AB}\times\mathbf{b}_{oa}\cdot\boldsymbol{\alpha}_B(v_a)\cdot\mathbf{T}\cdot\mathbf{b}_{oa}]v_a\boldsymbol{\mu}_{oa}^2, \tag{1.81}$$

where $\boldsymbol{\alpha}_B(v_a)$ is the polarizability tensor of group B at the frequency v_a; and $\mathbf{b}_{oa}$ is a unit vector along the direction of the transition. This predicts that if a transition owes its optical activity solely to the coupling with other electric dipole transitions, the circular dichroism will be proportional to the oscillator strength. The factor in brackets depends only on molecular geometry and the polarizability of the neighboring group. A similar expression will be obtained for any electric dipole transition in terms of the polarizability tensors of the neighboring groups. Note that R_{oa} varies inversely as the square of the inter-group distance R_{AB}.

The substitution of actual molecular values into (1.81) has proved useful in correlating optical activity with structure, but the agreement so often seen with experiment must be regarded as fortuitous in view of the approximations made.

If two transitions in neighboring groups have nearly equal frequencies, the denominator in (1.80) indicates that the role of other transitions in group B may be neglected. A similar calculation on the effect of group A on the transition, $o \rightarrow b$, gives a result identical with (1.80) with the opposite sign. This predicts that the circular dichroism of nearly degenerate electric dipole transitions will consist of two adjacent extremes of opposite sign. This effect has been observed in many instances, particularly in polypeptides.

The singularity at $v_a = v_b$ warns us that the problem has not been correctly treated by degenerate perturbation theory. When this is done the singularity disappears, but other problems remain. If the observed circular dichroism consists of two peaks of opposite sign with a clear-cut separation, we can be sure that $[V_{oa;\,ob}/h(v_b - v_a)] \ll 1$, and nondegenerate perturbation theory applies. When the two bands coalesce into a single S-curve (1.80) no longer applies. There are only two extremes which have any physical meaning. First, there is the dispersion outside the absorption band, which will be found proportional to the geometric factors in (1.80) with the frequency factor, $[v^2 v_a^2/(v_a^2 - v^2)^2]$. In principle the calculated rotational strength may be correlated with the observed dispersion; in practice this is hard to do because of the large overlap among bands owing to the gradual asymptotic behavior of the dispersion function. Second, in analogy with nonoverlapping bands, an attempt can be made to relate the magnitudes of the areas under half the band to some calculated molecular parameter. In this regard the only observable quantity is the dichroism itself. It is theoretically unsound to attempt a resolution of a degenerate S-shaped circular dichroism curve into the difference of two displaced bell-shaped curves with the intention of relating the resolved areas to the individual rotational strengths of the two components. The areas so obtained are incommensurate with those of nonoverlapping bands.

In (1.53) we found that we could relate the limiting form of the dispersion to the rotational strength, $Im(\mathbf{\mu}_{oa} \cdot \mathbf{m}_{ao})$, by assuming no overlapping bands. The dichroism was given the form, $\theta(\omega) = \sum_a \theta_a \delta(\omega - \omega_a)$. If now we assume the existence of a degenerate band such that

$$\theta_a(\omega) = \theta_a[\delta(\omega - \omega_a) - \delta(\omega - \omega_a - \Delta)], \tag{1.82}$$

the contribution to $\phi(\omega)$ will be

$$\phi_a(\omega) = \frac{\omega^2}{2\pi} \frac{\theta_a}{\omega_a} \left[\frac{1}{\omega_a^2 - \omega^2} - \frac{1}{(\omega_a + \Delta)^2 - \omega^2} \right] \simeq \frac{\omega^2 \theta_a}{\pi} \frac{\Delta}{(\omega_a^2 - \omega^2)^2}. \tag{1.83}$$

The equal and opposite rotational strengths are given by (1.80). If ω_a and ω_b differ by some small quantity, the resulting limiting form of the dispersion is

$$\phi_a(\omega) = \text{const } \omega^2 \omega_a^2 \mu_{oa}^{(A)} \times \mathbf{R}_{AB} \cdot \mu_{oa}^{(B)} \mu_{oa}^{(A)} \cdot \mathbf{T} \cdot \mu_{oa}^{(B)} \frac{1}{(\omega_a^2 - \omega^2)^2}. \tag{1.84}$$

From the last two equations we may infer that

$$\theta_a \Delta = \text{const } \omega_a^2 \mu_{oa}^{(A)} \times \mathbf{R}_{AB} \cdot \mu_{oa}^{(B)} \mu_{oa}^{(A)} \cdot \mathbf{T} \cdot \mu_{oa}^{(B)}.$$

This varies as $(1/R_{AB}^2)$. So long as the two components of the band are absolutely separated, any combination of θ_a and Δ with the same product will lead to the same result for the dispersion outside the absorption region. This will be equally true for arbitrarily large θ_a's and small Δ's and, within certain limits, smaller θ_a's and large Δ's.

If degenerate perturbation theory is applied to the above system,

$$\Delta = 2\mu_{oa}^{(A)} \cdot \mathbf{T} \cdot \mu_{oa}^{(B)} \sim \frac{1}{R_{AB}^3} \tag{1.85a}$$

$$R_a = \text{const } \frac{v_a}{c} \mu_{oa}^{(A)} \times \mathbf{R}_{AB} \cdot \mu_{oa}^{(B)} \sim R_{AB}. \tag{1.85b}$$

It follows that both θ_a and R_a vary as R_{AB}; in fact, they are connected by the relation (1.55). This resolution into nonoverlapping components leads to individual rotational strengths which increase with the distance of separation. We may then expect that the process of resolving an overlapping S-curve into individual Gaussian or bell-shaped components will result in the same behavior. It is doubtful whether the area under the actual resolved curves would vary in this manner, owing to many other complicating factors, particularly if we attempted only to obtain the best fit for an arbitrary Δ instead of the value in (1.85).

If the simple oscillator theory were valid it would provide us with rotational strengths that increase with distance. The way out of the difficulty is to recognize that, while (1.84) is the correct limiting form of the dispersion, (1.82) is not the correct form of the dichroism even though it leads to (1.84). There is in fact an infinite variety of analytical forms for $\theta(\omega)$ which leads to the correct limiting form. In fact, any analytic odd function of ω that vanishes at $\omega = \omega_o$ and has one maximum and one minimum will do.

For example, the function

$$\theta_a = A(\omega_a^2 - \omega^2)[e^{-a(\omega - \omega_a)^2} + e^{-a(\omega + \omega_a)^2}], \tag{1.86}$$

has the necessary properties. The constant A may be related to the limiting form (1.84) through the Kronig-Kramers relation (1.53). Experimentally A may be found from the best fit of (1.86) to the data. If the dispersion curve is available, the above procedure may be tested for self-consistency by the degree of its fit to the equation predicted by (1.53). If necessary, the empirical functions may be employed until a completely consistent result is obtained.

The most accurate procedure will probably be a graphical one, but we may attain a feeling for the method by observing that once an analytical form such as (1.86), along with its Kronig-Kramers transform has been established to fit both the observed dispersion and dichroism curves, only a few experimental numbers equal to the number of undetermined constants are needed to completely specify the system. In the above function this number is two. One relation may be obtained by finding the area under the positive half of the curve. Error may well be minimized by taking the other from the dispersion curve. Dispersion curves tend to overlap more, and the evaluation of areas is unsuitable: however, other quantities such as half-widths and differences between extrema may prove useful. If a graphical method is not used, several alternatives are recommended as a check.

For example, the value of ϕ at $\omega = \omega_a$ is readily found from (1.53):

$$\phi(\omega_a) = -\frac{\omega^2}{2\pi} \int_0^\infty A[e^{-a(\omega' - \omega_a)^2} + e^{-a(\omega' + \omega_o)^2}]\,d\omega'$$

$$\cong -\frac{\omega^2}{2\pi}\sqrt{\frac{\pi}{a}}\,A. \tag{1.87}$$

The limiting form of the dispersion will always coincide with that calculated from the Rosenfeld equation and an identification of calculated quantities with suitably chosen experimental parameters can be made. These quantities will always have a reasonable distance dependence. This procedure could serve as a stopgap until a more comprehensive theory of absorption is developed. Simple models indicate that for such systems the dichroism curve may be scaled to fit the derivative of the ordinary absorption curve.

6 INTRAMOLECULAR FIELDS

In the preceding section two simple models were discussed, a one-electron or local dissymmetry theory and a coupled oscillator or nonlocal dissymmetry mechanism. For instructive or empirical purposes these approximations seem adequate to cover most instances. They have the advantage of simplicity, and when prudently treated have enough flexibility to simulate actual behavior and provide expressions from which the rotary strengths of several compounds may be correlated with a few parameters.

In (1.70) we obtained a perturbation expression for a symmetrical chromophore in the presence of a dissymmetric field. One of the disadvantages of this formulation is that it requires a knowledge of excited states generally in the far ultraviolet, for which there are little data. In most cases the transition whose optical activity is being studied lies below all the transitions which give nonvanishing terms in (1.70). Rather than struggling with crude estimates to excited-state function, we can make a summation with an average energy denominator. A typical term will become

$$\sum_{s \neq n} \frac{V_{sn} \mathbf{\mu}_{os} \cdot \mathbf{m}_{no}}{E_s - E_n} \cong \frac{1}{\bar{E} - E_n} \sum_{\substack{s \neq o \\ \neq o}} V_{sn} \mathbf{\mu}_{os} \cdot \mathbf{m}_{no} + \frac{1}{E_o - E_n} V_{on} \mathbf{\mu}_{oo} \cdot \mathbf{m}_{no}$$

$$= \frac{1}{\bar{E} - E_n} \left[\sum_s V_{sn} \mathbf{\mu}_{os} \cdot \mathbf{m}_{no} - V_{nn} \mathbf{\mu}_{on} \cdot \mathbf{m}_{no} \right]$$

$$+ \left(\frac{1}{E_o - E_n} - \frac{1}{\bar{E} - E_n} \right) V_{on} \mathbf{\mu}_{oo} \cdot \mathbf{m}_{no}$$

$$= \frac{1}{\bar{E} - E_n} (\mathbf{\mu} V)_{on} \cdot \mathbf{m}_{no} + \left(\frac{1}{E_o - E_n} - \frac{1}{\bar{E} - E_n} \right) V_{on} \mathbf{\mu}_{oo} \cdot \mathbf{m}_{no},$$

$$(1.88)$$

where $\mathbf{\mu}_{on} \cdot \mathbf{m}_{no}$ equals zero by hypothesis, and the quantum mechanical matrix summation rule has been used. The term for $s = o$ must be treated separately, since the denominator is of opposite sign.

Equation (1.70) then becomes

$$\mathbf{\mu}'_{on} \cdot \mathbf{m}'_{no} \cong - \frac{1}{\bar{E} - E_n} \left[(\mathbf{\mu} V)_{on} \cdot \mathbf{m}_{no} + \mathbf{\mu}_{on} \cdot (V \mathbf{m})_{no} - \mathbf{\mu}_{oo} \cdot \mathbf{m}_{no} V_{on} \right]$$

$$- \frac{1}{\bar{E} - E_o} \left[(\mathbf{\mu} V)_{on} \cdot \mathbf{m}_{no} + \mathbf{\mu}_{on} \cdot (\mathbf{m} V)_{no} - \mathbf{\mu}_{nn} \cdot \mathbf{m}_{no} V_{on} \right]$$

$$+ \frac{V_{on}}{E_n - E_o} (\mathbf{\mu}_{oo} - \mathbf{\mu}_{nn}) \cdot \mathbf{m}_{no}. \qquad (1.89)$$

If reasonably good wave functions for the ground and excited state are known along with a good estimate to V, we might attempt an evaluation of (1.89) with a mind to assessing the importance of this term rather than expecting a meaningful mathematical result.

As an example of the use of this formula, consider two basic types of transition, an allowed magnetic dipole transition of the type, $\psi_{2px} \to \psi_{2py}$, and an allowed electric dipole transition represented by $\psi_{2px} \to \psi_{3d_{xz}}$. With m_z taken as $-i\hbar(\partial/\partial\phi)$,

$$m_z \psi_{2px} = i\frac{e\hbar}{\partial mc}\psi_{2py} \tag{1.90a}$$

$$m_z \psi_{2py} = -i\frac{e\hbar}{\partial mc}\psi_{2px} \tag{1.90b}$$

$$m_z \psi_{3dxz} = i\frac{e\hbar}{\partial mc}\psi_{2pyz}. \tag{1.90c}$$

For the magnetic dipole transition (1.89) gives

$$R_{on}(M) = Re[\mathbf{\mu}'_{on}(M) \cdot \mathbf{m}'_{no}(M)] = -\frac{e^2 h}{2mc}\left[\frac{1}{\bar{E}-E_n} + \frac{1}{\bar{E}-E_o}\right]$$

$$\times N \int xyzVe^{-ar}\,d\tau. \tag{1.91}$$

since the permanent dipole terms are zero at the origin. If only a point charge Q located at (X, Y, Z) is considered as a perturbation, the lowest order, lowest symmetry terms in the expansion of the potential are given by

$$V = \frac{-Qe^2}{\sqrt{(x-X)^2 + (y-Y)^2 + (z-Z)^2}}$$

$$= -Qe^2 \sum_k (-1)^k \frac{1}{k!}\left(x\frac{\partial}{\partial X} + y\frac{\partial}{\partial Y} + z\frac{\partial}{\partial Z}\right)^k\left(\frac{1}{R}\right)$$

$$= -Qe^2\left[\frac{1}{R} + \frac{xX + yY + zZ}{R^3} + \frac{3(xyXY + xzXZ + yzYZ)}{R^5}\right.$$

$$\left. + \frac{15}{R^7}xyzXYZ + \cdots\right], \tag{1.92}$$

where $R = \sqrt{X^2 + Y^2 + Z^2}$. The lowest nonvanishing term is the last, and (1.91) becomes

$$R_{on}(M) = (Qe^2)\left(\frac{e^2 h}{2mc}\right)\left[\frac{1}{\bar{E}-E_n} + \frac{1}{\bar{E}-E_o}\right]\left[N\int x^2 y^2 z^2 e^{-ar}\,d\tau\right]\left(\frac{15\gamma_x \gamma_y \gamma_z}{R^4}\right). \tag{1.93}$$

Except for Q and the direction cosines, $\gamma_x, \gamma_y, \gamma_z$, all the quantities in this expression are positive.

The result may be summarized: If a right-handed coordinate system is established with $\mathbf{i} \times \mathbf{j} = \mathbf{k}$, the sign of the rotatory strength for a magnetic dipole transition $x \to y$ is governed by that of the product $Q\gamma_x\gamma_y\gamma_z$. Note that $\mathbf{i}$ and $\mathbf{j}$ may be chosen along either direction. A reversal in direction of $\mathbf{i}$ or $\mathbf{j}$ requires a reversal in direction of $\mathbf{k}$, since $\mathbf{i} \times \mathbf{j} \cdot \mathbf{k} = 1$. The product, $\gamma_x\gamma_y\gamma_z$, is unaffected by a change in sign of any pair of direction cosines. Within the framework of this simple model the sign is unambiguously related to conformation.

A similar calculation for the electric dipole transition, $\psi_{2px} \to \psi_{3dxz}$, gives

$$R_{on}(E) = (Qe^2 \left(\frac{eh}{2mc}\right) \mu \left[\frac{1}{\bar{E} - E_n} - \frac{1}{\bar{E} - E_o}\right] \left[N' \int x^2 y^2 z^2 e^{-a'r} \, d\tau\right] \frac{15\gamma_x\gamma_y\gamma_z}{R^4},$$

$$(1.94)$$

where μ is the electric moment of the transition. This leads to the rule: In a right-handed coordinate system the sign of an electric dipole transition, which may be approximated by the hydrogen like transition, $x \to xz$, is governed by that of the product $Q\gamma_x\gamma_y\gamma_z$.

These results apply only to the effects of a dissymmetric point charge distribution without electron exchange. This will be a serious defect in the case of electric dipole transitions, where the coupled oscillator mechanism must always be considered. Sometimes both effects are present when a chromophore is perturbed by a polarizable group with a large charge. The preceding discussion indicates that if only point charges and coupled dipole oscillations are considered, the rotational strengths of electric and magnetic dipole transitions will have the form,

$$R_{on}(E) = \sum_k \frac{A_k}{R_k^2} + \sum_k B_k \frac{Q_k}{R_k^4} \tag{1.95a}$$

$$R_{on}(M) = \sum_k C_k \frac{Q_k}{R_k^4} \tag{1.95b}$$

where the parameters B_k and C_k depend only on the direction cosines of the perturbing group, and A_k depends both on the location of the group and its orientation in space. More general expressions would take into account other powers of $1/R$ as well as constant and exponential terms.

If, instead of a pure $2p_y$ function, we had chosen a distorted $2p_y$ function polarized along the z-axis, the integral, $\int (xy)z\psi_{2px} \psi'_{2py} \, d\tau$, would no longer vanish; and the $1/R^5$ term in (1.92) would make the lowest order contribution with the result,

$$R'_{on}(M) = (Qe^2)\left(\frac{e^2h}{2mc}\right)\left[\frac{1}{\bar{E} - E_n} + \frac{1}{\bar{E} - E_o}\right]$$
$$\times \left[N \int xyz\psi_{2px} \psi'_{2py} \, d\tau\right] \frac{3\gamma_x\gamma_y}{R^3}. \tag{1.96}$$

Both (1.93) and (1.96) lead to the same qualitative result in the so-called back octants, but (1.93) predicts a reversal in sign as the perturbing group passes through the transition's plane of polarization. In carbonyl derivatives (1.96) has been found useful in correlating experimental data. A study of compounds with substituents in front octants indicates that a quadrant rule, and not an octant rule, applies. On the basis of the above considerations this is not at all unreasonable, since the π^* orbital is most certainly not a pure $2p$ orbital on oxygen and is more than likely polarized toward the carbon atom.

A substantial advance in the formulation of empirical expressions can be made by the next reasonable approximation to the coulombic field of a molecule by the division into negative line charges along the bonds and positive or negative point charges at the nuclei. In effect this will approximate the many center integrals occurring in an exact treatment of the problem by their lowest order terms.

At this stage of quantum mechanical development one of the greatest contributions that can be made by optical rotation theory is the derivation of expressions for correlating the behavior of optically active molecules. In the important field of biochemistry, progress is already being made with the use of such expressions in predicting the behavior of large molecules from the properties of small ones.

Part 2 THE FARADAY EFFECT

7 GENERAL DESCRIPTION

The ordinary electromagnetic spectrum of a substance is obtained by investigating the behavior of randomly polarized light traversing the medium. The data obtained are put into two categories, infrared and ultraviolet, which give general information about the nuclear and electronic structure of the molecule. As we have seen, the use of circularly or plane-polarized light makes it possible to delineate subtle properties of a dissymmetric molecule. This technique may be used in the solid, liquid, or gas phase.

When molecules are arranged in the orderly pattern of a crystal, X-rays can give quite precise information on geometry of the molecules. If all molecules had a single equilibrium geometry and formed perfect crystals, the task of conformational analysis would lie almost entirely within the province of X-ray diffraction. Many complex biomolecules take on an entirely different conformation depending on their environment; some of them decompose when one tries to put them into the solid state. For this reason both optical activity and x-ray diffraction comprise a useful complementary study of molecular conformations in the various phases of matter.

From quantum theory we know that the presence of an external electromagnetic field introduces an extra perturbing term in the Hamiltonian of a molecule,

$$H' = i\,\frac{e\hbar}{mc}\sum_i \mathbf{A}(\mathbf{r}_i)\cdot\nabla_i + \frac{e^2}{2mc^2}\sum_i |\mathbf{A}(\mathbf{r}_i)|^2 + \sum_i e\phi(\mathbf{r}_i), \qquad (1.97)$$

where $\mathbf{A}(\mathbf{r}_i)$ is the vector potential of the field at the position of the ith electron and $\phi(\mathbf{r}_i)$ is the scalar potential. In radiation problems the time-dependent part of ϕ is set equal to zero because at low radiation intensities the charge density of the resonating molecule is constant. That is the justification for setting $\nabla\cdot\mathbf{A} = 0$, since the conservation of charge requires $\nabla\cdot\mathbf{A} + (1/c)(\partial\phi/\partial t) = 0$. This is probably a good approximation for weak transitions but warrants closer examination for electric dipole transitions, particularly when there is group degeneracy.

In general we can perturb a molecule with an arbitrary electromagnetic field consisting of time-dependent, time-independent, uniform, and nonuniform components. The degree of which $\mathbf{A}$ and ϕ can be adjusted to fit arbitrary functional forms subject to charge continuity requirements depends on the ingenuity of the experimenter. A summary of some of the common combinations is given in Table 1.2.

In addition, many other combinations have been used, such as multiple irradiation, in which light of two different frequencies is employed. The special effects that are observed when two identical groups are in a molecule may also be in evidence when beams of different frequency are used. In this section we investigate the properties of induced optical activity.

A major distinction between the Faraday and Kerr effects must be made. The Faraday effect arises from the propagation of circularly polarized light with different phase velocities and may be compared with natural optical activity; whereas the Kerr effect arises from macroscopic anisotropy rather than from induced modifications in the motion of the atomic currents. A further distinction must be made from natural optical activity. The situation may be summarized as follows.

In a naturally optical active medium dissymmetric currents effect the propagation of circularly polarized waves with different indices of refraction. This leads to the reciprocally related phenomena of optical rotatory dispersion and circular dichroism. If the medium is also anisotropic there will be certain directions, the optic axes, for which the transmission is much like the isotropic case. For propagation along any other directions the situation is complicated by birefringence. In its simplest form linear birefringence, in contrast to optical rotation, is caused by the propagation of plane-polarized waves with different phase velocities. In general the waves transmitted are elliptically polarized.

Table 1.2

Field	Designation	Data
Uniform periodic	I.R., U.V., radio, x-ray, Raman spectroscopy	Characteristic group and crystal properties
Uniform periodic with constant magnetic field		
Perpendicular, high frequency	Zeeman effect	Spin-orbit coupling, generally for atoms
Perpendicular, radio frequency	Nuclear and electron spin resonance	Environmental characteristics of interacting groups and molecular charge density
Parallel, polarized, high frequency	Faraday effect	Symmetry properties of excited states
Uniform periodic with constant electric		
Unpolarized	Stark effect	Splitting of energy levels
Polarized	Kerr effect	Molecular polarizability tensors

In an isotropic optically active medium, whether natural or magnetically induced, a plane wave will be resolved into two oppositely polarized circular waves with different phase velocities. These components recombine upon leaving the medium to produce a rotated plane wave. For an arbitrary direction in an anisotropic medium a pair of elliptically polarized waves are propagated with different phase velocities. Any incident radiation must be resolved into these two components. If the incident beam has exactly the ellipticity of one of these solutions, it will be propagated without change, as would a circularly polarized ray through an isotropic optically active medium. For any other polarization the incident beam will be converted into an elliptically polarized beam, whose principal axis is continuously rotated upon passage through the medium. The principal axis of the emergent elliptically polarized light may be determined by optical means. If the incident light was plane polarized, the angle between the two directions can be interpreted as an angle of rotation.

If, in addition, the medium is optically active, there will be an additional contribution to the angle of rotation due to the dissymmetry. At present there seems to be no straightforward technique for separating the combined effects of dissymmetry and anisotropy for propagation along directions other than the optic axis. If the medium is isotropic, pure optical rotation occurs; if it is anisotropic and inactive, we observe pure linear birefringence.

In the Faraday effect both effects are possible. If the light ray is parallel to the magnetic field, only magnetically induced optical rotation is observed, because this direction is an axis of symmetry of the system. If the two directions are not parallel, the effect of birefringence will be superimposed on the magnetically induced rotation. A more detailed analysis shows that the two phenomena are really different aspects of the same thing—the modification of the molecular currents in the presence of a constant magnetic field.

The Kerr effect is the result of pure birefringence, if the molecules themselves are naturally inactive. The electric field establishes an axis of symmetry for the anisotropic system. No birefringence effects are found for light parallel to this axis. When the direction of the ray is perpendicular to the electric field, the maximum effects of the induced anisotropy are observed in the resulting birefringence. Plane-polarized light whose direction of polarization is perpendicular or parallel to the electric field will pass through the medium unchanged. Light with any other direction of polarization will be converted to elliptical polarization with the principal axis rotated with respect to the original direction of polarization. This angle is also called an angle of rotation and is often confused with that of natural or magnetically induced optical activity.

A major distinction between the two is that only for the Kerr effect is there a direction for which no rotation is seen; and one necessarily deals with elliptical polarization even in the absence of absorption. We shall confine ourselves to a discussion of the Faraday effect, which is exhibited by all atoms and molecules in gas, liquid, and solid phases.

8 QUANTUM MECHANICAL DERIVATIONS OF GENERAL EQUATIONS

From (1.21*a,b*) the induced electric and magnetic dipole moments were found by averaging over all molecular orientations relative to the fixed vectors of the electromagnetic field. In the presence of a constant force field the averaging takes a different form. The two principal effects of such fields are a change in the molecular wave functions and a change in their statistical distribution function. As it happens, the Kerr effect and the Faraday effect individually represent these two extremes. For the low fields generally employed a constant electric field most strongly affects the distribution function by orienting polar molecules along a preferred direction. The Kerr effect should not be confused with the Stark effect for which the field is strong enough to cause a noticeable perturbation of the wave functions.

On the other hand, for diamagnetic molecules a magnetic field has a negligible effect on the distribution function but appreciably changes the wave functions. The field brings about a perturbation term, $H' = -\mathbf{H}_o \cdot \mathbf{m}$, which can be rationalized in terms of the classical energy of orientation of a magnetic dipole, where $\mathbf{m}$ is a quantum mechanical operator. The perturbed functions for the ground and excited state are

$$\psi'_n = \psi_n + \sum_{s \neq n} \frac{(\mathbf{H}_o \cdot \mathbf{m})_{sn} \psi_s}{h\nu_{ns}} \tag{1.98a}$$

$$\psi'_o = \psi_o + \sum_{t \neq o} \frac{\mathbf{H}_o \cdot \mathbf{m}_{to} \psi_t}{h\nu_{ot}}, \tag{1.98b}$$

where $\nu_{ns} = (E_s - E_n)/h$. Even though the molecules are not oriented by the field, their wave functions depend on their orientation relative to the field. This can be seen from the above equations by noting that $\mathbf{H}_o$ is fixed and $\mathbf{m}_{sn}$ depends on the orientation of the particular molecule. We are entitled to average (1.21) over all orientations with equal weighting of molecular orientations but unequal weighting of wave functions and energies. In the expressions for the induced electric and magnetic moments will occur terms of the type,

$$\boldsymbol{\mu}'_{on} \boldsymbol{\mu}'_{no} \cdot \mathbf{A}_{\text{eff}} = \left\{ \boldsymbol{\mu}_{on} \boldsymbol{\mu}_{no} + \sum_{s \neq n} \frac{\boldsymbol{\mu}_{on}(\mathbf{H}_o \cdot \mathbf{m}_{ns})\boldsymbol{\mu}_{so} + \boldsymbol{\mu}_{os}(\mathbf{H}_o \cdot \mathbf{m}_{sn})\boldsymbol{\mu}_{no}}{h\nu_{ns}} \right.$$

$$\left. + \sum_{t \neq o} \frac{\boldsymbol{\mu}_{on}(\mathbf{H}_o \cdot \mathbf{m}_{to})\boldsymbol{\mu}_{nt} + \boldsymbol{\mu}_{tn}(\mathbf{H}_o \cdot \mathbf{m}_{ot})\boldsymbol{\mu}_{no}}{h\nu_{ot}} \right\} \cdot \mathbf{A}_{\text{eff}}. \tag{1.99}$$

As before, $\langle \boldsymbol{\mu}_{on} \boldsymbol{\mu}_{no} \cdot \mathbf{A}_{\text{eff}} \rangle = \frac{1}{3} |\boldsymbol{\mu}_{on}|^2 \mathbf{A}_{\text{eff}}$. The averaging of the rest of the expression will lead to terms of the form, $\langle \mathbf{A}_1(\mathbf{A}_2 \cdot \mathbf{B}_1)(\mathbf{A}_3 \cdot \mathbf{B}_2) \rangle_{\text{av}}$, where $\mathbf{A}_1, \mathbf{A}_2, \mathbf{A}_3$ are fixed relative to one another in an arbitrary orientation with respect to the fixed plane determined by the vectors, $\mathbf{B}_1$ and $\mathbf{B}_2$. The averaging is over all relative orientations; $\mathbf{A}_1, \mathbf{A}_2, \mathbf{A}_3$ will be identified with the appropriate matrix elements of a molecule and $\mathbf{B}_1, \mathbf{B}_2$ with the electromagnetic field vectors.

In dealing with any such average between two sets of vectors internally fixed in orientation a useful relation is

$$(\mathbf{A}_1 \cdot \mathbf{B}_1)(\mathbf{A}_2 \cdot \mathbf{B}_2)(\mathbf{A}_3 \cdot \mathbf{B}_3)_{\text{av}} = \tfrac{1}{6}(\mathbf{A}_1 \cdot \mathbf{A}_2 \times \mathbf{A}_3)(\mathbf{B}_1 \cdot \mathbf{B}_2 \times \mathbf{B}_3), \tag{1.100}$$

from which it follows that

$$\langle \mathbf{A}_1(\mathbf{A}_2 \cdot \mathbf{B}_1)(\mathbf{A}_3 \cdot \mathbf{B}_2) \rangle_{\text{av}} = \tfrac{1}{6}(\mathbf{A}_1 \cdot \mathbf{A}_2 \times \mathbf{A}_3)\mathbf{B}_1 \times \mathbf{B}_2. \tag{1.101}$$

The use of (1.98*a,b*) and (1.101) in (1.21) gives

$$\langle \mathbf{\mu} \rangle = [\alpha \mathbf{E} + f_1 \mathbf{H}_o \times \mathbf{E}] + f_2 \mathbf{H}_o \times \dot{\mathbf{E}}$$

$$+ [\gamma \mathbf{B} + f_3 \mathbf{H}_o \times \mathbf{B}] + \left[-\frac{\beta}{c} \dot{\mathbf{B}} + f_4 \mathbf{H}_o \times \dot{\mathbf{B}} \right] \quad (1.102a)$$

$$\langle \mathbf{m} \rangle = [\kappa \mathbf{B} + f_1' \mathbf{H}_o \times \mathbf{B}] + f_2' \mathbf{H}_o \times \dot{\mathbf{B}}$$

$$+ [\gamma \mathbf{E} + f_3' \mathbf{H}_o \times \mathbf{E}] + \left[\frac{\beta}{c} \dot{\mathbf{E}} + f_4' \mathbf{H}_o \times \dot{\mathbf{E}} \right], \quad (1.102b)$$

where the effective field quantities are understood, and $\mathbf{E} = -(1/c)\dot{\mathbf{A}}$, $\mathbf{B} = \mathbf{V} \times \mathbf{A}$. The quantities, $\mathbf{B}$ and $\mathbf{E}$ have been used because only they can be derived from a single vector potential. Throughout the entire medium the electromagnetic field at any point is represented by a scalar potential ϕ and a vector potential $\mathbf{A}$. In radiation problems it is generally assumed that ϕ is independent of time. One should always remember, in considering effective fields, that only for time-independent constant fields can the magnetic and electric fields be specified independently. In the following discussion only the propagation of light parallel to the constant magnetic field will be of interest; and again it may be assumed that $\mathbf{A}_{\mathrm{eff}} = S\mathbf{A}_{\mathrm{av}}$.

Not all of the eight parameters, $f_1 \cdots f_4, f_1' \cdots f_4'$, are needed to describe the most important application, where $\mathbf{H}_o$ is parallel to the light beam. The other quantities, α, κ, β, and γ have been discussed in Part 1; they describe the behavior of the medium in the absence of external fields other than the radiation.

To see what types of waves are propagated under the above circumstances we must solve the wave equation

$$\mathbf{V} \times \mathbf{B} = \frac{1}{c} \dot{\mathbf{E}} + \frac{4\pi}{c} \dot{\mathbf{P}} + 4\pi \mathbf{V} \times \mathbf{M}, \quad (1.103)$$

where $\mathbf{E} = -(1/c)\dot{\mathbf{A}}_{\mathrm{av}}$ and $\mathbf{B} = \mathbf{V} \times \mathbf{A}_{\mathrm{av}}$. In (1.102*a,b*), as in Part 1, it will be assumed that $\mathbf{A}_{\mathrm{eff}} = S\mathbf{A}_{\mathrm{av}}$. The substitution of (1.102) into (1.103), when all vectors are written in terms of $\mathbf{A}_{\mathrm{av}}$, gives

$$-(1 - 4\pi NS\kappa)\mathbf{V}^2\mathbf{A} = -\frac{1}{c^2}(1 + 4\pi NS\alpha)\ddot{\mathbf{A}}$$

$$= -\frac{1}{c^2}(1 + 4\pi NS\alpha)\ddot{\mathbf{A}} - 8\pi \frac{NS\beta}{c^2} \mathbf{V} \times \ddot{\mathbf{A}} + 4\pi NS\left\{ -\frac{f_2}{c^2} \mathbf{H}_o \times \ddot{\mathbf{A}} \right.$$

$$- f_2'(\mathbf{H}_o \cdot \mathbf{V})\mathbf{V} \times \dot{\mathbf{A}} + \frac{1}{c}(f_3' - f_3)(\mathbf{H}_o \cdot \mathbf{V})\dot{\mathbf{A}}$$

$$+ \frac{f_3}{c} \mathbf{V}(\dot{\mathbf{A}} \cdot \mathbf{H}_o) + \frac{1}{c}(f_4' - f_4)(\mathbf{H}_o \cdot \mathbf{V})\ddot{\mathbf{A}} + \frac{f_4}{c} \mathbf{V}(\ddot{\mathbf{A}} \cdot \mathbf{H}_o) \right\}. \quad (1.104)$$

Various vector identities have been used to obtain this equation, along with the relation, $\mathbf{V} \cdot \mathbf{A} = 0$. In the expressions for the magnetic rotatory parameters it happens that f_1 and f_1' are zero and $f_4' = f_4$; this reduces the number of parameters to five. Furthermore, if the direction of propagation is along $\mathbf{H}_o$, $\mathbf{A} \cdot \mathbf{H}_o = 0$; and the terms in f_3 and f_4 will be eliminated. These terms will describe the birefringence for nonparallel propagation. The effect of the magnetic field will now be described by three parameters, f_2, f_2', and $f_3' - f_3$. To the degree of approximation used here we may certainly set $\kappa = 0$.

With these modifications the substitution of the trial function, $\mathbf{A} = \mathbf{A}_o e^{-i\omega(t - nz/c)}$, with $\mathbf{H}_o = H_o \mathbf{k}$ leads to the equation,

$$[(n^2 - \bar{n}^2) - 4\pi NSH_o(f_3' - f_3)n]\mathbf{A}_o$$

$$= 8\pi NS\beta \frac{i\omega n}{c} \mathbf{k} \times \mathbf{A}_o - 4\pi NSi\omega H_o[f_2 + n^2 f_2']\mathbf{k} \times \mathbf{A}_o, \quad (1.105)$$

where $\bar{n} = \sqrt{1 + 4\pi NS\alpha}$ is the average index of refraction in the absence of the magnetic field.

This is the magnetic modification of (1.28). The term in $(f_3' - f_3)$ makes a small correction to $\bar{n}$ and may be neglected. By analogy with the result in (1.33) we conclude that the total rotation is given by

$$\phi = \phi_{\text{ord}} + \phi_{\text{mord}},$$

where

$$\phi_{\text{ord}} = \frac{16\pi^3 v^2 NS\beta}{c^2} \quad (1.106a)$$

$$\phi_{\text{mord}} = \frac{8\pi^3 v^2 NS}{c} H_o \left[\bar{n} f_2' - \frac{f_2}{n}\right]. \quad (1.106b)$$

This result differs from the usual presentation because of the different assumptions on the nature of the effective field. The assumption that the effective electric and magnetic fields may be specified in a manner inconsistent with the requirement that $\mathbf{E} = -(1/c)\dot{\mathbf{A}}$ and $\mathbf{B} = \mathbf{V} \times \mathbf{A}$ is certainly wrong; but the relation between the average field and the effective field may be more complicated than a simple constant linear proportionality.

Caution again must be urged against the use of the Lorentz correction. In discussions of the Faraday effect this often leads to the factor, $[(n^2 + 2)/3]^2$, in the dispersion. Such expressions should be used only if they provide a clear-cut empirical correlation of the data, for they currently have no sound theoretical basis.

The quantum mechanical expressions for the parameters, f_2 and f_2', may now be written

$$f_2 = -\frac{1}{3\pi h}\sum_n \frac{1}{v_{on}^2 - v^2} Im\left\{\sum_{s\neq n}\frac{\mu_{on}}{hv_{ns}}\cdot \mathbf{m}_{ns}\times\mu_{so} + \sum_{t\neq o}\frac{1}{hv_{ot}}\mu_{on}\cdot \mathbf{m}_{to}\times\mu_{nt}\right\}$$

(1.107a)

$$f_2' = -\frac{1}{3\pi h}\sum_n \frac{1}{v_{on}^2 - v^2} Im\left\{\sum_{s\neq n}\frac{1}{hv_{ns}}\mathbf{m}_{on}\cdot \mathbf{m}_{ns}\times\mathbf{m}_{so} + \sum_{t\neq o}\frac{1}{hv_{ot}}\mathbf{m}_{on}\cdot \mathbf{m}_{to}\times\mathbf{m}_{n}\right\}.$$

(1.107b)

An analysis of the induced radiation discloses another distinction between natural optical activity and the Faraday effect. In the case of natural optical activity both the induced magnetic and electric moments make equal contributions to that part of the radiation field responsible for optical activity. If parallel electric and magnetic dipole transitions have nearly equal frequencies, they will borrow strongly from each other in a reciprocal manner, making equal and opposite contributions to the circular dichroism. While the magnetic transition makes a smaller contribution to the absorption spectrum, it enters on an equal basis in the circular dichroism.

On the other hand, these two types of transitions play much the same role in the Faraday effect as they do in absorption. The parameters, f_2 and f_2', separately describe the effects of induced electric and magnetic dipoles. The first is seen to depend on the product of one magnetic and two electric dipole moments, and the second on three magnetic moments. For the same reason that the magnetic oscillator strength, m_{on}^2, makes a negligible contribution to the total absorption of a symmetry-forbidden electric dipole transition in comparison with its vibronic induced electric oscillator strength, μ_{on}^2, we may conclude that $f_2' \ll f_2$ for all transitions.

If there are degenerate transitions, (1.98a,b) must be modified by the use of the proper zeroth-order combination of the functions. In a magnetic field the proper linear combinations for double degeneracy are $\psi_1 \pm i\psi_2$. The final result of such a calculation may be inferred from (1.107a) by taking the limit as $v_{ns} \to 0$ for a pair of transitions, $o \to n$ and $o \to s$. We obtain

$$\lim\left\{\left[\frac{1}{v_{on}^2 - v^2} - \frac{1}{v_{os}^2 - v^2}\right]\frac{\mu_{on}\cdot \mathbf{m}_{ns}\times\mu_{so}}{h(v_{os} - v_{on})}\right\} = \frac{2v_o\mu_{on}\cdot \mathbf{m}_{ns}\times\mu_{so}}{h(v_o^2 - v^2)^2}.$$

(1.108)

$$v_{os} \to v_{on} = v_o.$$

The functions, ψ_n and ψ_s, are now degenerate and the result in (1.108) is valid only if we take those linear combinations for which $\mu_{on}\cdot\mu_{so} = 0$.

This leads to an additional term in the dispersion of

$$f_2(\text{degen}) = -\frac{2}{3\pi h} \sum_n \frac{v_{on}}{h(v_{on}^2 - v^2)^2} \, \mathbf{\mu}_1(n) \cdot \mathbf{m}_{12}(n) \times \mathbf{\mu}_2(n) \qquad (1.109)$$

where $\mathbf{\mu}_1(n)$ and $\mathbf{\mu}_2(n)$ are the two perpendicular degenerate transition moments and $\mathbf{m}_{12}(n)$ is the magnetic moment between these two states. For higher degeneracy it is best to generalize by use of the proper linear combinations. For spherical or cylindrical symmetry it can be shown that $\mathbf{\mu}_1 \cdot \mathbf{m}_{12} \times \mathbf{\mu}_2 = |\mathbf{\mu}_1|^2(e\hbar/2mc)$. Owing to the orthogonality of the spherical harmonics, $\mathbf{m}_{ns} = 0$ for all pairs of states with a different principal quantum number; therefore, only terms of the type (1.109) contribute to the magnetic rotation of diagmagnetic atoms. Since e, the electronic charge, is negative, inspection of (1.106) and (1.109) leads to the conclusion that the rotation of diamagnetic atoms on the low-frequency side is negative.

A certain point of confusion may be encountered in the literature owing to the existence of two conventions for measuring optical rotation—the physicist's and the chemist's. The physicist's measurement is taken from the viewpoint of an observer looking along the direction of the wave, and the chemist's from that of an observer looking in the opposite direction as the light shines into his eye. The two lead to opposite signs. In the investigation of natural optical activity, chemists have predominated, and their convention has been used in the presentation of data on particular molecules.

The contributions of chemists comprise a fractional part of the literature on the Faraday effect, and a mixture of conventions will be found. The one used in this article is, of course, the chemist's; but we must establish which convention is employed in a given discussion.

The results of simple classical models are worth remembering when we attempt to extract information from the Faraday effect. An isotropically bound electron will exhibit a negative rotation on the low-frequency side of resonance. An anisotropically bound electron will also exhibit negative rotation for the lowest frequency but positive for the highest with the middle of either sign depending on the relative values of the three frequencies. This result is also obtained from the quantum theory for the harmonic oscillator.

9 PROPERTIES OF THE EQUATIONS

The general method for reporting magnetic optical rotatory dispersion (MORD) data is by means of the Verdet constant, V, defined by the relation

$$\phi = VLH_o, \qquad (1.110)$$

where L is the path length and H_o is the magnetic field. The term "positive rotation" refers to observations made with a positive magnetic field along the axis of wave propagation. Unfortunately for the investigator interested in both natural optical activity and the Faraday effect, the physicist's convention has been used in reporting the values of most Verdet constants.

As with natural optical activity, a magnetic rotational strength may be defined in terms of the area under the dichroism curve which, in turn, may be related by the Kronig-Kramers transform to the numerators in the expression for the Verdet constant in nonabsorbing regions,

$$V = \sum_n \frac{V_n}{v_{on}^2 - v^2}. \tag{1.111}$$

Again, degenerate transitions must be considered separately. Attempts to define V_n in terms of (1.108) and (1.109) will lead to a constant that increases with decreasing magnetic field strength. Any correlation of theory and experiment inside absorption regions will best be made in terms of the areas under the individual positive and negative portions of the S-curve degenerate dichroism band.

Like all electromagnetic response parameters, the magnetic rotation parameters satisfy a sum rule, that is, $\sum_n V_n = 0$. This may be seen from (1.21) by considering the sum,

$$Im \sum_{n \neq o} \mu_{on} \mu_{no} = Im |(\mu\mu)_{oo} - \mu_{oo} \mu_{oo}| = 0. \tag{1.112}$$

This applies to the system before averaging over molecular orientations, and will accordingly apply to crystals as well as to gases and liquids.

All molecules with thermally accessible degenerate states are paramagnetic. This means that their permanent magnetic dipoles may be oriented by the magnetic field. The averaging process leading to (1.107) must be modified by the inclusion of a weighting factor. For excited states this will make only a small correction to the Boltzmann factor, $e^{-E_n/kT}$; but for the ground state a substantial effect will be seen. If for the ground state $E_o = 0$, in the presence of $\mathbf{H}_o$ the energy will become

$$E_o' = -\mathbf{H}_o \cdot \mathbf{m}_{oo}.$$

The averaging of $\mu_{on} \mu_{no} \cdot \dot{\mathbf{E}}$ must now take the form,

$$Im[\langle \mu_{on} \mu_{no} \cdot \dot{\mathbf{E}} e^{\mathbf{H}_o \cdot \mathbf{m}_{oo}/kT} \rangle_{\mathrm{av}}]$$

$$\cong Im \left[\langle \mu_{on} \mu_{on} \cdot \dot{\mathbf{E}} \rangle_{\mathrm{av}} + \frac{1}{kT} \langle \mu_{on} (\mathbf{H}_o \cdot \mathbf{m}_{oo}) \mu_{no} \cdot \dot{\mathbf{E}} \rangle_{\mathrm{av}} \right]$$

$$= \frac{1}{6kT} Im(\mu_{on} \cdot \mathbf{m}_{oo} \times \mu_{no})(\mathbf{H}_o \times \dot{\mathbf{E}}), \tag{1.113}$$

where the approximation

$$e^{\mathbf{H}_o \cdot \mathbf{m}_{oo}/kT} \cong 1 + \mathbf{H}_o \cdot \mathbf{m}_{oo}/kT, \tag{1.114}$$

has been made. This leads to a paramagnetic contribution given by

$$f_2(\mathrm{para}) = -\frac{1}{3\pi h} \sum_n \frac{1}{(v_{on}^2 - v^2)kT} I_m |\mu_{on} \cdot \mathbf{m}_{oo} \times \mu_{no}|. \tag{1.115}$$

In this expression $\mu_{on} \times \mu_{no} \neq 0$; because in the presence of the magnetic field the correct linear combinations of the degenerate functions are complex due to existence of currents. For spherically and cylindrically symmetrical systems it follows from the properties of spherical harmonics that this term is always negative leading to a positive contribution to the Verdet constant. It seems likely that this will also be true for many degenerate systems of lower symmetry, but there does not appear to exist any general proof of the fact.

The anticipated behavior for the lowest strong electric dipole transition may be inferred by summing the first term of $(1.107a)$ with an average denominator $h\nu_n$. This leads to the result,

$$\sum_{s \neq n} \frac{\mu_{on}}{h\nu_{ns}} \cdot \mathbf{m}_{ns} \times \mu_{so} = \frac{1}{h\nu_n} \mu_{on} \cdot \sum_s \mathbf{m}_{ns} \times \mu_{so} = \frac{1}{h\nu_n} \mu_{on} \cdot (\mathbf{m} \times \mu)_{no}. \quad (1.116)$$

From elementary vector analysis it follows that

$$\mathbf{m} \times \mu = \frac{e^2}{2mc} (\mathbf{r} \times \mathbf{p}) \times \mathbf{r} = \frac{e^2}{2mc} [x\mathbf{p}x + y\mathbf{p}y + z\mathbf{p}z - \mathbf{r}(\mathbf{p} \cdot \mathbf{r})], \quad (1.117)$$

where the ordering of the noncommuting factors has required us to write $x\mathbf{p}x + y\mathbf{p}y + z\mathbf{p}z$ instead of $r^2\mathbf{p}$.

Use of the fundamental commutation rules gives

$$\mathbf{p} \cdot \mathbf{r} = \mathbf{r} \cdot \mathbf{p} - 3i\hbar \quad (1.118a)$$

$$x\mathbf{p}x + x\mathbf{p}y + z\mathbf{p}z = r^2\mathbf{p} - i\hbar\mathbf{r} \quad (1.118b)$$

This gives the relation,

$$\mathbf{m} \times \mu + \mu \times \mathbf{m} = 2i\left(\frac{e\hbar}{2mc}\right)\mu. \quad (1.119)$$

If the ground state orbital is totally symmetric or nearly so, $\mathbf{m}\phi_0 = 0$ and $(\mathbf{m} \times \mu)_{no} = 2i(e\hbar/2mc)\mu_{ro}$, from which it follows that

$$f_2 = -\frac{2}{3\pi h}\left(\frac{e\hbar}{2mc}\right)\frac{1}{h\bar{\nu}} \mu_{on}^2 \frac{1}{\nu_{on}^2 - \nu^2} \quad (1.120)$$

Since f_2 is of opposite sign to ν and e is negative, this predicts a negative sign for the lowest-allowed electric dipole transition of a diamagnetic molecule whose ground state orbital has nearly spherical symmetry (i.e., $s \to p$ type transitions).

If the excited state orbital is of the s type, the above term vanishes and the ground state perturbation term must be used. This leads to the result

$$f_2 \cong \frac{1}{3\pi h} \frac{1}{h\bar{\nu}_0} \mu_{on} \cdot (\mu \times \mathbf{m})_{no} \frac{1}{\nu_{on}^2 - \nu^2}.$$

$$= \frac{2}{3\pi h}\left(\frac{e\hbar}{2mc}\right)\frac{1}{h\bar{\nu}_0} \mu_{on}^2 \frac{1}{\nu_{on}^2 - \nu^2} \quad (1.121)$$

The opposite is expected in this case ($p \to s$ type transitions). In the event that neither ground nor excited state orbital approaches spherical symmetry, no simple deductions can be made.

The Faraday effect of paramagnetic species is rich in detail and somewhat complex in nature. Electron spin plays an important role and must be given special treatment; (1.115) may be used for spin degeneracy as well as for spatial degeneracy, but there are believed to be other terms of importance. It is generally found that at room temperature the paramagnetic effects are an order of magnitude larger than the diamagnetic effects.

According to (1.115) the paramagnetic contribution should be proportional to the product of the permanent magnetic moment and the oscillator strength of a transition. We can use this fact in assessing the relative paramagnetic and diamagnetic contributions.

10   USES OF THE FARADAY EFFECT

Finally we shall summarize the areas in which the Faraday effect is most likely to give important information about molecules. At present the wave mechanical theory of complex molecules can only provide functions best suited for calculating particular properties. The most widely investigated but least sensitive is the energy. Once reasonable agreement with experiment has been found, the next most sensitive quantities, the oscillator strengths, are investigated; the results are generally not as good as those for the energies. Even poorer agreement is found for transition moment directions in compounds of low symmetry. Owing to its extreme sensitivity to conformation, the poorest agreement of all would be expected for natural optical activity. The Faraday effect comes somewhere between transition dipole moment directions and optical activity.

In terms of matrix elements the following progression in sensitivity seems likely: H_{nn}, $|\mu_{on}|^2$, μ_{on}, $\mu_{on} \cdot \mathbf{m}_{ns} \times \mu_{so}$, $\mu_{on} \cdot \mathbf{m}_{no}$. In general the Faraday effect depends on both direction and magnitude of transition dipole moments. Unlike natural optical activity, the triple product of matrix elements is dependent on the properties of individual groups rather than subtle features of their interaction.

The insensitivity of $|\mu_{on}|^2$ to molecular environment is strikingly demonstrated through the additivity rule for molar refractivities. A similar situation has been encountered in the Verdet constants for a series of similar compounds, but the correlation is not quite so striking.

We can never expect a simple group additivity rule for Verdet constants or magnetic rotational strengths because both the magnitudes and directions of transition dipole moments govern the Faraday effect. Second, we are dealing with a signed quantity. The sum of the areas for the dichroism curve must be zero, whereas the total area of the absorption curve must obey the Kuhn-Thomas sum rule. The additivity of refractivities outside absorption regions may be regarded as a consequence of this rule. Inside an absorption band the refractivity is dominated by a single frequency and the rule might break down.

The Verdet constant is the sum of dispersion terms in $(V_n)/(v_{on} - v^2)$, where numerators sum to zero. This sum will be much more sensitive to changes in v_{on} and V_n than a sum of terms with positive coefficients.

The Faraday effect can answer many specific questions; for example, if an absorption band is thought to be degenerate, the question may be settled by studying the magnetic circular dichroism of the band. A composite band will display the characteristic S-shaped degenerate behavior.

A most fruitful area of research seems to be the behavior of symmetry forbidden transitions. Much vibronic fine structure is observed with intervening sign changes. All of this will be related to the symmetry of the contributing vibrational modes. The data so obtained may provide a valuable supplement to the information furnished by infrared, Raman, and neutron spectroscopy.

References

Condon, E. U., *Rev. Mod. Phys.*, **9**, 432 (1937).

Condon, E. U., W. Altar, and H. Eyring, *J. Chem. Phys.*, **5**, 753 (1937).

Eyring, H., H. Liu, and D. J. Caldwell, *Chem. Rev.*, **68**, 525 (1968).

Eyring, H., J. Walter, and G. E. Kimball, *Quantum Chemistry*, Wiley, New York, 1944.

Groenewege, M. P., *Mol. Phys.*, **5**, 541 (1962).

Kauzmann, W. J., J. E. Walter, and H. Eyring, *Chem. Rev.*, **26**, 339 (1940).

Kirkwood, J. G., *J. Chem. Phys.*, **5**, 479 (1937).

Kramers, H. A., *Proc. Acad. Sci. Amst.*, **33**, 959 (1930).

Landau, L. D., and E. M. Lifshitz, *Electrodynamics of Continuous Media*, Pergamon, Oxford, 1960.

Lowry, T. M., *Optical Rotatory Power*, Longmans, Green, London, 1935.

Tobias, I., and W. Kauzmann, *J. Chem. Phys.*, **35**, 538 (1961).

Chapter **II**

OPTICAL ROTATION — EXPERIMENTAL TECHNIQUES AND PHYSICAL OPTICS

Wilfried Heller
Revised by Henry G. Curme and Wilfried Heller

51

---

1 DEFINITIONS

Polarimetry deals with the quantitative investigation of all optical phenomena involving polarized light. It has become customary, however, to restrict the term "polarimetry" to quantitative investigations of the two following changes that may occur in polarized light when it interacts with optically anisotropic matter: (*a*) a change in the direction of vibration, that is, a rotation of a linear vibration perpendicular to the direction of propagation; and (*b*) a change in the state of vibration, that is, a transition of a linear vibration into an elliptic or circular vibration or vice versa. Substances that produce neither change are called optically *isotropic*, which implicitly defines the term *anisotropic*. A large number of phenomena are thus excluded in the customary definition of polarimetry, although their study calls for polarimetric methods, for example, the polarized fluorescence of dye solutions irradiated with unpolarized light. The organic chemist, however, is accustomed to an even narrower definition of polarimetry which excludes everything except quantitative investigations of the change in the direction of vibration of linearly polarized light during its passage through optically anisotropic liquids or solutions. This definition of polarimetry arises from the fact that, when polarimetry was developed a little over one hundred years ago, the optical phenomenon just mentioned was the only one involving polarized light known in organic liquids or solutions. The present chapter is limited to a discussion of the fraction of polarimetry defined by this tradition.

The change in the direction of vibration of linearly polarized light during passage through anisotropic matter is called " rotatory polarization," or " optical rotation " if the effect is due to anisotropic refraction. If it is caused by anisotropic absorption or anisotropic scattering, it should not be called "optical rotation," but "dichroism" (see p. 67). Substances that cause rotatory polarization are said to exhibit rotatory power. Substances that possess a "natural" rotatory power are called "optically active." A rotatory power may also be induced in a naturally isotropic substance by a magnetic field (Faraday effect). We shall deal primarily with the measurement of natural rotatory power.

The effect of rotatory power is shown in Fig. 2.1. A beam of linearly polarized light traveling perpendicular to the plane of the paper toward the

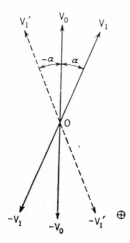

Fig. 2.1.   Optical rotation.

reader intersects the plane at O. The beam vibrates along $-V_0OV_0$, but will vibrate along $-V_1OV_1$ or $-V_1'OV_1'$ after traversing an optically active substance. Rotation α is positive if it is clockwise $(-V_0OV_0 \rightarrow -V_1OV_1)$, negative if the direction of the vibration is changed to $-V_1'OV_1'$. Substances that cause a positive rotation are called dextrorotatory or dextrogyrate, and substances causing a negative rotation, levorotatory or levogyrate.

In the following, the direction of vibration will be indicated by a single symbol only. V_0, for example, will stand for $-V_0OV_0$. In order to keep the figures as simple as possible, only half of a vibration will be reproduced, for instance, OV_0 instead of $-V_0OV_0$. This half represents the amplitude, a, of a vibration. The value of a is proportional to the square root of the light intensity, I.

Most authors use the terms "plane polarized," "elliptically polarized," and "circularly polarized." We use the term "linearly polarized" instead of "plane polarized" because the principle of projection involved is then identical for the three types of vibration. The terminology is essentially the same as that introduced by Brewster [1], who distinguished between rectilineal, elliptical, and circular polarization.

Symbols

A, analyzer

B, brightness (B_S of light source or optical field; $B_{Eq.}$, equality in two fields)

D, diaphragm

E, extinction (OE, direction of extinction in analyzer)

G, glass

I, intensity

 I_i, of beam incident upon analyzer

 I', after passing through light-absorbing substance

 I_0, emerging from analyzer in absence of optically active substance

 I_1, emerging in presence of optically active substance

 I_0, I_1, same if optically active substance absorbs or scatters light

L, lens

M, mirror

N, nicol, general meaning (see p. 79)

P, polarizer

Pr, prism

Pt, polarimeter tube

Q, quartz

R, light ray

S, light source

T, transmission (OT, direction of maximum transmission through analyzer)

V, vibration; direction of vibration (in abbreviation of OV)

V_0, direction of vibration of beam emerging from polarizer

V_1, after traversing optically active substance

V_1', after traversing optically active substance which absorbs or scatters light

V_1, vibration of linearly polarized beam incident upon analyzer, without specification of changes in V_0

a, amplitude (meanings of subscripts to a are analogous to those of subscripts to V)

i, angle of incidence

l, l', layer thickness

n, refractive index

r, angle of reflection (as subscript to R, resultant ray)

t, time

v, velocity of light (c, in vacuo)

x, optic axis of uniaxial substance

α, angle of rotation, in degrees, due to natural rotatory power (polarizability)

[α], specific rotation

{α}, intrinsic rotation

β, angle at apex of quartz wedge, cut perpendicular to optic axis

γ, angle between optic axis of aniaxial crystal and direction of vibration of incident linearly polarized light beam

δ, angle at apex of birefringent crystal wedge

ε, one-half of the half-shade angle

ζ, angle by which the position of half-shade equality deviates from position of crossed nicols (ζ', increment of ζ, due to loss in intensity by reflection)

η, angle between direction of transmission of analyzer and direction of vibration of incident beam

ϑ, angle of rotation due to linear dichroism

λ, wavelength ($λ_F$, extinct by analyzer; $λ_{Eq.}$, equally bright in two spectra compared after passage of light through analyzer)

ν, frequency, in sec^{-1}

π, 180°

ρ, angle of mechanical rotation ($ρ_P$, of polarizer; $ρ_A$, of analyzer; $ρ_{Sp}$, of spectrograph)

σ, angle between direction of extinction of analyzer and direction of vibration of incident beam

τ, range of angles of rotation of optically active substance between 6500 and 4500 A

φ, angle of rotation, in radians (cgs units)

ϕ, phase difference

χ, angle of rotation due to magnetic rotatory power (Faraday effect)

ω, numerical reading on circular scale of polarizer or analyzer

Ω, rotivity

↔, vibration in the paper plane or in perspective

×, vibration perpendicular to paper plane

x

↔, optic axis in the paper plane or in perspective

x ●, optic axis perpendicular to paper plane

O→←O, direction of propagation in paper plane or in perspective

⊕, direction of propagation perpendicular to paper plane and toward reader

O, direction of propagation perpendicular to paper plane and away from reader

2 PHYSICAL OPTICS OF ROTATORY POLARIZATION

Rotatory polarization in the case of optical activity was analyzed by Fresnel [2]. Linearly polarized light may be considered as the resultant of two coherent circular components of opposite sense of rotation, the d- and the l-component. The term *coherent* implies that the two components are able to interfere with each other. Furthermore, it is assumed that the velocities of the two components in the optically active substance differ. Figure 2.2 illustrates the case of an isotropic liquid. A linearly polarized beam of monochromatic light travels along the plane of the paper in the direction of the arrow. Rotating around the direction of propagation are the two circular components into which the beam is assumed to be resolved after it enters the liquid. The velocity of the two components in the liquid, $v_d = v_l = v_2$, differs from that in air, v_1, according to

$$v_1 : v_2 = n_2 : n_1 \qquad (2.1)$$

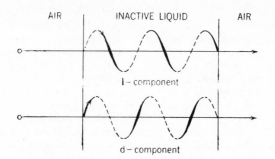

Fig. 2.2. Schematic resolution of a beam of linearly polarized light into two circular components. Optically inactive liquid. The actual phenomenon is visualized by a super-position of upper and lower drawings. Vibrations behind the plane of the paper are indicated by broken lines.

where n_1 is the refractive index of air and n_2 that of the liquid. The refractive index is defined by

$$n = \frac{c}{v} \quad \therefore n_1 = \frac{c}{v_1} \quad \text{and} \quad n_2 = \frac{c}{v_2} \tag{2.2}$$

where c is the velocity of light in a vacuum. Since the frequency, v, the number of vibrations per second, is constant and since

$$v = \frac{v}{\lambda} \tag{2.3}$$

it follows that

$$\lambda_1 : \lambda_2 = v_1 : v_2 = n_2 : n_1 \tag{2.4}$$

where λ_1 and λ_2 are the wavelengths in air and in the liquid, respectively. Consequently, the light waves are contracted in the liquid.

In Fig. 2.3, the isotropic liquid is replaced by an optically active one. This time, $v_d \neq v_l$, that is, the optically active liquid has two refractive indices,

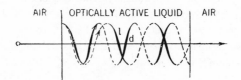

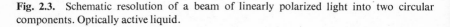

Fig. 2.3. Schematic resolution of a beam of linearly polarized light into two circular components. Optically active liquid.

n_d and n_l. This implies that λ_d differs from λ_l and that the planar projections of the two components will get out of phase, as shown in Fig. 2.3. Before discussing the significance of this getting out of phase, we consider the recombination of the two components leaving the isotropic liquid. In Fig. 2.4, the direction of propagation is perpendicular to the paper plane. We study the

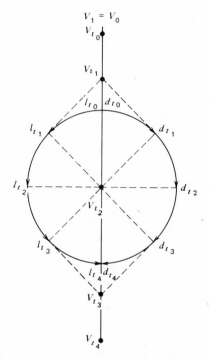

Fig. 2.4. Result of interference between two coherent d- and l-components if both travel equally fast. Optically inactive liquid.

recombination, in a given plane perpendicular to the direction of propagation, during one half-period, that is, during the time required by the light to advance by one half-wavelength, and we divide this period into four intervals, defined by the times t_0 to t_4. Evidently the resultants of the circular components fall on a straight line and the recombination results in a linear vibration, V_1, in the same direction as V_0. In Fig. 2.5, the case of an optically active liquid, the planar projections of the two circular components have reached a phase difference, ϕ, of $\pi/4$, that is, 45° upon arriving in the plane considered, the d-component being ahead of the l-component. The recombination leads again to a linear vibration, V_1, but the latter is rotated by 22.5° with respect to the initial vibration, V_0. The rotation, φ, amounts therefore to one-half

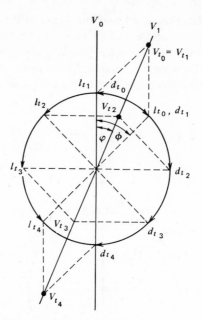

Fig. 2.5. Result of interference between two coherent d- and l-components if the d-component travels faster. Optically active liquid.

of the phase difference, ϕ, of the planar projection of the two circular components. It is evident from the preceding equations and figures that the angle

$$\phi = \left(\frac{2\pi l'}{\lambda_0}\right)(n_l - n_d) \tag{2.5}$$

and the rotation

$$\varphi = 0.5\phi = \left(\frac{\pi l'}{\lambda_0}\right)(n_l - n_d) \tag{2.6}$$

where both the layer thickness, l', and the vacuum wavelength, λ_0, are expressed in cm. The optical rotation is generally referred to a layer thickness of 10 cm. Consequently

$$\alpha_0 = 5\phi = \left(\frac{10\pi l}{\lambda_0}\right)(n_l - n_d) \text{ radian} \tag{2.7}$$

where l is the layer thickness in decimeters. On expressing the rotation in degrees, as is customary,

$$\alpha = \left(\frac{1800l}{\lambda_0}\right)(n_l - n_d) \text{ degrees} \tag{2.8}$$

We see that the optical rotation is caused by and is the experimental expression of a finite difference between the refractive indices n_l and n_d. (The rotation is, according to (2.6), positive if $n_l > n_d$ and negative if $n_l < n_d$.) The circular birefringence

$$(n_l - n_d) = \frac{\varphi \lambda_0}{\pi l'} \tag{2.9}$$

is rarely considered by the experimental man, although it is basically more significant than α which is proportional to it. One reason is that the numerical values of the circular birefringence are always very small. It is readily seen that a rotation of 0.1° by a 10-cm layer—measured at 5461 Å—is caused by a refractive index difference in the ninth decimal (2.8×10^{-9}). The molecular anisotropy responsible for optical rotation is therefore very small indeed.

It is to be noted that the theoretical proportionality between ϕ—and therefore of α—and layer thickness was verified at an early date by Biot [3].

Fresnel's concept is supported by an interesting experiment of Fleischl [4]. A linearly polarized beam is split into a left and a right circular component by refraction through a series of alternating prismatic layers of a dextrorotatory and a levorotatory liquid. This experiment duplicates (for the case of liquids) a result obtained by Fresnel on optically active crystals [5]. According to the writer, another formal analysis of rotatory polarization is possible. It is, for some purposes, more advantageous. The arrows shown in Fig. 2.6 give the

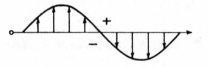
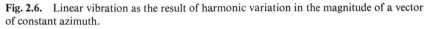

Fig. 2.6.   Linear vibration as the result of harmonic variation in the magnitude of a vector of constant azimuth.

direction and the relative magnitude of the electric vector of linearly polarized light, which travels in the direction of the horizontal arrow and vibrates in the plane of the paper. The electric vector maintains its azimuth, but varies harmonically between a maximum positive and negative value. An attempt is made in Fig. 2.7 to express a circular vibration by a vector quality. To that effect, the circular vibration is considered as equivalent to the resultant of two orthogonal coherent linear vibrations which have a phase difference of 90° (see pp. 74–76). It follows from the figure—which shows only the formation of the d-component—that rotatory polarization can be characterized as the result of a difference in the rate of rotation of two electric vector resultants of opposite sense of rotation. This concept leads at once to an easy understanding of the fact (p. 76) that a right circular or left circular vibration,

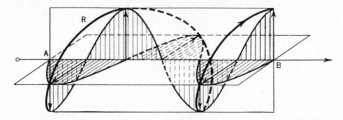

Fig. 2.7. Circular vibration by rotation of a resultant vector of constant magnitude. The arrows perpendicular to the horizontal direction of propagation show the electric vector of either of the two coherent orthogonal linear vibrations at the instant of maximum value. The helix, that is, circular vibration (heavy print), results from interference of the two linear vibrations due to their phase difference of 90°.

important for the investigation of the Cotton effect (p. 66), can be obtained from linearly polarized light by sending the light beam through a linearly anisotropic substance (pp. 70 *et seq.*) which produces a phase difference of $n \times 360° - 90°$ and $n \times 360° + 90°$, respectively, between two coherent orthogonal linear vibrations, n being any integral number or zero.

The arrows perpendicular to the horizontal direction of propagation show the electric vector of either of the two coherent orthogonal linear vibrations at the instant of maximum value. The helix, that is the circular vibration (heavy print), results from interference of the two coherent linear vibrations due to their phase difference of 90°.

3 PREREQUISITES FOR MOLECULAR ROTATORY POWER

Molecular Dissymmetry

Pasteur [6] discovered the principle which makes it possible to predict what molecules can be optically active. Each of the models of right and left circular vibration in Fig. 2.2 is the mirror image of the other; they cannot be brought into coincidence around whatever axis one may turn them. Similarly, according to Pasteur, a compound will be optically active if its structure cannot be brought into coincidence with its mirror image. Pasteur arrived at his principle after discovering: (1) that optically active compounds can exist in two enantiomorphic modifications—"optical antipodes or isomers"—with a rotatory power of identical value but opposite sign; and (2) that crystals of optically active compounds may have a shape exactly like the mirror image of crystals of the respective antipode.

Structural formulas that satisfy Pasteur's principle were given by van't Hoff [7] and Le Bel [8], who conceived of organic molecules as three-dimensional structures, the four valences of carbon extending from the center to the corners of a tetrahedron. The simplest molecule which, according to van't Hoff, would be optically active, is symbolized in Fig. 2.8. The four substituents, R_1, R_2, R_3, and R_4, at the carbon atom, C, must all be different from each other in order that structures (a) and (b) cannot be brought into coincidence. A carbon atom thus substituted is called *asymmetric.*

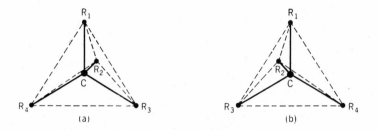

Fig. 2.8. Van't Hoff's tetrahedral molecule and its theoretically simplest two optical antipodes. R_1, R_2, R_3, and R_4 represent unequal substituents.

Pasteur's requirement for optical activity, namely, that a molecule and its mirror image are not superimposable, will be fulfilled whenever a molecule has neither a plane nor a center of symmetry and possesses no "improper" rotation axis. An improper rotation axis is present whenever a rotation about an axis, followed by a reflection in the plane perpendicular to that axis, results in a configuration identical to the original. A molecule and its nonsuperimposable mirror image are called *enantiomers.* Enantiomers may have an axis of symmetry, in which case they are said to be dissymmetric; those having no axis of symmetry are asymmetric. Enantiomers may only be distinguished physically through their interaction with some standard which is, itself, dissymmetric (or asymmetric)—a fundamental principle underlying all physical and chemical methods of identifying or separating enantiomeric pairs. For example, the fact that two beams of right and left circularly polarized light, respectively, are related to each other like mirror images of an object permits their usage (as dissymmetric standards) to distinguish between enantiomers Therefore, optical activity arises from the physically different interactions between each member of the enantiomeric pair (the optical isomers) with each member of the dissymetric standard (the circularly polarized components of the incident light). A few examples, of molecules capable of existing as enantiomers are shown below, namely, a spiro compound (I) [9], an allylenic compound (II) [10], and a compound in which steric hindrance prevents internal rotation about the *x*-axis (III) [11–13]. Furthermore, it is obvious, that even in single-ring systems optical activity can be achieved in a relatively simple fashion if they are nonaromatic. Consider, for instance, 3-methyl-cyclopentanone (IV). Substitution of 2 H by O in cyclopentane eliminates the

(I)

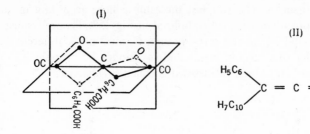

(II)

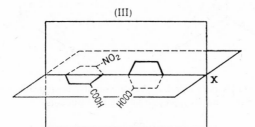

(III)

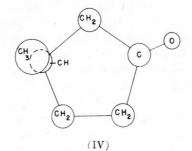

(IV)

center of symmetry and all planes of symmetry except that which coincides with the plane of the ring and that which is normal to it. The additional replacement of any one H by CH_3, for example, in the 3-position eliminates these planes of symmetry also and the resulting molecule is therefore optically active.

A particularly interesting case of molecular dissymmetry is that due to a helical structure such as found in polypeptides, proteins, polynucleotides, and other biological macromolecules [14]. Since most of these compounds also produce optical activity in the uncoiled state, the formation of a helical conformation, which, on its own account, would produce optical activity, leads to an increment or decrement of optical activity, depending on whether the sign of the helix effect is identical with or opposite to the effect that would be observed in absence of a helical structure [15–17]. Because of the complexity of the optical effects in these cases, it has become customary not to determine the rotation at a single wavelength but to determine its dispersion over a wide spectral range. This subject is covered in more detail in Chapter V.C. Although an intramolecular helix is probably the most fascinating type of helical structure—it may also be obtained synthetically with the aid of suitable catalysts (" stereospecific " polymers, such as high-density polystyrene)—intermolecular helices are also possible. Reusch [18] actually constructed a related model in the nineteenth century. He stacked up thin lamellae of biaxial mica so that the direction of the principal section in the plane of the leaflets changed from leaflet to leaflet by 60°. Thus on forming a right-handed macroscopic screw for the sequence of the optically privileged directions, dextrorotation of linearly polarized light was obtained. For the reverse screw levorotation resulted. Sohnke [19] showed that the rotation observed agreed quantitatively with that calculated and that it agreed also quantitatively with that observed for quartz, more so the thinner the lamellae of mica. Because of the extremely weak birefringence involved, however, Sohnke failed to duplicate Fresnel's experiment of splitting the polarized beam into two components upon passage through mica. An equivalent but spontaneous screwlike intermolecular arrangement of molecular axes has recently been considered as the cause of the extraordinarily large rotatory power of mesophases [20] (see page 154). An investigation on the interaction of microwaves with macroscopic models of helices [21] is also very interesting in this connection.

A molecule may contain several asymmetric carbon atoms without being dissymmetric. Tartaric and mesotartaric acid are classic examples. The structure of tartaric acid is dissymmetric; it exists in two optically active modifications. The structure of mesotartaric acid is not dissymmetric; it is inactive. The structures are shown below in simplified planar projection [22], the broken line indicating the plane of symmetry.

COOH	COOH	COOH	COOH
HCOH	HOCH	HOCH	HCOH
HOCH	HCOH	---+---	---+---
		HOCH	HCOH
COOH	COOH	COOH	COOH
d	*l*	Inactive	

The presence of an improper rotation axis will also render a molecule optically inactive. An example is spirane B which is optically inactive as the result of the S_4-axis (i.e., a rotation of $\pi/2$ radians followed by a reflection in a plane perpendicular to the axis of rotation).

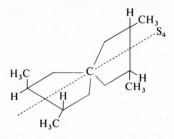

The spatial relationship of the bonds constituting a molecule defines its *absolute configuration*, whereas rotations about these bonds result in different *conformations*. The utility of optical rotation lies in its great sensitivity to slight changes in absolute molecular configuration or conformation. There is no better example of the practical application of optical activity to the elucidation of absolute configuration than the work of Djerassi and his associates [23] in their studies of the steroids. They demonstrated how relatively minor changes in substituents may completely change the observed rotatory dispersion. From this work evolved the "octant rule" regarding the Cotton effect (see p. 65 associated with the 2900 Å electronic transition of the carbonyl chromophore. Often the energy barriers separating various conformational isomers, or *conformers*, are too low to permit their physical separation. However, the development of rules such as the octant rule permits an interpretation of the observed optical activity when one conformation is predominant. An example is (+) *trans*-6-chloro-3-methylcyclohexanone, whose optical rotatory dispersion exhibits a "negative" Cotton effect at room temperature but changes to a "positive" Cotton effect when the temperature is lowered to − 192°C [24]. The octant rule predicts that conformation (**1**), will exhibit a "negative" Cotton effect whereas (**2**) should have a "positive" Cotton effect. Therefore, (**1**) is conformationally more stable at high temperature while (**2**) predominates at low temperatures.

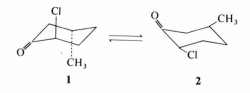

Before the development of the theory of rotatory power, it was not possible to state which of the two basic configurations of an optically active molecule is dextrorotatory or levorotatory. Fischer's assignment illustrated above for the case of tartaric acid was purely intuitive and not supported by evidence. It is remarkable that recent calculations, on the basis of the theory of Kuhn

[25], of Eyring [26], and of Kirkwood [27] showed conclusively that Fischer's assignment of absolute configurations was correct. Direct experimental checks by means of x-ray diffraction confirmed this [28]. This method is at present the only generally applicable experimental method for correlating absolute configurations with optical rotation whenever deductions by means of the now secure Fischer convention are not possible. (For a discussion and review of the x-ray method see ref. 29.)

Anisotropic Absorption

If a molecule has the structural dissymmetry defined by Pasteur's principle, it exhibits rotatory power due to the inequality of the refractive indices, n_l and n_d (see p. 58). This inequality of n_l and n_d is, in turn, a corollary of an even more fundamental phenomenon, optically active absorption, discovered by Cotton [30]. The relationship between the two effects will be discussed briefly on the basis of the classic theory of refraction and absorption.

Refraction results from the superposition of a primary light wave entering the refracting substance and a secondary "scattered" wave created by the vibration of dipoles that are induced in the substance by the primary wave. The dipole vibration is a function of the frequency of the primary wave, and of the natural (*eigen*) frequency of the dipoles. If both frequencies are identical, resonance occurs and absorption results. The dispersion (spectral variation of the refractive index) within a given spectral region depends, therefore, on the location of the nearest absorption band, as shown in Fig. 2.9 [31]. The

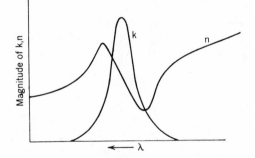

Fig. 2.9. Variation of refractive index in the spectral region of an absorption band.

refractive index, n, increases rapidly as one approaches the band from longer wavelengths and varies anomalously with the wavelength in the region of absorption (k = absorption coefficient). There is a large class of (uniaxial) substances that have two refractive indices, an "extraordinary" refractive index, n_e, in one direction, and an "ordinary" refractive index, n_o, in all orthogonal directions (see p. 71). Biot [32] found on turmaline crystals that this "linear birefringence" $(n_e - n_o)$ was accompanied by "linear dichroism," that is, by unequal absorption of light vibrating along extraordinary and ordinary directions. Thus, n_e is associated with an absorption coefficient, k_e, and similarly n_o depends on k_o. Similarly, the optical rotation, that is, the circular birefringence, $(n_l - n_d)$ is accompanied by circular dichroism $(k_l - k_d)$. In other words, n_l is associated with an absorption coefficient k_l and, correspondingly, n_d is associated with k_d. Cotton was the first to anticipate and actually to prove experimentally this relationship, which is illustrated schematically in Fig. 2.10. The whole set of phenomena illustrated by this figure is known as the Cotton effect.* It will be noted from an inspection of Fig. 2.10 that the circular dichroism reaches its maximum value at a wavelength where the optical rotation passes through 0. (The inversion of the sign of the optical rotation illustrated here rather by the circular birefringence curve, often does not take place exactly at the wavelength where the circular dichroism reaches its

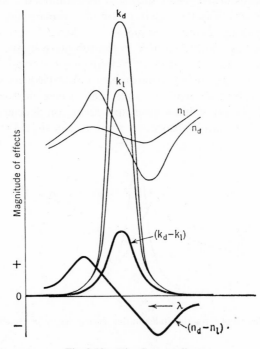

Fig. 2.10.   The Cotton effect.

* Because of an unfortunate misunderstanding of the classical definition of the Cotton effect, it became customary during the last 10 years to restrict the designation Cotton effect to the anomalous variation of the optical rotation with wavelength rather than to the spectral variation of optical rotation *and* circular dichroism, the latter being defiined by the difference $(k_l - k_d)$.

maximum but may differ slightly). The discovery of the circular dichroism and its interrelation with the circular bierefringence by Cotton was of fundamental importance for the development of the theory of optical activity [34].†

If the spectral location of the nearest optically active absorption band and the sign of the circular dichroism are known, the sign of α in the neighborhood of the band is given by Natanson's rule [35], which states that the less absorbed circular component is less refracted at larger wavelengths and more refracted at shorter wavelengths (see Fig. 2.10). A positive rotation, α, in the visible indicates, therefore, that the absorption band dominating the sign of the rotation, and located at shorter wavelength is characterized by a stronger absorption of left circular light.

It follows from the preceding discussion that the determination of α for a single wavelength is of limited significance, and that far more information on an optically active compound can be obtained by computing the dispersion of α, and particularly, by studying, at the same time, the spectral variation of the circular dichroism whenever the latter is sufficiently strong to be measurable with reasonable precision.

Since the circular dichroism leads to ellipticity if one operates with incident linearly polarized light, the simultaneous determination of the optical rotation in the respective spectral range may become very difficult if the circular dichroism is very large, unless ellipticity compensators are used (see Chapter V.C. The intensity of the optical field will then vary much less with the polarizer-analyzer setting than in absence of circular dichroism. The worst possible situation would arise if the circular dichroism were large enough to lead to circularly polarized light. In this case, it would be impossible to determine the rotation of the incident linearly polarized beam since the intensity of light transmitted through the analyzer would be independent of its setting.

4 OPTICAL ROTATIONS NOT DUE TO NATURAL ROTATORY POWER

Linear Dichroism

An optically active substance rotates the direction of vibration of linearly polarized light. Rotation observed during the passage of a linearly polarized beam through a liquid or a solution is, however, not necessarily a proof of

---

† It is useful to emphasize the difference between the effects of linear anisotropy and of circular anisotropy. Linear birefringence leads, in the very general case, to the transformation of linearly polarized light into elliptically polarized light. On the other hand, linear dichroism leads to a rotation of the plane of polarization of incident linearly polarized light. Thus the optical effects of linear birefringence and of circular dichroism are the same. Similarly, the optical effects of circular birefringence (optical rotation by optically active substances) and of linear dichroism are the same.

optical activity, a fact which is often overlooked and which deserves some attention. Rotation may also be caused by linear dichroism. This, the linear analogue to circular dichroism, is the effect of an anisotropic absorption in linearly anisotropic substances. (The elementary optics of such substances is presented on pp. 70 *et seq.*) In Fig. 2.11, V_o is considered the

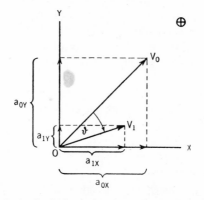

Fig. 2.11   Optical rotation due to linear dichroism.

resultant of two coherent orthogonal linear components, V_{ox} and V_{oy}. This is similar to the decomposition of a linear vibration into two coherent circular components of opposite sense of rotation (Figs. 2.3 and 2.5) in the case of circularly anisotropic substances. The absorption coefficients, k_x and k_y, of the dichroic substance are unequal, so that the amplitudes, a_{ox} and a_{oy}, of the two components are reduced to a_{lx} and a_{ly}, respectively, during the passage of the beam through the substance. The recombination of the two components leads then to a resultant linear vibration, V_l, which is rotated by the angle ϑ, with respect to V_o. In Fig. 2.11, an angle V_oOX of 45° is assumed. If the angle is either smaller or larger, the rotation is smaller. It would be zero for an angle of 0° or 90°. This gives us the criterion for distinguishing between a rotation due to natural rotatory power and a rotation due to linear dichroism: angle α is independent of the direction of vibration of the incident beam, while angle ϑ varies with this direction. The safest practical method of distinguishing between the two effects is the rotation of both polarizer and analyzer, in crossed position, in the same sense and to the same extent. The light transmission will remain unaltered in the case of rotatory power. However, if linear dichroism causes the rotation, then rotating the polarizer and analyzer will lead to changes in the brightness of the field. In the case of the extinction method (Fig. 2.27 and p. 99), for example, the transmission changes between zero and a maximum, twice during a complete rotation of polarizer and analyzer, by 360°. Linear dichroism is unlikely, in the absence of an external

orienting factor, unless the anisotropic substance represents a supermole-
cular structured colloid or a mesophase (liquid crystal) in which the spontan-
eously oriented elements are anisometric [36]. One can predict that the effect
will occur, for example, in protein structures, particularly in the ultraviolet.
Such structures may also be optically active. A linear dichroism is particularly
likely, however, if linearly dichroic molecules or particles are oriented by an
external force. The danger of confusing linear dichroism with rotatory power
is therefore particularly high in a transverse electric or magnetic field, or in
streaming solutions [37, 38]. A linear dichroism may be due not only to
anisotropic true absorption but also to anisotropic light scattering [39]. The
color of the substance is therefore no reliable criterion for the presence or
absence of dichrosim.

Faraday Effect

Parallel to the lines of force of a magnetic field, a "magnetic rotatory
power" known as the Faraday effect is observed in any liquid or solution [40].
Just as rotatory power in a spectral region close to an optically active absorption band is accompanied by circular dichroism, the Faraday effect is accompanied by a magnetic circular dichroism (inverse Zeeman effect) under
similar conditions. The Faraday effect, like the related Zeeman effect [41],
results from an influence of the magnetic field upon the electron movements
within the atoms, and the sign of the effect depends on the direction of the
magnetizing current. In most instances the rotation, α, has the same direction
as the magnetizing current, as illustrated by Fig. 2.12. The effect is then called

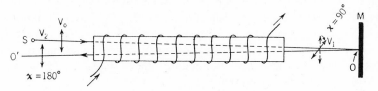

Fig. 2.12.   The Faraday effect.

positive. It follows from Fig. 2.12 that the dependence on the direction of the
current leads to two characteristic properties of the Faraday effect. First, if
the beam is reflected by a mirror, M, and returns through the substance, the
effect observed at O' will be double since the direction of the current is the
same both times. Second, to the observer at O, in the absence of M, the rotation appears as positive, which it is according to the definition given above.
However, if we interchange the light source, S, and the position of the observer at O, the rotation appears negative, although it is still positive according
to the definition. Consequently, the apparent, not the true, sign of the Faraday

effect depends on the direction of observation, while the magnitude of the effect is independent of it. The situation is different in the case of natural rotatory power. Here, the sign of the rotation is independent of an interchange between light source and observer. It follows implicitly that the result of reflection is different here. The rotation due to natural rotatory power is annulled if the light beam is reflected and thus forced to travel twice through the substance in opposite directions.

5 TECHNIQUE OF POLARIMETRIC MEASUREMENTS

Polarimetric measurements of rotatory power are concerned with the direct or indirect determination of the angle of rotation, α. While this operation appears simple, any polarimeter for precision measurements applies quite a number of optical phenomena. The chemist interested in precision measurements is therefore likely to encounter difficulties if he is not familiar with the functioning of the various parts of a polarimeter; to give this information is the first aim of this chapter. A number of excellent polarimeters are on the market, but some problems require the construction of special equipment; to furnish the basic knowledge necessary for such a task is the second aim of this chapter.

Introduction to Elementary Optics of Linearly Anisotropic Substances

We have seen that optically active substances respond differently to a right and a left circular vibration (circular anisotropy). There is a second, much larger, class of substances which respond differently to two orthogonal linear vibrations (linear anisotropy). Such substances are of paramount importance in polarimetric technique, where extensive use is made of the phenomenon of linear double refraction. This effect, and some resulting phenomena, will be outlined below, to the extent required for the discussion of polarimetric measurements of rotatory power.

The cause and the optical consequences of linear double refraction may be visualized most easily on the example of "morphically" double refracting (birefringent) substances. Figure 2.13 represents a rectangular rod which consists of a large number of isotropic solid sheets separated from each other by a gas or liquid. Both the thickness of the sheets and their mutual distance are small compared with the wavelength of light. According to Wiener [42], such a rod will exhibit linear double refraction. If light is propagated along X, that is, perpendicular to the sheets, the rate of propagation will be the same for a vibration $\parallel Y$ and for a vibration $\parallel Z$. If light is propagated along Y or Z, however, the rate of propagation will vary with the direction of vibration. If the direction of vibration is $\perp X$, the rate will be equal to that

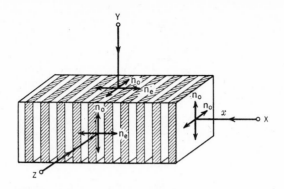

Fig. 2.13. Double refraction of a morphically anisotropic substance.

found in the case of propagation along X. It will be smaller if the direction of vibration is parallel to X. Since the rate of propagation of a linearly polarized beam depends on the refractive index in the direction of vibration, the stratified rod can be characterized by two different refractive indices, one, n_e, for vibrations parallel to X and another, n_o, for vibrations perpendicular to X. The X-direction, a privileged direction, is called the "optic axis," x. The principal "extraordinary index" is n_e, and n_o is the "ordinary index." If $n_e > n_o$, the birefringent substance is called "uniaxial positive" and the double refraction is called positive because $(n_e - n_o)$ is positive. If $n_e < n_o$, the substance is called "uniaxial negative," and the double refraction is called negative. The term "uniaxial" indicates that the substance has one optic axis. A linearly anisotropic uniaxial substance is optically characterized by the direction of the optic axis, the ordinary refractive index, and the principal extraordinary refractive index. Substances that have two optic axes, that is, three different refractive indices, are not of interest here.

The distance covered, in a certain very small period of time, by a light impulse created in the origin, is plotted in Fig. 2.14. The origin is assumed to be in the interior of an anisotropic substance. Only a propagation parallel or perpendicular to the optic axis will be considered. Light propagated along the optic axis, which coincides in the figure with X, can vibrate only perpendicular to X. Consequently, there will be only one rate of propagation in the X-direction (ordinary velocity, v_o, determined by n_o). Light propagated in the Y- or Z-direction may vibrate either parallel or perpendicular to the optic axis. Consequently the light propagated in these directions will consist of two components, one propagated with the ordinary velocity and the other propagated with a "principal extraordinary" velocity, v_e. It follows that any point reached after a certain time by the light impulse lies on one or the other of two wave surfaces. One of the surfaces is that of a sphere, that is, identical with the

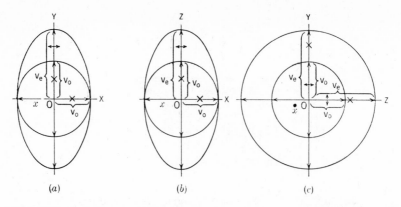

(a)   (b)   (c)

Fig. 2.14. Wave surfaces of uniaxial negative substance ($n_e < n_o$) if light emanates from center of substance. (*a*), (*b*), and (*c*) are cross sections through the three-dimensional scheme of wave surfaces. The ellipse is exaggerated for the sake of demonstration; it actually is almost a circle. V represents light path per unit time (velocity).

wave surface of an isotropic substance of refractive index n_o. The other surface is that of an ellipsoid. The ellipsoid is obviously tangent to the sphere at the points of intersection with the optic axis, which is the symmetry axis of revolution of the ellipsoid. In Fig. 2.14 $v_e > v_o$; in Fig. 2.15 $v_e < v_o$.

We now study the propagation of a pencil of parallel natural light incident upon a uniaxial negative crystal. There are three possibilities of orientation for the optic axis with respect to the crystal face through which the beam enters: the axis and the face may form an angle of 90°, of 0°, or of $\neq 90° \neq 0°$. The direction and the rate of propagation of light in the crystal shall be established for all three cases.

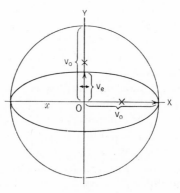

Fig. 2.15. Wave surfaces of uniaxial positive substance ($n_e > n_o$). As in Fig. 2.14, the ellipse is much closer to being a circle than the figure shows. For V, see Fig. 2.14.

By restricting the discusion to perpendicular incidence, which is sufficient for our needs, it can be postulated that the light travels in that direction in which it penetrates deepest into the crystal within a certain time, taking the distance from the face as the criterion. The deepest penetration and the direction of propagation are then defined by the point at which a plane parallel to the face of the crystal is tangent to the wave surface (Fig. 2.16). The double-headed arrow gives the direction of the optic axis. If the optic axis is perpendicular to the face of the crystal (I), the direction of propagation leading to

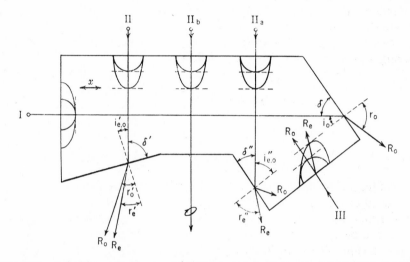

Fig. 2.16. Refraction, reflection, and interference phenomena of a light beam traversing a uniaxial crystal.

deepest penetration is in the direction of the optic axis. The light beam is therefore propagated without deviation and at the rate v_o, that is, the beam is propagated as in an isotropic substance. If the optic axis is parallel to the face of the crystal (II), the light will again travel through the crystal without deviation, but this time there will be two components. They travel along the same path, but at different rates, v_o and v_e. If the optic axis forms an angle $\neq 0°$ and $\neq 90°$ with the face of the crystal (III), the beam will be split up into two components, one propagated without deviation at the rate v_o, the other propagated under deviation at the rate v_e. The latter component is called the "extraordinary" ray, R_e, and the former, the "ordinary" ray, R_o. The direction of vibration of the ordinary ray is perpendicular to the optic axis; the direction of vibration of the extraordinary ray is in a plane parallel to the optic axis.

The next important problem is that of refraction and reflection, respectively, at the point at which the pencil of light reaches the end of the crystal. According to Snell's law of refraction

$$n_e \sin i = n_1 \sin r_e \; ; \; n_o \sin i = n_1 \sin r_o \qquad (2.10)$$

$$\therefore \frac{\sin r_o}{\sin r_e} = \frac{n_o}{n_e} \qquad (2.11)$$

where n_1 is the refractive index of air, and i and r are the angles of incidence and refraction, respectively. In Fig. 2.16, the beam is refracted by the angle r_o. In case II, it is split up into divergent rays, the ordinary and the extraordinary. As regards reflection back into the crystal, the critical angle of incidence leading to total reflection, i_{Cr}, is reached if:

$$\sin i_{eCr} = \frac{n_1}{n_e} \; ; \quad \sin i_{oCr} = \frac{n_1}{n_o} \qquad (2.12)$$

$$\therefore \frac{\sin i_{eCr}}{\sin i_{oCr}} = \frac{n_o}{n_e} \qquad (2.13)$$

The gradual reduction in angle δ (Fig. 2.16) leads therefore to three characteristic situations in the case considered: (a) two divergent linearly polarized components emerge (II); (b) only one linearly polarized component emerges, the second being totally reflected (IIa); (c) both components are totally reflected, a case not shown in the figure. The case $\delta = 90°$, that is, a plane-parallel crystal plate (IIb), is discussed below.

If the incident light is not natural but is linearly polarized, the phenomena discussed thus far by means of Fig. 2.16 remain unchanged but the physical relationship of the ordinary and extraordinary components becomes different. When natural light is split into two orthogonal linearly polarized components, these components are incoherent, just as two circular components are incoherent if they are obtained from natural light. However, if a linearly polarized beam is split into two orthogonal linearly polarized components, one vibrating parallel to the optic axis, the other vibrating perpendicular to it (Fig. 2.17), the components are coherent, just as two circular components are coherent if they are obtained from linearly polarized light. This difference leads to a characteristic difference in the behavior of an original natural and of an originally linearly polarized beam if $\delta = 90°$ (Fig. 2.16 IIb). In the former case, the two components, traveling along the same path, behave in every respect as natural light. However, if the original beam is linearly polarized, then the coherent orthogonal components will interfere. The resultant beam is therefore generally elliptically polarized, and only under very particular conditions will it be

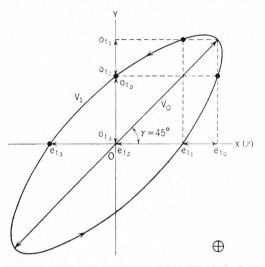

Fig. 2.17. Elliptical polarization (Case II*b* of Fig. 2.16).

linearly or circularly polarized. This is explained by Fig. 2.17, in which it is assumed that the passage through the plane-parallel uniaxial negative crystal plate has led to a path difference of $\frac{1}{8}$ wavelength (phase difference of $-45°$). No discussion of the figure appears to be necessary, since it is drawn according to the same principle as Figs. 2.4 and 2.5. Figure 2.18 gives the variation in type and state of vibration with the phase difference if $\gamma = 45°$. If $\gamma < 45°$, the picture does not change characteristically except for phase differences of $-90°$ and $-270°$. For these, elliptical vibrations with the long axis parallel to x are obtained instead of circular vibrations, as demonstrated in one case by a dashed curve. If $\gamma = 0$ or $= 90°$, the linearly polarized beam remains obviously unaltered. The types of vibration resulting from the various phase differences are the same, as above, if the optic axis coincides with the Y-axis, or if the crystal is uniaxial positive with the optic axis parallel to the X-axis. The sense of rotation of the elliptical and circular vibrations is then reversed, however.

Figure 2.18 suggests the procedure in measurements of the Cotton effect if one is interested in obtaining the absolute values of k_d and k_l by means of a photometer. A birefringent plate with a phase difference of $\pm 90°$, under an angle, γ, of $45°$, produces left circular or right circular light, respectively, from linearly polarized light depending on whether its optic axis coincides with the X-axis or Y-axis. Because of the dispersion of double refraction, measurements with a single plate are restricted to one wavelength. Far better in this respect are Fresnel rhombs [43]. Other methods of measuring the Cotton effect are discussed elsewhere in this volume.

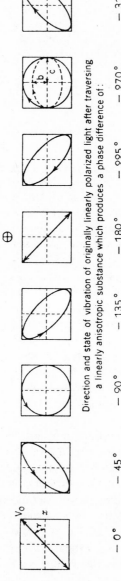

Direction and state of vibration of originally linearly polarized light after traversing
a linearly anisotropic substance which produces a phase difference of:

— 0°   — 45°   — 90°   — 135°   — 180°   — 225°   — 270°   — 315°   — 360°

Fig. 2.18. Effect of linearly anisotropic substance upon linearly polarized light.

Basic Parts of a Polarimeter

Polarizer

The polarizer produces linearly polarized light from natural light.

POLARIZERS BASED ON DOUBLE REFRACTION

Cases II, IIa, and III of Fig. 2.16 lead to the construction principles of the most important types of polarizers. The polarizers that do not affect the direction of propagation of the light are used most frequently. In these, corresponding to case IIa, one component of the light is totally reflected. Deviation of the transmitted component is prevented by combining two equally cut half-prisms of a uniaxial crystal. We shall first discuss the working conditions of prisms if used with parallel light and with an angle of incidence of 0°. In the Glan prism [44], the two half-prisms are separated by a thin layer of air. Figure 2.19 exaggerates the thickness of this layer and the parallel displacement of the emerging beam in order to show the principle. The totally reflected ordinary ray is absorbed by black paint on the lateral face of the double prism.

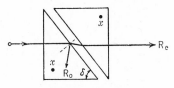

Fig. 2.19. Glan prism.

The Glan prism and other polarizing prisms based on total reflection are often made of calcite, a uniaxial negative crystal with particularly strong double refraction (n_e, 1.4865; n_o, 1.6584; D line). Calcite is soft, and any scratch will seriously impair the polarizing quality of a prism by the scattering of light. In cleaning the faces of calcite polarizers, only lens paper or chamois skin should be used. Loose dust should be brushed off with a very soft camel's hair brush. Some progress has been made in growing large, optically faultless, and even more strongly double refracting crystals of sodium nitrate [45] (n_e, 1.336; n_o, 1.587; D line). Polarizing prisms of this material can be made much shorter than calcite polarizing prisms with the same field and aperture. They are thus of very definite interest, notwithstanding (a) their solubility in water and (b) the difficulties involved in finding a cementing substance of sufficiently low refractive index $n \sim n_e$.

Other doubly refracting crystals used in ultraviolet polarimetry include quartz ($n_e = 1.6114$; $n_o = 1.6003$ at 2503 Å) [46], ammonium dihydrogen phosphate ($n_e = 1.479$; $n_o = 1.524$ at 5893 Å) [47], and magnesium fluoride ($n_e = 1.4148$; $n_o = 1.4021$ at 2537 Å) [48]. The ammonium dihydrogen phosphate polarizers in the Cary instrument [49] are immersed in cyclohexane contained in sealed Invar tanks, that have fused silica windows.

The beam emerging from a Glan prism is slightly displaced (but parallel) with respect to the incident beam. Rotation of the prism around the direction of light propagation as the axis causes the polarized beam to wander along a circle. This imperfection may be negligible if the air interlayer is very thin. More important is the twofold reflection of the transmitted polarized ray at the air layer, which results in a loss of polarized light. The main deficiency is the small field of the Glan prism—a deficiency existing for all polarizers with air interlayer. This means that perfect polarization can be obtained only if the incident beam represents parallel light and if the angle of incidence is 0°. The permissible maximum deviation from this condition is ±4° for the Glan (see also the discussion on pp. 79 et seq.).

Both disadvantages are eliminated in the Glazebrook or Glan-Thompson polarizing prism [50] (Fig. 2.20). Here, the two half-prisms are cemented together; Canada balsam is often used for this purpose. The refractive index

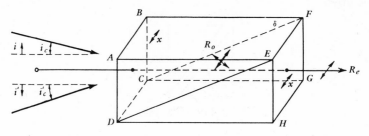

Fig. 2.20. Glan-Thompson prism.

of the latter (1.55) is so close to n_e of calcite (1.49) that the loss of intensity of the transmitted extraordinary ray is much smaller than with the Glan prism. On the other hand, n of Canada balsam is also much closer to n_o of calcite (1.66) than is n of air. According to (2.12), δ, which is (90° − i) here, must therefore be reduced in order to maintain total reflection of the ordinary ray. Consequently, the ratio of length to width is much larger for Glan-Thompson prisms than for Glan prisms of equal width. The internal reflection of the extraordinary ray can be completely avoided by using linseed oil ($n = 1.486$) between the two half-prisms. With this intermediate, the prisms can be approximately 10% shorter than prisms cemented with Canada balsam. Optical companies now use resins which combine cementing qualities with a refractive index close to that of linseed oil. The optic axis, parallel to the ground and polished face of the crystal, through which the light enters, is parallel to AB. It may be parallel to AD instead which, however, is slightly less advantageous (Ahrens prism) [51]. In some instances, the axis is parallel to AC or BD. The optic axis in calcite crystals is strongly oblique with respect to any of the natural faces. A crystal must therefore be cut down considerably for a Glan-Thompson prism, which consequently is expensive.

The nicol prism [52] shown in Fig. 2.21 is much cheaper. The faces of the natural crystal are less corrected, leaving the optic axis oblique with respect to the frontal face, $ABCD_0$. In addition, this face is oblique to the lateral faces, and the angle of incidence of the light beam is $\neq 0°$. Two characteristic properties are the result: the beam is split into two components immediately upon entering the nicol prism; and both the extraordinary and the ordinary ray are deflected.

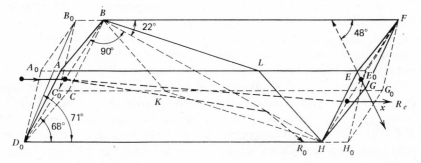

Fig. 2.21. Nicol prism with angles and other dimensions drawn out of proportion to show principles of construction: $A_0 B_0 C_0 D_0$ and $E_0 F_0 G_0 H_0$, faces of original calcite crystal; $ABCD_0$ and $EF_0 GH$, faces of crystal after correction; $BLHK$, plane along which corrected crystal is cut and recemented.

The nicol prism has two disadvantages which exlcude its use in precision measurements. (1) The light beam emerging from the polarizer moves in a circle as the polarizer is rotated, a defect evident from Fig. 2.21, and much stronger than with the Glan prism (Fig. 2.19.) (2) If used as an analyzer, in combination with a polarizing nicol and a very bright light source, complete extinction cannot be brought about for the whole optical field, but only for a limited section of it, and the locus of this section within the field varies with the rotation of the analyzer around the true position of "crossed nicols." The meaning of statement (2) may become fully clear only after the discussion of the polarizer–analyzer combination (pp. 88 *et seq.*). According to Bruhat, the latter deficiency of the nicol prism introduces an uncertainty of several minutes in the determination of the angle of rotation [53]. If used in a polarizing microscope, the nicol prism introduces astigmatism, a third deficiency which is unimportant, however, at low magnifications and irrelevant in those polarimetric methods not involving the use of a microscope.

It has become customary to call birefringent polarizers and analyzers of any type "nicols," provided they are composed of two half-prisms with equally oriented optic axes. We shall use the term in this sense.

Under actual working conditions the light incident upon a Glan-Thompson prism may be slightly convergent or divergent, so that the angle of incidence is 0° for a fraction of the beam only. However, only a limited deviation from 0° is permissible. This may be discussed by means of Fig. 2.20. We assume that the two half-prisms are separated by air and vary the angle of incidence in a plane perpendicular to $ABCD$ and parallel to $AD(BC)$. For an angle of incidence, i_c, both the extraordinary and ordinary components are transmitted; if the angle of incidence is i'_c, both components are totally reflected at the diagonal plane inside the prism. For intermediate angles, the extraordinary component is transmitted, and the ordinary component is totally reflected. The critical angle—which cannot be exceeded if polarization is to be complete—is characterized by the subscript c. The sum $(i_c + i'_c)$ defines the permissible deviation from normal incidence. Actually, the two half-prisms are not separated by air, but by Canada balsam. The angle i'_c is then obviously far larger, another important advantage of Glan-Thompson prisms over Glan prisms. If the refractive index of the cement equals n_c, then i' has no significance, since total reflection of the extraordinary component is impossible. The "field" of the polarizing prism, that is, the permissible solid angle of the light cone, then depends exclusively on i_c which increases with decreasing δ and therefore with increasing ratio of length to width (AE/EH) of the prism. i_c is calculated on the basis of (2.10) and (2.12). For example, the angle $2i$, defining the cone of a convergent beam, must not exceed 4° if the ratio AE/EH is 2.15. If the prism were cemented with Canada balsam, the permissible angle $2i_c$ would be smaller; and, in order to make it 4°, the ratio AE/EH would have to be 2.81.

The most important practical conclusions are the following: (a) The incident beam must not be more convergent or divergent than is compatible with the polarizer field; otherwise, partially natural light will be transmitted. It is therefore wrong to concentrate a beam upon the polarizer in the futile hope of getting "more intensity" in the apparatus. (b) If one recements a polarizing prism with Canada balsam—which must be free from dust in order to exclude depolarizing light scattering—i_c will be smaller than indicated by the manufacturer if the prism was cemented with a resin of refractive index closer to n_e. (c) The polarizer must be lined up properly. Even with a beam of parallel light, one will not obtain completely polarized light if the angle of incidence is $> i_c$. In order to test the proper alignment of the polarizer, of the light source, or of the condensing lens in front of the light source, a sheet of stiff black paper with a small hole is placed between condensing lens and polarizer. The light reflected from the face of the polarizer must be centered at the hole and must remain there upon rotation of the polarizer.

The aperture, $ABCD$, of a polarizer is determined by its length, since the ratio AE/EH is constant for constant i_c, and consequently large apertures require long polarizers. If, for example, one wants to increase the aperture of a square prism from 1 to 2 cm^2, while maintaining the same i_c, then the length of the prism must be doubled. Glan-Thompson polarizers with larger apertures are therefore very costly; if a large aperture is required, an Ahrens-Bénard [54] prism (Fig. 2.22a) may be used. The square prism has the aperture of a square Glan-Thompson prism twice as long; therefore, a much smaller crystal can be used for its construction. Light scattering by the apex of the central wedge, if Canada balsam is used, produces interference fringes which may disturb the homogeneity of the optical field in bright illumination. Nevertheless, the prism is much used in polarizing microscopes. A theoretically better solution is the " sensitive strip " polarizer of Jamin [55] (Fig. 2.22b),

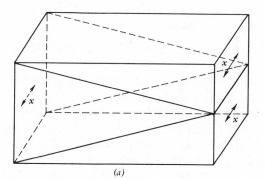

(a)

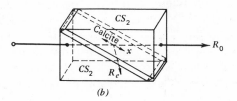

(b)

Fig. 2.22. Calcite polarizers with large aperture. (a) Ahrens-Benard prism; (b) Jamin-Brace "sensitive strip" polarizer.

in which the construction of a Glan-Thompson prism is inverted. The two half-prisms are replaced by carbon disulfide and the layer of Canada balsam is replaced by a strip of calcite. Under these conditions, the ordinary ray is transmitted. Brace [56] substituted monobromonaphthalene for carbon disulfide. This type of polarizer can be constructed at a comparatively low cost considering the large aperture obtained. A disdavantage is the absolute necessity of having a perfectly dust-free liquid; otherwise, only a partial polarization can be achieved.

For certain purposes, polarizers consisting of a single birefringent prism, simple "double image" prisms, are preferable. Polarization is then accompanied by a deflection of either polarized component. The most suitable type is shown in Fig. 2.23. Generally, the more refracted component is used. The interesting feature of these polarizers is their simultaneous action as a dispersing prism.

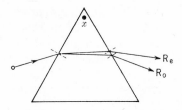

Fig. 2.23.   Double-image polarizing and dispersing prism.

When both the ordinary and the extraordinary components are to be used, two types of polarizers are particularly practical: (*a*) A single birefringent crystal, like calcite, under normal incidence (Fig. 2.24*a*) transmits the ordinary ray without deviation. The extraordinary ray is displaced laterally, but runs parallel to the ordinary ray after leaving the crystal. The calcite crystal should be three inches long in order to give satisfactory separation of ordinary and extraordinary rays [57]. (*b*) A composite "double image" birefringent prism, for example, a Wollaston prism [58] (Fig. 2.24*b*) displaces ordinary and extraordinary rays nearly symmetrically from the original direction of propagation. Closely related to the Wollaston prism are the Rochon prism [59] and the Senarmont prism [60], in which one of the components is not deviated. In order to increase the angular separation of the two components, double image prisms have been constructed with three prisms (Abbe prism) [61] or even four individual prisms (Ahrens prism) [62], one or two of the latter consisting of glass.

POLARIZERS BASED ON OBLIQUE REFLECTION
FROM ISOTROPIC PLANE-PARALLEL PLATES

The use of calcite and other birefringent materials is limited to spectral regions where these substances do not absorb light strongly. In regions where they do, polarization by reflection from isotropic surfaces is used. This alternate method is therefore of interest primarily in the infrared and would also become of major interest in the far ultraviolet if and when polarimetry should be extended into the hitherto neglected Schumann region of the spectrum.

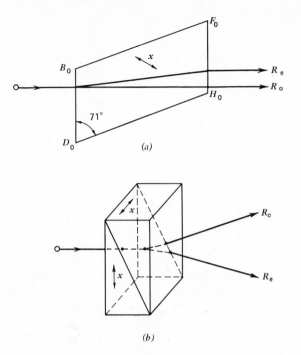

(a)

(b)

Fig. 2.24. Double-image polarizing prisms: (a) natural birefringent crystal (for the meaning of $B_0 \, D_0 \, F_0 \, H_0$, see Fig. 2.21); (b) Wollaston prism.

For certain polarimetric work, to be specified at the end of this section, it is and will continue to be of interest also in the intermediate spectral range.

Polarization by reflection from isotropic surfaces of isotropic materials is governed by the Fresnel equations

$$I_\perp = \left(\frac{1}{2}\right) I_0 \left[\frac{\sin^2(i - r)}{\sin^2(i + r)}\right] \tag{2.14}$$

$$I_\parallel = \left(\frac{1}{2}\right) I_0 \left[\frac{\tan^2(i - r)}{\tan^2(i + r)}\right] \tag{2.15}$$

Here I_0 is the intensity of the incident beam and $I_\perp$ and $I_\parallel$ are the reflected intensities. The intensity $I_\parallel$ is reflected from the component of the incident beam $[(1/2)I_0]$ which vibrates in the plane of incidence and $I_\perp$ refers to the intensity reflected from the orthogonal component of I_0, which vibrates perpendicular to the plane of incidence (the plane of incidence is the plane containing SFF' in Fig. 2.25). The angles i and r are the angles of incidence and

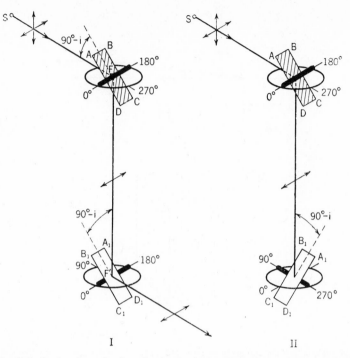

Fig. 2.25. Polarization by reflection (back of mirrors shaded): I, parallel reflectors; II, crossed reflectors.

refraction, respectively. It is apparent that $I_\perp$ is always in excess except if $i = r = 90°$. In view of Snell's law,

$$n_M = \frac{\sin i}{\sin r} \tag{2.16}$$

where n_M is the refractive index of the reflector, it is also clear that this preponderance of $I_\perp$ increases with n_M. Finally, it follows that $I_\parallel = 0$, if

$$\tan i = n_M \tag{2.17}$$

Equation (2.17) defines "Brewster's angle" of incidence for which the reflected light is 100% polarized. Table 2.3 gives Brewster's angle for various substances. Therefore, while perfect polarization can be obtained by reflection of a parallel beam of light under Brewster's angle, from a perfectly plane surface, the intensity of the polarized beam obtained is unfortunately quite modest. Using, for instance, the data for very heavy flint glass in Table 2.1, it follows from (2.14)–(2.18) that the polarized beam has an intensity slightly less than 16% of that of the incident beam. By contradistinction, a polarizing

Table 2.1.   Brewster's Angle, i_B, for Various Reflectors

Substance	Refractive Index (5890–5896 Å)	i_B
Fused quartz	1.458	55°33′
Lucite	1.50 ± 0.01	56°19′ ± 11′
Light crown glass	1.517	56°36′
Crown glass	1.520	56°40′
Flint glass	1.575	57°35′
Polystyrene	1.595 ± 0.003	57°55′ ± 3′
Heavy flint glass	1.650	58°47′
Very heavy flint glass	1.890	62° 7′
Selenium	2.54[a]	68°32′

[a] Applies to near infrared; not to 5890 Å.

prism yields an intensity in excess of 40%. This difference becomes serious—by amplification—on using a polarizer–analyzer combination (see p. 88). This weakness can be remedied partially by using a pile of transparent plane parallel plates all set at Brewster's angle. For infrared polarimetry, where polarization by reflection is more or less mandatory, Se has found special consideration, particularly on account of its relatively high transmittancy in that spectral range. It has also the additional advantage of a high refractive index [63]. By reflection on two consecutive Se-layers—each of them cast on glass—the reflected intensity was raised to 27% of I_0 and the degree of polarization achieved was stated to be practically identical with that obtained with a nicol prism, at least between 0.7 and 1.7 μ, the spectral range used for the comparison [63].

A corollary of the polarization by reflection is that the light transmitted through a transparent reflector also is partially polarized, the degree of polarization being largest for Brewster's angle. The excess polarized light in the transmitted beam obviously vibrates in the plane of incidence (parallel to *SFF′*), but the excess is clearly so small that the degree of polarization of the beam transmitted through a single reflector is generally of no practical interest. The excess of polarized light can be increased, however, by multiple reflections on successive transparent reflectors. One derives easily from the Fresnel equations the relation

$$\left(\frac{I_\perp}{I_\parallel}\right)_N = \left(\frac{I'_\perp}{I'_\parallel}\right)^N \tag{2.18}$$

Here the quantity on the left is the ratio of intensities transmitted through N

consecutive perfectly transparent plates all set at Brewster's angle and the quantities with the superscript ', at the right, pertain to the intensities transmitted through a single plate. Thus, a pile of eight glass plates (assuming $n_M = 1.50$) should yield transmitted light with a degree of polarization of 93%. This principle has been taken advantage of for infrared polarimetry. A pile of six Se-films was found to give a polarization of better than 98% within the tested spectral range from 2 to 14 μ [64]. The authors, who give explicit directions for the preparation of these delicate piles, claim a surprisingly high transmittancy. A pile of five Se-films is stated to transmit 47% of I_0.

Polarization by reflection will, in general, be of major interest only when conventional polarizing prisms fail, that is, outside of the visible and near ultraviolet range of the spectrum. This does not apply necessarily if the polarimetric investigations do not require a 100% polarization. In that case, polarizers built from a pile of transparent plates set at Brewster's angle will —on using either the reflected or transmitted beam—yield a preponderantly polarized beam of useful intensity of any desired aperture. This unlimited aperture—the practical limit is set by the breadth of the incident parallel beam—is a decisive advantage compared to polarizing prisms whose cost is prohibitive if apertures in excess of 1 cm^2 are desired. In contradistinction to Polaroids (see following section), which also can be obtained with impressively large apertures, the present type of polarizers will show a relatively minor spectral variation of the degree of polarization.

It should be noted that the effect of partial polarization by reflection must be taken into account also in prism polarimeters if they are preceded by a monochromator. Thus, 31% of the light emerging from a double monochromator with glass prisms is linearly polarized. In order to take full advantage of the brightness of the light source, it is therefore imperative to arrange the polarizer so that its direction of transmission is crossed with respect to the width of the exit slit of the monochromator. The technique used for identifying the direction of transmission is described on page 90.

POLARIZERS BASED ON DICHROISM

A third principle of polarization may be applied at certain wavelengths if no precision is required in the measurements. Not only the refractive indices but also the absorption coefficients are different parallel and perpendicular to the optic axis of uniaxial crystals. This effect of "linear dichroism" has already been mentioned (pp. 67 *et seq.*). If the two coefficients differ very strongly, partially polarized light is obtained from incident natural light. A tourmaline crystal, for example, cut perpendicular to the optic axis, absorbs almost entirely the component vibrating perpendicular to the optic axis if the layer is 1 mm thick. Herapath [65] found that the biaxial crystals of iodoquinine sulfate exhibit a still stronger dichroism. A 0.1-mm layer of herapathite absorbs almost entirely one of the two components. In order to use these crystals as polarizers, Herapath [66] tried to grow large specimens. However, another procedure proved more practical, namely the orientation of a large number of colloidal herapathite crystals. The polarizing effect is then essentially the same as that of a large single crystal. Since 1934 [67] such herapathite polarizers have been produced on a large scale. The colloidal crystals usually oriented by streaming, are kept oriented by embedding them in plastics.

More recently, herapathite seems to have been replaced by still more dichroitic crystals or molecules. Commercially, these polarizers are known primarily under the name of " Polaroid Polarizing Filters." The most recent models of Polaroid filters are claimed to produce an almost complete polarization between approximately 5200 and 6800 Å with a loss in intensity of the less absorbed component of approximately 20%. Other Polaroid filters are designed for use in the ultraviolet and the infrared. The most important weakness of polarizers based on dichroism is that they do not and never will achieve a 100% polarization, and that they are not usable at all outside of the limited spectral region defined [68]. Some of the optical properties of Polaroids have been investigated and reviewed and the reader is referred to these papers for further information [68, 69]. Of some interest also is the combination of Polaroids with other imperfect polarizing devices. Thus Gallup [70] follows a sheet of black Carrara glass, set at Brewster's angle for red light, with a Polaroid J-filter.

Analyzer

The analyzer is used to determine the direction of vibration of linearly polarized light. Any polarizer may be used as an analyzer if the incident light is linearly polarized. Let us assume that in Fig. 2.20 (p. 78) the incident light is linearly polarized. The prism then acts as an analyzer. No light is transmitted through the prism if the direction of vibration is perpendicular to the optic axis (direction of extinction). However, if the direction of vibration is parallel to the latter, all the light—disregarding losses by reflection—is transmitted (direction of maximum transmission). The transmission of light changes between a maximum and zero if the direction of vibration rotates through 90°. The same change is obviously observed if, instead of the vibration, the prism rotates around the direction of propagation as the axis. During a full rotation of 360° two maxima of transmission and two complete extinctions are observed. An analyzer can therefore be characterized quite generally by two orthogonal directions perpendicular to the incident light beam, the direction of maximum transmission and the direction of extinction. The angle σ, between a vibration, V_i, of amplitude a_i, of a light beam incident upon the

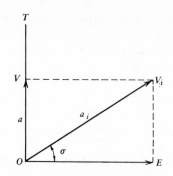

Fig. 2.26. Action of analyzer upon incident linear vibration: OT, direction of maximum transmission of analyzer; OE, direction of extinction of analyzer.

analyzer, and the direction of extinction OE (Fig. 2.26) determines the amplitude, a, of vibration, V, of the beam emerging from the analyzer:

$$a = a_i \sin \sigma \qquad (2.19)$$

The intensity, I, of the light transmitted through the analyzer and the brightness of the analyzer field, B, are proportional to the square of the amplitude:

$$\therefore \ I = KI_i \sin^2 \sigma \qquad (2.20)$$

$$B = B_S \sin^2 \sigma \qquad (2.21)$$

where I_i is the light intensity of the beam incident upon the analyzer and B_S is the brightness of the analyzer field if $\sigma = 90°$. Factor K takes into account the loss of intensity by reflection at and in the analyzer. In order to simplify, the factor K shall be considered as equal to 1.0. Equation (2.20) represents a modification of Malus' law:

$$I = KI_t \cos^2 \eta \qquad (2.21a)$$

where $\eta = (90° - \sigma)$. Figure 2.27 shows the variation of $I (B)$ with σ within the limits of $0°$ to $90°$.

Polarizer-Analyzer Combination

The polarizer furnishes linearly polarized light from natural light. The analyzer serves to determine the change in the direction of vibrational when an optically active substance is placed between polarizer and analyzer. In Fig. 2.28, the "nicols are crossed," that is, no light passes through the analyzer. If either prism were rotated by 90°, "parallel nicols," one would have the case of maximum transmission. Upon bringing an optically active substance between crossed nicols, light is transmitted because of the rotation of V_0 to V_1 (Fig. 2.1). Extinction can be reestablished by rotating the analyzer by the

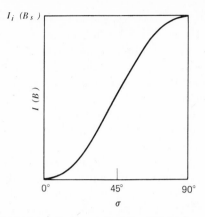

Fig. 2.27.   Variation of light intensity transmitted through analyzer, with angle σ between vibration of incident beam and direction of analyzer extinction.

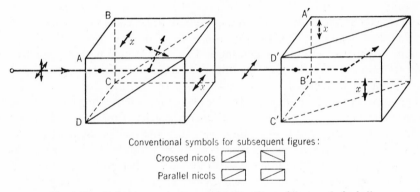

Conventional symbols for subsequent figures :
Crossed nicols
Parallel nicols

Fig. 2.28.   Polarizer–analyzer combination. Case of "crossed nicols."

angle ρ_A. This angle is, for the present type of operation, identical with the optical rotation of the substance, α. Thus, if the reading of the analyzer is ω_0 in the absence of the optically active substance and ω_1 in its presence (each time with the analyzer set for complete extinction):

$$\alpha = \rho_A = (\omega_1 - \omega_0) \qquad (2.22)$$

The polarizer–analyzer combination will function satisfactorily if the only change in V_0 is its rotation by an optically active substance. If in addition, the state of polarization is changed, by transformation of V_0 into an elliptical vibration, complete extinction cannot be reestablished (see Figs. 2.17 and 2.40). It is then necessary to introduce additional accessories, specifically, an ellipticity compensator and an ellipticity half-shade. These techniques go beyond the prescribed scope of this chapter, and they are therefore not discussed. Ellipticities, excepting those due to a Cotton effect (see the remarks on pp. 66 and 67), can generally be avoided or eliminated, however, in measurements of a rotatory power. One possible source of ellipticity are lenses that are rarely free of stress double refraction, except very expensive small lenses as used in condensers and objectives of polarizing microscopes built for precision measurements. It is therefore imperative to avoid the use of lenses or prisms between polarizer and analyzer. Another source of ellipticity is unsuitable or improperly handled cover glasses of polarimeter tubes (pp. 93 *et seq*).

For measurements of the rotatory power itself, it is unnecessary to know the direction in which a prism will transmit linearly polarized light, unless the influence of the electric or transverse magnetic field upon rotatory power is to be studied. It is very desirable, however, to know this direction if the polarizer is preceded by a monochromator (p. 86). The direction of the diagonal plane within a polarizing prism does not necessarily give the desired information. It is, however, simple to remove the polarizer or analyzer from the polarimeter and to point it toward a windowsill or a desk from which sky light is reflected. Upon rotating the prism in front of the eye, a position of minimum brightness is found. According to Fig. 2.25, in light partially polarized by reflection on a surface, the predominant component vibrates normally to the plane of incidence. Consequently, when the prism is in a position of minimum transmission, the direction of extinction is horizontal. An alternate procedure not requiring the removal of a prism from the polarimeter is as follows: the nicols are first crossed. Then the analyzer is rotated by 90° in order to obtain parallel nicols as accurately as possible. A third polarizing prism, "calibrated" as outlined, is set in front of the polarizer or between analyzer and observer and rotated until extinction is reestablished. It follows from this and from an earlier disussion (p. 86) that the direction of maximum transmission of the polarizer should be perpendicular to the width of the slit of the preceding monochromator.

Defects of polarizing devices are particularly pronounced when the latter are combined into a polarizer–analyzer system. Using Fresnel's formula, we find that the maximum intensity transmitted through an analyzing Glan-Thompson prism preceded by a polarizing Glan-Thompson prism is 42.65% of the intensity of the original unpolarized beam. The only loss, beside the obvious loss of 50%, is caused by fourfold reflection. In Polaroids, we have, in addition the loss by absorption. Transmission through two parallel HN22 Polaroids, recommended for "extreme extinction density" is 30% at 7000 Å falling to approximately 20% at 4500 Å [69]. In the case of polarization by reflection (pp. 82, *et seq*.), a combination of two reflectors [both consisting of very heavy flint glass (Table 2.1)] which corresponds to parallel nicols (Fig. 2.25 I), furnishes only about 2.5% of the original intensity. For crossed Glan-

Thompson prisms, the extinction will be complete even with the sun as the light source. It will be practically complete for a set of "crossed" reflectors. Starting out with position I in Fig. 2.25, the position of "crossed reflectors" is reached by rotating mirror $A_1 B_1 C_1 D_1$ by 90° around the incident beam, as the axis, without changing angle i (Fig. 2.25 II). The extinction will be fair for Polaroids unless one uses a light source of low brightness. But, even then, crossed Polaroids designed for work in the visible will transmit an appreciable amount of red and violet radiation.

Graduated Circle

The graduated circle for measuring the rotation of the analyzer in a manual polarimeter is divided into degrees and decimal fractions of degrees or minutes. In instruments for sugar analysis, the divisions may be in terms of sugar concentration. Regarding the latter instruments and the field of saccharimetry, the reader may be referred to the special literature [71]. The graduated circle is fitted with one vernier or, for precision measurements, with two verniers. Most commonly available commerical instruments allow the reading of 0.01°, or of 0.5′ and the safe estimate of 0.005°. The reading on a precision scale may be still more accurate with a pair of parallel index lines. By turning a supplementary graduate drum, the borders of that etched line on the main scale whose exact position is to be determined are made to coincide with the lines. This may allow a reading of 0.001° (3.6″). Commercial American polarimeters with this additional feature are also available. They allow a reading of 0.002°. The working principle of the index lines follows from Fig. 2.29 II, while the vernier principle is shown by Fig. 2.29 I. Instead of refining available commercial instruments in this manner, graduated circles may be used as manufactured for geodesic instruments which allow the direct reading of small multiples of seconds. Before installing such a high-precision circle in a polarimeter, one should be sure of a sufficiently high quality of the mechanical parts of the instrument. Another method of reading a rotation in small fractions of a minute is based on the use of the gear principle. A toothed circle, which this time does not need to be graduated, is rotated by a slow-motion gear coupled with a counter.

A precise graduation of the circle and of the vernier is a necessary but not sufficient prerequisite for precision measurements. (1) Any measurement which is to be accurate within less than two minutes must be made by means of two verniers 180° apart in order to eliminate errors due to eccentricity. (2) The instrument must have a sufficiently sensitive "null point" device (pp. 100–111). (3) The actual precision depends largely on the care with which the instrument is handled and operated, on the care in keeping the temperature constant, and on the sensitivity of the recording instrument, for example, the eye.

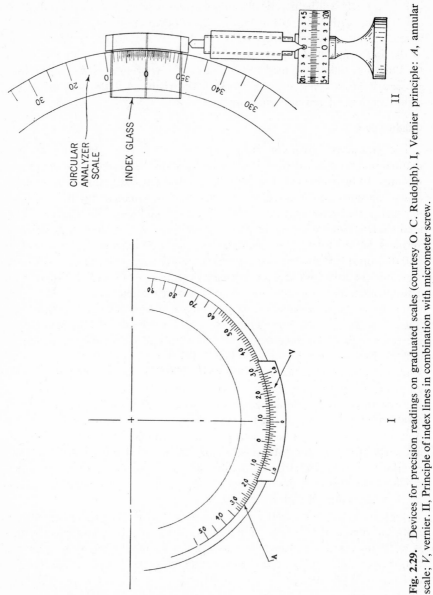

Fig. 2.29. Devices for precision readings on graduated scales (courtesy O. C. Rudolph). I, Vernier principle: *A*, annular scale; *V*, vernier. II, Principle of index lines in combination with micrometer screw.

Light Sources and Light Filters for Measurements in Visible Spectrum

For measurements in monochromatic light, there is a choice between two possibilities: one may use a monochromatic light source in connection with a monochromator of low dispersion, or with a set of filters; or one may use a light source that gives a continuous spectrum and combine it with a monochromator of high dispersion. In general, monochromatic light sources emit several wavelengths strongly. However, they limit the measurements to these wavelengths, a disadvantage most pronounced for sodium lamps which in the visible have a strong emission only at 5895.9 and 5889.9 Å, a doublet generally called the D line. For this reason and also because of the frequent increase of α with decreasing wavelength, it is, in principle, preferable to work with a mercury arc, which emits easily visible rays at 5790.7, 5769.6, 5460.7, 4916.0, and 4358.3 Å.

A sodium lamp of the commerical type should be used with a light filter in order to exclude a faint continuous spectrum, weak sodium lines present beside the D line, and lines due to foreign gases in the lamp. A mercury arc, and particularly the bright high-pressure mercury lamps [72], must also be operated with light filters or a monochromator. Interference filters [73] are very satisfactory for isolating fairly widely separated lines, since they can transmit 35 % of an incident line with a passband of 150 Å. It is often necessary to use other filters in series with these to exclude undesired orders. Colored glass filters [74] can be used for this, or can themselves be used in combination to isolate lines. Filters of dyed gelatin [75] are optically very satisfactory but some do not withstand prolonged intense illumination. Liquid solutions are also used as filters [76].

In regard to light sources with a broad emission spectrum, the sun, with 150,000 international cd/cm^2, is the brightest source by far, but its use, by means of a heliostat, is clearly limited. Tungsten and zirconium arcs have been popular but some movement of the arc restricts their use. Tungsten ribbon filament and tungsten coil filament lamps are more steady and satisfactory sources. The tungsten-halogen lamps and particularly xenon [77] arc lamps are widely used in both visible and ultraviolet work [78].

Polarimeter Tubes

Polarimeter tubes are covered at both ends by plane-parallel glass or fused silica disks which are sealed on or held in place by means of holders. For a tight fit without strain, an elastic ring is inserted between disk and holder. The tube must be placed in the light path with the reflecting surfaces of both disks strictly normal to the beam; otherwise the beam will be deflected and the light flux in the analyzer reduced or rendered inhomogeneous. In commercial polarimeters, the tube holders take care of the adjustment. If the disks are birefringent due to strain, they will transform linearly polarized light into elliptically polarized light (pp. 70 *et seq.*) and complete extinction between crossed nicols will be impossible. Since the strain is generally not uniform across the disks, the brightness of the field viewed through crossed nicols will not be uniform either. This will lead to incorrect zero points on using a half shade. Even isotropic glass disks may cause trouble if they are subjected to stress, particularly if they are fastened to the tube with a screw cap. If the latter is tightened too much, the glass disk may become birefringent. Differences in magnitude and nonuniformity of strain in consecutive series of measurements may therefore lead to annoying apparent changes in zero point. These changes can be avoided by preventing too tight a fit of the disks. A simple optical test to this effect has been described by Streuli [79] The disks are tested by crossing the nicols in the absence of the polarimeter tube and inserting the tube filled with distilled water. If the field remains completely dark, it is, nevertheless, advisable to make an additional test by rotating the tube around the light path as the axis by about 20 and 45°. If the field remains uniformly dark the disks are practically free of double refraction. Since double refraction may be caused by pressure on the disks, it is advisable to determine, in a separate experiment, with how many turns the disk holders may be screwed in without causing noticeable double refraction. The pressure should be released at once after the detection of an effect, in order to avoid incomplete relaxation. The disks may also produce a rotation. It is therefore necessary to test, in addition, whether the angular reading found for the position of extinction is the same in the absence and in the presence of the disks. King [80] has shown that even very thin surface films may cause problems if they are oriented. One may produce a troublesome oriented surface film by simply running a finger across the glass disk since even a clean finger has a slight amount of grease on its surface. The problem is, of course, most severe when high-precision work in the ultraviolet is attempted. King also has shown that interactions between birefringence in the sample and in the polarimeter can amplify the effects of sample birefringence.

The disk holders of commerical demountable polarimeter tubes usually have a narrow opening to serve as a diaphragm in order to avoid reflection of the beam on the tube walls. These, and other diaphragms, should be painted black to avoid stray light. For precision measurements, commercially available tubes with a water jacket for thermostated water are used, because such measurements require that the liquid in the polarimeter tube be at a constant and exactly known temperature. A special water-jacketed tube developed by the National Bureau of Standards [81] makes it possible to study the rotatory power at any temperature between 0 and 100°C without the risk of leakage or of a distortion of the optical field through temperature gradients. A special cap, to prevent fogging of the windows at low temperature, has also been developed [81]. Instead of using water-jacketed polarimeter tubes, the polarimeter itself may be placed in a constant temperature cabinet [82] or, even better, in a thermoconstant room.

One-piece cells of fused silica have come into wide use for work in the visible as well as in the ultraviolet. When such cells are properly annealed by the manufacturer, stress birefringence is practically eliminated.

For measurements of optical activity of substances available in amounts too small for a standard-size polarimeter tube, micropolarimeter tubes have been developed. Capillary tubes that fit snugly into a standard size polarimeter tube are particularly convenient [83]. Capillary tubes may not require more than 1.8–4.6 mg of substance in 0.64–1.35 g of solution [84]. If the rotatory power is very strong, even much less than 1 mg may then be sufficient. Thermostatic control of micropolarimeter tubes is particularly simple because they may be made of drawn stainless steel surrounded by a water jacket, and the whole assembly need not exceed the dimensions of an ordinary polarimeter tube. A thermostated micropolarimeter tube, 12.1 cm long and 0.3 cm in diameter has been described by Kacser and Ubbelohde [85]. An ultramicrocell has been made available by the Perkin Elmer Corporation [86], 10 cm long with a volume of 38 μl; this is designed for use with a laser light source. A device by means of which a polarimeter tube, made of hard rubber can be agitated in a horizontal and circular motion has been described by Straub [87] for use with protein solutions.

In order to determine the optical length of a polarimeter tube, one may measure the rotation of a liquid or a solution of known strong rotatory power at a constant and accurately known temperature. The rotatory power of nicotine in ethyl alcohol is so strong that the thickness of a layer can be determined with an error $< \pm 0.01$ mm. Data are given in Table 2.2. The

Table 2.2.   Optical Rotation of Nicotine in Ethanol[a]

Nicotine, g per 100 g Solution	Ethanol, g per 100 g Solution	α
100.00	0	$-161.42°$
90.09	9.91	$-141.16°$
74.93	25.07	$-110.62°$
59.93	40.07	$- 83.63°$
45.08	54.92	$- 59.49°$
30.03	69.97	$- 37.32°$
14.96	85.04	$- 17.46°$

[a] From data of H. Landolt, *Ann.*, **189**, 241 (1877). The boiling point at 74.5 mm Hg of the nicotine used was 246.6–246.8°. d_4^{20} of ethanol was 0.7957. The value of α was obtained at 20°C for the D line, with a layer of 9.992 cm.

variation of α with the concentration is practically linear. The optical length of a polarimeter tube varies slightly with temperature. If the optical length, that is, the layer thickness, l, is determined at 20°C, it may be calculated for any temperature, T, from the equation:

$$l_T = l_{20}[1 + \mu(T - 20)] \tag{2.23}$$

where μ is the linear expansion coefficient of the material. The value of μ for a glass tube is 8×10^{-6}.

Visual Polarimetry without Compensator in Monochromatic Light

Some Elements of Physiological Optics

Since the eye is the recording instrument in visual polarimetry, it is necessary, in order to use it to best advantage, to be familiar with some of its relevant characteristics.

RELATIVE VISIBILITY

The sensitivity of the eye for a given wavelength can be measured in terms of visibility at this wavelength, which is defined by the ratio of luminosity to radiant power of the light source, that is, by lumen/erg sec^{-1}. The extreme limits of visibility reached experimentally, under optimum conditions, are 8350 and 3650 Å, respectively. The visibility at the extremes of the given range is very small compared with that in the middle. It is a million times smaller at 3650 Å than at 5600 Å. For brightnesses corresponding to that of diffuse daylight, the maximum visibility—photopic vision—is at 5550 Å, as shown in curve I of Fig. 2.30 which gives averages compiled by Gibson and Tyndall [88]. The data are given in terms of relative visibility; that is, the maximum visibility at 5550 Å is set equal to 1.0.

At low brightness levels when the eye is said to possess scotopic, or dark vision, the wavelength of maximum sensitivity is shifted toward the blue (Purkinje phenomenon). The transition from daylight to scotopic vision begins at about a brightness of 0.1 ft-L, (foot-Lambert) which corresponds to the brightness of a white surface in a room receiving barely enough daylight for reading or writing without strain to the eyes. Figure 2.30 (curve II) shows the spectral distribution of sensitivity of the eye when the brightness of the field observed is, at best, 56 μft-L [89–91], which corresponds to the brightness of the clear night sky in absence of the moon. The maximum sensitivity is this time reached at 5110–5113 Å, the latter value being generally accepted as the standard for scotopic vision.

In polarimetry, we are frequently confronted with brightnesses below the critical value for which curve I (Fig. 2.30) holds. If, therefore, measurements

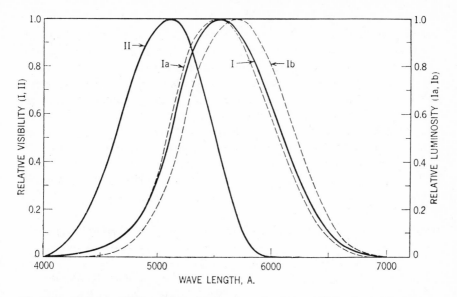

Fig. 2.30. Relative visibility and luminosity. Curve I, relative visibility for daylight vision (data of Gibson and Tyndall): Ia, relative luminosity of sun in zenith recorded at sea level; Ib, relative luminosity of commercial tungsten-filament lamp; II, relative visibility for scotopic vision (data of Luckiesh and Taylor).

are to be made at a single wavelength selected by the observer, a wavelength within the range of highest sensitivity should be chosen. Even at a high brightness, the eye is definitely more sensitive to the green line of the mercury arc than to the yellow sodium lines of equal absolute brightness; and at low brightnesses the differences are much greater. A mercury arc is therefore preferable to a sodium lamp if the radiant power of the mercury arc at 5461 Å is equal or slightly inferior to the radiant power of the latter at the D line. Particularly for sources with continuous spectrum, it should be kept in mind, however, that the relative visibility, though it is the decisive factor, is not the only factor to be considered in the choice of the most advantageous wavelength. If the radiant power of the light source varies strongly with wavelength, the actual impression of brightness in photopic vision, that is, luminosity, may reach its peak at a wavelength somewhat lower or higher than 5550 Å. Curves *Ia* and *Ib* give the variation in luminosity with wavelength for the sun—observation at sea level with the sun in the zenith—and for a commercial gas-filled tungsten-filament lamp, respectively. Detailed calculations of luminosities of a radiator within the range of color temperatures corresponding to that of tungsten lamps employed in experimental work (2000–3120° K) have been carried out by Skogland [92].

ADAPTION OF THE EYE

At a given wavelength, the eye is more sensitive toward a newly introduced brightness, B_{II}, the lower the brightness, B_I, to which it was previously adapted. A certain time elapses before the eye ceases to be dazzled if $B_{II} \gg B_I$, and before B_{II} is perceived if $B_{II} \ll B_I$. In the latter case, full sensitivity to B_{II} requires quite some time. The case $B_{II} \ll B_I$ is the common situation in polarimetry. The eye is first adapted from daylight to the lower brightness of the polarimeter field and later is readapted to it after an action at higher brightness, as during the reading of a scale or the recording of results on paper. Table 2.3 gives data of Blanchard [93] showing the time, t, required before the eye, previously adapted to a brightness B_I, is just able to perceive a much lower brightness, B_{II}.

Table 2.3. Rate of Adaptation of the Human Eye in Various Spectral Regions[a]

	White				Blue	Green	Yellow	Red
B_I :	0.1	1.0	10	100	0.1	0.1	0.1	0.1
t				$-\log B_{II}$				
0 sec	2.79	2.20	1.60	0.90	2.82	2.69	2.61	2.32
2 sec	4.13	3.27	2.53	2.00	4.36	4.39	4.17	2.98
5 sec	4.50	3.79	3.08	2.46	4.91	4.82	4.41	3.37
10 sec	4.75	4.15	3.54	2.64	5.27	5.11	4.65	3.57
1 min	5.32	5.06	4.61	3.84	5.81	5.56	5.09	3.80
2 min	5.52	5.22	4.83	4.12	6.00	5.70	5.24	3.92
5 min	5.68	5.52	5.22	4.76	6.23	5.80	5.39	4.02
10 min	5.70	5.68	5.59	5.38	—	—	—	—
60 min	6.06	6.04	6.01	5.97	—	—	—	—

[a] Data of Blanchard. Symbols in this table are explained in the text. Both B_I and B_{II} are brightnesses, expressed in millilamberts (1 mL $= 0.000354$ cd/cm^2).

In the light of these facts, the polarimeter should be operated in a darkened room, with a shielded light source, and measurements should not be started before 10 minutes after darkening. The necessary periodic illumination of scales and notebook should be as low as possible without strain for the eyes. Small light bulbs, operated at a voltage lower than that indicated at the mounting and properly shielded, should be installed near the area to be illuminated in order to allow only indirect light to reach the eye. The eye will then become almost completely readapted to dark within 10 seconds. If a weakly absorbing substance exhibits strong rotatory power, the high intensity transmitted through crossed nicols may blind the eye. It is then advisable to rotate the analyzer to near-extinction and to wait with the final measurement until the eye has recovered its sensitivity.

The sensitivity of a well-adapted eye is very often underrated. According to Hecht [94], the energy necessary for the production of the minimum visual effect, by means of an eye well adapted to the dark, is 5–8 light quanta. To that effect, it is necessary that 54–148 quanta are incident upon the cornea of the eye, the balance being lost before the light reaches the visual purple of the retina. In order to produce a detectable grain of silver on a photographic plate, 50–100 quanta are necessary [95]. In order to liberate one photoelectron from a photosensitive layer, 15–50 quanta are necessary; and 100,000 quanta per second are required in order to produce a steady photoelectric current. The sensitivity of the dark-adapted eye is therefore not surpassed by these artificial recording devices.

ABILITY OF THE EYE TO DISTINGUISH BETWEEN BRIGHTNESSES

The simplest method of measuring a rotation would be by rotation of the analyzer until complete extinction is established. If the sensitivity of the eye were the only important physiological factor, it would be possible to achieve a remarkably high accuracy in such measurements. A minute deviation of the analyzer from the position of extinction would be detectable. However, success in finding the position of complete extinction would require that the eye also *remember quantitatively* the various degrees of brightness perceived during a series of trials with analyzer positions slightly off the position of the true minimum. Unfortunately, small differences in *gradually varied sensation* which are involved in the present instance are not remembered well, whereas *fluctuations in sensation* near the absolute limit of visibility of the human eye are registered with such a remarkably high sensitivity that sudden differences, of the order of magnitude of 10 quanta, can be detected and estimated by particularly well-trained persons. The error in finding the position of complete extinction is generally defined by an uncertainty of not less than 3 minutes on the graduated scale. In cheap instruments, the graduated circle may not allow a better reading. It is then superfluous to use a principle other than that of extinction. For precision measurements, however, the principles of the half-shade or of the fringes are used which do not require the recollection of brightnesses.

In the methods based on the half-shade principle, the determination of minimum brightness is replaced by a determination of equal brightness in adjacent fields. According to the psychophysical law of Weber-Fechner, the eye is able to distinguish between two brightnesses, B_I and B_{II}, if their logarithmic difference is not smaller than 1–2%, that is, if

$$\log(B_I/B_{II}) \sim \log 1.01 \sim 0.004 \tag{2.24}$$

This law holds for moderate brightnesses but not for brightnesses that dazzle the eye adapted to diffuse daylight, or for brightnesses equivalent to, or lower than, the brightness of a white surface in moonlight. The decline in the differentiating power of the eye with declining brightness is more pronounced at longer wavelengths. If, at a given low brightness, the error in differentiation is 30% at 5050 Å, it rises to 80% at 6050 Å.

In some of the methods using the fringe principle, the brightness or darkness of a stationary fringe is varied until it equals that of the field, that is, the fringe disappears. From the point of view of physiological optics, the problems are the same as in the half-shade method. In other methods, one observes the linear displacement of a fringe or of fringes in a contrasting field. In this case, additional phenomena of physiological optics are involved. Only a few practical aspects can be discussed here. The fringes encountered in polarimetry are generally without sharp boundaries. In order to measure a displacement accurately, it is necessary that the boundaries *appear* as sharp as possible. If the fringes are broad, it is, in addition, desirable that they *appear* narrow, because the position of the center of a fringe is often the criterion in fringe measurements. As discussed extensively by Helmholtz [96] the impressions of both sharpness and narrowness are furthered by adapting the eye well, by focusing as sharply upon the fringe as possible, and by making the objective dark-white contrast—or color contrast—as strong as possible (without blinding the eye) through use of as powerful a light source as possible.

An interesting, but fragmentary, discussion of the precision of the half-shade and fringe methods has been made by Yves Le Grand [97].

Half-Shade Method

The half-shade is a device which transforms complete extinction, in the case of crossed nicols, into equal illumination of two adjacent fields. It will be seen that both fields then generally appear very dim. This effect has led to the coining of the term "half-shade," or "penumbra," either corresponding to the French *pénombre*. Since no shadow of anything is involved, the often preferred, but incorrect, term "half-shadow" shall be disregarded. The term "isophotostatic" device [98] is too cumbersome.

The half-shade principle is explained by Fig. 2.31, which illustrates the action of two adjacent Glan-Thompson prisms, A and B. Dotted lines across the prisms represent the directions of extinction for linearly polarized light. The inner circle is the circumference of a rotatable mounting for the two prisms. The pointer protruding from the rotatable mounting serves to indicate on the outer, fixed, circle the positions of the mounting when the fields of prisms A or B are brightest (b) or completely dark (d), respectively. The symbols at the peripheral part of the outer circle refer to A, those at the inside, to B. In Fig. 2.31a, the dotted directions of extinction are parallel and the double

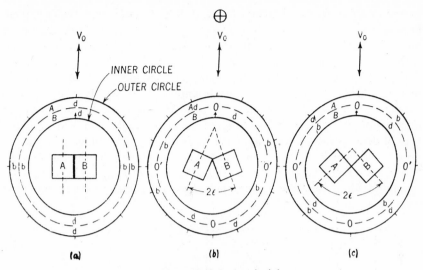

Fig. 2.31. Half-shade principle.

prism functions as a single analyzer. In Fig. 2.31b, where the directions of extinction form a "half-shade angle," 2ϵ, of 45°, a rotation of $-22.5°$ will lead to complete extinction in A, and a rotation of $+22.5°$ to complete extinction in B. The half-shade angle thus is the smaller of the two possible angles, 45° and $(180° - 45°)$, which characterize the angular difference for complete extinction in A and B. In the intermediate "zero" position, 0, both fields are equally dim.

The change in brightness in A and B with a rotation, ρ_A, of the rotatable analyzer mounting is illustrated by Fig. 2.32. It will be noticed that at right angles to 0, the brightness of the two fields is also equal, 0′, but far higher than at 0, and that 0′ is less suitable as a practical zero position (see Fig. 2.32). Depending on the angle, 2ϵ, the brightness of the double field in the zero position, 0, will vary between zero ($2\epsilon = 0$, Fig. 2.31a) and half the value at

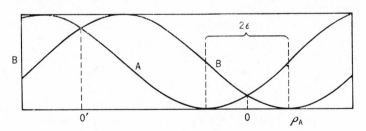

Fig. 2.32. Variation in brightness of the two analyzer fields with rotation, ρ_A, of analyzer mounting of Fig. 2.31. Half-shade angle of 45°.

maximum transmission ($2\epsilon = 90°$, Fig. 2.31c). In the latter case, the brightness in position 0 is equal to the brightness in position 0' (Fig. 2.31), that is, in monochromatic light, the position equivalent to crossed nicols is identical with that of parallel nicols.

Figure 2.33 shows that a small half-shade angle should be chosen for maximum precision. B_{Cr} is the critical brightness below which the eye no longer detects a brightness difference, ΔB, of 1–2%. B_{Eq} is the brightness at the intersection of the two curves—brightness equality in the two fields. For the sake of demonstration, ΔB is assumed to be 0.2 B_{Eq}. The range of rotation, $2\Delta\rho$, within which the eye cannot distinguish between the brightness of the two fields is much larger for a large half-shade angle (Fig. 2.33a) than for a small one (Fig. 2.33b). The smallest half-shade angle is set by the requirement

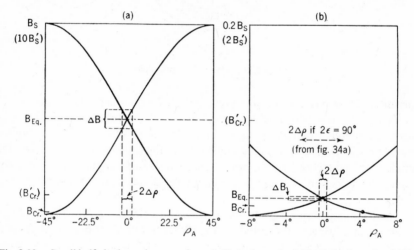

Fig. 2.33.   Small half-shade angle versus large half-shade angle: (a) half-shade angle of 90°; (b) half-shade angle of 16°.

that the brightness in the zero position, B_{Eq}, not be lower than B_{Cr}. If the maximum brightness of the analyzer field, B_S, is smaller, B'_S (less powerful light source), then the critical brightness may have the value B'_{Cr}. It is obvious from Fig. 2.33 that the large half-shade angle (Fig. 2.33a) will give more accurate results than the small angle (Fig. 2.33b). Hence, the half-shade angle should be variable if the brightness of the analyzer field is variable, that is, if the transparency of the substance investigated or the layer thickness varies, if the wavelength is to be varied, or if the brightness of the light source changes.

The precision of the half-shade method follows from Table 2.4, which gives data calculated by Landolt [99] on the assumption that ΔB, perceptible to the eye, is 0.02 B_{Eq} for all the half-shade angles. This assumption is

Table 2.4. Landolt Data for Half-Shade Angle, 2ϵ, and Uncertainty, $\pm\Delta\rho$, in Determination of Optical Rotation

2ϵ	$\pm\Delta\rho$
0.5°	4″
1°	9″
2°	18″
4°	36″
6°	54″
8°	72″
10°	1.5′

certainly not fulfilled for the two smallest half-shade angles, excepting possibly the case in which the sun is the light source. The data are expressed in terms of $\Delta\rho$ which is the physiological uncertainty involved in the determination of the rotation, ρ [99–101]:

$$\pm\Delta\rho = \frac{\epsilon}{4}\frac{\Delta B}{B_{\mathrm{Eq}}} \qquad (2.24a)$$

It follows from this equation that the uncertainty, $\Delta\rho$, is proportional to the half-shade angle as long as the Weber-Fechner law holds, that is, as long as $\Delta B/B_{\mathrm{Eq}}$ is a constant, and that it increases, for a given half-shade angle, if the Weber-Fechner constant increases (see pp. 99–100).

The simplest practical analyzer with half-shade effect is the Cornu modification of the Jellett prism [102–103]. It is made by sawing a Glan-Thompson prism in half, along AB in Fig. 2.34, grinding one of the halves down to AC and cementing the two parts. Since the half-shade angle is invariable, the Jellett-Cornu prism is satisfactory only if the light flux arriving in front of the

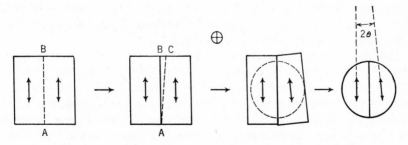

Fig. 2.34. Jellett-Cornu prism. Arrows indicate direction of extinction.

prism is approximately the same under all experimental conditions and, in addition, of the order of magnitude for which the fixed half-shade angle is most suitable. Jellett-Cornu prisms are recognized by the fine dividing line across their frontal face. Figure 2.35a gives the aspect of the optical field, which is usually made circular by a diaphragm, for the zero position (2), after rotation by $-\epsilon$ (1), and after rotation by $+\epsilon$ (3). The virtual image of one analyzer face is usually observed by means of a telescope.

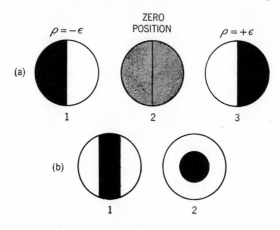

Fig. 2.35. Aspect of half-shade field: (a) changes in aspect during the search for field equality; (b) aspect of special half-shade fields, off position of field equality.

Double-image prisms (see pp. 82, *et seq.*) represent another type of simple analyzer with half-shade effect. The half-shade angle is 90° since the two emerging components are polarized at right angles. These prisms play an important role in nonvisual polarimetry, where large half-shade angles are desirable, particularly in photoelectric polarimetry, but may be without major interest as half-shade analyzers in visual polarimetry, where small half-shade angles are required.

In most cases, the half-shade effect is brought about by setting an appropriate device in front of a simple analyzing prism. The telescope is focused upon the device, not, as above, upon the analyzer. With the exception of a few cases, the field is the same as in Fig. 2.35a.

The best "compound analyzer," that is, analyzer with additional half-shade device, is the Lippich prism [104]. Half of the field of a simple analyzer, II (Fig. 2.36a), is covered by a preceding analyzer, I, of smaller size. The directions of extinction of the two prisms form a small angle (see Fig. 2.37). Figure 2.36b shows a commercially favored type of Lippich prism in which the field is divided into three parts by setting two small prisms in front of the main analyzer; the aspect of the field is, in this case, as shown in Fig. 2.35b,

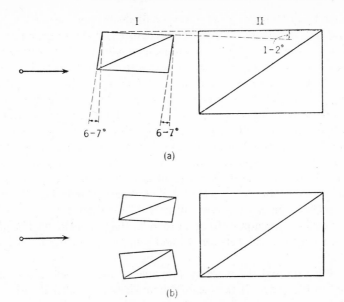

Fig. 2.36. Lippich prism: (*a*) double-field arrangement; (*b*) triple-field arrangement.

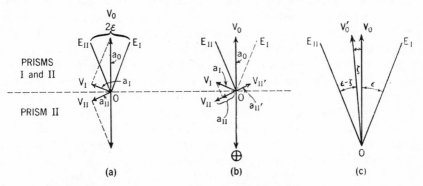

Fig. 2.37. Principle of Lippich half-shade method (double-field arrangement).

view 1, assuming a rotation by ϵ. The small prism of the two-field Lippich prism is arranged in a slightly slanting position, 1–2°, in order to obtain a well-illuminated and sharp dividing line of the two fields. For the same reason, the faces of the small prism are corrected by 6–7°. The half-shade angle can be varied by rotating—within a limited range—the small prism and the main prism independently one from the other. For this purpose, polarimeters fitted with a Lippich prism usually have a lever on the mounting.

In Fig. 2.37, it is assumed that the directions of extinction, OE_I and OE_{II}, of prisms I and II, respectively, form a half-shade angle, 2ϵ, of 45°, and that OE_I and OE_{II} are symmetrical with respect to V_0. If the Lippich prism could function like the Jellett-Cornu prism, field equality would be expected in this position; but such is not the case. To demonstrate this fact and the important reason for it, we propose to find the directions and amplitudes of the vibration, V_{II} (Fig. 2.37a), transmitted through the lower half of the prism II, which is not preceded by prism I, and of the vibration, V_I, which is transmitted through prism I, disregarding at first the subsequent transmission through prism II. The amplitudes are obtained according to the procedure used for Fig. 2.26. It is apparent that the amplitudes, a_I and a_{II}, are equal. V_I is now altered during the subsequent passage of the beam through prism II. The resulting vibration, $V_{II'}$ (Fig. 2.37b), is parallel to V_{II} but its amplitude, $a_{II'}$, is smaller than a_{II}. The two half-fields of the Lippich analyzer are therefore not equal in brightness. In order to make them equal a_I must be increased by increasing the angle between V_0 and OE_I to $V_0'OE_I$ by an increment, ζ (Fig. 2.37c).

The situation and the working principle of the Lippich prism are summarized in Fig. 2.37c. If the incident light vibrates parallel to either OE_I or OE_{II}, the respective half-field is completely dark. The vibration, V_0', which leads to field equality, deviates by the angle ζ from the position of symmetry with respect to OE_I and OE_{II}; that is, it deviates by the angle ζ from vibration V_0, which would be extinct if a simple analyzer were used. The half-shade effect of the Lippich prism may therefore be called "asymmetrical," while the half-shade effect of the Jellett-Cornu prism may be called "symmetrical." The asymmetry as such is of no importance in the measuring of rotations since ζ obviously does not vary with α. However, we should remember that ζ varies with the half-shade angle. A zero position determined for a given half-shade angle is therefore not strictly valid for another half-shade angle.

The Lippich prism has another peculiarity which, if overlooked, may cause errors. The light traversing both II and I is weakened more by reflection than that traversing I only. The difference changes the angle ζ by $\Delta\zeta$ to ζ' (not shown in Fig. 2.37), and $\Delta\zeta$ is likely to be even larger than ζ itself if the half-shade angle is small. As a result the zero position will vary with the wavelength because of the spectral variation in the reflecting power of the prisms. A zero reading taken in absence of the optically active substance holds, therefore, strictly only for the wavelength and the half-shade angle used.

Another, often used, compound analyzer with variable half-shade angle has been introduced by Laurent [105, 106]. It is sturdier and somewhat cheaper than the Lippich prism. The Laurent compound analyzer, shown schematically in Fig. 2.38, consists of a Glan-Thompson prism, II, half of which is preceded by a plane-parallel plate, I, of quartz cut parallel to its optic axis. In some cases, a circular strip of quartz is cemented onto a glass plate [107].

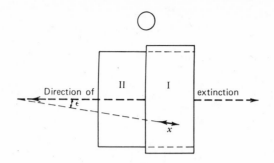

Fig. 2.38.   Laurent arrangement.

The aspect of the field is then that of Fig. 2.35*b*, view 2. This construction has some advantages.

Quartz is optically active toward light propagated parallel to its optic axis (see p. 71) but birefringent toward light propagated perpendicular to its optic axis. The half-shade effect of the Laurent compound analyzer is therefore based on the principle of double refraction. With the incident beam linearly polarized and the angle of incidence 0°, the light emerging from the quartz plate is linearly polarized only if the phase difference equals π (180°) or a multiple of π (see Fig. 2.18). For even multiples of π, the direction of the vibration emerging from the plate is identical with that of the original vibration. For a phase difference of π or an uneven multiple of π, the direction of vibration, V_1, of the beam emerging from the quartz plate differs from the original direction of vibration, V_0. V_1 and V_0 are symmetrical with respect to the optic axis. The angular displacement, 2ϵ, increases with the angle γ between V_0 and the optic axis up to an angle of 45°. The latter angle leads to the maximum angular displacement of 90° (Fig. 2.18). The quartz plate in the Laurent compound analyzer causes such a symmetrical displacement of the incident linear vibration. From the optical constants of quartz ($n_e = 1.553$; $n_0 = 1.544$) for the D line, it follows that a thickness of 32 μ or of an uneven multiple of 32μ is required in order to obtain a phase difference of π, or of an uneven multiple of π, respectively, in work with the D line.

Figure 2.39 demonstrates the result of the symmetrical displacement. The part of the Glan-Thompson prism covered by the plate is at the left; the other part of the prism is at the right of the vertical coordinate. In Fig. 2.39*a*, V_0 is parallel to the direction of extinction of the prism. *OE*; in Fig. 2.39*b* it is parallel to the optic axis of the quartz plate; and in 2.39*c* it is twice as far from *OE* as from *Ox* in terms of angular distance. Case *b* represents the position of field equality. The half-shade angle, 2ϵ, 45° in the example, is equal to twice the angle between *OE* and *Ox*. The half-shade effect is symmetrical, except for the disregarded asymmetry introduced by unequal reflections in the

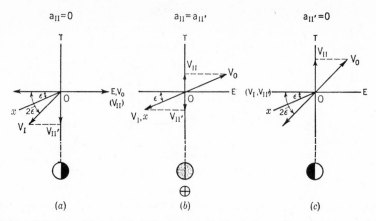

Fig. 2.39. Principle of Laurent half-shade method. Vertical coordinate represents dividing line between uncovered half-field (right) and quartz-covered half-field of Glan-Thompson prism. V_I, direction of vibration after light has traversed quartz plate and before it traverses prism; V_{II}, direction of vibration of light traversing prism only; $V_{II'}$, alteration of V_I by subsequent passage through prism.

two half-fields, and can be varied by rotating the quartz plate while the prism is kept in position.

The magnitude of double refraction changes with the wavelength. A quartz plate producing a phase difference of π at a given wavelength will cause a larger difference at shorter, and a smaller difference at longer wavelengths. This must be kept in mind when a Laurent compound analyzer is used. Let us assume in Fig. 2.40 that the phase difference is $(3/2)\pi$. If $\gamma \neq 45°$, the light emerging from the quartz plate will vibrate elliptically, V_I, the long

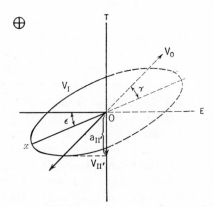

Fig. 2.40. Deficiency of Laurent arrangement if used with light of unsuitable wavelength. Changes for light beam that traverses prism only are omitted.

axis of the ellipse coinciding with the optic axis of the plate (Fig. 2.18). The analyzer transforms this elliptical vibration into a linear vibration, $V_{II'}$, of amplitude $a_{II'}$. For a given angle, γ, a linear vibration will be transmitted whatever the orientation of the ellipse with respect to the direction of extinction, OE. It will be impossible, therefore, to obtain complete extinction in that part of the analyzer field which is preceded by the quartz plate. A change of γ, on the other hand, changes the axial ratio, c/b, of the ellipse (Fig. 2.18). The ratio increases to infinity as γ decreases to zero. In the latter limiting case, the linear vibration passes the quartz plate unaltered. The phenomena just described lead to a particular kind of asymmetry of the half-shade effect. The zero position (Fig. 2.39b), characterized by field equality, remains unaffected by the state of vibration of V_I. In positions a and c (Fig. 2.39), however, the light transmission through the part of the prism preceded by the quartz plate (left-hand part of the drawings) will always be finite if V_I is an elliptical or circular vibration. A more detailed picture may be obtained by visualizing a Fig. 2.39c in which the left-hand part is replaced by the left-hand part of Fig. 2.40. Figure 2.41 gives the actual aspects of the field for

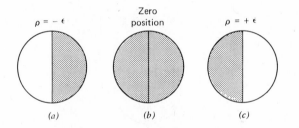

Fig. 2.41.   Variation in aspect of half-shade field if Laurent arrangement is used with light of unsuitable wavelength.

positions equivalent to those in Fig. 2.36. The accurate determination of the zero position is somewhat difficult because of the absence of pronounced contrast if the analyzer rotates from the position of field equality (position b) toward position c. The uncertainty in determining the position of field equality is therefore larger than at the wavelength, λ^*, for which the quartz plate is a "half-wave plate." Assuming that λ differs from λ^* by 500 Å, the error will be 50% larger than defined by (2.24). If the phase difference at λ^* is 3π, as is usual in commercial Laurent compound prisms using quartz, even a difference of $(\lambda - \lambda^*) = \pm 150$ Å will lead to a 50% increase in error. It is therefore preferable to use a quarter-wave mica plate instead of a commercial Laurent quartz plate.

The symmetrical displacement of the direction of vibration by a half-wave plate is equivalent to a rotation. Consequently, a half-shade effect can also be produced by superimposing upon half of the analyzer field an optically active substance, for example, a plate of quartz cut perpendicular to the optic axis. A number of such half-shade arrangements are on the market, for example, those designed by Macé de Lepinay and Wright. Their use is not advisable for precision measurements.

Much more preferable is the half-shade arrangement by Nakamura [108], a modification of the Soleil biquartz (pp. 117 *et seq.*). A thin plate of left quartz, Q_l, and an equally thin plate of right quartz, Q_d, are mounted, side by side, in front of the analyzer. Nakamura recommends a thickness of $\frac{1}{5}$ to $\frac{1}{25}$ mm (rotation for D line and 20°C 4.4 and 0.9°, respectively). The half-shade angle is therefore equal to twice the rotation of either plate. A half-shade angle of 2, 4, and 6° requires a biquartz 46, 92, and 138 μ thick, respectively (D line). Figure 2.42 shows the direction of vibration before and after the light beam

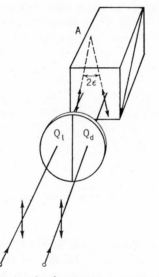

Fig. 2.42. Nakamura half-shade. Q_l, levorotatory quartz plate; Q_d, dextrorotatory quartz plate.

has traversed the twin plate. This compound analyzer has an advantage over the Lippich and the Laurent compound analyzers in that the half-shade is symmetrical, and that the loss of light by reflection is the same in both half-fields. On the other hand, the half-shade angle of the biquartz is invariable at a given wavelength, and increases with decreasing wavelength because of the dispersion of rotatory power. This latter quality is an advantage over the Jellett-Cornu prism—which has a completely invariable half-shade angle—if one uses a light source the brightness of which decreases toward the short wavelengths.

Half-shade arrangements based on the principle of dichroism are also possible. If the two half-prisms of a Jellett-Cornu are replaced by two tourmaline crystals or by two Polaroids with their optic axis inclined toward each other at an angle 2ϵ, a half-shade effect is obviously obtained. This principle for obtaining a triple field half-shade arrangement is incorrectly called " Lippich triple field." While precision measurements are excluded because of the incomplete polarizing effect of Polaroids, the sensitivity of Polaroid half-shade arrangements may approximate that of the classical half-shade devices if light of the proper wavelength is used (see p. 87).

In a compound analyzer, the separating line of the two half-fields rotates with the rotation of the analyzer. Some manufacturers avoid this by using a compound polarizer instead of a compound analyzer. The half-shade producing device (e.g., the small prism in the Lippich arrangement) is then mounted not in front of, but behind, the polarizer—that is, between the polarizer and the optically active substance.

Fringe Method

Prior to the extensive development of photoelectric polarimetry, "fringe methods" were quite popular. They consist of producing one or several dark fringes in an optical field, which can be taken as a quantitative argument for the magnitude of an optical rotation. These methods, though out of favor at present, are intrinsically very valuable, and one can safely predict that they will be favored again when they are properly automated. Since they are of particular interest for measurements of the dispersion of optical rotation, their principles are further discussed below.

Visual Polarimetry with Compensator in Monochromatic Light

The angle, α, can be measured not only by a rotation of the analyzer or polarizer, but also by compensation with a second optically active substance of known rotatory power of a sign opposite to that of the substance investigated. The layer thickness of this compensator is varied while polarizer and analyzer remain crossed. Generally, quartz is used for compensation. The use of an optically active liquid in a cell of variable thickness is possible, in principle, but is not practical. Since a 1-mm layer of quartz causes, at 20°C, a rotation of $\pm21.7°$ at 5893 Å and of $\pm25.5°$ at 5461 Å, a variation of less than 1 cm in the thickness of the plate will compensate any rotation ordinarily occurring in liquids in the visible spectrum. The problem of varying the layer thickness has been solved in a simple manner by the Soleil [109] double wedge. A plane-parallel plate of quartz cut normally to the optic axis is cut diagonally into two wedges, and the ends of the wedges are ground off (Fig. 2.43). By shifting one movable wedge, Q_l^*, laterally, while keeping the other, Q_l, in position, the rotation of the compensator is varied between a greatest and least value. The narrower the diaphragm, D, in front of the wedge, the further the movable wedge can be shifted in either direction. In order to obtain a zero rotation, a plane-parallel plate of quartz of opposite sign of rotation is added to the double wedge. In Fig. 2.43, the double wedge consists of levorotatory quartz, and the plate of dextrorotatory quartz. By giving to the latter the thickness of the double wedge in position I, the problem of compensating both positive and negative rotations is also solved. Compensation of a positive rotation with the compensator of Fig. 2.43 requires a shift downward to position III of the movable wedge. In position III, a negative rotation is compensated.

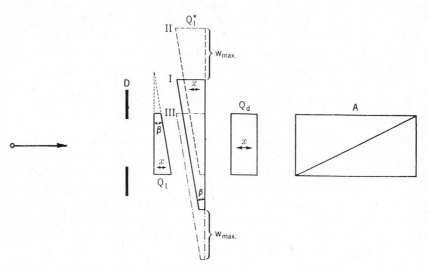

Fig. 2.43. Standard quartz-wedge compensator: I, zero position; II, compensation of negative rotation; III, compensation of positive rotation.

The rotation produced by the compensator is proportional to the lateral displacement, w, of the movable wedge, w being zero in position I. Implicitly, the displacement of the movable wedge required for compensating the rotation of a third substance is proportional to this rotation. The range of compensation increases with both w and β. The largest rotatory power, α_{max}, that can be compensated is therefore defined by:

$$\pm\alpha_{max} = \pm\alpha_Q w_{max} \tan \beta \qquad (2.25)$$

where α_Q is the rotation of quartz per unit length. On the other hand, the sensitivity of the compensator and the accuracy of compensation increase for a given Δw with decreasing β. The choice of β should therefore depend on the accuracy and the range of compensation which are required. For both high accuracy and a wide range , a very long movable wedge and a small compensator angle β should be used. The range of commercial instruments generally does not exceed 150° and may be considerably smaller. A slight disadvantage of the Soleil quartz wedge is a systematic error introduced in the measurements if the slanting face of the movable wedge is not completely planar throughout its entire length. This error was brought under control by a construction used in some polarimeters made by Schmidt & Haensch (shown in Fig. 2.44a). Both wedges, Q^*, can be displaced laterally; the others are stationary. The advantage of this construction is that a simultaneous displacement of both movable wedges in the same direction does not alter the zero position. One can therefore repeat a measurement by using other sections of the movable wedges and taking the mean value. The advantages of this procedure are great if the rotations to be compensated are small.

The movement of the movable wedge, or wedges, by means of rack and pinion, is recorded on a linear, usually horizontal, graduated scale. This, in commercial instruments, is mounted above the analyzer, close to the tele-scope. The scale may be divided in degrees and fractions of degrees permitting the direct reading of the compensated rotation, or it may be divided in frac-tions of 100 or marked in other units (e.g., give directly the concentration of sucrose).

There are a few other types of compensators that are different in detail but not in principle. Figure 2.44b shows, as an example, a Martens compensator.

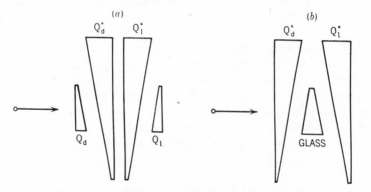

Fig. 2.44. Improved quartz-wedge compensators: (a) Schmidt and Haensch compensator; (b) Martens compensator; Q^*, movable quartz wedges.

It consists of two movable wedges and of one immobile glass wedge [110]. Polarimeters provided with compensators are mostly employed for sugar analysis. They are then called saccharimeters and are generally used in white light (see following section). Precision instruments are operated on the half-shade principle.

The data obtained with a polarimeter without compensator are indepen-dent of temperature except for the temperature effect exhibited by the optically active substance in the polarimeter tube (p. 167). A polarimeter with quartz-wedge compensator, however, is sensitive toward temperature changes. The variation with the temperature, T (in degrees C), of the optical rotation of quartz is given for the D line by:

$$\alpha_T = \alpha_{20} + \alpha_{20} \times 0.000143(T - 20) \tag{2.26}$$

where α_{20} stands for the optical rotation at 20°C. This correction factor is valid in the neighborhood of 20°C. It should be applied to the readings on the compensator scale—on the assumption that the latter had been calibrated for 20°—if a high precision in measurements of large rotations is desired. If, for example, the rotation measured at 25°C is 100°, the deviation of the actual rotation from that read on the compensator scale amounts to 7'.

Visual Polarimetry in White Light

Polarimetry in white light has the advantage over polarimetry in monochromatic light in that the light flux in the polarimeter is incomparably larger with a given light source. This makes possible the use of very small half-shade angles, and consequently a high accuracy in the measurements defined by an error possibly as small as 2–3″. However, the use of white light is limited to polarimeters with compensator and to the investigation of few substances; and even then the results have limited significance, although their practical value may be high.

The limitation of the use of white light to polarimeters with a compensator and to a few substances is due to the fact that the rotatory power varies with the wavelength. Instead of extinction or field equality, respectively, a colored field or difference in coloring in the two half-fields would be obtained in a polarimeter without a compensator. However, even the use of a compensator is limited, because its rotatory power obviously must have the same dispersion as the investigated substance. By a fortunate coincidence, the rotatory dispersion of quartz is so close to that of several sugars, including cane sugar, that quartz compensators and white light can be used in the polarimetry of such sugars.

Table 2.5 gives the variation of the rotatory power with wavelength for several substances. The optical rotations of all substances are arbitrarily set equal to unity at wavelength 6867 Å. Apparently turpentine is another of the few substances that could be investigated in white light using a quartz compensator. The table shows, on the other hand, that the dispersions of quartz and sugar are similar but not identical. The difference in the dispersions increases toward the violet. Polarimetry of sugars in white light ("saccharimetry") must therefore be restricted to sugar solutions which do not exhibit large rotations. In other words, the solution must not be too concentrated, the approximate limit being a concentration of 40%. Since the dif-

Table 2.5. Data of Landolt[a] for Relative Rotatory Dispersion of Several Substances

Substance	Wavelength, Å					
	6867	6563	5890–5896	5270	4862	4308
Quartz	1	1.09	1.38	1.75	2.08	2.70
Sucrose	1	1.11	1.40	1.78	2.13	2.77
Turpentine	1	1.09	1.36	1.71	2.03	2.60
Cholesterol	1	1.24	1.53	1.93	2.36	3.02

[a] H. Landolt, *Das optische Drehungsvermögen organischer Substanzen*, Vieweg, Braunschweig, 1898, p. 133.

ferences in dispersion are most pronounced in the blue and violet, one can improve the measurements and somewhat extend the range of usable sugar concentrations by employing a light filter of potassium dichromate solution, which eliminates the region of short wavelengths. For precision measurements it is advisable to have such a filter even in the case of small rotations. In some commercial instruments a colored glass filter is used as a substitute for the liquid filter.

The significance of results obtained in white light is limited by the fact that they represent the "mean" of rotations, the statistical weight of which varies with the wavelength. Factors that affect the significance of the "mean value" are the spectral sensitivity of the eye, the spectral distribution of the light source, and the dispersion of the rotatory power. Polarimetric measurements in white light have, therefore, mainly an analytical value in determining the concentration of substances of known rotatory power.

The polarizer and the analyzer must be suitable for white light. Any prism of the Glan-Thompson type, including compound prisms such as the Lippich or Jellett-Cornu, may be used. The polarization is uniform and complete for all wavelengths within the visible spectrum if the prisms are properly constructed and installed. Theoretically not as perfect, but practically satisfactory, are polarizers based on the principle of reflection. Since the refractive index of a reflector varies with the wavelength, Brewster's angle (2.17) is dependent on the wavelength, although very slightly. On the other hand, polarizers based on dichroism are unsuitable because of the variation of their polarizing power with the wavelength.

The Laurent prism is somewhat deficient, since a quartz plate can be a half-wave plate for one wavelength only (p. 108). Consider, for example, case *c* of Fig. 2.39. The field part of the analyzer covered by the quartz plate would be completely dark if the vibration emerging from the quartz plate were linear. Light of the wavelength for which the plate is actually a half-wave plate would be extinct, but light of any other of the neighboring wavelengths would be transmitted to some extent due to ellipticity. Consequently, the respective half-field would appear slightly colored in white light. On the other hand, it has been shown (pp. 108 *et seq.*) that the position of field equality (Figs. 2.39*b* and 2.41*b*) is not affected. The zero position is therefore independent of the wavelength, but the coloring of one of the fields, near the zero position, makes it difficult to locate the latter.

Far more disturbing is the color effect produced by those half-shade devices in which one half-field is covered by a plate of quartz cut perpendicular to the optic axis. Here, the coloring is due to the rotatory dispersion. At 6563 Å, the rotation produced by 1 mm of quartz is 17.31°. At 4308 Å, it is 42.59°, that is, almost 2.5 times larger. Consequently, there is no definite zero position. Much more favorable is the situation if a biquartz plate is used (Fig. 2.42).

We shall consider the case in which the largest rotation of the quartz—in the violet—is smaller than 90°. In Fig. 2.45, a thickness of 1 mm is assumed

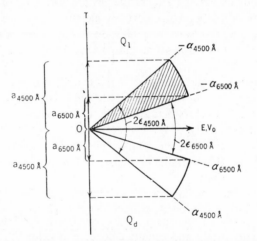

Fig. 2.45. Functioning of thin biquartz as half-shade device in white light. Both Q_d and Q_l are 1.0 mm thick.

for both the right and left quartz. The vibration, V_0, incident upon the bi-quartz, is parallel to the direction of extinction, OE. The shaded area repre-sents the range of rotations between 4500 and 6500 Å as produced by the left quartz; the symmetrical blank area in the lower quadrant refers to the right quartz. It follows from the figure, that is, from the projections onto the OT axis, that field equality—the case considered—is characterized by a bluish tint of equal shade and brightness in both half-fields. Upon rotating V_0 by a small angle, for example, by introducing an optically active substance, the tint will change to whitish in one half-field and become fuller in the other. It will change to a saturated color as soon as the shaded area overlaps OE. This type of half-shade is useful, but less sensitive than the following.

We assume that the rotation of both halves of the quartz plate exceeds $90°$ ($-90°$), but is smaller than $180°$ ($-180°$) in the violet (layer thickness >2 mm and <4 mm). If the nicols are crossed, the biquartz shows an equal hue in both half-fields. The tinge is dominated by the wavelength for which the rotation is exactly $90°$. The optical field is very bright because of the transmission of light of all wavelengths. Rotation of the analyzer by $90°$ makes the aspect of the field far more favorable physiologically (Fig. 2.46). No light

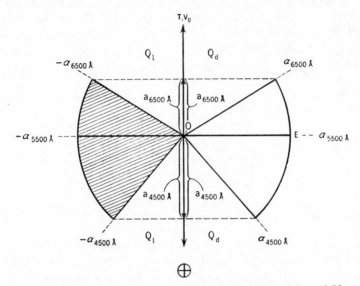

Fig. 2.46.   Soleil biquartz for polarimetry in white light. Both Q_d and Q_l are 3.75 mm thick.

is transmitted of that wavelength for which the rotation amounts to $90°$ and the two equally bright and equally colored half-fields have a saturated pure color which, upon a slight rotation of the analyzer, undergoes a very pronounced antibatic change in the two half-fields. The resulting contrast between the two half-fields is most strongly felt if the radiations which, in the position of field quality, are extinct or weakly transmitted belong to the yellow-green region (see Fig. 2.30). Field equality is then characterized by a purple, so-called " sensitive " tint. A minute rotation leads to a reddish color in one half-field and to a bluish color in the other. The biquartz operated on the principle of the sensitive tint and known as the Soleil biquartz [110a] was, for a long time, the favored half-shade for saccharimeters. It is usually 3.75 mm thick. The rotation is then $90°$ at ~ 5500 Å and $82°$ at the D line. The uncertainty in finding the position of field equality amounts to a few minutes. This is satisfactory if the effects to be measured are 30° or more. The Soleil biquartz is unsuitable, however, for precision measurements and for measurements of small effects. It is worthwhile to recall that field equality is reached when the nicols are *parallel*, while usually in half-shade methods field equality indicates that polarizer and analyzer are crossed or nearly crossed.

Visual Polarimetry by Fringe Methods

A fringe pattern will be produced if a plane parallel plate consisting of a wedge of left quartz and a wedge of right quartz is placed between crossed nicols. Figure 2.47 shows the experimental arrangement leaving out the polarizer. In the plane formed by points A, B, and C, the two rotations will compensate each other and cause a dark horizontal fringe in the center of the field between B and C while the upper and lower parts of the field are bright due to a negative and positive rotation, respectively.

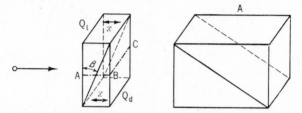

Fig. 2.47. Elementary fringe arrangement, with two optically active quartz wedges. Q_i: levorotatory quartz wedge; Q_d: dextrorotatory quartz wedge.

Figure 2.48 represents the appearance of the optical field for incident monochromatic light, if we disregard the inscriptions to the right-hand side of the figure. If the wedges are sufficiently thick, additional pairs of black fringes will be observed parallel and symmetrical to the fringe shown in Fig. 2.48 wherever the difference in rotation of the two wedges is 180° or multiples of 180°. If there are several such fringes, then they will be equidistant from each other, both in terms of rotation and linear distance. Insertion of an optically active sample of uniform thickness between the wedge and the analyzer or polarizer will cause a shift in the position of the fringes. The rotation of this sample can then be calculated from the magnitude of the shift or from the rotation of the analyzer needed in order to restore the fringe pattern to its original position. The precision obtainable by this method can be increased considerably by combining two inverse double wedges Q_1 and Q_2 such as shown in Fig. 2.49. The optically active substance then produces a shift of the fringes in opposite direction with respect to the zero position as shown schematically for the case of a single fringe in Fig. 2.50. This kind of arrangement is known as the Senarmont Fringe Method.

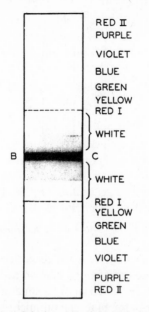

RED Ⅱ
PURPLE

VIOLET

BLUE

GREEN
YELLOW
RED I

WHITE

WHITE

RED I
YELLOW

GREEN

BLUE

VIOLET

PURPLE
RED Ⅱ

B   C

Fig. 2.48. Appearance of optical field in white light with double wedge of Fig. 2.47 between crossed nicols. Fringe is equivalent to zero fringe of a single quartz wedge, as considered in Fig. 2.68.

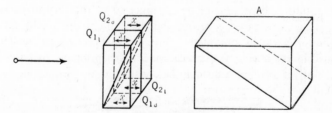

Fig. 2.49. Senarmont fringe arrangement with two pairs of quartz wedges.

If white light is used instead of monochromatic light, one observes in absence of an optically active substance with the experimental setup shown in Fig. 2.47 a dark central fringe and a colored field above and below the central fringe, the color sequence being symmetrical in the upper and lower part of the field. The sequence of colors is indicated at the right-hand margin of Fig. 2.48. On introducing an optically active substance now, the colors of the field will shift while the central fringe remains dark except for very strong rotation of the introduced optically active substance. (In the latter case, the central fringe will also assume coloration as well as the immediately adjacent field marked "white.") A very important analytical situation arises now if one places a dispersing prism into the light path. In absence of an optically

FIELD OF

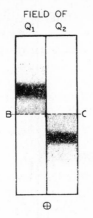

Fig. 2.50. Field aspect with Senarmont fringe arrangement.

active substance, one then obtains a fringe pattern shown schematically in Fig. 2.51. The field between the fringes, shown in white, is then colored, the color changing gradually from red at *B* to blue at *C*. Because the rotatory power of quartz is higher in the blue, the number of fringes is larger and their mutual distance smaller at *C* than at *B*. Introduction of an optically active substance results now in a displacement of all fringes, which, in addition, varies with wavelength. It is clear that from this record we can determine at once the rotatory dispersion of a substance. The rotatory dispersion thus obtained is an instantaneous record that has potentially major advantages over a photoelectrically obtained rotatory dispersion record which cannot be obtained instantaneously. Application of this principle by modern technology is not yet available and may be considered as overdue.

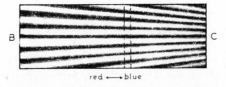

red ←—→ blue

Fig. 2.51. Spectropolarimetric record of Hussell-Nutting obtained by spectral resolution of the field in Fig. 2.48. Thicker quartz layer.

Another interesting fringe method (one which does not require quartz wedges) is that introduced by Fizeau and Foucault [111, 112]. Assuming that an optically active substance is placed between crossed nicols, the optical field will, in monochromatic light, obviously be completely dark, if the rotation is 0°, 180° or a multiple of 180°. Since this can occur only at a particular wavelength, the field will be colored on using white light. In that case that wavelength will be missing when the rotation is 180° or a multiple of 180°. If a dispersing prism is interposed now, one will get as a function of wavelength for the intensity transmitted through crossed nicols, a curve such as shown in Fig. 2.52, where λ_E represents the wavelength at which the rotation is exactly 180° or $n180°$, n being an integer number. The schematic drawing (a) in Fig. 2.52 shows the aspect of the fringe observed at λ_E. The similarities and differences of this fringe method compared to that discussed above (in Fig. 2.48) do not seem to require a discussion. The Fizeau-Foucault fringe method is, of course, particularly valuable if the rotations are very large, as in the case with liquid crystals. One then can expect several fringes throughout the spectrum, such as shown in Fig. 2.52a. Here again, one has the possibility of obtaining an *instantaneous* record of the dispersion of optical activity (ORD). This fringe method can be refined by using a Lippich prism instead of a simple analyzer. One obtains then two mutually displaced fringes (or fringe systems) as shown in Fig. 2.53, instead of the single fringe (or fringe system) shown in Fig. 2.52.

A series of other variants of the fringe principle have been developed by Perucca [113], Wiedemann [113a], Lommel [114], Landau [115]; and others. Particularly noteworthy is a device of Savart [115a]. Two birefringent quartz plates are brought into the light path in front of the analyzer with the optic axes inclined to each other. This device (described in detail in *Physical*

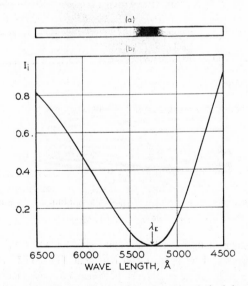

Fig. 2.52. Fizeau-Foucault fringe observed between crossed nicols in continuous spectrum of white light with optically active substance—6.55-mm layer of quartz—in the light path: (a) aspect of fringe; (b) intensity distribution in spectrum.

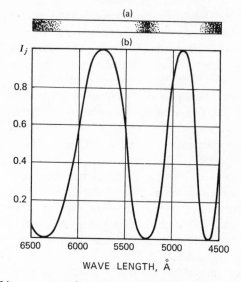

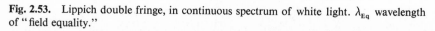

Fig. 2.52a.   Multifringe spectrum between crossed nicols with rotations $>360°$: (*a*) aspect of spectrum; (*b*) intensity distribution in spectrum. 19.65-mm layer of quartz.

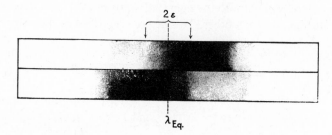

Fig. 2.53.   Lippich double fringe, in continuous spectrum of white light. λ_{Eq} wavelength of "field equality."

Methods of Organic Chemistry, 3rd ed., A. Weissberger, Ed., pp. 2228–2231) has the advantage of being extremely sensitive to small rotations, but it would not be easy, though not impossible, to adapt it to nonvisual techniques.

   The advantage of the various fringe methods referred to is briefly, as already mentioned, that they can provide an instantaneous record of the rotatory dispersion. This would be very valuable in all those cases where optical rotations change because of rapid changes in a system as a result either of reactions or system instability. Such fast changes in optical rotation or optical rotatory dispersion cannot be investigated at the present time with existing photoelectric methods. Suitably improved photographic methods could enable one to use the potential of these classical fringe methods.

Nonvisual Polarimetry

Survey of Available Techniques

Visual methods are generally limited to the visible region of the spectrum. Although they can be extended into the ultraviolet and infrared by means of fluorescence or phosphorescence effects [116–119], no precision is possible in such measurements. For precision work in the ultraviolet and in the infrared, nonvisual methods are used. On the other hand, the nonvisual methods are not restricted to the ultraviolet and infrared, but may be of advantage in the visible spectrum. There are essentially three principles of nonvisual polarimetric technique: photoelectric, photographic, and thermoelectric.

PHOTOELECTRIC TECHNIQUE

The photoelectric technique is by far the most convenient nonvisual technique. High amplification provided both by photomultiplier tubes and by modern external electronic amplification, combined with ingenious technical design, have made photoelectric polarimetry and spectropolarimetry a faster and more accurate method than either the photographic or visual methods. While the response of photoemissive surfaces is essentially limited to wavelengths < 15000 Å, various types of solid-state photodetectors can be used in infrared polarimetry, some of them to wavelengths as long as 10μ. The most commonly used infrared detecting cell used currently is lead sulfide but other materials in use include lead selenide, lead telluride, germanium, indium arsenide, and indium antimonide [120].

PHOTOGRAPHIC TECHNIQUE

The photographic plate can be used within the entire ultraviolet region, in the visible region, and in the near infrared up to approximately 12,000 Å. For most of the visible, the far ultraviolet, and the infrared, sensitized plates are necessary, as manufactured, for example, by Eastman Kodak. In the far ultraviolet—below 2800 Å, and especially below 2000 Å, where light absorption by gelatin is strong—photosensitive strata free from or low in gelatin (Schumann plates [121]) or ordinary photographic plates covered with a film of fluorescent oil [122, 123] are used.

The time-consuming operations involved in it are the main disadvantage of the photographic technique. If the compilation of α is based on photometric or microphotometric analysis of a photographic plate or film, a photometer or microphotometer is necessary. However, the photographic method has some advantages over other methods, which may make it useful in certain instances. One advantage is that the effect on a photographic plate is dependent not only on the intensity of illumination but also on the duration of exposure so that weak effects become measurable by long exposure. According to the empirical law of Schwarzschild [124]

$$K = It^p \tag{2.27}$$

where K is the optical density of the exposed and developed plate, I is the illumination, t is the time of exposure, and p is an empirical constant. Depending on the value of I, p varies between $\sim$0.9 and $\sim$1.2.

Another advantage of the photographic technique is the possibility of obtaining a nearly instantaneous polarimetric record which extends over a large spectral region. Furthermore, the photographic method is the most sensitive nonvisual method after the scintillation counter method. Finally, the photographic record is the only convenient means of analyzing rotations in excess of 360°, whether inside or outside the visible region [125].

THERMOELECTRIC AND BOLOMETRIC TECHNIQUES

Thermoelectric and bolometric techniques have proved satisfactory for polarimetry in the 1–10μ region and, at this writing, present a wider range of wavelength capability than that of the solid-state detectors.

Material Requirements for Polarimetry in the Ultraviolet

On its path through a polarimeter designed to operate in the visual region of the spectrum, the light may traverse the following substances: glass, calcite, Canada balsam or another binder, and air. Each of these substances, while transparent in the visible, becomes opaque in the ultraviolet. The absorption of Canada balsam, often used for cementing polarizing prisms, begins at 3400 to 2800 Å, depending on the brand. It is, therefore, advisable to use air as an interprismatic medium below 3500 Å (Glan prism, Fig. 2.19), or, if the polarimetric measurements are not extended below 2400 Å, glycerol. Still further in the ultraviolet, calcite also begins to absorb light strongly. A calcite polarizing prism of standard length is practically opaque slightly below 2400 Å. Prisms made of quartz, fluorite, ammonium dihydrogen phosphate, or other far-ultraviolet transmitting materials must then be used. The limit of transmission of pure samples of the former is approximately 1850 Å, that of ammonium dihydrogen phosphate 1850Å and that of fluorite 1550 Å. Because of the comparatively weak double refraction of both quartz and fluorite, it is difficult to obtain from these materials satisfactory polarizing prisms of the Glan type. Double-image prisms are therefore used [126]. If, finally, measurements are to be carried out below $\sim$1000 Å, polarization by reflection (see p. 82) is the only possible method. Glass lenses (flint or crown) used in front of the polarizer or behind the analyzer, and dispersing prisms made of glass, are unsuitable below 3500–3800 Å, depending on the thickness. Special ultraviolet-transmitting glasses may be used, however, as far down as 2700–3000 Å, while very much deeper penetration of the ultraviolet is possible through the use of fused silica. Plastics are often particularly unsuitable in the near ultra-violet. Most satisfactory are fused silica, quartz, and fluorite. A brief enumeration of comparative transparencies is given in Tables 2.6 and 2.7.

Table 2.6. Approximate Limits of Useful Light Transmission through Various Substances

Material	Approximate Shortest Wavelength, Å
Glass disks, of commercial polarimeter tubes 2–3 mm thick	3650
Film of Canada balsam in nicols	3400–2800
Microscope cover glass, 0.05 mm thick	2650
Calcite, 4-cm layer (= average length of polarizer + analyzer combined)	2400
Film of glycerol in nicol if used instead of Canada balsam	2300
Ammonium dihydrogen phosphate, 2-cm layer	1850[a]
Quartz, ~2-cm layer	1850
Fused silica, ~2-cm layer	1700[a]
Fluorite, ~2-cm layer (pure crystals)	1550[a]
Magnesium fluoride ~2-cm layer	1600[a]

[a] For 20% transmission of a 2-cm layer. A 3.6 mm layer of fluorite cuts off at about 1250 Å, and a 1.5 mm layer of magnesium fluoride at 1150 Å. Data from D. F. Heath and P. A. Sachet, *Appl. Optics*, **5**, 397 (1966).

The limiting wavelength given in Table 2.6 for the glass disks holds on the assumption that a powerful light source is used. The range of wavelengths given for Canada balsam signifies that the limiting wavelength depends on the brand. The mercury line with the shortest wavelength transmitted perceptibly from a mercury arc through a microscope cover glass of the specified thickness is 2654 Å. The light transmission through calcite of the specified length is, at the specified limiting wavelength, 20%; and it is already as small as 0.9% at the still shorter wavelength of 2310 Å. Very short nicol prisms made of calcite may be used down to 2200 Å, however, if the time of photographic exposure is long enough [127]. The light transmission of fluorite of the specified length is as high as 80% at 1860 Å.

In the far ultraviolet where silica and fluorite become opaque, lenses must be replaced by concave aluminized or silvered mirrors, and dispersing prisms by gratings. The use of glass for the cover glasses of polarimeter tubes is less serious, because they may be made very thin. Ordinary microscope cover glasses (about 0.05 mm thick) transmit sufficient light above approximately 2700 Å. At still shorter wavelengths, however, special glasses, quartz, fused

Table 2.7. Transmission of Special Ultraviolet Glasses and Comparison with Common Glasses

Wavelength, Å	Percent Transmission Through 10-mm Layer				
	Fused Silica[a]	Corex D (Corning 9700)[b]	Pyrex (Corning 7740)[b]	Hard Crown[c]	Dense Flint[c]
3303	95	75	48	65	10
3133	95	—	—	10	—
3081	95	52	5	—	—
2749	95	<1	—	—	—
2537	95	—	—	—	—
2250	90	—	—	—	—
2000	85	—	—	—	—
1850	70	—	—	—	—

[a] D. F. Heath and P. A. Sacher, *Appl. Optics*, **5**, 937 (1966).
[b] Corning Glass Works Catalog.
[c] B. K. Johnson, *Proc. Phys. Soc.* (*London*), **55**, 291 (1943).

silica, or fluorite windows must be used. Fluorite has the disadvantage of being very fragile in thin layers. If quartz is used, one of the disks must be *d*-quartz, the other *l*-quartz of exactly the same thickness in order to eliminate optical rotation. Fused silica is by far the most widely used material for cells used in the ultraviolet. Precautions against birefringence and dichroism [80] cited earlier in connection with visible polarimetry are even more important in the ultraviolet. The light absorption by air becomes an important factor at wavelengths below 1800 Å. Some gain in transmission can be achieved by working in nitrogen, rather than in air. Vacuum polarimeters must be used for the extreme ultraviolet.

Regarding light sources for ultraviolet polarimetry, continuous sources are preferred over line sources due to the modern trend toward recording spectropolarimetry. The hydrogen discharge lamps used in spectrophotometers have inadequate output for spectropolarimetry. Preferred sources are the high-pressure mercury arc, which emits a line spectrum* superimposed on a continuum, the tungsten-halogen lamp, and especially the xenon arc. Spectropolarimeters using the latter source have been designed to operate at 2000 Å or shorter wavelengths. The zirconium arc lamp is useful at wavelengths longer than 3000 Å.

* Principal lines in the mercury spectrum between 4100 and 2500 Å are 4046.8, 3663.3, 3650.2, 3125.6–3131.8, 3023.5, and 2536.5 Å.

Ultraviolet polarimetry is at present the most favored branch of polarimetry for research work since optically active absorption bands of organic substances are generally in the far ultraviolet. Consequently, the rotatory power generally reaches very high values in the intermediate ultraviolet. A few comparative data are given in Table 2.8. It is useful to add here that water as solvent may be used down to about 1850 Å.

Material Requirements for Polarimetry in the Infrared

As in the case of ultraviolet polarimetry, one of the principal problems in infrared polarimetry is the transparency of the polarizing prisms. Calcite may be used up to approximately 2μ. At still larger wavelengths, polarization by reflection is the preferred procedure.

Favored as light sources in the infrared are Nernst glowers and globars. Both of these sources are electrically heated and their emissivity over the range $1.5–15\mu$ is not far from that of black bodies at the same operating temperatures. Nernst glowers are made of rare earth oxides, and globars of silicon carbide. Tungsten lamps containing quartz windows can be used up to approximately 4μ. Alternate light sources for the near infrared are the well-known high-pressure mercury arc lamps. They furnish a strong continuum between 0.6 and 1.0μ. In view of the intense line spectrum emitted by these lamps at shorter wavelengths, stray light resulting from the latter should be eliminated by the use of suitable filters if a monochromator is used in conjunction with the polarimeter. Monochromatic radiation may be obtained from the continuous spectrum sources by means of appropriate dispersing prisms or gratings.

Water has a series of very strong absorption bands in the near infrared which makes the polarimetric investigation of aqueous solutions of optically active substances difficult or impossible above 0.9μ. The molecular extinction coefficients are 0.46, 1.30, 30.5, and 104 for 0.97, 12, 1.44 and 2.0μ, respectively. Carbon dioxide also has a series of absorption bands, so that it may be useful to conduct the measurements in air as free of carbon dioxide as possible [128].

Photoelectric Polarimetry

Demonstrations that optical rotation and, particularly, its variation with wavelength, that is optical rotatory dispersion in the ultraviolet gave information widely applicable in structural analysis [129–133] led to rapid advances in the design of polarimeters after about 1955. A number of good quality manual and fully automatic photoelectric polarimeters are now commercially available. Measurements can be made much faster and more easily with some of these instruments than with visually operated instruments, and the best photoelectric polarimeters can compete in accuracy of the results with those obtained with visual polarimeters [134].

Table 2.8. Specific Optical Rotations,[a] $[\alpha]$, in the Visible and Ultraviolet of Selected Organic Compounds in Aqueous Solution

Substance	Solvent	Grams in 100 ml Solution	Temperature, °C	$[\alpha]$ in Degrees at Wavelength in Å							
				5893	5461	5000	4000	3500	3000	2500	2000
Sucrose[b]	Water	26	20	66.53	78.34						
Sucrose[c]	Water	0.1	23–25			95.8	160		339	540	1340
d-Tartaric Acid[d]	aq. HCl pH 0.3	3.0	25	13.1	14.5	15.7	12.3	−7.9	−107.3		
10-Camphor	Water	1	25	22.1	29.0	41.6	127.4	331.1			
Sulfonic acid[e]	Water	0.2	25						1804	−2024	−3637

[a] See definitions, p. 159.

[b] NBS Circular 440, p. 82, 1942.

[c] S. England, G. Avigad, and I. Listowsky, *Carbohydrate Res.* **2**, 380 (1966).

[d] L. I. Katzin and E. Gulyas, *J. Phys. Chem.* **66**, 494 (1962). The optical rotatory dispersion curve of d-Tartaric acid shows a negative peak at 2280 Å, $[\alpha]_{2280}^{25} = -4300$. [L. I. Katzin and E. Gulyas, *J. Am. Chem. Soc.* **90**, 247 (1968).]

[e] DeL. F. DeTar, *Anal. Chem.* **41**, 1406 (1969). The optical rotatory dispersion curve of 10-camphorsulfonic acid shows a positive peak at 3050 Å, $[\alpha]_{3050}^{25} = 2003$; and a negative peak at 2700 Å, $[\alpha]_{2700}^{25} = -2501$.

Fundamentals

A photocell can give the absolute value of a light intensity but it cannot make possible a comparison between two light intensities in a single operation. Just the opposite holds for the human eye. This means (*a*) that the absolute variation of I with σ, as expressed by Fig. 2.27, is very important in photoelectric polarimetry, and (*b*) that true half-shade methods, as discussed in the section on visual polarimetry, are applicable only under special conditions, such as by using two photocells.

It is difficult to find two photocells with identical spectral sensitivity curves, so that the use of balanced photocells is generally possible for a single wavelength only, unless electronic compensating devices are used (done in several commercial spectrophotometers). In addition, the sensitivity curves might change with time and might change unequally, due to changes in the photosensitive surface layer, particularly if the incident light intensities are not very small.

It follows from (2.20) that:

$$\frac{dI}{d\sigma} = KI_i \frac{d(\sin^2 \sigma)}{d\sigma}, \tag{2.28}$$

that is, a photoelectric polarimeter based upon absolute intensity measurements is most sensitive if $\sigma = 45°$. While a number of polarimeters have been built on this design principle, factors other than sensitivity can affect the choice of this angle (see below).

The photoelectric current is not only dependent on the light intensity, that is, on the number of light quanta incident per unit time on the unit area, but also on the width of the beam, that is, on the total exposed area of the photosensitive layer. For a given light intensity, the photoelectric current can therefore be increased by selecting a wider aperture for the polarizing and analyzing prism.

The photosensitive layer of a photocell is occasionally anisotropic, that is, the photoelectric current may vary with the direction of vibration of the incident linearly polarized light. It is therefore advisable to rotate the polarizer rather than the analyzer. If it is not possible to rotate the polarizer, the photocell should be rotated together with the analyzer [135].

A Cotton effect does not seriously affect visual measurements of α as long as the ellipticity due to it is small (pp. 65–67). This does not hold if the measurement of α is based, as in some of the following photoelectric methods, on measurements of intensity differences instead of intensity balancing.

Photoelectric Equivalent of the Visual Extinction Method

An instrument that imitates the visual extinction method [136] ($\sigma = 0$) cannot be used for accurate measurements in view of (2.28). If the rotations are very large—in excess of 50°—the *relative* error may, however, be small enough for routine experiments or for exploratory experiments.

In this connection, the inverse of the extinction method should be mentioned. In visual polarimetry, it is impossible to determine the angle at which the transmittancy is highest because physiologically one would operate under the conditions of minimum sensitivity of the eye to intensity changes. In photoelectric polarimetry, this difficulty does not arise. Therefore, the inverse of the extinction method, that is, the determination of the polarizer or analyzer position for which the transmittancy reaches a maximum, is at least as good as, and in general far better, than the determination of the angle for which one reaches extinction. This method, like the preceding one, has the advantage of being largely independent of light absorption or light scattering.

The introductory statement regarding the precision of results thus obtained applies, of course, only to the technique whereby one searches *directly* for the analyzer (or polarizer) position giving minimum (or maximum) light transmission. In contradistinction, methods used to find this position *indirectly* are very powerful. They are discussed below.

One actual modern apparatus, based on measuring *directly* maximum transmittancy, deserves particular attention. Rank, Light, and Yoder [137] constructed a recording photoelectric polarimeter which operates on this principle, that is, by determining the position of maximum transmittancy through the analyzer.

In those cases where optical rotation is accompanied by circular dichroism, and this is invariably true in the spectral regions where the Cotton effect is very strong, the elliptic polarization may make it somewhat difficult to determine the analyzer rotation of either minimum or maximum transmission. Compensation of the circular dichroism is advisable in such cases.

Methods Based on Intensity Equalization Related to Half-Shade Methods

In a true half-shade method, two light intensities recorded *simultaneously* are matched by rotation of analyzer or polarizer. This is not possible in photoelectric polarimetry unless two photocells are used (for such an arrangement, see Fig. 2.66). With one photocell, the performance of a half-shade method may be approached by matching light intensities recorded in two *successive* experiments. Several principles can be used to achieve this.

METHOD OF SYMMETRICAL ANGLES

From a number of measurements, carried out with systematically varied analyzer or polarizer positions, the position of complete extinction can be interpolated with great accuracy, as demonstrated by the analogous photographic method of Bruhat [138, 139]. The following simple interpolation method is particularly convenient and accurate and only possible in photoelectric polarimetry. Two analyzer positions, defined by the readings ω_{oa} and

ω_{ob} on a graduated scale in the absence of the optically active substance and by ω_{1a} and ω_{1b} in the presence of the substance, are determined for which the photocell currents are equal. Figure 2.54 illustrates primarily the first pair of experiments.

$$\omega_{Eo} = \frac{\omega_{ob} + \omega_{oa}}{2} = \omega_{oa} + \frac{\rho_o}{2}, \tag{2.29}$$

and consequently

$$\alpha = \omega_{E_1} - \omega_{Eo} = \frac{\omega_{1b} + \omega_{1a}}{2} - \frac{\omega_{ob} + \omega_{oa}}{2} \tag{2.29a}$$

ω_E being the interpolated reading that would correspond to extinction.

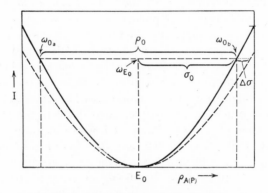

Fig. 2.54   Method of symmetrical angles.

For maximum sensitivity, the galvanometer deflection found for $\sigma_o = \pm 45°$ is restored, by rotation of analyzer or polarizer, after the optically active substance has been introduced. If, however, the latter strongly absorbs or scatters light, the restoration of the same deflection implies that $\sigma_l \neq 45°$ (it is $\sigma_o = \Delta\sigma$ in Fig. 2.54). It may then be advisable to choose smaller deflections in the second pair of experiments.

In Fig. 2.54, the fully drawn curve—valid in the presence of a light-absorbing, optically active substance—is made to coincide with the fully drawn curve at E_o. Actually, the former is displaced toward higher $\rho_{A(P)}$ values, as curve B in Fig. 2.65 is displaced with respect to curve A.

This method, which we have designated as the "method of symmetrical angles," was introduced by Kenyon and Dickes [140]. With an amplified photoelectric circuit, an accuracy of 0.01° was achieved in the red and of 0.1° in the violet. Rudolph [134] has compared the performance of a photoelectric polarimeter employing the method of symmetrical angles with a high-quality visual instrument. He concluded that the accuracy of both methods was close to 0.001° under optimum conditions. The visual method is considerably less convenient to use if the light intensity is very small because of the adaptation required in order to obtain an accuracy comparable to, or superior to, the photoelectric method which would be impractically long.

MODULATED NULL-POINT METHOD USING A SINGLE BEAM

The modulated null-point method is a very convenient practical variant of the method of "symmetrical angles" (p. 130). Here, the plane of polarization of the beam incident upon the analyzer is made to oscillate through a small angular range, which when the instrument is near balance includes the extinction angle. This is frequently brought about by oscillation of the polarizer. (The same could be achieved by oscillating the analyzer.) The principle of the method is illustrated in Fig. 2.55. The V_1 and V_2 represent, in column A, the extreme positions of the plane of polarization. The direction $\overline{V}$, the bisectrice of the angle $V_1OV_2 = 2\epsilon$, represents the time average position of the plane of polarization, and $\bar{\sigma}$ represents the time average angle between the position of the beam producing extinction, E and $\overline{V}$. The drawings in column B represent the variation of I_t, that is, of the intensity of the beam transmitted through the analyzer with the time, t. The situation aimed at is that of Case III, where the extinction direction of the analyzer coincides with the bisectrice of the angle of oscillation. Here the variation of I_t with time occurs with twice the frequency found in the other special case, BI, where E coincides with one of the extreme positions of the oscillating plane of polarization. While in these two cases the amplitude of the oscillating intensity is constant, all intermediate possibilities such as the one illustrated in BII show a periodic variation of the amplitude and therefore have a more complicated appearance. Since $BIII$ is therefore characteristically different from all other arrangements, defined by $\bar{\sigma} \neq 0$, it is electronically easy to arrive at an actual instrumental electronic output which will indicate when the extinction position coincides with $\overline{V}$. (One commonly used method is electronic filtering of the output to observe only that part of the signal varying at the fundamental frequency of oscillation of the beam. As seen in the set of sketches given under C, the output at this frequency is zero when the extinction position coincides with $\overline{V}$.) Several commercial photoelectric polarimeters are based upon the modulated null-point method. In some of them, the angular position of $\overline{V}$ is being changed by motor drive until, by virtue of the signal defined by $CIII$, coincidence of $\overline{V}$ and E is achieved.

Figure 2.56 shows a block diagram representing a number of single-beam null-point systems [141–149]. Although the order of components is often as shown, several choices are available in the position and nature of the modulator.

Light from the source, usually monochromatic, is polarized linearly. The

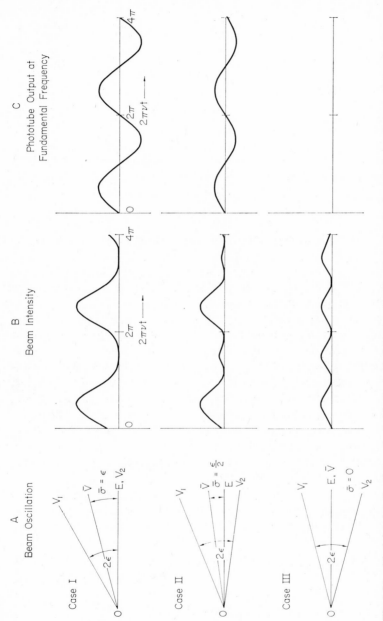

Fig. 2.55. Modulated null-point method. σ is the angle between the direction of extinction of the analyzer and the direction of vibration of the incident beam. $\sigma = \bar{\sigma} + \epsilon \sin 2\pi\nu t$, where $\bar{\sigma}$ is the time average of σ, $\epsilon = (1/2)V_1OV_2$, ν is the frequency of oscillation, and t is the time.

133

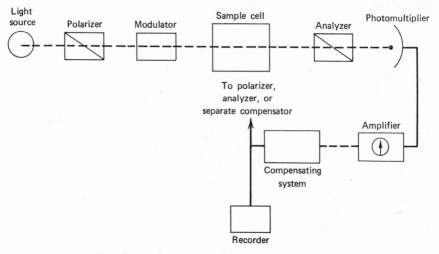

Fig. 2.56. Single-beam modulated null-point system.

light can be modulated through the half-shade angle 2ϵ, either by movement of the polarizer [149–151] or analyzer [144–148], or by a separate device inserted between polarizer and analyzer [141–143; 145–147].

In the simpler manual forms of the system [144, 149, 152] the analyzer is rotated until a current-measuring device indicates that the same current is produced by the phototube at each of the two half-shade positions. This is simply the method of symmetrical angles (see above). Higher relative sensitivity can be obtained by making the modulation periodic and amplifying, rectifying, and reading out the fundamental modulation frequency and adjusting $\bar{\sigma}$ manually until the phototube signal is zero [141, 147]. The operation can be made fully automatic through use of the amplified photocurrent to drive a compensator which restores the beam-analyzer position to null, and recording the analyzer position electrically [141–143, 145, 146, 148, 151] or by a mechanical counter [150].

The symmetry of this method permits the accurate measurement of rotation in samples showing circular dichroism, although the precision is of course smaller the stronger the dichroism. The best of these instruments permits the measurement of rotation to better than a millidegree in solutions that do not attenuate the light, or to a few millidegrees in samples showing optical densities as high as 2. Performance of this quality will not, however, be achieved at short wavelengths where light intensities are very low.

Modulating Devices for Single-Beam Instruments. While the angle σ can be varied with time by oscillating either the polarizer or analyzer, still other separate devices have been used to perform this modulation. Bürer, Kohler, and Günthard [146], in one of the early high-precision photoelectric polari-meters, used a rotating disc which inserted in the light beam alternately a levo- and dextrorotatory quartz plate. The apparatus of Gillham [141] accomplished the same effect by using alternately quartz and glass. Gillham [141, 153] introduced still another method—that of passing the polarized light through a Faraday cell, consisting of a transparent core to which an alter-nating magnetic field was applied by a solenoid wound on it. Materials used for this core have included lead-zinc borate glass [145, 153], water [147], and fused silica [142, 143]. The beam can also be modulated by the application of an electric field across an ammonium dihydrogen phosphate crystal [154] (Pockels effect).

It is easily demonstrated that maximum sensitivity is obtained in the modu-lated instruments when the beam incident upon the analyzer is modulated at $\pm 45°$ from the extinction position [141, 155]. In Fig. 2.57 the beam incident upon the analyzer is modulated through a half-shade angle of 2ϵ. OE is the extinction position for the beam at $90°$ to OA, the latter being the direction of maximum transmission of the analyzer. The difference in intensity of the light emerging from the analyzer for the two extreme positions of the plane of polarization, OV_1 and OV_2 is

$$\Delta I = KI_i[\sin^2(\bar{\sigma}+\epsilon) - \sin^2(\bar{\sigma}-\epsilon)]$$

$$= KI_i \sin 2\bar{\sigma} \sin 2\epsilon \tag{2.30}$$

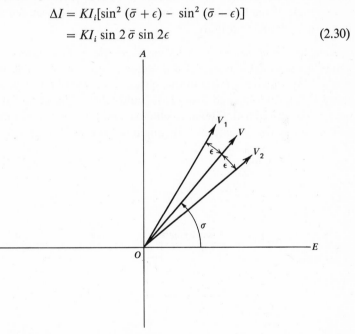

Fig. 2.57. Modulation of a plane-polarized beam incident upon an analyzer. The single headed arrow is used to indicate the directionality of σ.

The photoelectric system must use this difference to determine and compensate the value of $\bar{\sigma}$. The sensitivity, then

$$\frac{\partial \, \Delta I}{\partial \bar{\sigma}} = 2KI_i \sin 2\epsilon \qquad \text{when } \bar{\sigma} \approx 0. \tag{2.31}$$

This is clearly a maximum when ϵ, the modulation angle, is 45°. However, other considerations affect the modulation angle. As Rouy and Carroll [155] pointed out, the relative sensitivity, that is $(dlnI/d\bar{\sigma})$, increases as ϵ decreases although, of course, the absolute magnitude of the signal decreases as ϵ decreases. Amplification of the AC component of the modulated signal and removal of the first harmonic (which amounts to retaining only the signal due to a nonzero value of $\bar{\sigma}$) greatly improves the sensitivity of the measurement, and noise may become the limiting factor [141] (see below). Mechanical vibrations used to produce large values of ϵ may introduce noise which more than outweighs the beneficial effects of large values of $\partial I/\partial \bar{\sigma}$, and a polarimeter can show no better precision than shown by the stop positions with oscillating parts. In actual practice, modulation angles of $\pm 1°$ to $\pm 3°$ appear to be adequate in most instruments when light intensities are not too attenuated, regardless of the method of modulation used [141–145, 147–151, 156]. Modulation angle of the Faraday cells increases with shorter wavelength due to the increase in Verdet constant.

Detector-Compensator Systems for Single-Beam Instruments. Figure 2.58 shows a simplified diagram of a detector-compensator system used by Cary et al. [143], which is similar to the system used by Bürer, Kohler and Günthard [146], Rudolph and Bruce [148], and others. The signal from the photomultiplier is amplified, filtered to eliminate all but the fundamental frequency of the modulator, synchronously amplified, and used to drive a servo motor

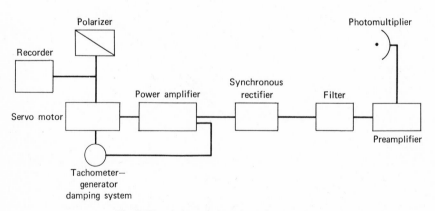

Fig. 2.58. A detector-compensator system.

to position the polarizer (or analyzer [146]) back to the crossed position. The angular position of the polarizer (or analyzer) is indicated on a recorder. An alternate compensation system was introduced by Gillham [141], who used the DC output to power a Faraday cell placed into the beam between polarizer and analyzer. Thus, the beam was rotated back toward the null position and the current passing through the cell at balance was a measure of the rotation of the sample. When used for compensation in a spectropolarimeter [142], the wavelength-dependence of the Faraday effect must be taken into account.

Angular Resolution of Single-Beam Instruments. Sensitivity of the electrical system to angular displacement can be a limiting factor in null-balance systems, especially in instruments that do not filter out from the phototube signal all but the fundamental modulation frequency [155]. However, modern photomultipliers, amplifiers, and electrical attenuating systems provide sufficient sensitivity in the null instruments that the combined interactions of sensitivity and noise may become the limiting factor in angular resolution. Performance of these instruments has been analyzed by a number of authors [143, 155–157]. These analyses point out that the signal/noise ratio is decreased either by factors that reduce the signal (absorbance, scattering, very small half-shade angle) or that increase the noise. Factors that increase the noise include fluctuations in the brightness (energy output) of the light source, failure of optical components to produce strictly plane polarized light, light scattering, circular dichroism in the test sample, and shot noise in the phototube (noise due to the random nature of the emission of electrons from the cathode). Cary et al. [143], examine the case of a pure sinusoidal modulation of the polarized beam

$$\sigma = \bar{\sigma} + \epsilon \sin 2\pi v t, \tag{2.32}$$

where σ is the instantaneous angle between the plane of polarization of the radiation incident upon the analyzer and the direction of extinction, $\bar{\sigma}$ is the angle by which the average plane of polarization of the radiation incident upon the analyzer departs from the crossed position, and ϵ is the peak amplitude of the sinusoidal angular modulation. The instantaneous phototube current is

$$i = kL(T_o + T_1 \sin^2 \sigma), \tag{2.33}$$

where L is the radiant flux delivered to the polarimeter system, T_o is the transmission of the system with polarizer and analyzer crossed, T_1 the additional transmission when the polarizer and analyzer are parallel, and k is the conversion constant of the photocell in amperes per unit flux. If $\bar{\sigma}$ is small and ϵ on the order of 6° or less, the photocurrent can be expressed as a DC component, a component at the fundamental modulation frequency, a first harmonic, and higher harmonics of much smaller amplitude. Peak amplitudes of the first three components are, respectively:

$$i_o = kL\left(T_o + \frac{T_1\epsilon^2}{2}\right) \tag{2.34}$$

$$i_1 = kLT_1\bar{\sigma}\epsilon \tag{2.35}$$

$$i_2 = \frac{kLT_1\epsilon^2}{2} \tag{2.36}$$

The desired information is present in the fundamental i_1. Presence of the i_2 component forces the system to detect a small current in the presence of a much larger one and, hence, for reasonable relative sensitivity i_2 must be very effectively filtered out. Even with effective filtering, the very low ratio of signal intensity to incident light intensity—$\bar{\sigma}\epsilon$ is 3×10^{-6} if $\epsilon = 6°$ and $\bar{\sigma}$ is $0.002°$—places great demands on the polarimeter design. Cary et al., further show that the ratio of DC output associated with the fundamental to root mean square shot noise in the phototube can be approximated by

$$\frac{i_{dc}}{i_{noise}} = \frac{\bar{\sigma}\epsilon}{\pi}\left[\frac{2kLT_1}{\varepsilon F\left(\dfrac{T_o}{T_1} + \dfrac{\epsilon^2}{2}\right)}\right]^{1/2} \tag{2.37}$$

where ε is the charge on the electron, and F is the noise bandwidth defined by a low-pass filter following demodulation. This indicates that it is important to maintain values of phototube conversion (k), light flux (L), and transmission through the system (T_1) as high as possible. Absorption and scattering, of course, reduce T_1, and ellipticity (nonlinear polarization of the light incident upon the analyzer caused by circular dichroism), and stray light increase T_0, decreasing the signal/noise ratio. It may be seen that the ratio is not a very strong function of ϵ, the modulation angle, especially when $T_1\epsilon^2/2 \gg T_o$.

While absorption, ellipticity due to circular dichroism, and light scattering do decrease the signal/noise ratio and therefore increase the minimum detectable rotation, it is important to note that these do not introduce systematic errors into readings of the null-type polarimeter. Rouy and Carroll [158], however, have shown that bias can be introduced into such a polarimeter through phase shift brought about by an overloaded amplifier coupled with a synchronously rectified signal that is not of simple harmonic form.

Cary et al. [143], have also shown that the polarimetric zero can be displaced when the intensity of the light source fluctuates with a frequency component equal to the fundamental modulation frequency. The amount of the shift is approximately

$$\bar{\sigma}' = \left(\frac{a}{8T_1\epsilon}\right)(4T_o + 3T_1\epsilon^2) \tag{2.38}$$

where a is the ratio of the peak amplitude of the lamp modulation at the fundamental frequency to the average lamp flux. If $(T_1\epsilon^2/2) \gg T_o$ and it is desired to restrict $\bar{\sigma}'$, the uncertainty in determination of the null point, to $0.005°$ when $\epsilon = 6°$, then (2.12) indicates that the lamp fluctuation at the modulation frequency must be restricted to about 0.1%.

The noise level in a well-designed polarimeter will be as low as a few tenths of a millidegree when the intensity of the light transmitted through the polarimeter is sufficiently large [143, 159] or one or two millidegrees when the transmission of the sample is $1–10\%$.

MODULATED NULL-POINT METHOD USING A DOUBLE BEAM

In the modulated null-point method discussed above, the plane of polarization of a single beam is allowed to oscillate between two extreme positions until the midpoint of the position is such that the output as a function of time is either constant or zero. The same result can be achieved by using a double beam if the plane of polarization differs for the two beams. One can then, by various methods (several of which will be discussed below), produce equality of intensity of the two beams transmitted through the analyzer in what amounts to a photoelectric imitation of the visual half-shade method (see p. 100). Automatic recording polarimeters described by Earl and Bernhardt [160] and by Levy, Schwed, and Fergus [161] projected the image of the half-shade from a Lippich analyzer onto a phototube. A rotary semicircular sector interposed between the tube and analyzer produced a sinusoidal alternating photocurrent as long as the two half-fields of the Lippich were of unequal intensity. In the Earl and Bernhardt instrument, the photocurrent was used to move a compensating wedge by motor drive across the field until balance was reestablished. Lecy, Schwed, and Fergus used the photocurrent to rotate the analyzer by motor until balance was established. The sensitivity of the instrument was given as $\pm 0.005°$.

In the instrument described by Daly [162], polarizer and analyzer are Wollaston prisms, each of which transmit two rays with orthogonal planes of polarization (see p. 82), so that two beams pass through the sample. With the plane of polarization of the analyzer at approximately $45°$ to that of the polarizer, each beam is split into two beams. The instrument photoelectrically compares the intensity of the pair of beams inclined at one angle to that of the orthogonal pair and a recording servo drives the analyzer until the two intensities are equal.

A Wollaston prism was used as analyzer in an instrument designed by Bruhat and Chatelain [163]. The prism split the incident beam into two beams, and the polarizer position was adjusted until the intensities of the two beams were equal; that is, when the beam was at an angle of $45°$ with respect to the direction of maximum transmission of the prism.

FLICKER METHOD

In the preceding methods based on intensity equalization, the principle of measurement was a search for the conditions under which the intensity of light transmitted through the analyzer oscillated harmonically with constant amplitude. In the flicker method, which is in principle related to the others, the condition of quasi-steady intensity output is approached. The basic difference in this method compared to the preceding ones is that in the present instance the polarizer or analyzer is allowed to rotate continuously, while in the preceding methods the rotation is limited to a certain angular range. The intensity of light transmitted through the analyzer and, with it, the output photocurrent, vary twice between zero and maximum if the analyzer is rotated by 360°. A continuous rotation of the analyzer therefore produces a fluctuating photocurrent. The introduction in front of the rotating analyzer of a stationary half-shade device, with a half-shade angle of 90° (e.g., of a Wollaston prism) makes this flicker phenomenon useful in polarimetry. (Figure 2.59 shows the apparatus of Schrönrock and Einsporn [164]). It is

Fig. 2.59. Scheme of photoelectric flicker apparatus of Schönrock and Einsporn.

assumed first that the beam incident upon the Wollaston prism, W, vibrates under an angle of 45° with respect to the orthogonal optic axes, that is, that the intensity of the two emerging components is equal. Instead of analyzing the intensity of either component separately, both components are sent through the rotating prism, A, and are received together by the photocell, C. According to Fig. 2.33a, the combined intensity of the two components is independent of the rotation of the analyzer, ρ_A. The 45° incidence upon the Wollaston prism thus manifests itself by a steady photocurrent which corresponds to $\frac{1}{4}I_i$ in Fig. 2.60, curve I. If an optically active substance is introduced which produces a rotation of 45°, only one component will be transmitted through the Wollaston prism, and the intensity, recorded by the photocell, will vary between $\frac{1}{2}I_i$ and 0 as the result of the analyzer rotation (curve II or IIa of Fig. 2.60). The polarizer is now rotated until the fluctuations in photocurrent cease. The rotation necessary—45° in the present example—gives the value of α directly.

The sensitivity of the method is defined by the angle $\Delta\rho_P$, by which the polarizer must be rotated in order to transform an intensity which is constant (curve I of Fig. 2.60) into one which changes periodically (curve III). The square of the amplitude, a^*, defines the smallest change in light intensity

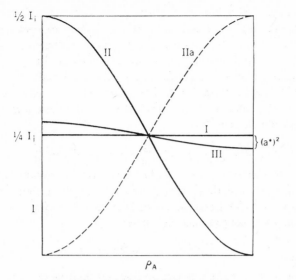

Fig. 2.60. Principle of flicker method.

which makes the fluctuations in photocurrent perceptible. For a given amplification factor, the sensitivity of the flicker method is obviously identical with that of a quasi half-shade method of the types previously discussed. A disadvantage of the flicker method is that the analyzer must be kept in continuous rotation at a proper and constant speed in order to prevent an erratic character in the period and amplitude of the vibrations. Schönrock and Einsporn [164] operated with 100 rpm, that is, a time of 0.3 sec was allowed for one period in the vibration of the galvanometer deflections.

It is clear that the presence of circular dichroism or turbidity will reduce the sensitivity of the flicker method but will not otherwise introduce systematic errors into determinations of rotation.

In a variant of the flicker method, Landegren [165] splits the beam after its passage through a polarizing Polaroid disk which rotates at 5000 rpm. To that effect a semitransparent mirror is set in front of the optically active solution, inclining it suitably with respect to the incident beam. The transmitted beam passes through the optically active solution and the stationary analyzer (also a Polaroid disk) and impinges upon phototube 1. The reflected beam impinges upon phototube 2. The phase shift of the signal received by the two phototubes is translated, by conventional electronic procedure, into a meter reading. By means of suitable adjustments, the meter reading can be made to indicate directly the rotation in degrees.

Another variant of the flicker method, a stroboscope polarimeter was also described by the same author [166]. Here the degree of rotation is read off visually by means of a fiduciary mark on the rotating polarizer using the illumination provided by a Strobotac whose flashes depend on the same phase shift. By means of this rather ingenious principle, the manipulations necessary in visual polarimetry are reduced to a minimum.

In conclusion, it may be added that the flicker method—based on a very old and well-known principle of flicker photometry—is without interest for visual polarimetry because it is equivalent to a half-shade method operated with a half-shade angle of 90°.

Methods Involving the Measurement of Intensity Differences

Provided that the test solution neither depolarizes nor elliptically polarizes an incident linearly-polarized beam, and assuming the validity of Lambert's law, the intensity of light passing through an analyzer after passing through a polarizer and test solution can be expressed as

$$I = I_o k_1 e^{-hl} \sin^2(\sigma_o + \alpha) \qquad (2.39)$$

In this expression, I_o is the intensity of light incident on the polarizer, k_1 expresses optical reflection losses in the system and therefore is less than one,

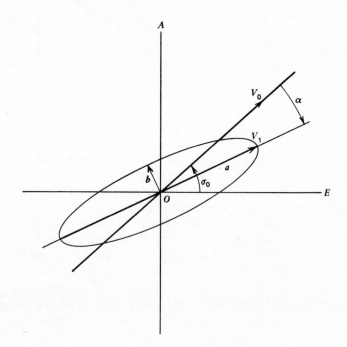

Fig. 2.61. Amplitude of elliptically polarized beam incident upon an analyzer. The single-headed arrows are used to indicate the directionality of σ_0 and α.

h is the absorption coefficient*, l the length of the polarimeter cell, σ_o the angle σ in the absence of the optically active substance, and α the rotation. If a scattering solution depolarizes part of the beam passing through it, or if the solution shows circular dichroism, thus elliptically polarizing the beam, (2.39) as it stands no longer can hold. The reason for the latter conclusion is shown in Fig. 2.61. A sample showing circular dichroism, as well as optical rotation, has been placed in the path of a beam originally linearly-polarized along OV_o. The result is rotation of the beam by an angle α and its simultaneous transformation into an elliptically polarized beam of major axis a and minor axis b. It is clear from the drawing that no position for the ellipse can be found within the AOE plane at which the sum of the squares of the projection of the axes a and b on OA (corresponding to the intensity passed by the analyzer) is zero. In such a case (2.39) clearly does not hold. Figure 2.62 shows

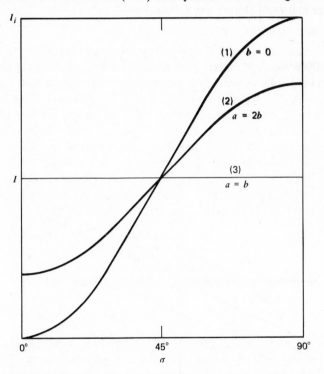

Fig. 2.62. Variation of intensity of (1) linearly polarized, (2) elliptically polarized, and (3) circularly polarized beam passing through analyzer with angle σ; a and b are semiaxes of ellipse.

* In the case of a solution that scatters light, part of h may represent an apparent absorption which is really turbidity.

the angular dependence of the intensity of light passing through the analyzer for (1) linearly-polarized light, (2) elliptically polarized light with $a = 2b$, and (3) circularly polarized light, that is, $a = b$.

As used in (2.39) and below, σ_o is the angle traversed on bringing the direction of extinction of the analyzer into coincidence with the plane of polarization of the polarizer, giving this angle a positive sign if the rotation is clockwise as viewed from the analyzer.

Measurement of the intensity of light passing through an analyzer when it is not in its crossed position can be used to calculate α by means of (2.39) only in the absence of depolarization or circular dichroism, and only if additional measurements are made to eliminate h. On the other hand, there is some possibility that information on depolarization and circular dichroism, as well as h, can be extracted from intensity measurements by means of a more complete analysis of the effects of these factors [167, 168].

SINGLE-ANGLE METHOD

The simplest and, at first thought, most inviting principle of photoelectric polarimetry consists of measuring the change in intensity of light transmitted through two nicols, as brought about by insertion of an optically active substance into the light path between polarizer and analyzer. This intensity is expressed by (2.39). The attractive feature of this method is that no rotation of either polarizer or analyzer is necessary. This also eliminates the need for graduated circles unless it is desirable to operate with various σ_o settings. Implicitly, the error in reading polarimeter scales is eliminated. The prime weakness of the method is that the effects of absorption, scattering, depolarization, or elliptical polarization must be determined separately unless they are zero. If this is the case and if σ_o is set at 45°

$$\sin 2\alpha = \frac{2I - k_1 I_o}{k_1 I_o} \tag{2.40}$$

and for small angles of rotation

$$\alpha = k'T - \tfrac{1}{2} \tag{2.41}$$
$$\scriptstyle \alpha \to 0$$

where T is the ratio of light transmitted through the analyzer after and before insertion of the optically active substance (quasi-transmittance). In the latter case, T varies linearly with α. This method was introduced long ago by Desains [169] in thermoelectric polarimetry and, later on, by Todesco [170, 171] into photoelectric polarimetry of the Faraday effect.

The method can, of course, be used with other σ_o values also, except that the range of near linearity between T and α is more limited as σ_o approaches 0° or 90° (see also Fig. 2.27).

SINGLE-ANGLE SWITCHOVER METHOD

The problem of combining a polarimetric measurement with an absorption measurement, on applying the method of fixed prisms, was solved first in an elegant fashion by Landt and Hirschmüller [172, 713]. The apparatus is shown schematically in Fig. 2.63.

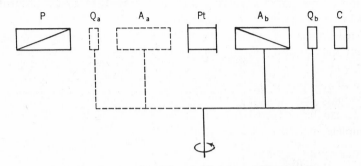

Fig. 2.63. Scheme of apparatus of Landt and Hirschmüller.

In a first pair of experiments, in the absence and the presence, respectively, of the optically active substance, the analyzer, A, is set between polarizer P and polarimeter tube Pt (position A_a). In a second pair of experiments, the analyzer is switched to its normal position A_b. From the two sets of data, the correct optical rotation follows, without interference of absorption. Instead of permanently setting the planes of polarization of polarizer and analyzer at a fixed angle, σ_o (most likely the best approach since σ_o is independent of wavelength in contradistinction of α_q), the authors had the nicols permanently crossed and inserted a quartz plate, Q, cut perpendicular to the optic axis and producing a rotation α_q of 34.65° for the D line. This quartz plate was mounted and switched together with the analyzer. The instrument was designed for technical measurements and fitted with a Herotar polarizer and analyzer [174, 174], products similar to the Polaroid.

Assuming a finite angle σ_o and that depolarization and elliptical polarization are negligible, in absence of quartz, the first set of experiments gives

$$I_a = I_o k_1 e^{-hl} \sin^2 \sigma_o \tag{2.42}$$

and the second gives

$$I_b = I_o k_1 e^{-hl} \sin^2 (-\sigma_o + \alpha). \tag{2.43}$$

Consequently, the relative quasi-transmittance*

$$S = \frac{I_b}{I_a} = \frac{\sin^2(\sigma_o - \alpha)}{\sin^2 \sigma_o}. \tag{2.44}$$

* On using the quartz, Q, with the nicols crossed, $S = \sin^2 (\alpha_q + \alpha)/\sin^2 \alpha_q$ since the direction of rotation produced by Q is the same in positions Q_a and Q_b.

If $\sigma_o = 45°(-\sigma_o = -45°)$, one would obtain,

$$\sin 2\alpha = \frac{(I_a - 2I_b)}{I_a} \tag{2.45}$$

which would, for small rotations, reduce to the extremely simple equation

$$\alpha \underset{\alpha \to 0}{=} \tfrac{1}{2} - S \tag{2.46}$$

A simple polarimetric device based upon Landt and Hirschmüller's principle was described by Crumpler, Dyre, and Spell [176]. It is fitted with Polaroids and is to be used in conjunction with a commercial photoelectric colorimeter. Reproducibility of rotation measurements with this instrument is about 0.03°.

ANGLE-PAIR METHOD

General Remarks:

The pair of experiments required for elimination of the effects of absorption can also be carried out in a different way. Two consecutive experiments, a and b, with angles σ_{o_a} and σ_{o_b}, respectively, yield the ratio

$$R = \frac{I_a}{I_b} = \frac{\sin^2(\sigma_{o_a} + \alpha)}{\sin^2(\sigma_{o_b} + \alpha)} \tag{2.47}$$

provided depolarization and elliptical polarization are small or zero. The rather unwieldy form of this equation can be simplified on making the two angles σ_o numerically equal but opposite in sign so that

$$R = \frac{\sin^2(-\sigma_o + \alpha)}{\sin^2(\sigma_o + \alpha)} = \frac{\sin^2(\sigma_o - \alpha)}{\sin^2(\sigma_o + \alpha)}. \tag{2.48}$$

A simple transformation yields

$$\alpha = \tan^{-1}\left[\left(\frac{1 \mp \sqrt{R}}{1 \pm \sqrt{R}}\right)\tan \sigma_o\right]. \tag{2.49}$$

Since the intensity ratio, R, may be expressed in terms of a quasi-relative absorption coefficient, g [177]

$$R = e^{-g} \tag{2.50}$$

one obtains, for *small* values of α, the relatively simple expression

$$\alpha \underset{\alpha \to 0}{=} \left(\frac{\tan \sigma_o}{4}\right)g = Kg = K'D. \tag{2.51}$$

If, in addition, σ_o is made small, this reduces further to

$$\alpha \underset{\substack{\alpha \to 0 \\ \sigma_o \to 0}}{=} \left(\frac{\sigma_o}{4}\right)g = K_1 g = K_1' D \qquad (2.52)$$

The constants K' and K_1' may be calculated or determined experimentally from the slope of D versus α plots, where D is the quasi-relative optical density ($\log_{10}$ base). It is apparent that K has the maximal value of 0.25 if σ_o has the values of $45°$ and $-45°$ (orthogonal angles), respectively, and approaches zero as $\sigma_o \to 0$. This is significant for the choice of σ_o in the angle-pair method. It shows that the intensity *ratio* R is more sensitive to changes in small α-values the smaller σ_o. On the other hand, the *intensity increments* or *decrements* (relative to I_o) accompanying changes in small α-values are largest if $\sigma_o = \pm 45°$ (2.28). The choice of the most promising σ_o-value for the angle-pair method and for related methods will therefore have to be made in the light of the sensitivity of response and quantitative reproducibility of response of the photoelectric device used. Obviously, the sensitivity of this and of related methods to changes in α will, at a given σ_o, vary also with α since (see Fig. 2.27) the change in R that results from a change in α will be maximal if $\alpha + \sigma_o \to 0$. This factor, that is, the magnitude of optical rotations encountered or expected, also should be taken into account on selecting the pair of σ_o-values most suitable for a given experiment or for a particular section within an experimental series involving wide variations of α.

Other functions besides the ratio I_a/I_b have been used for calculating α [155, 178, 179]. In particular, the quotient

$$\frac{I_b - I_a}{I_b + I_a}$$

has been shown mathematically [155, 178, 179] to be linear in $\tan \alpha$ over a wider range in α than the ratio I_a/I_b. Several workers have used intensity comparison methods to calculate α when working with solutions in which there was appreciable scattered or elliptically polarized light [167, 168, 180, 181], making appropriate corrections for these phenomena.

METHOD OF ORTHOGONAL ANGLES

If $\sigma_o = \pm 45°$, (2.49) simplifies to

$$\alpha = \tan^{-1}\left(\frac{1 \mp \sqrt{R}}{1 \pm \sqrt{R}}\right). \qquad (2.53)$$

The respective experimental method was introduced almost ninety years ago by Desains [182, 183] into thermoelectric polarimetry. It can be used also in photoelectric polarimetry. The apparatus needed is very inexpensive; no graduated circles are required. The method operates with a rotation of the analyzer by a fixed angle of 90°. The mechanical rotation, from one position to the other, can be carried out in rapid succession by two mechanical stops on the analyzer mounting.

In Fig. 2.64, X and Y define the plane of polarization of the analyzer in its two orthogonal positions. Remembering that I is proportional to the square of the amplitude, a, it follows from the figure that, quite generally,

$$\cos^2 \sigma_{o_b} = \frac{I''_{o_x}}{I''_{o_x} + I''_{o_y}} \tag{2.54}$$

$$\cos^2 \sigma_{1_b} = \frac{I_{1_x}}{I_{1_x} + I_{1_y}} \tag{2.55}$$

where $I''_o = k_1 I_o e^{-(h)l}$. If Lambert's law holds, (2.54) reduces for the case considered here to

$$\frac{I''_{o_x}}{I''_{o_x} + I''_{o_y}} = \tfrac{1}{2}. \tag{2.56}$$

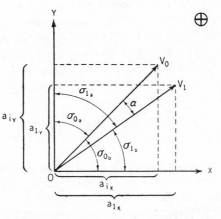

Fig. 2.64. Method of " orthogonal angles."

The optical rotation obviously

$$\alpha = \sigma_{1_b} - \sigma_{o_b}. \tag{2.57}$$

Equations (2.54), (2.55), and (2.57) were used by Desains. They are alternatives and equivalent to the single (2.53). As still another alternative, one may propose a quite general equation, which also follows at once from Fig. 2.64:

$$\cos \alpha = \frac{\sqrt{I''_{o_x} I_{1_x}} + \sqrt{I''_{o_y} I_{1_y}}}{\sqrt{(I''_{o_x} + I''_{o_y})(I_{1_x} + I_{1_y})}} \tag{2.58}$$

which, for $\sigma_{o_a} + \sigma_{o_b} = 0$, reduces to:

$$\cos \alpha = A\left(\frac{\sqrt{I_{1_x}} + \sqrt{I_{1_y}}}{\sqrt{I_{1_x} + I_{1_y}}}\right) \tag{2.59}$$

(where $1 = /A\, I''_{o_x} = I''_{o_y}$). If it is sufficient to determine relative α-values (values relative to a standard), this simplifies to

$$\cos \alpha_{rel} = \frac{\sqrt{I_{1_x}} + \sqrt{I_{1_y}}}{\sqrt{I_{1_x} + I_{1_y}}} \tag{2.60}$$

and finally to

$$\alpha_{rel} \underset{\sigma \to 0}{=} \frac{\sqrt{I_{1_x}} + \sqrt{I_{1_y}}}{\sqrt{I_{1_x} + I_{1_y}}} \tag{2.61}$$

if the optical rotations are small.

An instrument designed by Fordyce, Green, and Parker [184] splits the linearly polarized beam emerging from the polarizer into two orthogonal beams by using a Wollaston prism, and compares the intensities of these two beams.

METHOD OF SMALL ANGLES

A polarimetric attachment to be used in connection with a Beckman spectrophotometer, intended for and limited to an σ_o-value of $\pm 5°$, was made commercially available by Keston [185]. Its sensitivity obviously is maximal at α-values of about $5°$, decreasing towards larger α-values. This maximal sensitivity is given as $0.0015°$. An improved Keston-type device—also to be used in connection with a Beckman spectrophotometer—containing Glan-Thompson prisms instead of Polaroids and allowing a variation of the σ_o-values by virtue of retained graduated circles, was described by Kirschner [186]. Obviously (2.49) and, in the limiting case of very small rotations (2.51) apply.

DOUBLE-BEAM METHOD

Instead of producing a changeover from σ_o to $-\sigma_o$ by rotation of the analyzer (or polarizer) from one mechanical stop to another, or by the switchover method, one may split the light beam in two and insert in the path of each component both a polarizer and analyzer using a finite $\sigma_o(-\sigma_o)$-value. (Consequently, four polarizing prisms are needed.) The result obtained is identical with that achieved in the methods already mentioned. The double-beam method was introduced by Savitzky [187] who thus added a new variant to the use of double beams in photoelectric polarimetry. The respective device was intended for use in connection with Perkin-Elmer or Cary double-beam spectrophotometers.

Methods of Compensation Not Related to Half-Shade Methods

The first photoelectric method introduced into polarimetry [188–190] was based on the principle of compensation (Fig. 2.65). The polarimeter tube filled with an active reference liquid is brought into the light path, and the analyzer or polarizer is rotated until $\sigma_o = 45°$. The corresponding reading on the graduated scale is ω_o, and the intensity, I_r, recorded by the galvanometer is, except for reflectivity losses, $I_i/2$ in the absence of absorption (Curve A). After introduction of the optically active substance, the analyzer or polarizer is rotated until I_r is reestablished (transition from Curve A to B). The corresponding reading on the graduated scale is ω_1. The difference $(\omega_1 - \omega_o) = \alpha$.

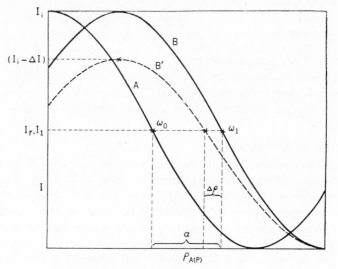

Fig. 2.65. Method of compensation.

A considerable advantage of this method is that all measurements are made in the region of maximum sensitivity of the I versus σ curve *regardless* of the magnitude of the rotation. A serious disadvantage of the method when used in the simplified form described is the falsification of results if the natural rotatory power is accompanied by light absorption, absent in the reference liquid, which reduces I_i by ΔI. Curve B' is then valid instead of B; that is, the apparent value of α differs by the decrement $\Delta\rho$ from the true value. For maximum accuracy, it is necessary that the refractive index of the optically active liquid or solution be as close as possible to that of the reference liquid or solvent in order to keep the reflectivity losses the same. Neither of these two problems arises if the method is used for measurements of a Faraday effect because the substance to be studied can be used in zero magnetic field as the reference.

For measurements of the natural rotatory power, von Halban designed an instrument (shown schematically in Fig. 2.66) which eliminates errors due to absorption, although it does not completely eliminate errors due to scattering and to circular dichroism. The light coming from a mercury arc is split into two beams by oblique incidence upon a quartz plate, Q. Under the selected incidence of 45°, the reflected beam carries 10% of the total light flux. One of these beams traverses polarimeter tube, Pt_1, the other Pt_2. These tubes are of the same length and are filled with an identical liquid. The two photocells, C_1 and C_2, are balanced to zero photoelectric current. After introducing the optically active liquid into both light paths, the balance is reestablished by rotating the analyzer, A, or polarizer, P. The reproducible maximum accuracy is given as 0.01° for the strong ultraviolet lines of the mercury arc. The monochromatic measurements of Mayrhofer [189] have covered the range down to 2536 Å. The method makes use of the photoelectric photometer of von Halban and Siedentopf [191] and has been altered in matters of detail by Ebert and Kortüm [135], who discuss the sources of error at length. The apparatus of von Halban gives, in the same set of experiments, both the rotatory power and the light absorption.

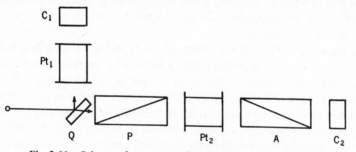

Fig. 2.66.   Scheme of apparatus of von Halban and Siedentopf.

Photographic Polarimetry

The main field of photographic polarimetry and spectropolarimetry is the ultraviolet, although photographic methods may be useful in the visible, particularly for spectropolarimetric measurements, and in the near-infrared.

Photographic polarimetry in monochromatic light may be carried out by adapting any of the visual methods. The main difference, except for substituting a photographic plate for the eye, and for the choice of the half-shade angle, is that visual methods generally call for the observation of a virtual image (e.g., of the face of the half-shade) by means of a telescope while a real image must be formed on the photographic plate. In the visible, photographic polarimetry with monochromatic light hardly has an advantage. In the ultraviolet or the near-infrared, photographic spectropolarimetry deserves preference over photographic polarimetry with monochromatic light unless the latter achieves a higher accuracy. This holds at present only for the methods of Bruhat and his collaborators [192, 193]. With either the extinction method or a half-shade method, a series of exposures is made on a single photographic plate for systematically varied analyzer positions. By making the images small enough and by shifting the plate after each exposure in the most economical manner, Bruhat and Pauthenier [192] obtained sixteen exposures on a single plate. The developed plate is analyzed by means of a microphotometer. The analyzer or polarizer position that corresponds to complete extinction or complete field equality, respectively, can be determined with an error which does not exceed 0.25'. In order to make the method sufficiently convenient, Bruhat et al., synchronized an automatic stepwise shift in the position of the photographic plate.

The optimum half-shade angle in photographic polarimetry—and also in photographic spectropolarimetry—is 90°. This is a significant difference compared with visual polarimetry. The reasons were discussed in the section on photoelectric polarimetry where the situation is analogous.

All the visual spectropolarimetric methods can be used in photographic spectropolarimetry. In addition, a few other methods based on principles explained earlier, but little suited for visual spectropolarimetry, are either most useful or very promising.

The sequence in the relative merits of spectropolarimetric methods is not the same in visual and in photographic spectropolarimetry. Methods that call for the direct coordination of the center of a fringe to a wavelength, or for the determination of the distance between the centers of two fringes, are unsuitable for visual precision measurements. Such determinations are easily carried out with precision by microphotometric analysis of photographic records. Consequently, spectropolarimetric methods, which are not suited for visual precision measurements, may be precision methods with the photographic technique.

Several methods of photographic spectropolarimetry involve the illumination of a polarizer-cell-analyzer combination with a source emitting a continuous or line spectrum, and analysis of the emergent light with a spectrograph. For details see *Physical Methods of Organic Chemistry*, 3rd ed., A. Weissberger, Ed., pp. 2260–2267.

Method Useful for Rotations of Less Than 180°

In the method of Cotton and Descamps [193, 194] the optical rotation of light of various wavelengths is determined by rotation of a spectrograph and analyzer during exposure, and noting on the photographic plate the angle of rotation where a particular wavelength shows extinction. This photographic method was based on the earlier method of visual spectropolarimetry devised by Fizeau and Foucault [111–114, 195].

Methods Useful Over a Wide Range of Rotations

Hussell [196, 197] and Nutting [198] inserted a plane parallel plate consisting of a wedge of right quartz and a wedge of left quartz between a polarizer
and analyzer. When illuminated with monochromatic light, dark fringes are
produced as described in the section on visual polarimetry by fringe methods.
Optical rotation of a sample can be calculated by measuring the shift in position of the fringes on a photographic plate caused by placing the sample in
front of the analyzer. The rotatory dispersion can be calculated by using white
light and observing these shifts as a function of wavelength with a spectrograph. Application of the fringe method to spectropolarimetry was preceded
by its use in visual monochromatic polarimetry [199–201].

A photographic method was devised by Landau [115] and is applicable to
both small and large rotations but is particularly useful for measuring rotations larger than 360°. Modifications and improvements on the method have
been made by Darmois [202], and by Lowry and collaborators [203–206].
For details the Third Edition of this chapter (pp. 2260–2267) should be
consulted.

Methods Particularly Applicable to the Study of Mesophases (Liquid Crystals)

The combination of very high rotation and ease of flow of certain types of
liquid crystals (mesophases) makes it possible to study their rotation with
relatively simple equipment. A drop of an optically active mesophase on a
glass slide will spread to a circular wedge if a planoconvex lens is set into the
drop (Fig. 2.67). A plane-parallel composite layer is thus obtained which is
free from optical distortion between the dotted lines of Fig. 2.67, if the refractive indices of the lens and of the mesophase are very similar. Between
crossed nicols in monochromatic light, the circular wedge exhibits the circular
analogue (Fig. 2.68b) to the fringes observed on a thick quartz wedge cut
perpendicular to the optic axis (Fig. 2.68a). Each of the circular fringes represents the geometrical locus of a phase difference of $n \times 180°$, n being 1 for
the innermost fringe, 2 for the second, and n for the nth fringe. The center of

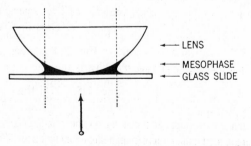

Fig. 2.67. Lehmann's arrangement for measuring rotatory power of mesophases (liquid
crystals).

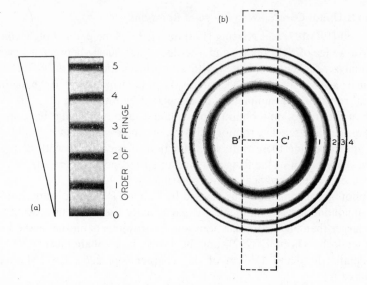

Fig. 2.68. Fringes between crossed nicols in monochromatic light produced (*a*) by single quartz wedge, (*b*) by circular liquid wedge of Fig. 2.67.

the fringe pattern should be a dark spot if the glass slide and the center of the lens were in optical contact. The absence of such a dark spot in actual experiments indicates that there is no optical contact, but there is instead a film of mesophase between the center of the lens and the glass slide. Because of the curvature of the lens, the fringes are narrower the larger their diameter. From the known curvature of the lens and the distance between the centers of two adjacent fringes, the optical rotation per unit layer can be determined. This simple and very elegant method of measuring the rotatory power of mesophases by means of a polarizing microscope is described by Lehmann [207]. By carrying out measurements successively at different wavelengths, the dispersion can be compiled.

Because of the great sensitivity of mesophases toward minute changes in temperature, Stumpf's [208, 209] spectropolarimetric modification of Lehmann's method for white light, which excludes the time element, is preferable. If a diaphragm restricts the passage of light to a rectangular section of the circular wedge between the dotted lines in Fig. 2.68*b*, spectral resolution will lead to a pattern as shown in Fig. 2.69. This is the photographic pattern obtained at 80°C with the mesophase of pure cyanobenzylidene aminocinnamate, which shows an anomalous dispersion of rotatory power in the visible. The rotation per millimeter layer is, with $-6000°$, relatively "small" at 6420 Å, in excess of $-16,000°$ at 6100 Å, in excess of $+30,000°$ at 5200 Å, and 21,000–26,000° between 5080 and 4460 Å, with a minimum at 4800 Å. The

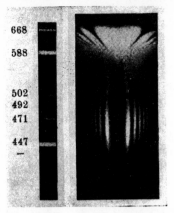

668

588

502
492
471

447
—

Fig. 2.69. Photographic record [208] of rotatory dispersion of a mesophase, as obtained by Stumpf's method with a continuous spectrum. λ in reference spectrum in millimicrons. Analysis of pattern shows changes in rotation with wavelength from $> -16,000°$ to $> +30,000°$ with an error in evaluation of $\pm 25°$. $B'C'$ runs horizontally.

optically active absorption band, responsible for the excessive rotations, reaches its peak between 5600 and 5800 Å. Within this region, the rotation goes through zero. The narrow spectral region of excessive values of $d\alpha/d\lambda$ appears washed out on the pattern. The rotation is determined by measuring, for different wavelengths, the distance between the centers of two adjacent fringes. The smaller the distance, the larger is the value of α. As in the method of Hussell and Nutting, the dispersion of the rotatory power is obtained in a single exposure.

Polarimetry in the Infrared

Thermoelectric and bolometric polarimetry and spectropolarimetry date back almost as far as visual polarimetry and spectropolarimetry. The first observations in the infrared were made in 1836 [210] and 1846 [211] on the natural rotatory power and on the Faraday effect, respectively. As early as 1849, Desains carried out quantitative measurements of either effect in the infrared by means of methods discussed on pages 144 and 148.

Thermoelectric and bolometric polarimetry are closely related to photoelectric polarimetry. In either case, measurements of a rotation are based on the recording of an electric current or of an electric potential as produced or changed by illumination or irradiation. All methods of photoelectric polarimetry are therefore directly applicable to thermoelectric and bolometric polarimetry. In fact, several methods developed in thermoelectric polarimetry and applicable to photoelectric polarimetry have been discussed in the section on photoelectric polarimetry because the latter has a more general interest. It is therefore sufficient to discuss a few important details here.

Ingersoll [212] adapted the method of orthogonal angles to measurements with a differential bolometer. The use of a differential bolometer or differential thermocouple is, in fact, less objectionable than the use of differential photocells. A Wollaston prism is employed as the analyzer. Each of the two emerging rays is incident upon one of the two bolometers, and instead of two intensities, I_X and I_Y, the intensity difference $(I_X - I_Y)$ is measured (Fig. 2.64). If $\sigma_{0_a} = \sigma_{0_b} = 45°$, the intensity difference is zero in the absence of the optically active substance. The galvanometer deflection obtained after introduction of the optically active substance is proportional to $I_{1X} - I_{1Y}$ (see Fig. 2.64). Since the numerical values of I_{1X} and I_{1Y} must be known for the calculation of α, a second differential measurement is necessary after rotating the polarizer by 90°. The number of measurements, three, is therefore not reduced compared with the original method of the orthogonal angles if the latter is operated with $\sigma_{0_a} = \sigma_{0_b} = 45°$.

Of particular technical interest is the infrared polarimeter designed by Meyer [213]. It was used for investigating the Faraday effect in the far infrared, but it could also be used for investigating the natural rotatory power. A scheme of the apparatus is given in Fig. 2.70. S represents a Nernst lamp and P a polarizing mirror. Other symbols are M_1, M_2, M_3—concave mirrors; A—analyzing mirror; D_1, D_2—diaphragms; L_1, L_2—sodium chloride lenses; R_m—radiomicrometer. The instrument parts between and including D_1 and D_2 represent a Wadsworth spectroscope.

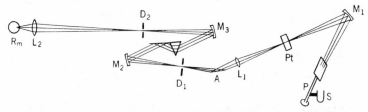

Fig. 2.70.   Meyer's apparatus for spectropolarimetry in the far-infrared.

Both the polarizer and the analyzer are reflecting mirrors. The rotation is measured by the extinction method: the polarizing mirror is rotated until the galvanometer deflection reaches a value of zero. In order to maintain the relative position of the source of radiation (Nernst lamp) with respect to the polarizing mirror, lamp and mirror are mounted together rigidly and rotated around the direction of the reflected beam as the axis. Meyer extended his measurements up to a wavelength of 8.85μ.

Thermoelectric and bolometric spectropolarimetry for dispersion measurements in the infrared may—like photoelectric spectropolarimetry—be based upon successive measurements at systematically varied wavelengths or upon scanning.

Two of the pioneers of infrared spectropolarimetry, Meyer [213] and Inger-soll [212], whose infrared methods were discussed above, followed their polarimeter with an infrared monochromator consisting of a prism-mirror combination of constant deviation of the Wadsworth [214] type. (This combination is preferable to the Pellin-Brocca prism usually found in monochromators and spectrographs, if the radiation is partly absorbed by the substance of which the prism is made.) Introduction of the method of Fizeau and Foucault into thermoelectric spectropolarimetry is due to Carvallo [215, 216], who used a double-purpose prism of the Joubin type [217] to serve both as an analyzer and as a dispersing prism. The polarizer was rotated until a fringe or its complement, a spectral range of maximum transmission, passed the thermopile. This method is equivalent to the original method of Fizeau and Foucault (p. 120), which is useful if the rotations are very large, that is, if the fringes are narrow. Lowry and Coode-Adams [218] improved the limited accuracy of results obtained by this method by making two series of extinction (fringe) and intensity maxima determinations. In the second series, the polarizer was rotated by 90° so that the maxima of the first series appeared as fringes and vice versa. By taking the mean values the centers of the maxima and minima, respectively, are obtained somewhat more accurately. Instead of rotating the polarizer and keeping the wavelength constant, in a given measurement, an arbitrary polarizer position may be chosen, for example, the one corresponding to crossed nicols, and the wavelength may be varied instead. An improved version of this alternate method has been used by Lowry and Snow [219], who measured the infrared rotation of quartz at regular spectral intervals of 100 Å between 18,000 and 27,000 Å.

By employing differential thermocouples or bolometers, additional methods could be introduced into thermoelectric spectropolarimetry, such as the Lippich modification of the method of Fizeau and Foucault (p. 121). Dongier [220] used such a differential method. Each of the two juxtaposed spectra (see Fig. 2.53) is focused upon one of the two thermoelements. The wavelength drum of the monochromator is rotated until the galvanometer deflection reaches zero. The wavelength fulfilling this condition is λ_{Eq}. Dongier used a Wollaston prism as the analyzer so that the mutual displacement, 2ϵ, of the fringes in the two juxtaposed spectra amounted to 90°. This angle is evidently the optimum condition for accuracy, just as in photoelectric polarimetry (p. 147).

The modern development of commercial infrared spectrophotometers has facilitated the construction of infrared spectropolarimeters. H. S. Gutowski [221] and Hediger and Günthard [222] were the first to take advantage of this. In both instances the polarimeters were combined with a Perkin-Elmer infrared spectrometer functioning as monochromator. In both instances also, Se-films were used as polarizer and analyzer (see also pp. 85 and 86); the latter authors used 6 films for each. Gutowski's range of measurements (on quartz) extended from 2.2 to 9.7μ, that of the latter authors (on organic solutions) from 0.75 to 4μ. The latter authors also incorporated in their instrument a recorder. The precision of the data obtained is given as 0.1°. A more detailed discussion of modern spectropolarimetry in the infrared is given by Parker in this volume.

Radiowave Polarimetry

Macroscopic models of optically active molecules rotate the plane of polarization of Hertzian waves, as shown by Lindman [223]. A more sensitive apparatus was described by Servant and Loudette [224]. Apparatus of this type should become extremely interesting beyond the purpose originally intended. They could be used in order to prove, once more, but quite directly, that the absolute configurations coincide with the convention established by Fischer (see p. 64). Secondly, by using macroscopic models of molecules of complex structure, a direct study of the effects of substitutions in optically active molecules would become possible. For a comparison of the results thus obtained with those expected in solution, one would have to take into account of course, that the solutes are dielectrics in contradistinction to the perfectly conducting macroscopic models necessary in this type of experiment. There is also the comparatively minor difference that the optically active solutes have random orientation, which reduces their effect.

6 STANDARD EXPERIMENTAL CONDITIONS AND STANDARD UNITS

The optical rotation exhibited by an optically active solid or liquid of a given dissymmetric structure depends on the layer thickness, the wavelength, and the temperature. The optical rotation exhibited by a solution of an optically active substance depends, in addition, on the concentration of the optically active solute, on the nature of the optically inactive medium and possibly also on the nature and concentration of optically inactive co-solutes. Finally, the rotatory power may vary with time (racemization, mutarotation). In order to make experimental data significant, it is therefore necessary to define fully the experimental conditions. In order to make them easily comparable with other data, it is important to collect them at standard conditions and to express them in terms of standard units defined below. There are, of course, cases where adhering to standard conditions is not compatible with the purpose of a particular investigation.

The standard for the layer thickness, l, introduced by Biot, is 10 cm. Since α is proportional to the layer thickness, α/l is a constant at constant temperature. The number of optically active molecules contained in the unit volume of a given optically active liquid or solid varies with changes in its density, ρ. A significant standard for the comparison of experimental results is therefore obtained by reducing the data to both standard layer and unit density. The respective quantity is the specific rotation

$$[\alpha] = \frac{\alpha}{l\rho} \quad (\text{degrees cm}^2 \text{ g}^{-1} \times 10) \tag{2.62}$$

where ρ is the density of the liquid or solid at the temperature of the experiment and l is to be expressed in decimeters. It is clear that the temperature dependence of $[\alpha]$, in contrast to that of α, is, in general, fairly small and reflects exclusively the effect of temperature upon the rotatory power of the individual molecules (see also pp. 167–169).

On multiplying α/l by one hundredth of the volume of the individual molecule, v, and by Avogadro's number, N, the molecular rotation

$$[M] = \left(\frac{\alpha}{l}\right)\left(\frac{vN}{100}\right) = \left(\frac{M}{100}\right)[\alpha] \quad (\text{degrees cm}^2 \times 10^{-2}) \tag{2.63}$$

is obtained which is another very useful standard. M is the molecular weight of the compound. The choice of the factor 100 like that of one decimeter for l in (2.62) is dictated by the desire to have quantities of a convenient order of magnitude.

The fact that $[M]$ is, by definition, proportional to the molecular volume explains to a large extent the finding that $[M]$ varies with the atomic dimensions of substituents not directly attached to an assymmetric C-atom [225].

The quantities $[\alpha]$ and $[M]$ are not cgs units. In addition, they refer to rotations in degrees. Numerical factors and coefficients contained in theoretical equations, on the other hand, are generally given in cgs units and refer, necessarily, to rotations in radians. The comparison of experimental and theoretical data requires, therefore, a conversion of units. Thus, it is useful to point out the numerical relationship between $[\alpha]$ and $[M]$, on the one hand, and the corresponding cgs units, on the other hand. Using φ as the basic symbol, since this is favored in theoretical work (in order to avoid confusion with the polarizability, α), one has for the specific rotation expressed in radians, per 1-cm layer,

$$[\varphi] = \left(\frac{\pi}{1800}\right)[\alpha] = 1.74533 \times 10^{-3}[\alpha] \text{ cm}^2 \text{ g}^{-1} \tag{2.64}$$

Since everything points to an increasing intimate contact between theory and experiment in the field of rotatory power, the need for these conversion factors is regrettable. On the other hand, the cgs units are quantities of practically inconvenient magnitude and it is for this reason that Biot selected the units now in use. This objection would easily be overcome by selecting the cgs units divided by 1000 as the practical units. Such units shall be proposed herewith as alternate practical units in which to express experimental data. It appears fitting to propose for these units the designation "Biot" and "Cotton" respectively. Thus the rotation in units of the Biot is

$$[B] = 1000[\varphi] = 1.74533[\alpha]$$

and in units of the Cotton is

$$[C] = 1.74533[M]$$

Whenever the rotation is determined or given for one wavelength only, the standard generally is the sodium doublet (D line; 5890, 5896 Å). An alternate standard, the green Hg line (5461 Å), has been favored increasingly without replacing the former, classical standard. The green Hg line is preferable to the D line for several reasons: (1) whenever the dispersion is normal, the rotations are larger at 5461 than at 5890 Å; (2) the green Hg line is practically a single line so that no problem arises if a high-dispersion monochromator is used; (3) the eye is more sensitive to green than to yellow; (4) relatively inexpensive high-pressure Hg lamps of high intrinsic brightness are commercially available while, in the early periods of polarimetry, the only convenient method of producing monochromatic light was contamination of the flame of a Bunsen burner with NaCl vapor.

Whenever the rotation is determined or given for one temperature only, the standard, classically, is 20°C. A large number of authors prefer the alternate standard of 25°C.

For the sake of brevity, wavelength and temperature are indicated as subscripts and superscripts, respectively, to $[\alpha]$ or $[M]$, no matter whether the two quantities refer to standard conditions or not. Thus $[\alpha]_D^{20}$ specifies that the specific rotation applies to the yellow sodium line and to a temperature of 20°C.

If the optically active substance is not a liquid or solid, but a solute, then the specific rotation is, in conformity with the definition given by (2.62), the rotation exhibited by a 10-cm layer of a solution containing, at the temperature of the experiment, 1 g of optically active solute per milliliter of solution. Since concentrations, c, are generally given in terms of g, per 100 milliliters of solution

$$[\alpha] = \frac{100\alpha}{lc} \tag{2.65}$$

and

$$[M] = \frac{M\alpha}{lc} = \left(\frac{M}{100}\right)[\alpha] \tag{2.66}$$

If the concentrations are given in terms of g per 100 g of solution, p, then, necessarily,

$$[\alpha] = \frac{100\alpha}{lp\rho_s} \tag{2.67}$$

and

$$[M] = \frac{M\alpha}{lp\rho_s} \tag{2.68}$$

respectively, where ρ_s is the density of the solution.

Surprisingly, it is quite uncommon to express concentrations in moles per liter, n. In that case one would have

$$[M] = \frac{10\alpha}{ln} \tag{2.69}$$

According to (2.69), the molecular rotation can be defined as the rotation, in degrees expected of a 1-m layer of a solution which contains 1 mole of optically active solute per liter.

Finally, one might consider the " molal " rotation

$$[M'] = \frac{10\alpha}{l\rho_s m} \tag{2.70}$$

where m is molality of the optically active solute.

It is generally not recognized that an $[\alpha]$-value, obtained at finite concentration of a solute in a given medium, is not necessarily a representative value. It would be if α were proportional to the concentration. While this is very often true in a good approximation, one cannot rely upon this proportionality without having made certain of it. This will be discussed in more detail in a later section.

There is, of course, no standard solvent, since the choice of the medium depends on the solubility of the optically active substance. When one can choose between several solvents, one will find invariably that the rotatory power varies from solvent to solvent. This may be due to one or several of the following differences: differences in (1) solvent-solute interaction, (2) solute-solute interaction, and (3) the refractive index of the solvent relative to that of the solute. In highly dilute systems, only the first and last factors enter. The true scope of factor (1) could then be ascertained by eliminating also factor (3). This can be done by considering, instead of $[\alpha]$ or $[M]$, rather the " rotivity," a quantity introduced by Gans [226]

$$\Omega = \frac{[\alpha]}{n^2 + 2} \tag{2.71}$$

where n is the refractive index of the solvent. The basic validity of this relation was verified experimentally by L. K. Wolff and H. Volkmann [227], by Rule and McLean [228], and by Beckmann and Cohen [231]. As one should expect, the value of Ω varies widely for solutes with dipole moment and also in polar solvents. The rotivity can, obviously, be a constant only if both solvent and solute are nonpolar. Deviations from constancy are in any event indicative of solvent-solute interaction provided the concentrations used are small enough. The rotivity of an optically active compound in the gas phase (absence of solute-solvent interaction)

$$\Omega_0 = \frac{[\alpha]}{3} \tag{2.72}$$

It has, therefore, been suggested [229] to replace the numerical definition of Ω by

$$\Omega' = \frac{3[\alpha]}{n^2 + 2} \tag{2.73}$$

According to this alternate definition, the rotivity in the gas phase, $\Omega'_0 = [\alpha]$. According to Rule and Chambers [230] and Beckmann and Cohen [231] Ω_0 is, in complete absence of solvent-solute and solute-solute interaction, related to Ω in solution by

$$\Omega = \Omega_0 + B\left[\frac{(\epsilon - 1)}{(\epsilon + 2)}\right] \tag{2.74}$$

where B is a constant for solvents of similar structure [232] and ϵ is the dielectric constant of the medium at the temperature of the experiment. For nonpolar solvents, one can write, in view of Maxwell's relation $\epsilon = n^2$,

$$\Omega = \Omega_0 + B\left[\frac{(n^2 - 1)}{(n^2 + 2)}\right] \tag{2.75}$$

For early reviews of the problem of the solvent effect see [229] and [233]. More recently Djerassi [233a] and co-workers have used optical rotatory dispersion to demonstrate changes in conformation in a variety of molecules with changes in solvent.

As stated earlier (p. 58) the optical rotation is due to the difference of the refractive indices n_d and n_l. In view of this, it is rather surprising that there has been, thus far, exceedingly little interest in reducing experimental α-values to numerical values of the basic phenomenon, circular birefringence. The conversion is easy by means of (2.8). Thus for solids and liquids,

$$(n_l - n_d) = \frac{\alpha \lambda_0}{1800 l} = \frac{[\alpha]\lambda_0 \rho}{1800} = \frac{[M]\lambda_0 \rho}{18M} = \frac{[\varphi]\lambda_0 \rho}{\pi} \tag{2.76}$$

and for solutes

$$(n_l - n_d) = \frac{[\alpha]\lambda_0 c}{180,000} = \frac{[M]\lambda_0 c}{1800M} = \frac{[\varphi]\lambda_0 c}{100\pi} \tag{2.77}$$

7 THE CONCENTRATION DEPENDENCE OF THE SPECIFIC ROTATION

The Intrinsic Rotation

The specific rotation, obtained in dilute solutions, is generally considered as a constant representative for the solute in a particular solvent at a particular wavelength and temperature. This is not entirely justified. Even in very dilute solutions, one finds, as already noticed by Biot, that

$$\frac{\alpha}{l} = Ac + Bc^2 \tag{2.78}$$

Consequently,

$$[\alpha] = a + bc \tag{2.79}$$

where $a = 100A$ and $b = 100B$. The truly constant value, independent of concentration, is therefore $[\alpha]_{c \to 0} \equiv \{\alpha\}$, a new quantity which we should like to introduce and to define as "intrinsic" rotation. The simplest possible concentration dependence of $[\alpha]$ is, accordingly, defined by

$$[\alpha]_\lambda^T = \{\alpha\} + bc \tag{2.80}$$

or

$$[\alpha]_\lambda^T = \{\alpha\} + b'p \tag{2.81}$$

This relationship is illustrated in Fig. 2.71 for nicotine in ethanol using data of Landolt [234]. The data follow the equation

$$-[\alpha]_D^{20} = 138.59 + 0.224p \tag{2.82}$$

In this particular case, therefore, the specific rotation determined in a 5% solution would differ from the intrinsic rotation, $\{\alpha\}$, by 0.8%. It follows from both the data of Landolt and from those of Winther [235] that the variation of $[\alpha]$ of nicotine with concentration is less in some solvents and considerably larger in others. Thus, for ethylene bromide, the difference between $[\alpha]$ and $\{\alpha\}$ can be estimated to be in excess of 5% at $p < 2\%$ (D line). In some solvents, a change in concentration at p-values $< 10\%$ is accompanied even by a change in sign of $[\alpha]$. Such variations from solvent to solvent, modest as they may be on the average, certainly impair a check on the validity of

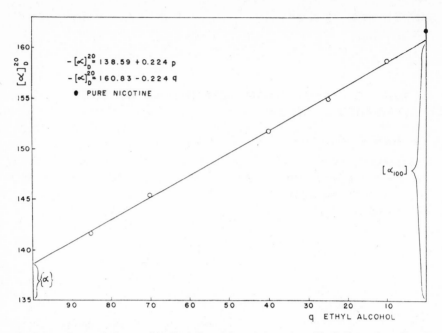

Fig. 2.71. Extrapolation to "intrinsic" rotation, $\{\alpha\}$, and to "bulk" rotation, $[\alpha_{100}]$. Nicotine in ethanol [236].

(2.71), which was verified experimentally by some authors [226, 227] but was questioned by others. The concentration dependence of $[\alpha]$ varies not only from solvent to solvent for a given solute, but also from solute to solute for a given solvent. Thus, $[\alpha]$ of a 5% solution of sucrose in water differs from $\{\alpha\}$ by less than 0.1% [see (2.83)] while that of a 5% solution of ethyl-tartrate in water differs from $\{\alpha\}$ by >3.5% [236]. The great majority of available data pertain to the concentration effect within the spectral range of normal dispersion. They are, therefore, undoubtedly, the minimal effects to be expected in these systems. The change in solute-solute interaction and in solute-solvent interaction with concentration is bound to lead to changes in the absorption of the optically active chromophores. These changes may be limited to the extinction coefficient but there may also be a slight spectral shift of the respective absorption band(s) in analogy to what is well known for optically inactive absorption bands. While such a shift will have a relatively small effect upon $[\alpha]$-values collected at a considerable spectral distance from the absorption band(s), profound changes in $[\alpha]$ are most likely within the range of anomalous dispersion. This is obvious from Fig. 2.10. There does not exist any literature to that effect, but some results by early authors, including Biot, support this reasoning. Thus, Biot noticed [237] that the concentration effect in aqueous solutions of tartaric acid varied with the color of the light. It was larger for violet than for blue, and larger for blue than for green or red. It is therefore to be expected that rotatory dispersion curves, in order to be fully significant and comparable, should be given in terms of $\{\alpha\}$ and not for an arbitrary, however small, concentration.

The relatively simple (2.80) and (2.81) will generally be sufficient if the concentrations used do not exceed a few percent. At larger concentrations, however, it will generally be necessary to consider more terms. Thus, the concentration dependence of sucrose between $c = 0$ and $c = 65$ is, in spite of its smallness, governed by the three-term equation [238]

$$[\alpha]_D^{20} = 66.456 + 0.00870c - 0.000235c^2 \tag{2.83}$$

In this particular case, the curve shows an inflection. In other instances, the third term may have the same sign, specifically,

$$[\alpha]_\lambda^T = \{\alpha\} + bc + dc^2 \tag{2.84}$$

Extrapolations to 100% Concentration

Extrapolation to zero solute concentration yields the intrinsic rotation, that is, the specific rotation free from solute-solute interactions. On the other hand, extrapolation to 100% concentration should yield the "bulk" rotation, that is, the specific rotation free from solvent interaction but incorporating maximal solute-solute interaction. This alternate extrapolated value, which shall be designated as $[\alpha_{100}]$, should, therefore, be identical for all $[\alpha]$ versus c curves of a given solute no matter what solvent is used. This is borne out by Fig. 2.72 taken from Winther's work [235]. The further expectation that the $[\alpha_{100}]$-values extrapolated from solution data are identical with experimental data obtained from the pure substance is also borne out by results of Landolt [234]. The $[\alpha]_D^{20}$-value of pure nicotine is $-161.55°$ as compared to $-160.83°$ and $-161.29°$, respectively, if extrapolated from an ethanol and water solution. (The first value is traced into Fig. 2.71.) The differences are not significant in view of the relatively large solvent concentration (10%) from which the extrapolation is carried out, relying on a constancy of slope [239]. It is worthwhile noting that extrapolation to $[\alpha_{100}]$ opens up a method for determining $[\alpha]$ of solids not obtainable in specimens large enough or transparent enough to allow a direct $[\alpha]$-determination.

It is clear that the difference between $[\alpha_{100}]$ and the "intrinsic rotivity" $\{\Omega\}$ (the Ω-value derived from $\{\alpha\}$) represents, except for a constant (solute-solute interaction) the true absolute solvent effect. For this reason, the comparative determination of $[\Omega]$ and $[\alpha_{100}]$ appears to be of major interest. The only systematic investigations in the literature along these lines are those by Winther and, particularly, the very careful work by Landolt [234] on nicotine,

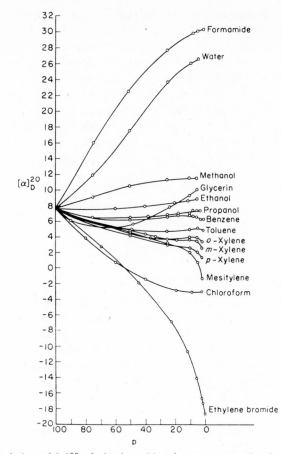

Fig. 2.72. Variation of $[\alpha]_D^{20}$ of nicotine with solvent concentration in various solvents [235].

camphor, turpentine, and ethyl tartrate [240]. Such double extrapolations will have to be limited to instances where enough material is available. This being the case, one will have to limit the investigations to spectral ranges distant from absorption.

If extrapolation to $c = 100$ are made by means of experimental values obtained between, say $c = 95$ and 97, the simple equation

$$[\alpha]_\lambda^T = [\alpha_{100}] + mq \tag{2.85}$$

will generally be sufficient. Here q is the concentration of the solvent, in g per 100 g of solution. If the concentration range to be covered quantitatively is larger, more terms are generally necessary, namely,

$$[\alpha]_\lambda^T = [\alpha_{100}] + mq + nq^2 + \cdots \tag{2.86}$$

8 THE TEMPERATURE DEPENDENCE OF THE SPECIFIC ROTATION

The effect of temperature upon the experimentally measured rotation consists of two parts: (a) that due to thermal volume changes, and (b) that due to changes in the rotatory power of the molecules themselves. The former effect is generally small. It can be eliminated by means of the equation

$$\alpha_\lambda^{T_0} = \alpha_\lambda^T[1 + \beta(T - T_0)] \tag{2.87}$$

where T_0 is the reference (standard) temperature, T is the experimental temperature, both in °C, and β is the volumetric expansion coefficient of the solution. In many cases, the expansion coefficient of the solvent may be sufficient for an elimination of the volume effect. For optimum precision, of course, the expansion coefficient of the solution should be used. This type of thermal change is automatically excluded on transforming α-values to $[\alpha]$-values.

Thermal changes of $[\alpha]$, that is, changes due to factor (b) may be due to several causes: (1) change in the solvation of the molecules; (2) changes in dipole-dipole interaction if both solute and solvent are dipolar; (3) changes in induction effects if either solvent or solute are dipolar; (4) changes in molecular association; (5) changes in the equilibrium of configurations or conformations of a given optically active compound; and (6) changes in chemical equilibria involving at least one optically active compound. Of considerable interest is finally (7) the effect of temperature upon the *rate* of configurational or chemical changes since $[\alpha]$ or Ω can then be used for the calculation of pertinent thermodynamic quantities. Of particular interest from the theoretical point of view is factor (5). The configurational change involved may not be drastic but subtle, such as the change due to lessening of a hindrance to free rotation about valence bonds as the temperature increases. The temperature dependence of optical activity is therefore a singularly sensitive and conclusive effect for investigating these phenomena whenever the contributions of the other factors to the optical changes can be eliminated or controlled by suitable procedures.

A fundamental investigation of such changes is that by Bernstein and Pedersen [230] concerned with *d*-secondary butyl alcohol. The schematic Fig. 2.73 illustrates the problem. On looking in the direction of the C—C* bond of this dissymmetric molecule (perpendicular to the plane of the paper)

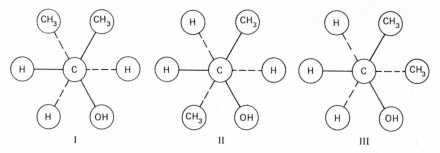

Fig. 2.73. The three stereoisomers of *d*-secondary butyl alcohol. C—C* axis ⊥ paper plane. Bonds of substituents below paper plane are dotted.

three rotational isomers can be readily recognized. The substituents bound to the C*-atom below the paper plane are identified by dotted bonds. A rotation, by 2π, requires the surmounting of three potential energy peaks of which one, that for direct passage from III to I, can be expected to be highest and that for direct passage from III to II as intermediate. One therefore would expect the concentration of isomer III to be the lowest in the equilibrium between the three configurations. Consequently, the contributions of the three to the [α]-value of the compound are bound to differ. It is clear that the thermal variation of Ω can yield both the actual equilibrium concentrations at a given temperature and the activation energies. Bernstein and Pedersen derive from the Ω-values extrapolated to high dilutions in nonpolar solvents the gas value-Ω_0 using to that effect (2.74). By means of the basic equations already available [241, 229, 231], the activation energy for configuration III was obtained as 803 ± 60 cal/mole at 20°C. The equilibrium concentrations were found to be: I and II, 42.35% (43.0%); and III, 15.3% (14.0%), the bracketed values referring to 70°C. Interesting, similar investigations were carried out by Volkenshtein [242] on methylvinylcarbinol. More recently, Djerassi [233a] and co-workers have used the temperature dependence of optical rotation and also of circular dichroism, extensively in the conformation analysis of organic molecules.

For all such investigations, it must be borne in mind that the variation of the primary quantity [α] with temperature is, in general, not very large at room temperature for nonpolar solutes in nonpolar solvents. This is illustrated by a few representative data, in Table 2.9, taken from Volkenshtein's publication [242]. Rather precise measurements and rigorous temperature control are therefore necessary if reliable data are to be obtained.

There are, on the other hand, cases where [α] varies quite extensively with temperature. Tartaric acid and its derivatives are classic examples. Here [α] may vary, at room temperature, by more than 10% per degree. Such a large temperature sensitivity is indicative of the existence of an equilibrium between

Table 2.9. Specific Rotation of Methylvinylcarbinol in Various Solvents at Several Temperatures in °C [242]; $\lambda_0 = 3340$ Å; $c = 0.05043 - 0.65358$

	15°	20°	35°	50°	55°
Pentane	207.5	200.6	180.2	161.2	142.8
Dioxane	213.7	208.2	192.3	180.8	171.1
Benzene	184.8	174.8	152.2	133.5	118.4
CCl$_4$	186.0	176.1	153.8	139.0	128.1
Cyclohexane	198.2	191.0	173.1	158.3	147.7
n-Hexane	195.0	181.4	141.0	113.8	93.7

two species differing appreciably in their rotatory power. Here, one may use the temperature effect for calculating the enthalpy and entropy differences between the two species applying the theory of rate processes. The case of tartaric acid esters has thus been treated by Eyring [229].

More straightforward are the appreciable thermal changes in the rate of racemizations of sterically hindered compounds treated by Volkenshtein [243] and, more recently in detail, also by Eyring [244]. Here again, enthalpies and entropies are obtained without difficulty from the theory of rate processes.

9 ANALYSIS OF MIXTURES

In absence of interaction, the specific rotations $[\alpha]_1$ and $[\alpha]_2$ of two optically active components in a mixture are additive, that is, the specific rotation of the mixture

$$[\alpha] = C_1[\alpha]_1 + (1 - C_1)[\alpha]_2 \qquad (2.88)$$

where C is the weight fraction and the subscripts refer to the two components, respectively. This mixture rule has practically useful implications. Considering first a mixture of two optical antipodes, it follows at once that

$$\frac{[\alpha]}{[\alpha]_1} = 2C_1 - 1 \qquad (2.89)$$

and

$$C_1 = \frac{[\alpha] + [\alpha]_1}{2[\alpha]_1}$$

$$C_2 = \frac{[\alpha] + [\alpha]_2}{2[\alpha_2]} \qquad (2.90)$$

and

$$\frac{[\alpha] - [\alpha]_1}{[\alpha] + [\alpha]_1} = \frac{C_2}{C_1} \tag{2.91}$$

if we associate with $[\alpha]_1$ the specific rotation of the dextrorotatory component. Since all these equations obviously apply at any wavelength it follows furthermore from (2.89) that

$$\frac{[\alpha]_{\lambda_1}}{[\alpha]_{\lambda_2}} = \frac{[\alpha]_{1\lambda_1}}{[\alpha]_{1\lambda_2}} = \frac{[\alpha]_{2\lambda_1}}{[\alpha]_{2\lambda_2}} = \text{constant} \tag{2.92}$$

meaning that the dispersion ratio in these mixtures is a constant independent of the mixture ratio of the components. If, therefore, mixtures with strongly different $[\alpha]$-values exhibit exactly the same dispersion ratios, it is at once clear that one is dealing with mixtures of optical antipodes and their concentrations can then be determined easily by means of (2.90).

If the two components of a mixture are not related, it follows easily from (2.88) that

$$C_1 = \frac{[\alpha] - [\alpha]_2}{[\alpha]_1 - [\alpha]_2}$$

$$C_2 = \frac{[\alpha]_1 - [\alpha]}{[\alpha]_1 - [\alpha]_2} \tag{2.93}$$

It is, therefore, again easy to determine the concentration of both components provided the specific rotations of the pure single-component solutions, $[\alpha]_1$ and $[\alpha]_2$ are known and that there is no interaction between the solutes in the mixture. On combining (2.93) we obtain

$$\frac{[\alpha]_1 - [\alpha]}{[\alpha] - [\alpha]_2} = \frac{C_2}{C_1} \tag{2.94}$$

Validity of all these equations is obviously independent of the wavelength selected. Therefore

$$\frac{[\alpha]_{1\lambda_1} - [\alpha]_{\lambda_1}}{[\alpha]_{\lambda_1} - [\alpha]_{2\lambda_1}} = \frac{[\alpha]_{1\lambda_2} - [\alpha]_{\lambda_2}}{[\alpha]_{\lambda_2} - [\alpha]_{2\lambda_2}} \tag{2.95}$$

One thus can, as pointed out by Darmois [245], base upon (2.95) a graphical test of interesting implications. It is illustrated in Fig. 2.74. It is clear, on the basis of elementary geometrical consideration, that the straight lines connecting $[\alpha]_{\lambda_1}$ with $[\alpha]_{\lambda_2}$, $[\alpha]_{1\lambda_1}$ with $[\alpha]_{1\lambda_2}$ and $[\alpha]_{2\lambda_1}$ with $[\alpha]_{2\lambda_2}$, respectively, must all meet in a single point 0. In Fig. 2.74, which is a reproduction of Darmois' original figure, one of the wavelengths is the D line and the other is the Hg

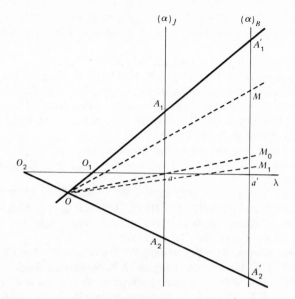

Fig. 274. The "Darmois" diagram [245]; abscissa, λ; ordinate, $[\alpha]$.

line at 436 mμ. The respective specific rotations are denoted as $(\alpha)_J$ and $(\alpha)_B$, respectively. The straight lines of the pure components are $A_1 A_1'$, and $A_2 A_2'$, respectively, assuming that their rotations have opposite signs. The symbol M characterizes mixtures of the two components. The existence of a common point of intersection for the various curves, at 0, is a quick and rapid proof that the respective mixture is a two-component mixture. In contradistinction noncoincidence of the intersection of the various curves would indicate either the presence of a third component or an interaction between the two components.

References

1. D. Brewster, *A Treatise on Optics*, 1st Am. ed., Carey, Lea, and Blanchard, Philadelphia, 1833, p. 191.
2. A. Eresnel, *Ann. Chim. et Phys.*, **28**, 147 (1825).
3. J. B. Biot, *Mém, Prem. Classe Inst. France*, **13**, 218 (1812).
4. E. von Fleischl, *Ann. Physik. Chem.*, **24**, 127 (1885).
5. A. Fresnel, *Ann. Chimie et Phys.*, **28**, 147 (1825); *Pogg. Ann.*, **21**, 276 (1831); see also G. Meslin, *Compt. Rend.*, **152**, 1666 (1911).
6. L. Pasteur, *Ann. Chim. et Phys.*, **24**, 442 (1848).

7. J. H. van't Hoff, *La chimie dans l'espace*, Rotterdam, 1874; or *The Arrangement of Atoms in Space* (trans. and ed. A. Eiloart), Longmans, Green, London, 1898.

8. J. A. Le Bel, *Bull. Soc. Chim.*, **22**, 337 (1874).

9. For example, W. H. Mills and C. R. Nodder, *J. Chem. Soc.*, **117**, 1407 (1920); **119**, 2094 (1921); W. H. Mills, *Trans. Faraday Soc.*, **26**, 431 (1930).

10. R. Maitland and W. H. Mills, *J. Chem. Soc.*, **1936**, 987.

11. W. H. Mills, *Chem. Ind. (London)*, **45**, 884 (1926); W. H. Mills, and K. A C. Elliott, *J. Chem. Soc.*, **1928**, 1291.

12. F. Bell and J. Kenyon, *Chem. Ind. (London)*, **45**, 864 (1926).

13. J. F. Hyde and R. Adams, *J. Am. Chem. Soc.*, **50**, 2499 (1928), and subsequent papers by R. Adams.

14. M. Perutz, *Nature*, **167**, 1053 (1951); W. L. Bragg and M. F. Perutz, *Proc. Roy. Soc. (London)*, **A213**, 425 (1952); L. Pauling and R. B. Corey, *Proc. Roy. Soc. (London)*, **B141**, 21 (1953); *Proc. Natl. Acad. Sci. U.S.*, **39**, 253 (1953); *J. Am. Chem. Soc.*, **72**, 5349 (1950).

15. See, for example, R. Simpson and W. Kauzman, *J. Am. Chem. Soc.*, **75**, 5139 (1953). K. Linderstrom-Lang and J. A. Schellman, *Biochim. et Biophysic. Acta*, **15**, 156 (1954). C. Cohen, *J. Biophys. Biochem. Cytol.*, **1**, 203 (1955). P. Doty and T. Nishihara, in *Recent Advances in Gelatin and Glue Research*, Pergamon, New York, 1958, p. 92. Y. Tanomura, K. Sekiya, and K. Imamura, *J. Biol. Chem.*, **237**, 3110 (1962). A. N. Glazer and N. S. Simmons, *J. Am. Chem. Soc.*, **87**, 2287 (1965). C. Tanford, R. H. Pain, and N. S. Otchin, *J. Mol. Biol.*, **15**, 489 (1966).

16. W. Moffitt, *J. Chem. Phys.*, **25**, 467 (1956); W. Moffitt, D. Fitts, and J. G. Kirkwood, *Proc. Natl. Acad. Sci. U.S.*, **42**, 33 (1956); I. Tinoco, *Advan. Chem. Phys.*, **4**, 113 (1962); R. W. Woody and I. Tinoco, *J. Chem. Phys.*, **46**, 4927 (1967).

17. J. T. Yang, *Poly-α-Amino Acids*, G. D. Fasmau, ed., Marcel Dekker, New York, 1967, Chap. 6.

18. E. Reusch, *Pogg. Ann.*, **138**, 628 (1869).

19. L. Sohnke, *Mathem. Ann. IX*, **504** (1876); *Pogg. Ann. Ergänzungsband, VIII*, **16** (1878); *Z. Krist.*, **13**, 229 (1888).

20. H. de Vries, *Acta Cryst.*, **4**, 219 (1951).

21. M. H. Winkler, *J. Phys. Chem.*, **60**, 1656 (1956).

22. E. Fischer, *Ber.*, **24**, 2683 (1891); **27**, 3189 (1894).

23. C. Djerassi, *Optical Rotatory Dispersion*, McGraw-Hill, New York, 1960; see also P. Crabbé, *Optical Rotatory Dispersion and Circular Dichroism in Organic Chemistry*, Holden-Day, San Francisco, 1965. T. M. Lowry, *Optical Rotatory Power*, Longmans, Green, London, 1935. W. Kuhn, *Trans. Faraday Soc.*, **26**, 293 (1930). E. Brand, E. Washburn, B. F. Erlanger, E. Ellenbogen, J. Daniel, F. Lippman, and M. Schen, *J. Am. Chem. Soc.*, **76**, 5037 (1954).

24. C. Djerassi, *Proc. Chem. Soc.*, **314** (1964).

25. W. Kuhn, *Z. Physik. Chem.*, **B31**, 23 (1935).

26. E. U. Condon, W. Altar, and H. Eyring, *J. Chem. Phys.*, **5**, 753 (1937).

27. W. W. Wood, W. Fickett, and J. G. Kirkwood, *J. Chem. Phys.*, **20**, 561 (1952).

28. J. M. Bijvoet, A. F. Peerdeman, and A. J. Van Bommel, *Nature*, **168**, 271 (1951), *Proc. Konink. Ned. Acad. Wetenschap*, **B54**, 16 (1951).

29. A. J. Van Bommel, *Chem. Weekblad*, **48**, 988 (1952); R. Pepinsky, *Record Chem. Progr.* (Kresge-Hooker Sci. Lib.), **17**, 145 (1956).

30. A. Cotton, *Compt. Rend.*, **120**, 989, 1044 (1895); *Ann. Chim. et Phys.*, **8**, 347 (1896).

31. For details on these and other elementary optical phenomena, see: G. S. Monk, *Light, Principles and Experiments*, McGraw-Hill, New York, 1937; R. W. Wood, *Physical Optics*, Macmillan, New York, 1934, J. Valasek, *Elements of Optics*, McGraw-Hill, New York, 1932.

32. J. B. Biot, *Bull. Soc. Philomath*, Paris (1815), p. 26.

33. L. D. Landau and E. M. Lifshitz, *Electrodynamics of Continuous Media*, Addison-Wesley, Reading, Mass., 1960.

34. Attention should be drawn to the fact that it is not necessary that k_l and k_d reach their maximum at the same wavelength although this is generally assumed at present. Their maxima could differ slightly in spectral location. This should lead to and may explain an occasionally observed lopsidedness of, and inflections within, anomalous rotatory dispersion curves.

35. L. Natanson, *J. Phys.*, **8**, 321 (1909).

36. W. Heller, in R. Abegg, *Handbuch der Anorganischen Chemie*, Vol. 4, Hirzel, Leipzig, 1935, Part III, **2B**, p. 860.

37. J. Kunz and R. G. LaBaw, *Nature*, **140**, 194 (1937).

38. E. B. Ludlam, A. W. Pryde, and H. Gordon-Rule, *Nature*, **140**, 194 (1937).

39. W. Heller, *Rev. Mod. Phys.*, **14**, 406 (1942).

40. M. Faraday, *Phil. Mag.*, **28**, 294; **29**, 153 (1846).

41. P. Zeeman, *Proc. Acad. Sci. Amsterdam*, **5**, 41 (1902).

42. O. Wiener, *Abhandl. Math.-Phys. Klasse Sächs. Akad. Wiss. Leipzig*, **32**, 509 (1912).

43. A. Fresnel, *Ann. Chim. Phys.*, **28**, 147 (1825); W. Kuhn and E. Braun, *Z. Physik. Chem.*, **B7**, 292; *ibid.*, **B8**, 445 (1930).

44. P. Glan, *Ann. Physik. Chem.*, **1**, 351 (1877); *Repertorium Exptl.-Physik.* (*Carl's*), **16**, 570 (1880); **17**, 195 (1881).

45. See, for example, B. K. Johnson, *Proc. Phys. Soc.* (*London*), **55**, 291 (1943).

46. International Critical Tables.

47. T. A. Davis and K. Vedam, *J. Opt. Soc. Am.*, **58**, 1446 (1968).

48. D. L. Steinmetz, W. G. Phillips, M. Wirick, and F. F. Forbes, *Appl. Opt.*, **6**, 1001 (1967).

49. H. Cary, R. C. Hawes, P. B. Hooper, J. J. Duffield, and K. P. George, *Appl. Optics*, **3**, 329 (1964).

50. R. T. Glazebrook, *Phil. Mag.*, **10**, 247 (1880); **15**, 352 (1883). S. P. Thompson, *ibid.*, **12**, 349 (1881); **15**, 435 (1883); **21**, 476 (1866).

51. Constructed by C. D. Ahrens. See S. P. Thompson, *Phil. Mag.*, **21**, 476 (1886).

52. W. Nicol, *Edinburgh New Phil. J.*, **6**, 83 (1828); **16**, 372 (1834).

53. G. Bruhat and M. Hanot, *J. Phys.*, **3**, 46 (1922).

54. H. Bénard, *Recherches Inventions*, **1**, 229 (1920).

55. J. Jamin, *Compt. Rend.*, **68**, 221 (1869).

56. D. B. Brace, *Phil. Mag.*, **5**, 161 (1903).
57. W. Heller, and G. Quimfe, *Phys. Rev.*, **51**, 382 (1942).
58. W. H. Wollaston, *Trans. Roy. Soc. (London)*, **A110**, 126 (1820).
59. A. Rochon, *J. Phys.*, **72**, 319 (1811).
60. H. de Senarmont, *Ann. Chim. Phys.*, **50**, 480 (1857).
61. See W. Grosse, thesis, Univ. of Kiel, 1886.
62. C. D. Ahrens, *Phil. Mag.*, **19**, 69 (1885).
63. A. H. Pfund, *J. Opt. Soc. Am.*, **37**, 558 (1947).
64. A. Elliot, F. J. Ambrose, and R. Temple, *J. Opt. Soc. Am.*, **38**, 212 (1948).
65. W. B. Herapath, *Phil. Mag.*, **3**, 161 (1852).
66. W. B. Herapath, *Phil. Mag.*, **6**, 346 (1853).
67. E. H. Land, Brit. Pat. No. 412,179 (June 18, 1934).
68. M. Grabau, *J. Opt. Soc. Am.*, **27**, 420 (1937); L. R. Ingersoll, J. Winans, and E. Krause, *J. Opt. Soc. Am.*, **26**, 233 (1936); J. Strong, *J. Opt. Soc. Am.*, **26**, 256 (1936).
69. E. H. Land and C. D. West, *Colloid Chem.*, **6**, 160 (1946); E. H. Land, *J. Opt. Soc. Am.*, **41**, 957 (1951). (Pamphlet FT 3374A, Polaroid Corp., Cambridge, Mass., 1967.)
70. J. Gallup, *Bull. Am. Ceram. Soc.*, **23**, 38 (1944).
71. F. J. Bates and associates, "Polarimetry, Saccharimetry and the Sugars," *Natl. Bur. Standards*, U.S. Circ. No. 440 (1942).
72. General Electric Vapor Lamp Co., Hoboken, New Jersey.
73. Baird-Atomic, Inc., Cambridge, Mass.; Bausch and Lomb Optical Co., Rochester, New York; Orial Optics Corp., Stamford, Conn.
74. Corning Glass Works, Corning, New York.
75. Wratten Filters, Eastman Kodak Co., Rochester, New York.
76. R. W. Wood, *Physical Optics*, Macmillan, New York, 1942, p. 15. F. Weigert, *Optische Methoden der Chemie*, Akadem. Verlagsgesellschaft, Leipzig, 1927, p. 62. A. Stähler, E. Thiede, and F. Richter, in *Handbuch der Arbeitsmethoden in der Anorganischen Chemie*, Vol. II, part 2, de Gruyter, Berlin, 1925, p. 1530.
77. W. Baum and L. Dunkelman, *J. Opt. Soc. Am.*, **40**, 782 (1950).
78. A wide variety of commercially available lamps is described in *Illuminating Society Handbook*, 4th ed., John E. Kautman, Ed., Illuminating Engineering Society, New York, 1968.
79. H. Streuli, *Mitt. Lebensm. u. Hyg.*, **42**, 79 (1951).
80. R. J. King, *J. Sci. Inst.*, **43**, 924 (1966).
81. F. J. Bates and associates, *loc. cit.* [71], pp. 103–105.
82. S. D. Gardiner, *Intern. Sugar J.*, **54**, 183 (1952).
83. D. Smith and S. A. Ehrhardt, *Ind. Eng. Chem., Anal. Ed.*, **18**, 81 (1946).
84. M. Békésy, *Magyar Chem. Folyóirat*, **49**, 114 (1943); *Biochem. Z.*, **312**, 103 (1942).
85. H. Kacser and A. R. Ubbelohde, *J. Soc. Chem. Ind. (London)*, **68**, 135 (1949).
86. Bodenseewerk Perkin Elmer and Co., GMBH/Uberlingen, West Germany.
87. J. Straub, *Chem. Weekblad*, **31**, 465 (1934).
88. K. S. Gibson, and E. P. T. Tyndall, *Natl. Bur. Standards*, *U.S. Sci. Paper* No. 475, **19**, 131 (1923–1924).

89. M. Luckiesh and A. H. Taylor, *Illum. Eng.*, **38**, 189 (1943); A. H. Taylor, *Illum. Eng.*, **38**, 89 (1943).

90. K. S. Weaver, *J. Opt. Soc. Am.*, **27**, 36 (1937); S. Hecht and R. E. Williams, *J. Gen. Physiol.*, **5**, 1 (1922).

91. N. I. Pinegin, *Nature*, **155**, 20 (1945).

92. J. F. Skogland, *Natl. Bur. Standards*, U.S. Misc. Pub. No. 86 (1929).

93. J. Blanchard, *Phys. Rev.*, **11**, 81 (1918).

94. S. Hecht, *Am. Scientist*, **32**, 159 (1944); see the extensive literature references given there, and *J. Opt. Soc. Am.*, **32**, 42 (1942).

95. K. O. Kiepenheuer, *Z. Physik*, **107**, 145 (1937).

96. H. Helmholtz, *Handbuch der Physiologischen Optik.*, Voss, Leipzig, 1867, especially pp. 135, 323.

97. Y. Le Grand, *Rev. Opt.*, **12**, 145 (1933).

98. N. Deer, *Intern. Sugar J.*, **37**, 421 (1935).

99. H. Landolt, *Das Optische Drehungsvermögen Organischer Substanzen.*, Vieweg, Braunschweig, 1898, pp. 304, 305.

100. A. Wüllner, *Lenrbuch der Experimentalphysik* (Vol. IV, *Die Lehre von der Strahlung*), Teubner, Leipzig, 1899, pp. 966–967.

101. See particularly L. Chaumont, *Ann. Phys.*, **4**, 61 (1915); **5**, 17 (1916).

102. J. H. Jellett, *Researches in Chemical Optics*, Dublin Univ. Press, 1875. See also *Brit. Assoc. Advancement Sci. Rept.*, **29**, 13 (1860).

103. A. Cornu, *Bull. Soc. Chim.*, **14**, 140 (1870).

104. F. Lippich, *Sitzber. Akad. Wiss. Wien. Math.-Naturw. Klasse, Abt. IIa*, **91**, 1059 (1885).

105. H. Laurent, *J. Phys.*, **3**, 183 (1874).

106. U. Gayon, *J. Phys.*, **8**, 164 (1879).

107. H. Heele, according to E. Gumlich, *Z. Instrumentenk.*, **16**, 269 (1896).

108. S. Nakamura, *Centr. Mineral, Geol.*, **1905**, 267.

109. H. Soleil, *Compt. Rend.*, **21**, 426 (1845).

110. F. Martens, *Z. Instrumentenk.*, **20**, 82 (1900).

110a. H. Soleil, *Compt. Rend.*, **20**, 1805 (1845); **24**, 973 (1847); **26**, 162 (1848).

111. H. Fizeau and L. Foucault, *Compt. Rend.*, **21**, 1155 (1845).

112. O. J. Broch, *Repetorium der Physik* (Davis), **7**, 113 (1846); *Ann. Chim. Phys.*, **34**, 119 (1852).

113. E. Perucca, *Nuovo Cimento*, **17**, 1 (1940).

113a. G. Wiedemann, *Die Lehre von Elektrizität*, Vieweg, Braunschweig, 1893–1898.

114. E. Lommel, *Ann. Physik. Chem.*, **36**, 731 (1889).

115. S. Landau, *Physik. Z.*, **9**, 417 (1908).

115a. F. Savart, discussed by J. C. Poggendorf in *Ann. Physik. Chem.*, **49**, 292 (1840); J. Müller, *Ann. Physik. Chem.*, **35**, 261 (1835).

116. J. L. Soret and E. Sarasin, *Compt. Rend.*, **95**, 636 (1882); **84**, 1362 (1887); **83**, 818 (1876); **81**, 610 (1875); *Arch. Sci. Phys. Nat.*, **8**, 5, 97, 201 (1882).

117. J. Duclaux and P. Jeantet, *J. Phys. Radium*, **2**, 156 (1921).

118. G. Bruhat and M. Pauthenier, *Rev. Opt.*, **6**, 163 (1927).

119. A. Hussell, *Ann. Physik. Chem.*, **43**, 498 (1891).

120. For more detailed information on phototubes and photodetectors see manufacturers data sheet, for example the RCA Tube Manual.

121. See, for example, E. C. C. Baly, *Spectroscopy*, Vol. II, Longmans, Green, London, 1927, p. 1380.

122. G. R. Harrison, *J. Opt. Soc. Am.*, **11**, 113 (1925).

123. J. Duclaux and P. Jeantet, *J. Phys. Radium*, **2**, 156 (1921).

124. K. Schwarzschild, *Astrophys. J.*, **11**, 89 (1900).

125. For further information on photographic technique, see J. Strong, *Procedures in Experimental Physics*, Prentice-Hall, New York, 1938, pp. 449–493, and on photographic theory, T. H. James, *The Theory of the Photographic Process*, 3rd ed., Macmillan, New York, 1966.

126. See, for example, W. Kuhn, *Ber.*, **62**, 1727 (1929).

127. S. Landau, *Physik. Z.*, **9**, 417 (1908).

128. For experimental details on work in the infrared see: G. B. B. M. Sutherland, *Infrared and Raman Spectra*, Methuen, London, 1935; J. Lecomte, *La Spectra Infrarouge*, Presses Univ. France, Paris, 1928; C. Schaefer and F. Matossi, *Das Ultrarote Spektrum*, Springer, 1930; W. W. Coblentz, *loc. cit.* [141]; and R. B. Barnes, R. C. Gore, U. Liddel, and Van Zandt Williams, *Infrared Spectroscopy*, Reinhold, New York, 1944.

129. W. Kuhn, *Trans. Faraday Soc.*, **26**, 293 (1930).

130. T. M. Lowry, *Optical Rotatory Power*, Longmans, Green, London, 1935.

131. E. Brand, E. Washburn, B. F. Erlanger, E. Ellenbogen, J. Daniel, F. Lippmann, and M. Schen, *J. Am. Chem. Soc.*, **76**, 5037 (1954).

132. C. Djerassi, *Optical Rotatory Dispersion*, McGraw-Hill, New York, 1960.

133. P. Crabbé, this volume.

134. H. Rudolph, *J. Opt. Soc. Am.*, **45**, 50 (1955).

135. L. Ebert and G. Kortum, *Z. Physik. Chem.*, **B13**, 105 (1931).

136. E. J. B. Wiley, *J. Sci. Instr.*, **20**, 74 (1943).

137. D. H. Rank, J. H. Light, and P. R. Yoder, *J. Sci. Instr.*, **27**, 270 (1950).

138. G. Bruhat, *Ann. Phys.*, **3**, 246 (1915).

139. G. Bruhat and M. Pauthenier, *Rev. Opt.*, **6**, 163 (1927).

140. J. Kenyon, *Nature*, **117**, 304 (1926).

141. E. J. Gillham, *J. Sci. Instr.*, **34**, 435 (1957).

142. E. J. Gillham and R. J. King, *J. Sci. Instr.*, **38**, 21 (1961).

143. H. Cary, R. C. Hawes, P. B. Hooper, J. J. Duffield, and K. P. George, *Appl. Optics*, **3**, 329 (1964).

144. B. R. Malcolm and A. Elliott, *J. Sci. Instr.*, **34**, 48 (1957).

145. J. W. Gates, *Chem. Ind.*, **190** (1958).

146. T. Bürer, M. Kohler, and H. H. Günthard, *Helv. Chim. Acta.*, **41**, 2216 (1958).

147. M. Billardon and J. Badoz, *Compt. Rend.*, **248**, 2468 (1959).

148. H. Rudolph and R. Bruce, *J. Opt. Soc. Am.*, **49**, 1127 (1959).

149. O. C. Rudolph and Sons, as described by A. N. James and B. Sjoberg, in *Optical Rotatory Dispersion* by C. Djerassi, McGraw-Hill, New York, 1960, pp. 18–26.

150. Bodenseewerk Perkin-Elmer and Co. GMBH/Uberlingen.

151. Japan Spectroscopic Company, Ltd

152. G. Bruhat and A. Guinier, *Rev. Opt.*, **12**, 396 (1933).

153. E. J. Gillham, *Nature*, **178**, 1412 (1956).

154. H. Takasaki, *J. Opt. Soc. Am.*, **51**, 462 (1961).

155. A. L. Rouy and B. Carroll, *Anal. Chem.*, **33**, 594 (1961).

156. T. Bürer, *Proc. 4th Intern. Meeting Mol. Spectr.*, Bologna, 1959.

157. T. Bürer and H. Günthard, *Helv. Chim. Acta.*, **43**, 810 (1960).

158. A. L. Rouy and B. Carroll, *Anal. Chem.*, **38**, 1367 (1966).

159. R. J. King, *P.S.G. Bul.*, **16**, 487 (1965).

160. W. O. Bernhardt, *Proc. Am. Sugar Beet Technologists*, **5**, 547 (1948).

161. G. B. Levy, *Anal. Chem.*, **23**, 1089 (1951); G. B. Levy, P. Schwed, and D. Fergas, *Rev. Sci. Instr.*, **21**, 693 (1950); G. B. Levy, *Biochem. J.*, **57**, 50 (1954).

162. E. F. Daly, *Proc. 4th Intern. Meeting Mol. Spectr.*, Bologna, 1959.

163. G. Bruhat and P. Chatelain, *Rev. Opt.*, **12**, 1 (1933).

164. O. Schonrock and E. Einsporn, *Physik. Z.*, **37**, 1 (1936).

165. G. F. Landegren, *Rev. Sci. Instr.*, **26**, 502 (1955).

166. G. F. Landegren, *Rev. Sci. Instr.*, **26**, 578 (1955).

167. F. Woldbye and S. Bagger, *Act. Chem. Scand.*, **17**, 817 (1963).

168. A. L. Rouy and B. Carroll, *Anal. Chem.*, **37**, 96 (1965).

169. de La Provostaye and P. Desains, *Ann. Chim. Phys.*, **27**, 232 (1849); **10**, 267 (1850).

170. G. Todesco, *Nuovo Cimento*, **5**, 376 (1928).

171. E. Perucca, *Atti Reale Accad. Lincei, Classe Sci. Fis. Mat. Nat.*, **7**, 733 (1928).

172. E. Landt and H. Hirschmüller, *Deut. Zuckerind*, **62**, 647 (1937).

173. E. Landt, H. Hirschmüller, and W. Bechstein, German Pat., assigned to Schmidt and Haensch, No. 691,441, April 30, 1940.

174. F. Bernauer, *Fortschr. Mineral. Krist. Petrog.*, **19**, 22 (1935).

175. M. Haase, *Zeiss Nachr.*, **2**, 55 (1936).

176. B. Crumpler, W. H. Dyre, and A. Spell, *Anal. Chem.*, **27**, 1645 (1955).

177. P. M. Gallop, *Rev. Sci. Instr.*, **28**, 209 (1957).

178. B. Carroll, H. B. Tillam, and E. S. Freeman, *Anal. Chem.*, **30**, 1099 (1958).

179. L. D. Kahn, R. C. Calhoun, Jr., and L. P. Witnauer, *J. Appl. Polymer Sci.*, **8**, 439 (1964).

180. B. Carroll and T. J. Quigley, *J. Appl. Polymer Sci.*, **9**, 1905 (1963).

181. A. L. Rouy, B. Carroll, and T. J. Quigley, *Anal. Chem.*, **35**, 627 (1963).

182. P. Desains, *Compt. Rend.*, **62**, 1277 (1866).

183. P. Desains, *Compt. Rend.*, **84**, 1056 (1877).

184. W. B. Fordyce, J. Green, and A. C. Parker, *Proc. Biochem. Soc.*, **68**, 33 (1958).

185. A. Keston and J. Lospalluto, *Fed. Proc.*, **12**, 229 (1953).

186. M. J. Albinak, D. C. Bhatnager, S. Kirschner, and A. J. Sonnessa in *Advances in the Chemistry of the Coordination Compounds*, S. Kirschner, ed., Macmillan Co., New York, 1961.

187. A. Savitzky, W. Slavin, and R. E. Salinger, paper presented at Pittsburgh Conference on *Analytical Chemistry and Applied Spectroscopy*, March 3, 1959. Also, same authors, *Advances in Molecular Spectroscopy*, Mangine, Ed., Pergamon, New York, 1962, *Proc. 4th Intern. Meeting Mol. Spectr.*, Bologna, 1959.

188. H. von Halban and K. Siedentopf, German Patent No. 386,537 (1922).

189. K. Mayrhofer, Thesis, Univ. of Wurzburg (1924).

190. H. von Halban, *Nature*, **119**, 86 (1927).

191. H. von Halban and K. Siedentopf, *Z. Physik. Chem.*, **100**, 208 (1922).

191a. G. Bruhat, *Ann. Phys.*, **3**, 246 (1915).

192. G. Bruhat and M. Pauthenier, *Rev. Opt.*, **6**, 163 (1927). This paper contains, a valuable review of photographic methods in the ultraviolet.

193. A. Cotton and R. Descamps, *Compt. Rend.*, **182**, 22 (1926).

194. R. Descamps, *Rev. Opt.*, **5**, 481 (1926).

195. F. Lippich, *Sitzber. Akad. Wiss. Wien, Math.-Naturw. Klasse Abt. IIa*, **91**, 1070 (1885); **85**, 307 (1882).

196. A. Hussell, *Ann. Physik. Chem.*, **43**, 498 (1891).

197. R. S. Minor, *Ann. Physik. Chem.*, **10**, 581 (1903).

198. P. G. Nutting, *Phys. Rev.*, **17**, 1 (1903).

199. H. de Senarmont, *Ann. Chim. Phys.*, **28**, 279 (1850).

200. F. Savart, discussed by J. C. Poggendorf in *Ann. Physik. Chem.*, **49**, 292 (1800).

201. E. Perucca, *Nuovo Cimento*, **17**, 1 (1940).

202. E. Darmois, *Ann. Chim. Phys.*, **22**, 247, 495 (1911).

203. T. M. Lowry, *Proc. Roy. Soc. (London)*, **A81**, 472 (1908); **A212**, 261 (1912).

204. T. M. Lowry and W. R. C. Coode-Adams, *Proc. Roy. Soc. (London)*, **A226**, 391 (1927).

205. T. M. Lowry and M. A. Vernon, *Proc. Roy. Soc. (London)*, **A119**, 706 (1928).

206. T. M. Lowry and H. K. Gore, *Proc. Roy. Soc. (London)*, **A135**, 13 (1932).

207. O. Lehmann, *Ann. Physik.*, **2**, 649 (1900); **18**, 796 (1905).

208. F. Stumpf, *Ann. Physik.*, **37**, 351 (1912).

209. F. Stumpf, *Physik. Z.*, **11**, 780 (1910).

210. J. B. Biot and Melloni, *Compt. Rend.*, **2**, 194 (1836).

211. Wartmann, *Compt. Rend.*, **22**, 556, 745 (1846).

212. L. R. Ingersoll, *Phys. Rev.*, **23**, 489 (1906); *Phil. Mag.*, **11**, 41 (1906).

213. U. Meyer, *Ann. Physik.*, **30**, 607 (1909).

214. F. L. O. Wadsworth, *Phil. Mag.*, **38**, 337 (1894); *Astrophy. J.*, **1**, 232 (1895).

215. E. Carvallo, *Ann. Chim. Phys.*, **26**, 113 (1892).

216. See also G. Moreau, *Ann. Chim. Phys.*, **30**, 433 (1893).

217. P. Joubin, *Ann. Chim. Phys.*, **16**, 78 (1889).

218. T. M. Lowry and W. R. C. Coode-Adams, *Proc. Roy. Soc. (London)*, **A226**, 391 (1927).

219. T. M. Lowry and C. Snow, *Proc. Roy. Soc. (London)*, **A127**, 271 (1930).

220. R. Dongier, *Ann. Chim. Phys.*, **14**, 331 (1898); *J. Phys.*, **7**, 637 (1898); *Bull. Soc. Franc. Phys.*, **105**, 1 (1898).

221. H. S. Gutowski, *J. Chem. Phys.*, **19**, 438 (1951).

222. H. J. Hediger and H. H. Günthard, *Helv. Chim. Acta*, **37**, 1125 (1954).

223. K. F. Lindman, *Ann. Physik.*, **63**, 621 (1920); **69**, 270 (1922); **74**, 541 (1924); **77**, 337 (1925). See also B. Y. Oke, *Proc. Roy. Soc. (London)*, **A153**, 339 (1936).

224. R. Servant and P. Loudette, *Compt. Rend.*, **231**, 1052 (1950).

225. D. H. Brauns, *Rec. Trav. Chim.*, **69**, 1175 (1950).

226. R. Gans, *Ann. Physik.*, **79**, 547 (1926); **27**, 164 (1924).

227. L. K. Wolff and H. Volkmann, *Z. Physik. Chem.*, **B3**, 139 (1929); H. Volkmann, *ibid.*, **B10**, 161 (1930).

228. H. G. Rule and A. McLean, *J. Chem. Soc.*, **1400** (1932).

229. W. J. Kauzmann, J. E. Walter, and H. Eyring, *Chem. Revs.*, **26**, 339 (1940).

230. H. G. Rule and A. R. Chambers, *Nature*, **133**, 910 (1934); see also H. G. Rule, *J. Chem. Soc.*, **1931**, 674; H. Bernstein and H. Pedersen, *J. Chem. Phys.*, **17**, 885 (1949).

231. C. O. Beckmann and K. Cohen, *J. Chem. Phys.*, **4**, 784 (1936); **6**, 163 (1938).

232. N. Gutzwiller, *Helv. Phys. Acta*, **18**, 497 (1945).

233. E. U. Condon, *Rev. Mod. Phys.*, **9**, 432 (1937).

233*a*. C. Djerassi, *Proc. Chem. Soc.*, 314 (1964); P. Crabbé, *Optical Rotatory Dispersion and Circular Dichroism in Organic Chemistry*, Holden-Day, San Francisco, 1965.

234. A. Landolt, *Ann.*, **189**, 241 (1877).

235. C. Winther, *Z. Physik. Chem.*, **60**, 563 (1907).

236. Calculated from data of A. Landolt [234].

237. J. B. Biot, *Mém. Acad. Sci.*, **15**, 280 (1838).

238. H. Streuli, *Mitt. Gebiete Lebensmi, u. Hyg.*, **42**, 79 (1951).

239. The nicotine data by Landolt and Winther differ (compare Figs. 2.71 and 2.72), indicating that the compounds differed. This does not affect the conclusions drawn here.

240. Landolt's $[\alpha_{100}]$-values for camphor are the only ones to vary widely with the solvent which is undoubtedly due to the very insecure extrapolation from $p \sim 50$–60% to 100%.

241. J. G. Kirkwood, *J. Chem. Phys.*, **5**, 487 (1937).

242. N. M. Baghenov and M. V. Volkenshtein, *Zh. Fiz. Khim.*, **28**, 1299 (1954).

243. M. V. Volkenshtein, *Uspekhi Khim.*, **13**, 234 (1944).

244. F. W. Cagle, Jr., and H. Eyring, *J. Am. Chem. Soc.*, **73**, 5628 (1951).

245. E. Darmois, *J. Phys.*, **5**, 225 (1924); *Rev. Gén. Sci.*, **23**, 670 (1922); *Compt. Rend.*, **147**, 195 (1908).

General References

POLARIMETRY. ROTATORY POWER IN GENERAL

Born, M., and E. Wolf, *Principles of Optics*, Pergamon, New York, 1959.

Brode, W. R., "Optical Rotation of Polarized Light by Chemical Compounds," *J. Opt. Soc. Am.*, **41**, 987 (1951).

Bruhat, G., *Traité de Polarimétrie*. Editions de la revue d'optique théorique et instrumentale, Paris, 1930.

Eyring, H., J. Walter, and G. E. Kimball, *Quantum Chemistry*, Wiley, New York, 1944.

Kauzmann, W., *Quantum Chemistry*, Academic, New York, 1957.

Lowry, T. M., *Optical Rotatory Power*, Longmans, Green, London, 1935; Dover, New York, 1964.

Mathieu, J. P., *Activité Optique Naturelle*, Handbuch der Physik Band XXVIII, Spektroskopie II, S. Flügge, ed., Springer-Verlag, Berlin, 1957.

Stone, J. M., *Radiation and Optics*, McGraw-Hill, New York, 1963.

ROTATORY POWER AND CHEMICAL CONSTITUTION

Balfe, M. P., "Optical Activity, Its Study, Terminology and Uses," *Sci. Progr.*, **38**, 459 (1950).

Bush, C. A., and J. Brahms, "Optical Activity in Single-Strand Oligonucleotides," *J. Chem. Phys.*, **46**, 79 (1967).

Condon, E. U., "Theories of Optical Rotatory Power," *Rev. Mod. Phys.*, **9**, 432 (1937).

Condon, E. U., W. Altar, and H. Eyring, "One Electron Rotatory Power," *J. Chem. Phys.*, **5**, 753 (1937).

Crabbé, P., *Optical Rotatory Dispersion and Circular Dichroism in Organic Chemistry*, Holden-Day, San Francisco, 1965.

Djerassi, C., *Optical Rotatory Dispersion*, McGraw-Hill, New York, 1960.

Freudenberg, K., *Stereochemie*, Deuticke, Leipzig (1933), reproduced by Edwards Bros., Ann Arbor, 1945, also, *Monatsch.*, **85**, 537 (1954).

Jirgensons, B., ed., *Optical Rotatory Dispersion of Proteins and Other Macromolecules*, Springer-Verlag, New York, 1969.

Kauzmann, W. J., J. E. Walter, and H. Eyring, "Theories of Optical Rotatory Power," *Chem. Rev.*, **26**, 339 (1940).

Kirkwood, J. G., "On the Theory of Optical Rotatory Power," *J. Chem. Phys.*, **5**, 479 (1937).

Klyne, W., "Optical Rotation," in *Determination of Organic Structures by Physical Methods*, E. A. Braude and F. C. Nachod, Eds., Academic, New York, 1955.

Kuhn, W., "Optical Rotatory Power," *Ann. Rev. Phys. Chem.*, **9**, 417 (1958).

Kuhn, W., and K. Freudenberg, "Näturlich Drehung der Polarisationsebene," in *Hand-und-Jahrbuch der Chemischen Physik.*, A. Eucken and K. L. Wolf, Eds., Vol. 8, Part III, Akadem. Verlagsgesellschaft, Leipzig, 1936.

Liehr, A. D., "Interaction of Electromagnetic Radiation with Matter. Part I. Theory of Optical Rotatory Power," *J. Phys. Chem.*, **68**, 3629 (1964).

Mason, S. F., "Optical Rotatory Power," *Quart. Rev. (London)*, **17**, 20 (1963).

Mathieu, J. P., "Les Théories Molécularies du Pouvoir Rotatorie Naturel," Centre national de la recherche scientifique, Paris, 1946.

Moffitt, W., "Optical Rotatory Dispersion of Helical Polymers," *J. Chem. Phys.*, **25**, 467 (1956).

Moffitt, W., D. D. Fitts, and J. G. Kirkwood, "Critique of the Theory of Optical Activity of Helical Polymers," *Proc. Natl. Acad. Sci. (U.S.)*, **43**, 723 (1957).

Moffitt, W., and A. Moscowitz, "Optical Activity in Absorbing Media," *J. Chem. Phys.*, **30**, 648 (1959).

Moffitt, W., R. B. Woodward, A. Moscowitz, W. Klyne, and C. Djerassi, "Structure and Optical Rotatory Dispersion of Saturated Ketones," *J. Am. Chem. Soc.*, **84**, 1945 (1962).

Rosenfeld, L., "Quantum Mechanical Theory of Natural Optical Activity of Liquids and Gases," *Z. Physik*, **52**, 161 (1928).

Snatzke, G., ed., *Optical Rotatory Dispersion and Circular Dichroism in Organic Chemistry*, Sadler, Philadelphia, 1967.

Tinoco, I., "Theoretical Aspects of Optical Activity," *Adv. Chem. Phys.*, **4**, 113 (1962).

Velluz, L., M. Legrand, and M. Grosjean, *Optical Circular Dichroism*, Academic, New York, 1965.

Chapter **III**

OPTICAL ROTATORY DISPERSION AND CIRCULAR DICHROISM

Pierre Crabbé and A. C. Parker

---

1 INTRODUCTION

A number of important breakthroughs in the development of sophisticated instrumentation for the investigation of the optical properties of optically active compounds have radically altered the nature and scope of optical rotatory dispersion and circular dichroism methods. Further advances in this field can be expected in three major directions: technologically for the improvement of existing instruments; experimentally in the examination of new chromophoric groupings and functions; and theoretically in the interpretation of the observed phenomena. Optical rotatory dispersion and circular dichroism have become the basis of widespread techniques for the study of chiral molecules. In many instances the information provided by these methods is as valuable as that from infrared or ultraviolet techniques. They do not give the wealth of data as, for example, nuclear magnetic resonance spectroscopy gives, but the stereochemical information given by rotatory dispersion and circular dichroism usually cannot be obtained easily by any other method.

Since several books, chapters, and review articles have appeared on various aspects of this topic, as well as a recent thorough survey of the applications in organic chemistry, this chapter will follow a different approach than this literature. The first section will be devoted to the instrumentation now available for optical rotatory dispersion (RD) and circular dichroism (CD) measurements. In the second section, the definitions commonly used in the RD and CD will be reviewed without any theoretical discussion, because this aspect is treated in detail in another chapter of this book. The influence of the nature of the solvent and the effect of temperature on RD and CD curves have been recognized as important factors that will be dealt with in a special section. Finally, there will be a review of the main functions and chromophoric groupings which can presently be examined by RD and CD, either for structural or stereochemical assignments. In each section, some examples, as well as leading references will be given in order to provide the reader with information about the scope of the methods. These sections are kept as brief and as concise as possible.

2 INSTRUMENTATION FOR SPECTROPOLARIMETRY

History and General Principles

This article covers both optical rotatory dispersion (ORD) and circular dichroism (CD) measurement, particularly for compounds that exhibit natural optical activity; when the activity is induced by an applied magnetic field (MORD and MCD), similar apparatus could be used, but with the addition of the means of applying that field. This, however, is the subject of a separate article in this volume.

All these studies are concerned with the variation of the particular chosen property with the wavelength (λ) of the exciting light. Therefore, it follows that the required measurements can be made by using a monochromator to provide light of the desired wavelength, followed by a suitable instrument— for example, a polarimeter when the studies are of ORD—to measure the required effect. Most instruments are just this combination. This order of instrument should not be reversed, even should the optical design so allow, because the greater intensity of the undispersed radiation vastly increases the risk of sample decomposition.

An ingenious apparatus used to measure both effects was described by Cotton in 1896 [1]. Because the only practical " detector " then available was the human eye, and the white light sources of the era were primitive, the limitations of such apparatus are obvious. In the last two decades the advent of suitable light sources and sensitive photomultipliers has enabled great strides to be made, though in the interim years following Cotton's paper descriptions of instruments continued to appear (e.g. [2, 3]), while Scherer in 1932 published the first paper on MCD [4]. Photographic means of detection also came to be used in this period so that measurements could be made beyond the visible spectrum.

The successful advent of the Rudolph photoelectric polarimeter in 1955 [5] marked the beginning of a new era. In conjunction with a monochromator and zirconium arc (later a xenon arc) it was exploited with great success by

Djerassi in a historic series of papers beginning in 1955 [6] followed by his monograph on the subject [7]. Other workers used similar arrangements of apparatus, perhaps with different monochromators (see, for example, Klyne [8]).

At about that time several people began considering the possibility of assembling automatic recording spectropolarimeters [9a, b, c]. These should match the performance and ease of use of contemporary recording absorption spectrophotometers. Several different systems were evolved and some of these are now available as complete manufactured instruments. Parallel developments took place in the CD field resulting in instruments, or adaptations for instruments, being marketed. A most comprehensive review of spectropolarimeters and instruments for measuring CD was presented by Woldbye at the NATO Summer School held at Bonn in September 1965 [10]. The last few years have seen the further development of some of these instruments for MORD and MCD measurements by adding the means of applying a strong magnetic field to the sample.

Such evolution requires a change of viewpoint compared with our chapter in the previous edition, which was primarily concerned to show how suitable optical elements could be assembled by the reader himself for ORD measurements, although the commercial instruments at that time just becoming available were mentioned. In the present article the optical elements required for both fields are first discussed together; then follow the principles of their assembly into complete instruments. Practical considerations regarding such things as cells and samples are also treated. Finally, the instruments now available are reviewed with short descriptions.

Optical Components

Polarizers

Reflection polarizers, birefringent polarizers, and dichroic polarizers have all been used for the ultraviolet (U.V.) spectral region, though reflection polarizers are of limited application because the polarizing angle varies with the wavelength of the incident light, and the deviated incident beam is usually inconvenient. Also there is considerable attenuation of the beam. In addition to these, piles of optically transmitting plates have been used as polarizers, mainly in the infrared [11–13] but a pile of lithium fluoride plates has been employed in the vacuum U.V. [14]. Dichroic polarizers, often known as "Polaroid," are now widely used for visible light and are extensively discussed by Shurcliffe [15]. They are now available for wavelengths down to about 270 nm [16], and it has been reported that specially prepared films of azo dyes can be useful as low as 215 nm [17]. If the above wavelength limitations are not important, then such sheet polarizers lend themselves to simple spectropolarimeters.

Birefringent polarizers are the most suited to U.V. use, and so have been the most extensively employed. Of the dozen or more types, the most useful in the U.V. are the three double-image prisms of Rochon, Wollaston, and Senarmont, the Glazebrook, and the Taylor modification [18] of the Glan Thompson prisms. These are illustrated in Fig. 3.1a where the optic axes

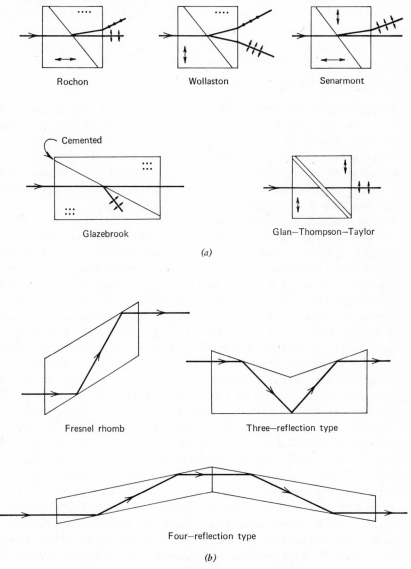

Fig. 3.1. (a) Polarizing prisms; (b) phase retarders.

are represented by dots and arrows where they are normal to and in the plane of the paper, respectively. Other polarizing prisms are discussed by W. Heller elsewhere in this volume. Naturally the materials from which the polarizers are constructed must be transparent to U.V. light, and so, where an air film is not required the components must either be assembled in optical contact, or a thin film of transparent contact fluid (e.g., "Nujol") must be used. The refractive index of the fluid should match that of the crystalline material used, so far as possible. Canada balsam does not have good transmission in the U.V. Crystalline quartz, calcite, and ADP (ammonium dihydrogen phosphate) are the most commonly employed materials.

Modulators for the Half-Shade Method

Visual polarimeters using the half-shadow device long ago superseded the method whereby a simple analyzer and polarizer, with the sample between them, were adjusted for extinction. It was found that the sensitivity of such a method—where two adjacent fields, offset each side of the null point, were adjusted for equal brightness—was for several reasons much superior to an adjustment for extinction, when the signal disappears at the point of measurement, just when it would be most useful to have it large.

For similar reasons, some version of the half-shade method is used in spectropolarimeters, but the choice of the specific half-shadow angle will be dictated by different considerations. If the light intensity is very weak, then 45° each side of the extinction position will give the most favorable sensitivity, but an examination of all the factors suggests the employment of a much smaller angle [19]. For a photoelectric instrument, the two beams can be separated in "time" or in "space." The former method possesses the advantage of requiring only a single detector; thus, the matching of photocell response over a wide range of spectrum and the overall stability requirements are greatly simplified. Even so, if the spectropolarimeter is to have a flat baseline over all the spectrum, then at any given wavelength the detector sensitivity must be the same for light in different states of polarization. This requirement is usually only partly satisfied, and some modern instruments incorporate means for obtaining a substantially flat baseline by electrical correction potentiometers, each covering only a short wavelength interval, similar to the method used in some spectrophotometers. At the same time, compensation can be arranged for baseline errors from other causes, for example, cuvette windows.

If the half-shadow method is to be used, means must be found to vary the azimuth of the polarized beam. Where polarizers of the double-image prism type are used, two mutually perpendicularly polarized beams of light are produced, which in themselves can be employed at a half-shadow angle of 45°, using a space separation method [20, 21]. The basic space separation can be turned into one of time separation if shutters, condensing optics, or things of this kind, are arranged so that both beams are imaged on the one detector.

Other devices produce changes in the state of polarization of the one beam. They include oscillating polarizers [5, 22], rotating levo- and dextrorotatory plates [19], magneto-optical rotation (Faraday effect) [23], and electro-optical rotation (Pockel effect) [24]. Oscillating polarizers, moved by hand or mechanically at moderate frequency, pose difficulties in their precision of mounting, but the rotating plates are perhaps optically more difficult to construct. If a dichroic polarizer is spun over and over in the beam on a mechanical axis at 45° to direction of orientation of its dichroic material, a modulation between +45° and −45° can be achieved. Still further methods have been devised where two distinct polarizers, set to different azimuths, are inserted in two already separated beams [25, 26].

The Faraday effect is probably the most extensively used method, in the form of a "Faraday cell" consisting of a solenoid surrounding a cylinder of glass or quartz, or a cuvette of distilled water. The beam travels through this parallel to the axis of the solenoid. Heating caused by the electrical power required to swing the beam azimuth the necessary amount is the principal design difficulty that has to be overcome.

The Pockel effect, using a crystalline plate, usually of ADP, has not found much use in spectropolarimeters, but comes into its own in the measurement of CD. Tagasaki [27] has described the use of two such modulators to extract both parameters of elliptically polarized light.

Quarter Wave Retarders

Simple quarter wave plates consisting of thin sections of uni- or biaxial crystals are suitable only for monochromatic radiation. Adjustable retarders (e.g., the Babinet-Soleil compensator) are sometimes used to cover a wider spectral range, but they are difficult to manufacture, have limited angular aperture, and multiple reflections can cause inaccuracies. The phase retarders most used are therefore either of the total internal reflection type (e.g., Fresnel rhomb) or those employing the electro-optical Pockel effect [24].

Types of retarder employing total internal reflection are illustrated in Fig. 3.1b. The Fresnel rhomb functions by combining two retardations of 45° at each internal reflection, and a rhomb of hard crown glass ($n = 1.5217$) varies in retardation by less than 2° over the visible spectrum [28]. For a single reflection the phase retardation is given by

$$\tan \tfrac{1}{2}\delta = \frac{\cos \theta (n^2 \sin^2 \theta - 1)^{1/2}}{n \sin^2 \theta}$$

and thus 45° retardation is obtained at angles of incidence of about 41° and 55°, the latter value being usually employed as here δ is less steeply dependent on θ. Residual strain in the rhomb material can affect the phase change, as can surface films caused in the polishing process. Ditchburn and Orchard reported errors of at least $4°$ due to this [29]. Silica rhombs are not as achromatic as the glass ones, but achromatism can be achieved by hard-coating the reflection surfaces with magnesium fluoride. Unfortunately, the constant retardation then becomes $82°$. An achromatic $90°$ is obtainable by using two films, one of high and one of low refractive index. The Fresnel rhomb suffers from the disadvantage of a parallel displacement of the beam, which then requires equal restoration to its path.

Another system, in effect two silica rhombs in series, with four reflections each of about $74°$, has been described by Kizel, Krasilov, and Shamraev [30]. It has good achromatism from 200 to 600 nm and is insensitive to small changes in angle of incidence. But because of the high angles of reflection the overall length must be 175 mm for a beam diameter of 10 mm, and so the usual residual birefringence could result in several degrees of error in the retardation. It is therefore unlikely that the inherent accuracy could be currently realized.

Pockel-effect retarders depend on the fact that when a Z-cut crystal of material, such as ADP, is subjected to an electric field in the direction of the natural optic axis (Z-direction), polarized light traveling along this axis will suffer retardation, the induced birefringence depending on the wavelength and the applied voltage V_z:

$$(n'_x - n'_y) = n_o^3 r_{eo} \frac{V_z}{d}$$

retardation

$$\delta_o = (n'_x - n'_y) \frac{2\pi d}{\lambda}$$

$$= \left(\frac{2\pi}{\lambda}\right) n_o^3 r_{eo} V_z$$

At a given wavelength, the retardation is directly proportional to the applied voltage and is independent of the thickness, d. In the above expression, the n's are refractive indices, and r_{eo} is the electro-optic coefficient.

Difficulty can be experienced in applying suitable electrodes to the plate, because obviously light must traverse them. The use of metallic grids or rings will reduce the effective aperture, while the nonuniform field between the spacings results in similar variations in retardation across the aperture. On the other hand thin films, such as evaporated gold, cause even greater transmission losses. Substantial errors in retardation can also be caused by an obliquity of the incident rays, even with rays having only 2° difference from the central ray incident angle [31].

Because of the possibility of using an alternating applied voltage the Pockel effect retarder can also be used as a modulator to produce a swing from right to left circularly polarized light.

Monochromators and Light Sources

These topics are covered in detail in articles elsewhere in this book, but it is desirable to call attention to certain requirements particularly important to polarimetric measurements using them.

The importance of lamps of higher intensity than usually adopted for spectrophotometry was early appreciated; zirconium arcs were used before the now popular xenon arcs became available. The lowest wavelength accessible was then about 290 nm. Early xenon arcs were notoriously unstable, particularly in the burning position of the arc. The importance of the positional stability and its influence on the polarimeter output due to changes in the consistency of the illumination at the various apertures of the instrument is discussed by Cary et al. [32]. In the case of the Osram lamp type XBO 501, these fluctuations have been investigated by Budde [33], but Wasserman [34] claims that the Hanovia type 901C lamp with an especially short arc (1.4 mm) shows little fluctuation.

Two methods that alleviate this problem considerably have been used by King and Gillam [35] and by this author. King and Gillam use a mirror in the slit-illumination system which is movable in the perpendicular and horizontal planes by small electromagnets, themselves energized by an amplified signal derived from four small photocells on which falls a subsidiary image of the arc. Intensity variations between these photocells control the mirror in such a way as to return the intensities to normal. This author has used (in one plane only) a similar arrangement of a split-cathode photocell, with the amplifier feeding an electromagnet acting on the arc directly. Since the arc itself is a flow of current, it is deflected in a direction mutually perpendicular to the magnetic and electric fields, restoring the illumination of the photocell by the subsidiary image to normal.

One spectropolarimeter [9b] has as its light source a high-pressure mercury arc. This gives a line spectrum superimposed on a strong continuum, but the lines apparently do not interfere with the functioning of the instrument.

Not only do these arc lamps have many times the U.V. output of the hydrogen or deuterium lamps used in spectrophotometers, but also their long wavelength output is disproportionately larger. Because rotation measurements in the U.V. are usually made in the vicinity of absorption bands, perhaps allowing a transmission of less than 1 %, it is therefore necessary to ensure that the stray light level of the monochromator is extremely low. The effect of stray light on accuracy is discussed by Jirgensons [36]. In practice, this means that a double monochromator has considerable advantage over the single one, which should almost invariably be used with a filter limiting the entry of the longer wavelengths into the measuring system. Prism instruments, because of their larger dispersion in the U.V., possess a certain advantage over gratings, because it is easier to obtain a more constant energy throughout the spectrum if the slits are programmed, without at the same time producing too large a variation in the spectral bandwidth. It is unnecessary to have instruments of large aperture because most of the polarimeter optical systems will accept beams of only limited angular aperture.

Sample Cuvettes

At first it would appear that any cell with fused silica windows should serve for polarimetry in the U.V., but certain practical considerations might restrict the choice of cell construction. If a cell contained a small optical rotation in its windows due to imperfections in the silica, then simply mounting it in such a way, permanently or kinematically, that it is always exactly in the same position in the beam would overcome the problem, as only a constant shift of the baseline would result. However, in a study of the effect of sample cell windows in precision polarimetry, King [37] has shown that errors over 0.01° can be produced by surface films, particularly if the surface is rubbed in one direction only. Thus it would appear advisable to run solvent baselines and then the sample successively, avoiding any intermediate handling of the cell.

Cells with fused-on end windows often exhibit considerable birefringence, while demountable cells can do the same due to strain produced when tightening. At the same time, cells with cemented-on windows will not suit all cases, because of limits they impose on the solvents that may be used. These points regarding strain assume even greater importance when low temperature measurements are attempted.

The effect of residual birefringence in the cells is magnified by any additional birefringence in the rest of the optical components, and these combined effects have also been examined by King [37]. In a well constructed polarimeter cell, the retardation produced by the windows should not exceed about 0.2°, and in a perfect polarimeter a variation of the baseline of a few millidegrees would be noticed. But King shows that the effect of two separate birefringences is multiplicative and errors of the order of 0.01° can easily arise in measuring the angle of the major axis of the elliptically polarized light. Even with a high-quality cell, the presence of instrumental birefringence results in the balance point varying with the cell orientation. With cell and instrument birefringences of 0.2° and 1.0°, respectively, the peak-to-peak variation is 3.5 millidegrees.

Detectors

Radiation detectors are more fully treated in the appropriate sections of this volume. For the visible and U.V. spectral regions, little further comment is required, as the photomultiplier will almost certainly be chosen. High photoelectric efficiencies are now being obtained with these tubes, and because of their own high amplifications, Schott noise in the photo current itself, due to the random nature of the arrival of light quanta and subsequent electron emission, imposes the ultimate limit of sensitivity available with practicable measuring times. Care should be exercised if the polarimeter system incorporates automatic gain control which varies the dynode voltages near the cathode, as in some tubes this causes alteration in the relative sensitivities to beams of different polarizations, but otherwise this is a very convenient method.

Though some photoemissive surfaces work a short way into the infrared (I.R.), it is most probable that any infrared spectropolarimetry will be done with the detectors now common in I.R. spectrometers: thermocouples, bolometers, Golay cells, and things of this kind. Rapid strides have been made in the development of photoconductive cells for use out to moderate wavelengths. Ultimate sensitivity of all these types of detector is limited by thermal (Johnson) noise and is independent, or nearly so, of the intensity of incident radiation. Similar considerations regarding their performance with respect to the state of polarization of the received radiation apply, of course.

Measurement of Rotation and Ellipticity

Basic Design Principles

An examination of published spectropolarimetric work yields the criteria that determine the design of a suitable instrument, but it is also necessary to derive an expression for the sample optical density required for optimum performance. That such an optimum exists will be apparent from the fact that the absorption of light by a substance increases exponentially, whereas the rotation of the plane of polarization varies only linearly with concentration. Because of the random emission of electrons by the detector caused by the random arrival of the light quanta, the mean fluctuation in the signal current (assuming no disturbance from other sources) increases as the square root of the current itself. If we assume that the detector is receiving a light signal in some manner proportional to the rotation of the sample, the expression for signal-to-noise ratio (S) must be

$$S = \alpha I_o^{1/2} c 10^{-cA/2} \tag{3.1}$$

where α = rotation of 1% solution, A = absorbance of 1% solution, I_o = intensity of incident light, c = concentration of solution (g/100 ml). By differentiation, S is a maximum when $cA = 2/\log_e 10 = 0.85$, and this is the absorbance of the sample of optimum concentration c.

In practice it is not possible to work continuously at this optimum, as the absorbance will vary with wavelength; nevertheless the expression is useful in arriving at a figure for the rotations that a successful spectropolarimeter must be able to measure. Determinations of rotatory dispersion curves made with samples having twice this optimum absorbance yield satisfactory results with our own instrument.

In the specific instance of cholest-4-en-3-one a solution with an absorbance maximum of about 1.5 in the ketone-absorption region would have a rotation maximum of about 0.5°. Though we cannot expect to cope with all values in any one design, it will be apparent that a recording instrument covering ranges upward from $\pm 0.05°$ for full-scale deflection, or a manual instrument capable of measuring 0.001°, will enable rotation measurements to be made somewhere near the optimum absorbance of most substances, except those that are very weakly rotating. For such substances measurements would have to be made at a higher concentration, or in a larger cell, and the precision of the measurements would be reduced owing to the increased noise level.

The precision of any measurements will also be influenced, through the signal/noise ratio, by the intensity of the incident light available for measurement; a tenfold increase in the brilliance of the light source illuminating the monochromator would permit measurements of the same precision with samples exhibiting a unit increase of absorbance. However, an intense light source is desirable for other and more important reasons.

Most photoelectric methods of measuring optical rotation rely on some variant of the half-shade method. Greatest sensitivity is obtained when the angle between the transmitted planes of vibration of the polarizer and analyzer is 45° (see, for example, the chapter by Heller in this volume). For a change of rotation of 0.001°, for example, which is the minimum detectable change in rotation desirable for spectropolarimetry, the intensity of the two light signals will change by 0.0035%.

Spectropolarimetry therefore requires about a hundredfold increase in the sensitivity of detection over that necessary for ordinary spectrophotometry, and this can be realized partly by replacing the usual hydrogen arc with a more intense source, and partly by operating with the largest monochromator slits consistent with the required resolution. There is also a tenfold loss of light in a sample whose absorbance is unity, and the polarimeter reduces the light still further (e.g., the polarizer itself transmits only 50% of the incident light). It is not possible with any known light source to obtain this increase of 1000 ($100 \times$ for sensitivity and $10 \times$ for sample); it is therefore necessary to increase the time required for measurement (the "time constant" or "damping" of a recording instrument), thus "integrating" the readings over a longer period.

The electrical signal produced by the polarimeter contains the information required to arrive at the optical rotation, that is, the orientation of the major axis of the elliptically polarized light leaving the sample. This can be used in two ways. The rotation can be extracted directly from the signal, or the signal can be used in some way to restore balance to the polarimeter. This latter can be done by means of a servomotor driving the polarizer or analyzer, or more conveniently by passing a current derived from the signal through another Faraday cell. This involves a very high degree of amplification because in this system of measurement a small residual out-of-balance signal is necessary to provide the balancing current. The residual error is thus inversely proportional to the gain. This method is successfully employed in some of the commercial instruments now on the market, and examples of both types are more fully discussed below.

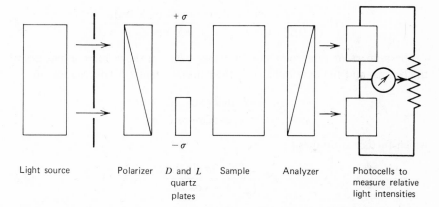

Light source   Polarizer D and L Sample   Analyzer   Photocells to
                                   quartz   measure relative
                                   plates   light intensities

Fig. 3.2. Illustration of the energy-measurement method of polarimetry. (After A. S. Keston, "Improvements in or relating to polarimetric apparatus," Brit. Pat. no. 715057, 1954.)

Direct use of the signal is attractive in that no further optical parts are required, or that no moving optical parts, with the associated problems of accuracy of setting, are called for. Its disadvantage is that, not being a null-balance system, it is sensitive to variations in the overall gain or output of the elements. The rotation can be derived as follows (see Fig. 3.2).

If the angle between the planes of polarization of the two beams is $2\sigma°$, it can be shown that the result of interposing a sample with a rotation $\Delta\alpha$ is to produce two beams of relative intensity, $I\cos^2(\sigma + \Delta\alpha)$ and $I\cos^2(\sigma - \Delta\alpha)$, where the initial intensity of each beam is $I_o \cos^2 \sigma$. I_o/I is the ratio of light received to light transmitted by the sample.

It is necessary to eliminate the intensity values I_o and I in any system for obtaining the rotation in terms of the energy of the two beams. The simplest method is to measure their ratio

$$R = \frac{\cos^2(\sigma + \Delta\alpha)}{\cos^2(\sigma - \Delta\alpha)} \tag{3.2}$$

$\Delta\alpha$ is then derived from the equation

$$\tan \Delta\alpha = \frac{(1 - \sqrt{R})(\cot \sigma)}{1 + \sqrt{R}} \tag{3.3}$$

A simpler equation relating $\Delta\alpha$ to the intensities of the beams can be derived by taking the ratio

$$R^* = \frac{I \cos^2(\sigma - \Delta\alpha) - I \cos^2(\sigma + \Delta\alpha)}{0.5I \cos^2(\sigma - \Delta\alpha) + 0.5I \cos^2(\sigma + \Delta\alpha)} \tag{3.4}$$

The numerator represents the difference between the relative intensities of the beams, and the denominator is their mean intensity. This reduces to

$$R^* = \frac{2 \sin 2\sigma \sin 2 \Delta\alpha}{1 + \cos 2\sigma \cos 2 \Delta\alpha} \tag{3.5}$$

which further simplifies to

$$R^* = \frac{2 \sin 2\sigma \sin 2 \Delta\alpha}{1 + \cos 2\sigma} \tag{3.6}$$

where $\Delta\alpha$ is small, from which

$$\sin 2 \Delta\alpha = \frac{R^*(1 + \cos 2\sigma)}{2 \sin 2\sigma} \tag{3.7}$$

In the special case when $\sigma = 45°$, an even greater simplification is possible.

$$\sin 2 \Delta\alpha = \frac{R^*}{2} \tag{3.8}$$

For small values of angle the sine varies linearly with the angle itself. Thus the ratio R^* is a linear function of the rotation.

Many workers, before the advent of commercial instruments, used variations of these methods (e.g. [20, 25, 26, 38]).

Rouy and Carroll [42, 43] have dealt comprehensively with all the methods of extracting the optical rotation from the signal by intensity measurements, and they have also discussed the recovery of the CD information by similar methods.

To measure CD, attachments that produce circularly polarized light can be inserted into the two beams of a recording spectrophotometer, which then measures the relative beam intensities in the normal way. One such device is described by Woldbye and Bagger [43] and is illustrated in Fig. 3.3. Light passing through a linear polarizer is circularly polarized by the Fresnel rhomb, and after passing through the sample cell is returned to its original path by the mirrors. A pair of these is needed, with the rhombs set to give right or left circularly polarized light. The difference in absorption of the two oppositely polarized beams is then measured by conventional use of the recording system.

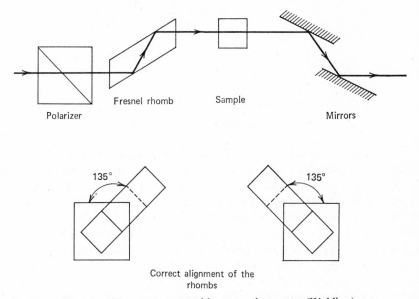

Fig. 3.3. CD measurement with spectrophotometer (Woldbye).

Circular dichroism attachments are now also offered by several manufacturers to use with their spectropolarimeters, using a quarter wave retarder. The method is ellipsometric, being based on analyzing the elliptically polarized light produced when plane-polarized light traverses an optically active sample. It can be shown that when a quarter wave retarder is introduced into the optical path of a spectropolarimeter, with the optical elements in the order illustrated in Fig. 3.4, then the output rotation recorded by the instrument is a direct measure of the beam ellipticity, whence the circular dichroism. King [44] gives the following expression:

$$\theta = 33.0 \ cd(\epsilon_l - \epsilon_r)\cos \delta - \phi \sin \delta$$

c and d being the concentration and path length of the sample, ϵ_l and ϵ_r are the molecular extinction coefficients of left and right circularly polarized light, and $90 + \delta$ is the actual retardation of the nominally 90° retarder. Therefore, if δ is not zero, the term $\phi \sin \delta$ appears, proportional to the rotation. Thus there is the necessity for achromatic, 90° retarding rhombs, discussed earlier.

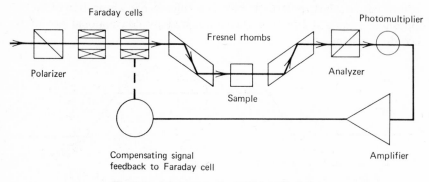

Fig. 3.4.   Measurement of CD by ellipticity.

An ingenious CD attachment making use of a relatively thick retardation plate was described by Holzwarth, Gratzer and Doty [45] and has been marketed by Rehoveth Instruments Ltd., to fit Cary 11, 14, and 15 spectrophotometers. The plate is of such a thickness that the retardation changes continuously from + to −90° each 8 nm. A recording of absorption made with this plate fitted therefore traces a rapidly alternating curve, with the CD appearing as the envelope illustrated in Fig. 3.5. The limited resolution of this type of recording may not always be acceptable.

Direct measurement of the CD can be made with automatic instruments using Pockel-effect modulators. The relative transmissions of the right and left circularly polarized light through the sample can then be compared as required. Complete instruments using this principle are described below.

The accuracy of all measurements of optical activity can be affected by many causes. In addition to stray light and residual birefringences, which have been mentioned already, there are such causes as sample scattering and depolarization at optical interfaces. High optical densities can cause errors due to increase in the noise levels, as well as the stray light effect. It is impossible to generalize on sources of error, but Rouy and Carroll [46] have

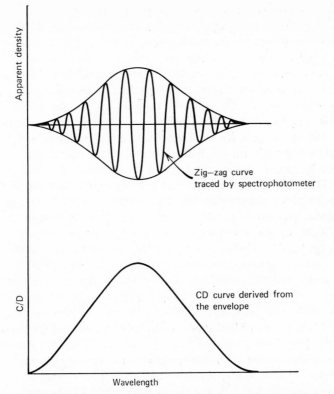

Fig. 3.5. Use of multiple quarter-wave plate to record CD.

discussed some interesting cases. Reference to the original papers describing the design of instruments will sometimes indicate within what limits measurements can be relied on.

INFRARED SPECTROPOLARIMETRY

Early work in this field parallelled that in the U.V. The history of infrared (I.R.) polarimetry is well documented by Heller [47]. Most early measurements were confined to the spectral range below 2.7 μ, where quartz optics could be used, but Meyer [48] used reflection polarizers to make measurements up to 8.8 μ. Lowry and others [49, 50] improved these early methods which, however, are now of mainly historical interest. Further progress awaited the appearance of modern I.R. spectrometers. Then both Gutowski [51] and Hediger and Günthard [52] combined polarimeters with Perkin-Elmer spectrometers to produce the first "modern" I.R. spectropolarimeters. Gutowski used the wavelength range 2.2 to 9.7 μ for measurements on quartz, while Hediger and Günthard, with organic solutions, used 0.75 to 4 μ.

Both these instruments had Se film polarizing elements. The precision obtained was about 0.1°, inferior to that of contemporary U.V. apparatus. Quite recently, Wyss and Günthard have described an I.R. spectropolarimeter [53] with about seven times this precision, and this instrument has been chosen as an example for description in the next section. The factors that control the precision are discussed in the paper by these authors.

In principle, any of the methods used at the shorter wavelengths are applicable equally to the I.R. if suitably transparent optical materials are chosen, though in practice such materials might not be suitable for any extended wavelength range. Several new polarimeter designs have been prompted by recent work on the Faraday effect in the I.R. Although MORD and MCD are treated in separate articles herein, brief mention of the polarimetric principles is appropriate here because they are in essence applicable to nonmagnetic measurements. At the same time, they illustrate the application in the I.R. of the above methods of photoelectric polarimetry.

As in the U.V., Faraday cell modulators are quite commonly used. For example, in addition to the above three references, Robinson [54] has described an instrument using a modulator of Schott SF 6 glass up to a wavelength of 1.9 μ. A cooled PbS detector enabled measurement of the rotation, by manually balancing the analyzer, with an accuracy of 3 to 6 millidegrees.

Instead of this type of half-shade modulation, Craig et al. [55] and Pidgeon and Smith [56] both employed analyzers set at 45° to the polarized light incident on the sample, and then measured the intensity changes caused by the presence of the sample. Craig et al., used a $\frac{1}{2}$mW He-Ne maser, itself giving polarized light of 3.39 μ, a Ge analyzer, and an InAs photovoltaic cell. A resolution of $\frac{1}{2}$° is quoted. Pidgeon and Smith used a double-beam system and measured the intensity difference by mechanical beam switching, achieving in this way 0.01 % accuracy. Their polarizers were stacks of "Polythene" plates and they used PbS and cooled InSb photocells to 3 and 5.5 μ, respectively, with a thermocouple up to 25 μ. The same instrument was used for ellipsometry, but the actual method used for this was applicable only to the study of *magnetic* circular dichroism.

Spectropolarimeters

The following descriptions of spectropolarimeters, while perhaps *not exhaustive*, serve to illustrate the many different combinations of basic elements that can be employed. A glance at their specifications will show that the distinct limitations of each method have been satisfactorily overcome.

An early spectropolarimeter was the nonrecording instrument designed by *Rudolph* [5]. In this instrument a polarizer was rocked through a small angle by an electric motor at a low frequency. After passing through the sample and analyzer, the light was measured by a photoelectric cell and the current generated was observed with a galvanometer. The analyzer was rotated by hand until the galvanometer deflections for the two positions of the polarizer were equal. Later a recording version of this instrument was produced in which the analyzer was rocked, while the polarizer was turned by a servomechanism to maintain balance [57]. A mechanical linkage from this moved the recorder pen while the monochromator automatically traversed the spectrum. An improvement to the Rudolph polarimeter using a Faraday cell of water as modulator was described by Foss [58].

Daly [21] has described a spectropolarimeter developed from the original design of *Fordyce, Green, and Parker* [20]. Like the latter, two Wollaston double-image prisms were used as polarizer and analyzer, but whereas the latter instrument masked off one beam from the polarizer, the former used all four images, which were measured in two groups by means of an oscillating shutter in the focal plane. Polarizer rotation balancing was also employed in this instrument, while the other one used the sum–difference ratio method.

Bürer, Kohler and Günthard [9a]: the order of optical parts in this instrument is polarizer, sample cell, modulator which is a rotating disc with half-shade plates, analyzer, and photomultiplier. The signal derived from this is used to rotate the analyzer to restore balance.

Gillam and King [35]: (Now produced and marketed by *Bellingham & Stanley* in association with *Bendix Ericsson* as the *Polarmatic 62.*) This is illustrated in Fig. 3.6. A novel principle is employed, where the two prisms of the double monochromator are made to serve also as polarizer and analyzer. The prisms are of crystalline quartz instead of the usual fused silica, and the monochromators are so coupled that the wavelength of the E-ray passed by one is that of the O-ray passed by the other. The polarimeter lies between the two halves, and Faraday cells are used for modulation and balancing. Because of this, the original instrument had a calibration (in rotation) that varied with wavelength according to the Verdet constant, but an "electrical cam" is used to compensate this effect. The wavelength range is 588 to 208 nm, rotation ranges from 0.01° to 5° for full scale, and sensitivity 0.2 to 1.9 millidegrees. Sample density limit is 2 at 240 nm.

Billardon and Badoz [9b] devised a spectropolarimeter based on Faraday cells containing water for both modulation and balance. Polarizer and analyzer are both Glazebrook prisms and the light source is a *Philips* SP 500 high-pressure mercury arc. Development of this instrument has led to the *Spectropol 1* produced by *Fica*. In the latter form, polarizers of ADP crystal are employed, followed by a Faraday cell for modulation and a second one for balancing. A double monochromator is employed, and twelve potentiometers allow zero correction over the whole wavelength range of the instrument, 200–600 nm. Rotation range is from 0.05° to 2° for full scale; sensitivity 2 millidegrees from 600 to 250 nm, rising to 5 millidegrees at 200 nm.

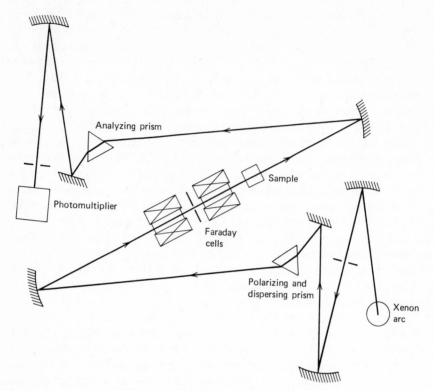

Fig. 3.6. Spectropolarimeter of Gillam and King.

Another spectropolarimeter using a Faraday cell modulator is the *Carl Zeiss REPM 12*. The wavelength range is 215–800 nm, rotation range 0.1° to 25° full scale, sensitivity 5 millidegrees.

Cary et al. [32], *Cary Instruments:* The polarizer and analyzer of this instrument are double-image prisms—Sennarmont and Rochon, respectively. The optical layout of the polarimeter is illustrated in Fig. 3.7. A Faraday cell modulator of fused silica is used, and the curved front surface of this also provides focusing. The beam is reflected from the rear surface of the Faraday cell, so effectively the cell has twice the path length. A lithium fluoride–fused silica achromatic condenser is used in front of the polarizer, and the unwanted beam is masked off before the sample cell. The output signal is used to restore balance by rotating the polarizer via an amplifier and servomotor. The double monochromator of this instrument has prisms. The wavelength range is 185 to 600 nm, rotation ranges 0.02° to 2° full scale, sensitivity is 0.5 millidegrees with most favorable conditions, and sample densities of 2 at 188 nm can be used, with higher densities elsewhere.

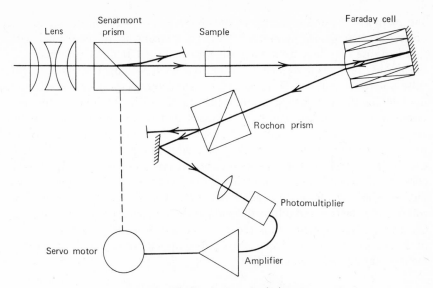

Fig. 3.7. H. Cary's spectropolarimeter.

Perkin-Elmer's P22, and the recently introduced P23 spectropolarimeters both employ an oscillating polarizer, similar to Rudolph's principle. The analyzer is rotated by a servomotor to restore balance. They are both calcite Glan prisms. A double grating monochromator is used [59]. The two instruments differ essentially only in their wavelength ranges, 220 to 600 nm for the P22, and 185 to 700 nm for the P23. Rotation range is in steps from 0.1° to 2° for full scale, or from 0.04° continuously variable. Sensitivity is better than 2 millidegrees, and densities of 2.5, 4, and 1.5 can be used at 220, 300, and 600 nm, respectively. The *Japan Spectroscopic Company's* instrument also uses an oscillating polarizer, this time a Rochon prism. Its range is also 185 to 700 nm, density accommodated up to 2, sensitivity 0.5 to 1 millidegree.

Wyss and Günthard I.R. spectropolarimeter: a "Globar" source and Perkin-Elmer 99G monochromator supply the radiation, chopped at 13 Hz. Both polarizer and analyzer are constructed from Ge plates, and modulation (at 1 Hz) is by rocking the polarizer through 8°. The detector is a Perkin-Elmer thermocouple. Measurement and compensation are by servodriven analyzer. Sensitivity is 15 millidegrees at 3.3 μ. This is lower than was anticipated, but a possible explanation is that drifts can occur in the very low frequency modulation cycle, itself dictated by thermocouple response speed.

Circular Dichroism Instruments

The measurement of CD has been tackled in two ways. First, there are attachments for double-beam spectrophotometers or for spectropolarimeters and methods of using such attachments have been covered in the section on polarizers. It remains, therefore, to describe instruments of the second type, that is, those that have been integrally designed for CD measurements.

Tagasaki [27] has described a method by which both parameters of elliptically polarized light can be measured simultaneously, using two Pockel cell modulators of ADP. The principle is illustrated in Fig. 3.8. The modulators, oriented as shown, are driven by voltages that differ 90° in phase angle, $E \sin \omega t$ and $E \cos \omega t$. It can be shown that the signal transmitted by the photocell contains three terms, the first being proportional to the average transmitted light flux. The other two terms, in $\sin \omega t$ and $\cos \omega t$, respectively, contain the orientation and ellipticity of the light passing the sample and are distinguished by phase-sensitive rectification. The signals are then used to rotate the polarizer and adjust the Babinet-Soleil compensator. The settings of these optical elements thus record simultaneously the rotation and CD of the sample. In this particular paper, the ellipticity of reflected light from a metallic surface was being studied.

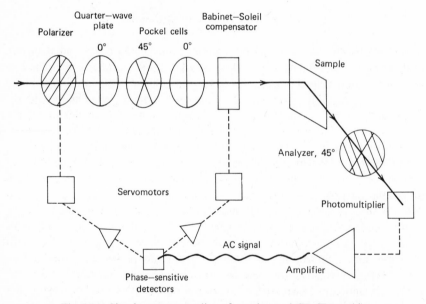

Fig. 3.8. Simultaneous recording of rotation and CD (Tagasaki).

Grosjean and Legrand [60] described the instrument now marketed as the *Dichrograph* by *Roussel-Jouan*. After being linearly polarized by a Rochon prism, the light traverses a Pockel cell of ADP, which modulates it alternately into left and right circularly polarized light. The alternating voltage used for this is varied with wavelength to keep the retardations correct at $\pm 90°$. After passing through the sample, both the alternating and steady components of the light signal are measured by the photomultiplier and appropriate amplifiers (Fig. 3.9). The ratio of these two voltages, which varies linearly as a function of the differences of the molecular extinction coefficients, is then recorded. The ranges of this instrument are from about 0.02 to 0.04 density units for full scale, and noise level and reproducibility for a sample of average density are 0.0002 and 0.0001. The maximum density is 2.

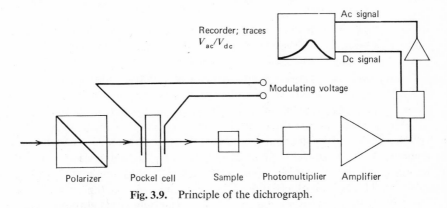

Fig. 3.9. Principle of the dichrograph.

3 DEFINITIONS

The specific rotation of optically active substances $[\alpha]_\lambda$ is a function of the nature and the concentration of the active compound, the solvent, the wavelength of the incident light, and the length of the cell [7].

The molecular rotation $[\Phi]$ is defined in (3.13):

$$[\Phi] = \frac{[\alpha] \cdot M}{100} \tag{3.13}$$

where M is the molecular weight of the optically active substance.

The rotatory power is associated with the presence of chromophores (light absorbing groups) which are either chiral (inherently dissymmetric) or in which the asymmetric vicinity makes the transitions optically active (inherently symmetric but asymmetrically perturbed chromophores).

The variation of optical activity with the wavelength gives an *optical rotatory dispersion curve*. For a compound which is optically active but contains no chromophore, that is, which does not absorb light at the wavelengths used in the examination, the optical activity decreases as the wavelength increases. A *plain positive* or *plain negative* RD curve is obtained, depending upon whether it rises or falls with decreasing wavelengths.

In the case of a compound presenting one or several optically active absorption bands within the spectral range under observation, the RD curve is *anomalous* and shows one or several peaks or troughs in the spectral region in which the chromophores absorb light.

Outside the region where optically active absorption bands are observed, Drude [61] proposed an expression relating the optical activity with the wavelength of the incident light. The first term of the Drude equation is

$$[\Phi] = \frac{K}{\lambda^2 - \lambda_o^2} \tag{3.14}$$

where K is a constant depending on the molecular weight of the optically active compound, λ is the wavelength of the incident light, and λ_o is the wavelength of the nearest absorption maximum.

If the compound possesses an active chromophore, absorbing between 200 and 700 nm, the RD curve will present a peak or a trough. The curve shows a *Cotton effect*, named after the discoverer of the phenomenon [62].

Figure 3.10 reproduces the RD, CD, and U.V. curves of *N*-(5,5-dimethyl-2-cyclohexen-1-on-3-yl)-gitingensine (**2**) [63]. The dimedone condensation compound of the steroidal alkaloid gitingensine (**2**) contains two chromophores absorbing above 200 nm namely the vinylogous amide at C-3 which shows its U.V. absorption band at 293 nm and the γ-lactone whose n-π^* transition appears at approximately 217 nm. The dimedonyl chromophore is optically active and shows a positive Cotton effect. The RD curve is characterized by a *peak* at 302 nm. The *trough* appears at lower wavelength (274 nm). The point λ_o (290 nm) of rotation $[\Phi] = 0$, where the curve inverts its sign, corresponds roughly to the wavelength of the ultraviolet absorption band (see Fig. 3.10). The vertical distance between the peak and the trough (*a* in Fig. 3.10) is called the *molecular amplitude*. It is defined as the difference between the molecular rotation at the extreme (peak or trough) of longer wavelength $[\Phi]_1$ and the molecular rotation at the extreme of shorter wavelength $[\Phi]_2$ divided by 100, as shown in (3.15):

$$a = \frac{[\Phi]_1 - [\Phi]_2}{100} \tag{3.15}$$

Whereas the dispersion effect of an optically active chromophore is attributed to a difference in speed between the oppositely circularly polarized beams of light, the *circular dichroism effect* is due to the fact that the right circularly polarized ray is differently absorbed from the left circularly polarized beam.

The differential dichroic absorption is defined by (3.16):

$$\Delta\epsilon = \epsilon_L - \epsilon_R \tag{3.16}$$

in which ϵ_L and ϵ_R are the molecular extinction coefficients for the left and right rays.

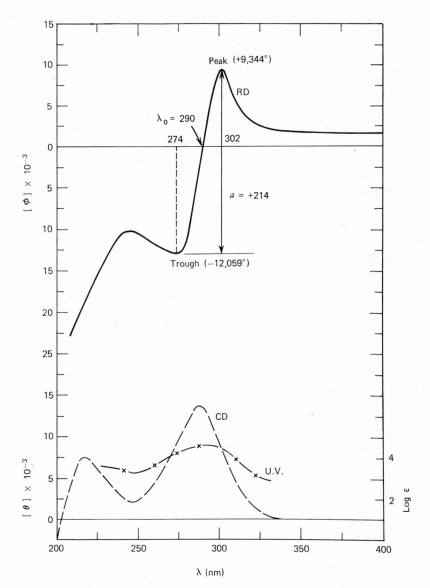

Fig. 3.10. RD, CD, and U.V. curves of *N*-(5,5-dimethyl-2-cyclohexen-1-on-3-yl)-gitingen-sine (**2**).

Another common unit is the *molecular ellipticity* $[\theta]$ of a CD curve. It is related to the differential dichroic absorption $\Delta\epsilon$ by (3.17):

$$[\theta] = 3300 \cdot \Delta\epsilon \qquad (3.17)$$

Figure 3.10 clearly shows that both RD and CD curves exhibit a positive Cotton effect in the 290 nm region. Thus, the sign of the Cotton effect is the same by both methods.

A positive Cotton effect is also associated with the lactone chromophore in **2** and is clearly observed by CD (see Fig. 3.10).

Some functions (e.g., a conjugated ketone such as cholest-4-en-3-one (**1**)) exhibit a multiple Cotton effect RD or CD curve. In this type of RD curve two or more peaks and troughs are observed. The corresponding multiple Cotton effect CD curve shows various positive and/or negative maxima [7].

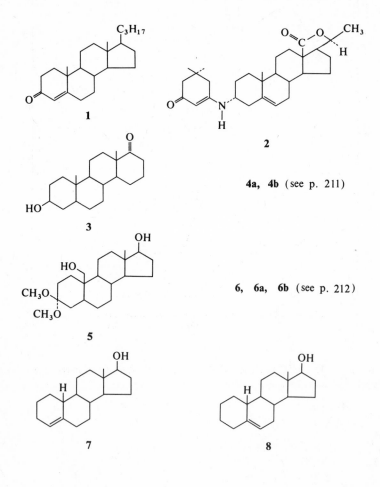

4a, 4b (see p. 211)

6, 6a, 6b (see p. 212)

While the Cotton effect associated with an optically active absorption band manifests itself by a CD curve and an anomalous RD curve, two major differences between these techniques are noteworthy. An optically active compound devoid of absorption in the wavelength range under examination will not exhibit any CD curve. However, despite the lack of Cotton effect, such a compound will present a plain RD curve, since the rotational contribution of more distant absorption bands gives rise to a background effect or skeleton effect [7]. Sometimes Cotton effect RD curves can be substantially affected by the skeleton effect; this is particularly true in the case of the RD curves of compounds exhibiting a weak Cotton effect (e.g., the lactone group in compound **2**), which is superposed on a strong skeleton effect of opposite sign.

Such a situation is clearly illustrated in Fig. 3.11 which shows that the weak negative Cotton effect associated with the 17a-keto-chromophore of 3β-hydroxy-D-homo-5α-androstan-17a-one (**3**) is better observed by CD than by RD, where it is partially masked by the background curve. In such a case, CD will be the technique of choice for the quantitative evaluation of the Cotton effect.

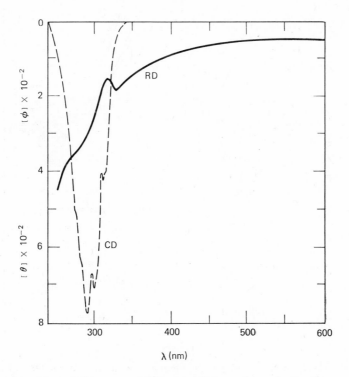

Fig. 3.11. RD and CD curves of 3β-hydroxy-D-homo-5-α-androstan-17a-one (**3**).

4 SOLVENT AND TEMPERATURE EFFECTS ON RD AND CD CURVES

Solute-solvent interactions manifest themselves in many spectroscopic measurements, and the choice of solvent for RD and CD curves is extremely important [7, 64]. The understanding of these effects is far from complete. Moreover, the technique of variable temperature RD and CD has been

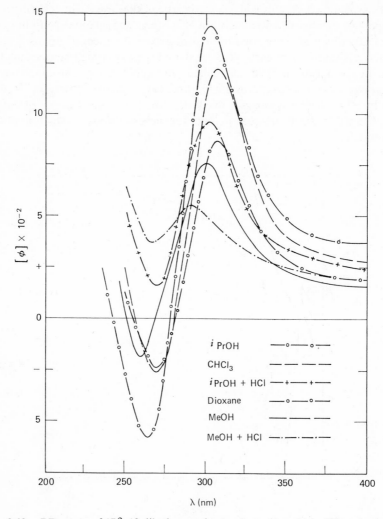

Fig. 3.12. RD curves of 17β, 19-dihydroxyandrostan-3-one (**4a**$\rightleftharpoons$**4b**) in different solvents [22].

successfully applied to study conformational equilibria, including the effect of free or hindered rotation.

Methanol is a polar solvent, which is transparent at low wavelengths and, therefore, useful for the examination of numerous chromophores. It is appropriate for the examination of ketal-formation of ketones, a study that can provide valuable information on the stereochemical vicinity and steric hindrance around the carbonyl group (4). A recent reinvestigation of this reaction has shown that it is a ketal and not a hemiketal which is usually formed when a ketone dissolved in methanol solution is treated with a trace of hydrogen chloride [65].

The influence of the polarity of the solvent on the Cotton effect associated with a saturated carbonyl group is illustrated in Fig. 3.12. 17β,19-Dihydroxyandrostan-3-one (4a) can exist in the free form and as an intramolecular hemiketal (4b). The equilibrium (4a $\rightleftharpoons$ 4b) is displaced according to the nature of the solvent. One should also bear in mind that in methanol containing HCl, besides 4a and 4b, very probably some 3-dimethoxy-ketal (5) is formed.

The variation of the Cotton effect with the dielectric constant of the solvent has been observed and commented upon [66] and examined from a theoretical point of view [67]. The dramatic effect of the nature of the solvent on the optically active transitions of unsaturated ketones has been discussed in

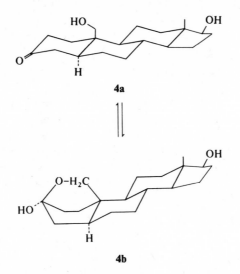

4a

4b

detail by Legrand and co-workers [7c].

The effects of temperature or solvent on conformational equilibria have been elucidated by RD and CD techniques [64b, 66, 68–70].

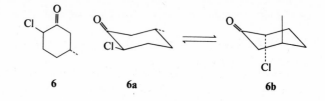

6   **6a**   **6b**

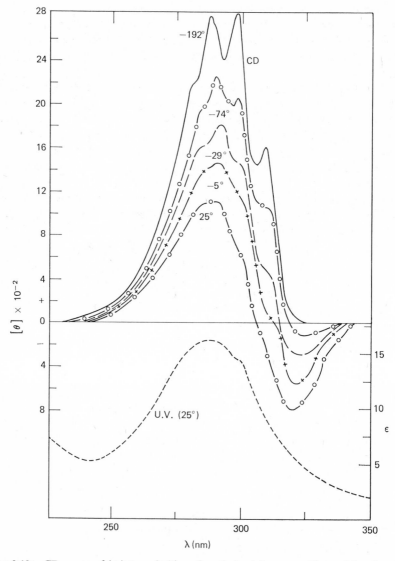

Fig. 3.13. CD curves of (+)-*trans*-2-chloro-5-methylcyclohexanone (**6**) at +25, −5, −29, −74, and −192°, as well as U.V. curve at +25° [84*b*].

Figure 3.13 shows the effects of temperature on the conformational equilibrium between the diequatorial (**6a**) and diaxial (**6b**) isomers of *trans*-2-chloro-5-methylcyclohexanone. The CD curves, obtained in ether-isopentane-alcohol in the temperature interval of $+25°$ to $-192°$, are reproduced in Fig. 3.13. The ultraviolet absorption band observed around 300 nm is a superposition of the separate bands of the individual conformers that absorb at different wavelengths. However, the overlapping is so great that the separate contributions are hardly recognizable. The presence of two conformations is clearly evident in the CD curves, because the separate contributions although severely overlapping, are of opposite sign [64b].

A particular source of complication arises from unsuspected solvent interactions, and Djerassi et al. [70d] have drawn attention to the need for caution in assigning configurations to chiral molecules, solely on the basis of the sign of the Cotton effect, in systems where the solute-solvent interactions are unknown. Combinations of conformational and solvational equilibria will show complex temperature variations. The presence of relatively widely separated CD extremes of opposite sign is usually indicative of solvation and/or conformational equilibria. It is noteworthy that RD and CD data can indicate solute-solvent interactions in media such as hydrocarbons which are ordinarily considered unlikely to participate in solvate formation [64b, 66].

5 FUNCTIONAL GROUPS

The octant rule for saturated ketones was the first successful attempt to correlate the three-dimensional structure of a chiral molecule with its optical properties. During the last decade a number of new rules have been proposed for the correlation of stereochemistry with optical activity, such as various extensions of the original octant rule, sector rules, and quadrant rules. Some of these propositions, either based on theoretical considerations or purely empirical, are now available for a variety of chromophoric groupings including saturated and unsaturated carbonyls, conjugated dienes, lactones, lactams, esters, amides as well as numerous derivatives of acids, alcohols, and amines. Recently octant, sector, and quadrant rules have been suggested for various nitrogen- and sulfur-containing chromophores. In this section, the optical properties associated with the most common functional groups and rules proposed will be reviewed. Table 3.5 gives a list of most of the functional groups and chromophoric derivatives that have been investigated by RD and CD.

Isolated Double Bonds

Allinger and Tai [71] have studied the ultraviolet properties of ethylenic compounds from the theoretical point of view and several workers have shown that the double bond is a chromophoric grouping which can be perturbed by the asymmetric surrounding. The U.V. absorption spectrum of mono-olefinic compounds presents a high-intensity band at $\lambda_{max} < 200$ nm, attributed to a π-π^* transition. Another band of lower intensity appears in the vapor phase at $\lambda > 200$ nm, which has been assigned to a π-σ^* transition. In solution, this longer-wavelength band is exhibited mainly by higher substituted ethylenic bonds.

RD and CD curves of several steroidal mono-olefins have been obtained [72, 73]. The relationship between the position of the double bond in the steroidal molecule and the sign of the Cotton effect seems to indicate that the latter depends on the asymmetric environment of the double bond. Indeed, 17β-hydroxyestr-4-ene (7) exhibits a positive Cotton effect, while its Δ^5-isomer (8) shows a negative Cotton effect [72].

According to Yogev and Mazur [72], the Cotton effect is due to a dissymmetric chromophore formed by the double-bond carbon atoms and their allylic quasi-axial hydrogens. The double-bond chromophore will exhibit a positive Cotton effect when the geometry is as in **A**:

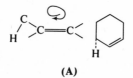

(**A**)

A negative Cotton effect will be shown in the case of a negative helix, as in **B**:

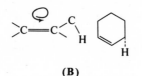

(**B**)

When more than one quasi-axial hydrogen atom is present in the allylic positions to the double bond, the Cotton effect may depend on the sum of the contributions of all quasi-axial hydrogens.

In the case of ethylenic compounds, one can use some chromophoric derivatives of the double bond to settle the configuration in its vicinity. Derivatives like osmic esters [74], trithiocarbonates (see below), organometallic complexes with platinum [75] and the like, have been prepared and their optical properties investigated.

Dienes

A careful analysis of the Cotton effect associated with 1,3-cyclohexadienes has indicated that the chirality imposed on such diene systems by structural and/or steric factors constitutes the major element of asymmetry responsible for the effect. It has been shown theoretically and confirmed experimentally that the sign of the Cotton effect of skewed cisoid dienes depends upon the sense of helicity of the diene system [76]. A strong positive Cotton effect associated with the lowest frequency cisoid diene π-π* absorption band (around 260–280 nm in polycyclic substances) indicates that the diene chromophore is twisted in the form of a right-hand helix (**C**). Conversely, a strong negative Cotton effect is indicative of a left-hand twist (**D**) [76].

(C)   (D)

Numerous optically active dienes have been shown to follow the "helicity rule for skewed dienes" [22, 76–79]; the rule seems generally applicable, provided that there is no interference by other factors as, for example, in the case of gliotoxin (**33**) [80] (see below).

During the total synthesis of the sesquiterpene lactone dihydrocustunolide (**9**), Corey and Hortmann [81] obtained the diene (**10**). This compound

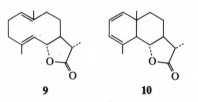

9   **10**

exhibits a strong negative molecular amplitude ($a = -244$) in agreement with the left-hand helix (**D**) formed by the diene chromophore.

The chirality rule for cisoid dienes has been invoked recently for the assignment of stereochemistry to a newly isolated acetylenic nor-sesquiterpene alcohol (**11**) [82]. Since dehydrochamaecynenol (**11**) presents a strong negative Cotton effect ($a = -421$), the absolute configuration indicated on formula **11** was proposed for this compound. According to the authors [82], nuclear magnetic resonance indicates that the conformation of the *cis*-fused hydronaphthalene system in **11** is nonsteroidal.

Several transoid dienes have also been examined by optical methods [83]; a rule has been proposed, and the optical properties of some homoconjugated dienes have been reported and commented upon [84].

Allenes

Little attention has been paid to the Cotton effect exhibited by optically active propadienes. This seems to be because only a small number of such allenes were available in the past and that the RD and CD instruments did not reach the low wavelength region where the chromophores absorb. Recently, several steroidal allenes have been prepared [85, 86] and their optical properties investigated. The CD data of compounds **12** to **15** are reported in Table 3.1.

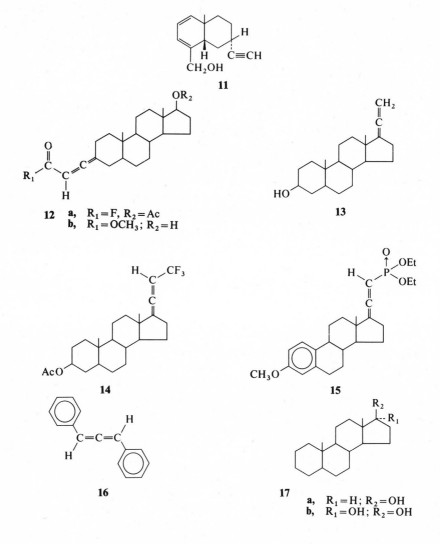

12 a, $R_1 = F, R_2 = Ac$
       b, $R_1 = OCH_3; R_2 = H$

17 a, $R_1 = H; R_2 = OH$
       b, $R_1 = OH; R_2 = OH$

Table 3.1. Cotton Effects of Some Steroidal Allenes

Compound	CD
12a	$[\theta]_{280} = +3000$ $[\theta]_{235} = +3990$ $[0]_{215} = -4290$
12b	$[\theta]_{274} = +3100$ $[\theta]_{235} = +3600$ $[\theta]_{214} = -9500$
13	$[\theta]_{243} = -1400$ $[\theta]_{225} = +1320$
14	$[\theta]_{253} = -600$ $[\theta]_{234} = +1600$
15	$[\theta]_{277} = -1300$ $[\theta]_{250} = +620$ $[\theta]_{232} = +5300$

All steroidal allenes investigated show at least one major optically active absorption band between 220 and 250 nm. The sign and the intensity of the Cotton effects are functions of the stereochemistry of the allene and of its surrounding, as well as of the nature of the substituents of the allene chromophore.

Both the allenyl acid fluoride (**12a**) and the corresponding methyl ester (**12b**) exhibit three Cotton effects. The long wavelength positive Cotton effect at approximately 280 nm is probably due to the conjugated carbonyl. The positive molecular ellipticity at 235 nm is attributed to the allene chromophore. The low wavelength (215 nm) strong negative Cotton effect is the summation of an optically active transition of the conjugated allene and the 17-acetate function (positive contribution) in **12a**. Indeed, compound **12b** devoid of the 17-acetoxy moiety shows a more intense negative molecular ellipticity at 214 nm.

In the three 17-steroidal allenes (**13** to **15**) a substantial Cotton effect also appears between 232 and 243 nm. In the estratriene derivative (**15**), the negative Cotton effect at 277 nm is allotted to the aromatic ring. In this respect it is worth mentioning that the absolute configuration of (+)-1,3-diphenylallene (**16**) has been deduced from its electronic absorption and circular dichroism spectra [87].

Alcohols

The alcohol grouping absorbs at wavelengths too low for available RD and CD instruments, but one can sometimes refer to plain RD curves to assign the configuration of the hydroxyl. Klyne et al. [88] have shown that whereas 17β-hydroxy-5α-androstane (17a) exhibits a plain positive RD curve, 17α-hydroxy-5α-androstane (17b) shows a plain negative curve. Plain RD curves have also been observed for isomeric allylic secondary alcohols [7b] as well as for tertiary hydroxyl groups [89].

Since the correct configuration will usually be more safely deduced from Cotton effect RD or CD curves than from plain RD curves, the investigation of chromophoric derivatives of alcohols may be advisable. Numerous derivatives of alcohols have been prepared, for example, esters (lactone, benzoate, xanthate, nitryloxy, etc.) (*vide infra*). From the Cotton effects the configuration can usually be assigned to asymmetric centers in the vicinity of the alcohol group. Figure 3.14 shows the Cotton effect RD and CD curves of the 2-iso-thiocyanato derivative of D- and L-2-amino-butanol methyl carbonate (18). Both enantiomers (18) exhibit three Cotton effects of increasing intensity from 350 to 200 nm. From longer wavelengths, a first rather weak Cotton effect, negative in the case of the D-isomer (D-18), positive for its antipode (L-18), appears around 340 nm; a second more intense Cotton effect of opposite sign is situated at approximately 255 nm; and the most intense Cotton effect, of the same sign, is observed in the 200 nm region (Fig. 3.14). The antipodal relationship between the D- and L-isomers is reflected in the signs of their CD and RD curves, as shown in Fig. 3.14 [90].

Figure 3.14 also indicates that CD will be preferred to RD for the quantitative evaluation of the Cotton effects exhibited by compounds presenting several optically active transitions.

Recently the CD data of cuprammonium complexes of diols and amino-alcohols have been reported [91]. The relationship between the chelate ring conformation and the sign of the Cotton effects has been used to determine the nature of the cuprammonium complexes of some acyclic ligands, particularly of carbohydrate derivatives. It should be mentioned that the optical properties of several carbohydrates and various of their derivatives have been reported for stereochemical assignments in this important class of active substances [92]. A dibenzoate chirality rule has been proposed for the determination of the absolute configurations of cyclic α-glycols (*vide infra*).

Saturated Ketones and Aldehydes

The experimental data on the Cotton effect associated with the saturated carbonyl chromophore accumulated primarily by the school of Djerassi [7] have culminated in the proposition of the octant rule [93], which permits establishment of the absolute stereochemistry from the sign and intensity of the Cotton effect. Conversely, the sign of the Cotton effect can be predicted

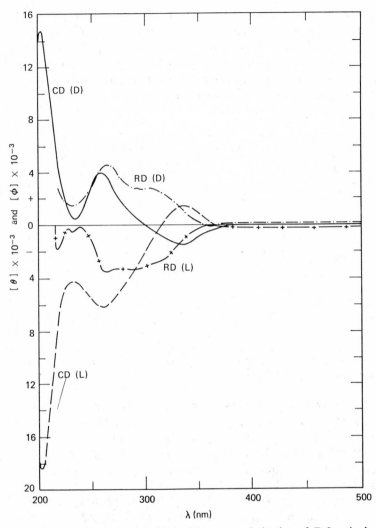

Fig. 3.14. RD and CD curves of the isothiocyanato derivative of D-2-aminobutanol methyl carbonate (**18**) and its L-isomer.

if the stereochemistry of the compound is known [94]. In spite of some relatively minor but pertinent questions [94–99] (sometimes called "failures" of the octant rule), which are still under debate and special cases in which other factors are interfering, over a thousand papers have now been published on successful applications of the octant rule, emphasizing its importance and its scope in chemistry [7d].

The octant rule expresses a relationship between the absolute configuration and conformation of the perturbing environment and the sign and, semiquantitatively, the intensity of the Cotton effect due to the long wavelength carbonyl n-π^* transition around 300 nm. The rule states [93] that three nodal planes of the n- and π^*-orbitals of the carbonyl group divide the molecular environment of the carbonyl into eight octants: four back octants and four front octants. Figure 3.15 represents the eight octants for the carbonyl in the chair conformation of the cyclohexanone ring.

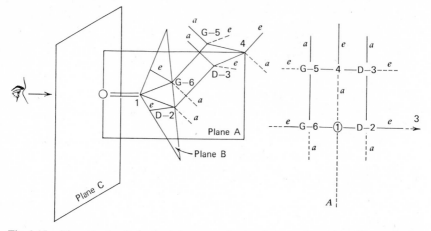

Fig. 3.15. The octant rule for cyclohexanone in the chair conformation. The projection, shown on the right, indicates the spatial orientation of the various substituents in the different octants, created by planes A, B, and C [4].

A group or atom, different from fluorine (because of the position of fluorine in the atomic refractivity and specific rotativity scale), situated in the upper-left or lower-right rear octant, relative to an observer looking at the molecule in the O=C direction induces a positive Cotton effect in the 300 nm region. A negative Cotton effect is produced by the corresponding substitution in the upper-right or lower-left back octant.

5α-Cholestan-3-one (**19a**) exhibits a positive Cotton effect ($a = +56$; $[\theta] = +4200$) around 300 nm. This has been interpreted in the light of the octant rule [93]. Indeed, the octant projection of this substance indicates that carbon atoms C-6, C-7, C-15, and C-16 lie in positive octants. Conversely, in coprostan-3-one, which is the 5β-isomer (**19b**), carbon atoms C-6, C-7, C-15, and C-16 make a negative contribution to the Cotton effect; thus both the molecular amplitude ($a = -27$) and the molecular ellipticity ($[\theta] = -1500$) are negative [93].

The Kronig-Kramers theorem [100] led to an expression which relates semiquantitatively the molecular amplitude of the RD curve to the dichroic absorption of the CD curve of a saturated ketone:

$$a = 40.28 \cdot \Delta\epsilon \qquad (3.18)$$

In terms of molecular ellipticity $[\theta]$ (3.18) becomes:

$$a = 0.0122 \cdot [\theta] \tag{3.19}$$

These expressions (3.18 and 3.19) are of obvious interest. It should be emphasized, however, that they were obtained for the n-π^* transition of a saturated carbonyl and should be used with caution for other chromophores.

The octant rule has been successfully applied to aliphatic optically active ketones and aldehydes, as well as to numerous mono- and polycyclic keto-derivatives belonging to all classes of organic compounds [7d]. For further details and/or examples, see refs [7, 93]. A "reversed" octant rule has been proposed for α-cyclopropyl and α-epoxyketones [101]. Recently, some exceptions of this rule have been reported [101b–104], and it may be that, by virtue of particular conformational and electronic factors, the "reversed" rule is not as general as anticipated.

The Unsaturated Keto Chromophore

α,β-Unsaturated ketones show two absorption maxima between 220 and 400 nm. The intense maximal absorption between 220 and 260 nm is associated with the π-π^* transition of the $C{=}C{-}C{=}O$ group. The less intense absorption band around 340 nm corresponds to the n-π^* carbonyl band. Both transitions are optically active in an asymmetric surrounding.

Figure 3.16 shows the RD and CD curves of the enantiomeric steroids, **20** and **21**, prepared from natural estrone [105]. Both transitions of the Δ^4-3-keto-chromophore are optically active. A more intense optical activity is associated with the π-π^* band than with the n-π^* band, which shows a multiple Cotton effect at approximately 340 nm. The antipodal relationship between the steroids **20** and **21** is reflected in their mirror image RD and CD curves (Fig. 3.16).

In α,β-unsaturated ketones one or both orthogonal reflection planes are lost, so that the octant rule is in general no longer applicable in its original form.

Extensions of the octant rule have been proposed for the n-π^* transition of α,β-unsaturated ketones [106–108]. These propositions seem to account for some experimental data, but should be used with extreme caution, for both theoretical [109] and experimental reasons [110, 111].

Recently, Ziffer and Robinson [112] have obtained the RD and CD curves of several α,β-unsaturated ketones. They discovered that the CD curves of a number of α,β-unsaturated ketones reveal the presence of a new strong optically active transition close to the π-π^* transition. The sign of the Cotton effect associated with the π-π^* transition observed by RD is sometimes obscured by the overlapping near transition. The CD curves of various conjugated ketones were obtained and the effect of the ring size on the sign and magnitude of the Cotton effects has been studied [113].

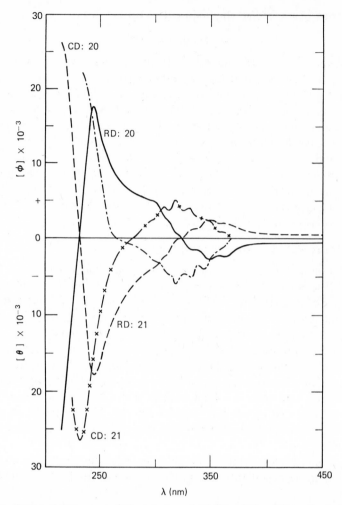

Fig. 3.16.   RD and CD curves of 13α-estr-4-ene-3,17-dione (**20**) and its antipode (**21**) [22].

The Cotton effects of halogen-substituted α,β-unsaturated ketones [7, 114] indicate that the substituent effect is consistent with a particular type but cannot be correlated with an octant rule.

The cyclopropenones can be considered as the extreme case of cyclic α,β-unsaturated ketones. Recently, several optically active cyclopropenones have been prepared in the steroid series [86, 115] and their optical properties have been investigated.

The ultraviolet spectrum of the 17α-cyclopropenonyl-androstane derivative (**22**) shows a maximum at λ_{max} 259 nm. This band is optically active and presents a negative Cotton effect, as shown in Fig. 3.17. Moreover, a weakly positive Cotton effect appears at 227 nm.

In the case of β,γ-unsaturated ketones it is known that with appropriate geometry the nonbounding n electrons on the carbonyl oxygen interact with the π electrons of the homoconjugated ethylene-carbonyl π-system [116]. It follows that the forbidden n-π^* transition borrows intensity from the allowed π-π^* transition. As a result, the extinction coefficient of the ultraviolet 300 nm band is enhanced. Moreover, several authors have noted that spectroscopic interactions between a carbonyl grouping and a β,γ-double bond are some-

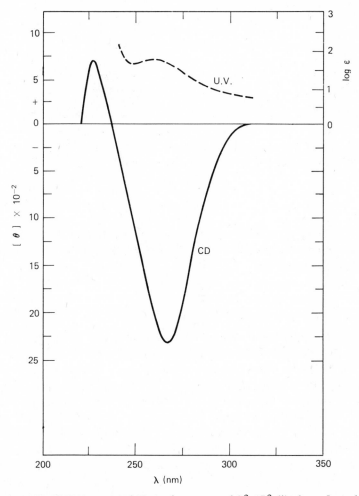

Fig. 3.17. CD and U.V. curves of 17α-cyclopropenonyl-3β, 17β-dihydroxy-5α-androstane (**22**).

times accompanied by high optical activity which increases at low temperature [7, 117–120].

Mislow, Moscowitz, and Djerassi [121] formulated the idea that the β,γ-unsaturated carbonyl system constitutes an inherently dissymmetric chromophore [122]. These authors proposed a modification of the octant rule for this chromophore. They suggested that the chirality of the β,γ-unsaturated keto chromophore may be discussed in terms of the geometric representations **E** and **F** [121].

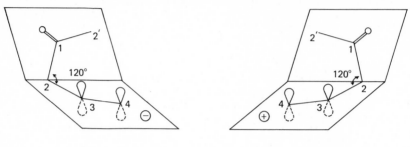

(E)   (F)

Two planes are defined by

and $C_2{-}C_3{-}C_4$ portions of the chromophore which intersect at a dihedral angle greater than 90° (usually about 120° in rigid structures). As indicated in **E** and **F**, the arrangement

$$C'_2{-}\overset{\overset{\displaystyle O}{\|}}{C_1}{-}C_2{-}C_3{-}C_4$$

assumes one of two enantiomeric conformations, one giving rise to **E**, a negative, and the other to **F**, a positive Cotton effect [121].

Steroids **23**, **24**, and **25**, which were obtained by photochemical addition of acetylene or allene to the corresponding Δ^{16}-20 keto-compounds [123], are interesting examples. These compounds exhibit respectively a strong positive (**23**; $[\theta]_{300} + 15,380$), an intense negative (**24**; $[\theta]_{295} = -18,180$), and a positive (**25**; $[\theta]_{295} = +5780$) Cotton effect. The examination of the geometry of the homoconjugated systems with molecular models indicates **23** to correspond to geometry **F**, while the β,γ-unsaturated chromophore of **24** has the same conformation as in **E**. Finally, as expected, the exomethylene isomer (**25**) shows the positive Cotton effect typical of 17α-substituted 20-keto-steroids (**22**).

Temperature-dependent CD studies have been undertaken with mobile β,γ-unsaturated carbonyl containing compounds [68c]. In some cases the CD temperature gradients have been interpreted in terms of preferred conformations.

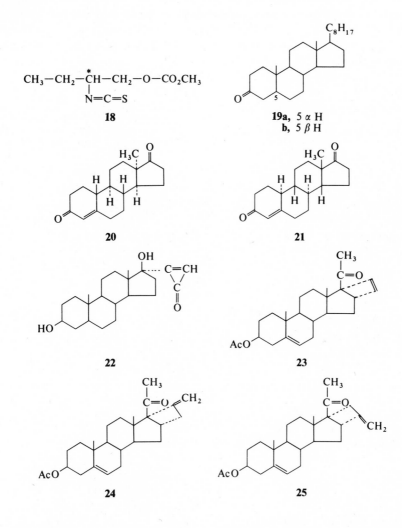

$$CH_3-CH_2-\overset{*}{C}H-CH_2-O-CO_2CH_3$$
$$|$$
$$N=C=S$$

18

19a, 5 α H
b, 5 β H

20

21

22

23

24

25

In summary, we can say that the n-π^* Cotton effects of nonconjugated ketones are low, as in saturated carbonyls, intermediate, as in axial α-halo-ketones, or cyclopentanones, or high, as in some β,γ-unsaturated ketones, depending on the extent to which the asymmetrically perturbing orbitals mix with the orbitals of the carbonyl grouping.

Carboxylic Acids

Klyne et al. [124] have shown that most optically active acids present a Cotton effect in the 210 nm region. The existence of a significant Cotton effect in this region must indicate an appreciable degree of conformational preference, even though free rotation about the carbon-carboxyl bond is formally possible. In some cases, a homoconjugation existing between the acid function and another chromophore leads to a bathochromic displacement of the Cotton effect [125].

For simple acids a correlation can be made between the sign of the experimental Cotton effect and the absolute configuration at the single asymmetric carbon atom. For more complex acids of known absolute configuration the preferred conformations of the carboxyl group have been discussed by Klyne and co-workers [124] in terms of the carboxyl sector rule, originally developed for lactones (see below).

The carboxyl sector rule, which is shown in Fig. 3.18, has been applied to numerous optically active acids [124–126]. The Cotton effects of some α,β-unsaturated acids have been reported by Weiss and Ziffer [127], who have focused their attention on the n-π^* transition of the carboxyl at approximately 250 nm.

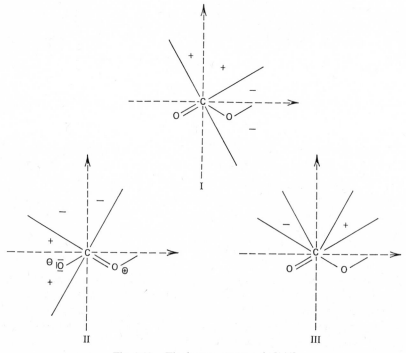

Fig. 3.18. The lactone sector rule [144].

Several dicarboxylic acids have been investigated [125]. It has been shown that monosubstituted succinic acids with the D configuration show positive Cotton effects when the substituent is alkyl, thioalkyl or bromine and chlorine. Conversely, hydroxy acids and amino acids of the same configuration exhibit opposite Cotton effects. Furthermore, the Cotton effect of substituted succinic acids of relatively simple structure is inverted in alkaline medium.

Recently, Montaudo and Overberger [128] have reported a case in which the analysis of the optical properties of some *trans*-dicarboxylic acids has allowed to suggest whether these molecules are fixed in a rigid conformation or are flexible systems. Although the U.V. spectra show only the n-π^* transition around 210 nm, the RD and CD spectra are dominated by the π-π^* transition [128]. This transition appears at 200–203 nm in the flexible systems but is shifted to 209–210 nm in rigid conformations. This red shift supports the hypothesis that the carboxyl groupings are coupled in the latter molecules [128]. This behavior, predicted by theory, has been observed in other similar instances [129]. The authors concluded that, if an optically active molecule possesses two identical neighboring chromophores, the optical techniques provide a way to test the conformational rigidity through application of the exciton theory [128, 129].

The anhydride, a derivative of dicarboxylic acid, is another function that can be investigated by optical methods. The optical properties of several active anhydrides have been described [130] and a Cotton effect is associated with the 220 nm U.V. absorption band.

Lactones

Klyne and his collaborators [124] have investigated the optical properties of lactones. As a result of this study a lactone sector rule has been proposed. The rule suggests that the space around the lactone group may be divided into sectors by means of planes meeting at the carboxyl carbon atom. The available data [124, 126, 131] show that the signs used for the ketone octant rule must be reversed for lactone sectors. Thus, atoms or groups in the back upper-right and lower-left sectors make positive contributions to the Cotton effect, while atoms situated in the back upper-left and lower-right sectors contribute negatively (see Fig. 3.18).

It is suggested that both carbon-oxygen bonds of a lactone have some double-bond character. Each carbon-oxygen bond of the lactone is considered in turn as a double bond, and the signs of the contributions made by the substituents in different octants are allocated according to the octant rule. If diagrams I and II in Fig. 3.18 are superimposed as in III, the signs of the contributions in some sectors cancel in varying degree, whereas in other sectors the contributions reinforce one another leading, on balance, to a positive contribution in the back upper-right sector, and a negative contribution in the back upper-left sector.

Wolf [132] has investigated the Cotton effect exhibited by several δ-lactones. The optical properties associated with this chromophore depend on the conformation of the δ-lactone ring. If the chirality of the lactone can be established, the sign of its Cotton effect will be deduced. Conversely, from the sign of the Cotton effect, one can ascertain the conformation of the ring system.

Recently, several investigators [133–135] have considered the influence of both the ring-chirality and the configuration of carbon atoms and substituents adjacent (Cα, Cβ) to the chromophore. Beecham [135] concludes that the sign of the n-π* Cotton effect in γ-lactones depends on the location of Cβ relative to the planar lactone system. The situation seems to be reminiscent of that in some bi- and polycyclic cyclopentanones, in which the out-of-plane carbon atoms of the five-membered ring have a dominant influence on the Cotton effect of the ketone [136]. Hence, there is evidence that the Cotton effect associated with the n-π* transition is determined both in sign and magnitude by interactions within the asymmetric ring [135].

Finally, Snatzke and Otto [135a] have shown that the rule proposed for the Cotton effect associated with the n-π* transition of α,β-cyclopropyl ketones [101] can be applied to α,β-cyclopropyl lactones.

Acetates, Amides, Lactams

Several authors have proposed to use the Cotton effect exhibited by the acetate for the assignment of the configuration to alcohols, mainly in the steroid field [137].

A "carboxyl sector rule" has been suggested to account for the experimental Cotton effects. However, a difficulty arises because most of the molecule usually falls in front sectors and it is only assumed that substituents in front sectors make contributions opposite to those in back sectors.

Rules have also been proposed for amides [138a], thioamides [138b], and lactams [138c] which allow us to correlate the sign of the Cotton effects with the configurations of these functions [138].

α-Hydroxy and α-Amino Acids

The work of Klyne and associates [124, 125] has shown that α-amino acids and α-hydroxy acids of L configuration show a positive Cotton effect around 215 nm, whereas their enantiomers D have a Cotton effect of opposite sign. Furthermore, in the case of α-amino acids, acidification of the medium in which the RD curve is being measured leads to an increment of the molecular rotation of the maximum and a bathochromic displacement of the Cotton effect by around 10 nm.

Gaffield [139] was able to measure the complete Cotton effect (RD) of some amino acids. All compounds examined in acid medium present their first extreme at approximately 225 nm, λ_o around 210–212 nm, and their second extreme in the 195–200 nm region. The molecular amplitudes depend on the size of the alkyl substituents; L-alanine, the most symmetric compound investigated [140], shows the lowest amplitude. Substitution of the alkyl chain progressively increases the intensity of the Cotton effect from L-valine to α-amino butyric acid.

More recently, several other groups of investigators have reported findings related to the amino acid chromophore [140]. While the situation is rather complicated in aromatic amino acids, aliphatic amino acids show a unique Cotton effect in the 210 nm region, the sign of which reflects the stereochemistry at the asymmetric center [139, 140]. Moreover, the exact wavelength where the Cotton effect appears, as well as its intensity, vary with the pH of the medium. RD and CD will therefore be of primary importance for assignment of configuration of amino acids, as well as α-hydroxy acids and derivatives [7, 141–149]. Of particular interest are the findings of Katzin and Gulyas [148], who found effects related to molecular structure and to the state of ionization of the species using the fully protonated form as reference states. The influence of vibrational fine structure on the absorption–CD relation has been shown, and some examples of the utility of comparing the U.V., RD, and CD data were mentioned.

Recently a series of derivatives of L-phenylalanine substituted in the aromatic ring have been investigated by RD [150]. The authors have shown temperature dependence of the parameters of the Drude equation. All aromatic amino acids examined exhibited positive Cotton effects corresponding to the $^1B_{2u} \leftarrow {}^1A_{1g}$ aromatic electronic transition which has a low rotational strength and which is structure dependent. The second aromatic transition seems to interfere with the n-π^* transition of the carboxyl group. The particular optical properties of β-aryl amino acids outside the region of the Cotton effects are ascribed to the vicinal effect of the aryl substituent. The chromophore most affected by this vicinal effect seems to absorb below 200 nm; it is conceivable that an optically active n-σ^* transition of the N—C_α bond is involved [150].

If the integral Cotton effect of the carboxyl chromophore in amino and α-hydroxy acids cannot be reached easily, one can refer to some of their derivatives in which the absorption band is shifted toward higher wavelengths [7]. Sjöberg [151] has discussed the optical properties associated with various derivatives of amino acids. Other recent studies [141–149, 152] have emphasized that in some cases appropriate derivatives of amino acids and amino alcohols will lead to safer conclusions.

In this respect the isothiocyanate derivatives of various amino acids have been prepared recently for examination of their optical properties [90]. Usually, these compounds show multiple Cotton effect curves. The isothiocyanate derivative of the aliphatic D-amino acids investigated exhibits a weakly negative Cotton effect around 250 nm, whereas the L-enantiomer presents a positive RD curve in this region. For example, the isothiocyanato derivative of D-alanine methyl ester (**26**) exhibits a weakly negative Cotton effect RD curve at about 250 nm, which is of opposite sign in the case of its L-isomer. Other chromophoric derivatives of the amino group are discussed in the following sections.

In the paragraph dealing with polyamino acids the fundamental role of RD and CD for assignment of configuration in this important class of natural products will again be emphasized. In the case of simple amino acids and

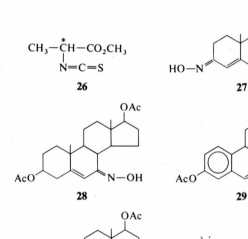

26   **27**

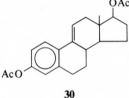   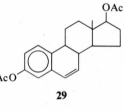

28   **29**

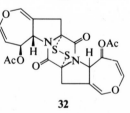   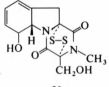

30   **31**

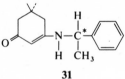

32   **33**

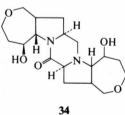

34   **35**

peptides, the Cotton effect can be measured either on the acids themselves, or on some esters, or also on some other appropriate chromophoric derivatives, or as is often the case, on ion complexes such as cupric or nickel (*vide infra*).

Oximes

The oxime of a saturated carbonyl is a chromophore which presents a relatively simple absorption pattern in the ultraviolet, so that it can easily be investigated by the optical methods. The sign and intensity of the Cotton effects reflect the stereochemistry in the vicinity of the chromophore. Table 3.2 gives the CD data of some oximes of saturated ketones [153].

Table 3.2. Cotton Effect of Some Oximes

Oxime of	Molecular Ellipticity
(−)-Menthone	$[\theta]_{197} - 14,250$
(+)-Camphor	$[\theta]_{215} + 1,910$
5α-Pregnan-3-one	$[\theta]_{210} + 7,160$
5β-Pregnan-3-one	$[\theta]_{195} + 2,700$
2,2-Dimethyl-17β-hydroxy-5α-androstan-3-one	$[\theta]_{214} + 6,500$
4,4-Dimethyl-17β-hydroxy-5α-androstan-3-one	$[\theta]_{221} - 5,940$

In Section 5 it was mentioned that α,β-unsaturated ketones exhibit two absorption bands between 230 and 350 nm; for example, the n-π* transition around 340 nm and the π-π* band in the 250 nm region. These transitions become optically active in an asymmetric surrounding (*vide supra*). The situation seems to be simplified in the case of the corresponding oximes. Indeed, the CD curves of compounds **27** and **28** exhibit one major Cotton effect, devoid of fine structure, around 240 nm, as shown in Fig. 3.19 (compare with the multiple Cotton effects in Fig. 3.16). Thus, it will be useful sometimes to refer to this derivative of the carbonyl in order to settle the configuration in the neighborhood of the chromophore, since Fig. 3.19 clearly indicates the chirality of the chromophore to be opposite in compounds **27** and **28**. Similarly, the Cotton effects of oximes of dienones are simpler than those of the ketones, and their sign also reflects the stereochemistry in the vicinity of the chromophore [7].

Consequently, the Cotton effect exhibited by an oxime will sometimes give safer stereochemical information than that of the parent ketone. It should also be mentioned that a difference of intensity has been observed in the

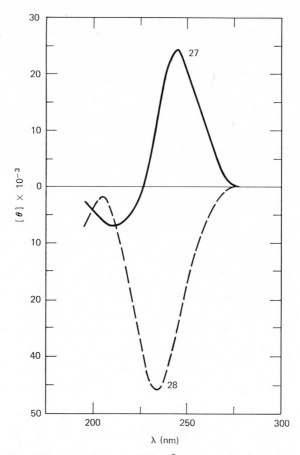

Fig. 3.19. CD curves of the steroidal α,β-unsaturated oximes, **27** and **28**.

Cotton effect of *syn-* and *anti-*oximes [154]. Moreover, in some instances it has been noted that the Cotton effects of oximes can be substantially affected by the nature of the solvent.

Aromatic Substances

In spite of the fact that the Cotton effects associated with optically active aromatic substances have been reviewed [7d, 155], few attempts have been made to relate the sign of the Cotton effects to the stereochemistry in the neighborhood of the aromatic chromophore or to the chirality of the molecule as a whole. Theoretical studies are as yet limited to a few groups of compounds, but empirical approaches have shown that from the sign and magnitude of the Cotton effects, the stereochemistry may often be deduced

[155]—that is, the absolute configuration of asymmetric centers and the conformation of rings or conjugated systems.

Empirical rules and some based on theory have been published for molecules containing the following aromatic chromophores: phenyl [156], tetrasubstituted phenyl [157], biaryl [158], phenylosotriazole [159a], and purine [159b]. In view of the recent publication of detailed surveys [7d, 155, 156] on the aromatic chromophores, only a brief mention of the aromatic quadrant rule will be made followed by a discussion of the styrene chromophore and the recently proposed benzoate sector rule. Furthermore, various aromatic chromophores are listed in Table 3.5.

The examination of the optical properties of various aromatic compounds of known absolute configuration has led to propose a quadrant rule, which allows us to predict the absolute stereochemistry of a molecule having an asymmetric center adjacent to the aromatic ring [156c].

The quadrant rule is shown in Fig. 3.20. The aromatic ring and the asymmetric benzylic carbon atom should be located in plane A, plane B being a symmetry plane [156c]. As in the octant rule [93], the aromatic quadrant rule states that substituents in the upper left and lower right quadrants make a positive contribution to the Cotton effect. Conversely, atoms situated in the upper right and lower left quadrants make a negative contribution to the Cotton effect [156c].

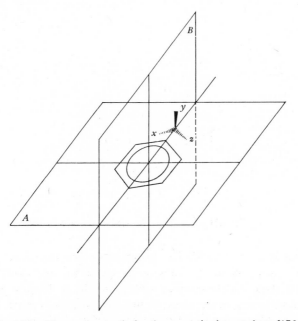

Fig. 3.20. The quadrant rule for the aromatic chromophore [176c].

An examination of the geometry of the styrene chromophore with molecular models shows that the chirality of the conjugated system in the Δ^6-aromatic steroid (**29**; $a = -236$) (right-handed helix; negative Cotton effect) is opposite to that in the $\Delta^{9(11)}$-compound (**30**; $a = +194$) (left-handed helix; positive Cotton effect). Thus, a strong negative Cotton effect associated with the 260–270 nm transition indicates that the styrene chromophore is twisted in the form of a right-hand helix. Conversely, an intense positive Cotton effect is indicative of a left-hand twist [160].

It has been shown previously that correlations between the sign of the Cotton effect and the absolute configuration of acetates have been proposed on the basis of similar rules for lactones and carboxylic acids. Nakanishi and collaborators [161] have found that the strong Cotton effect of benzoates attributed to π-π^* intramolecular charge transfer transition at about 225 nm allows prediction of the absolute configuration of a variety of cyclic secondary hydroxyl groups. The benzoate sector rule is represented in Fig. 3.21 [161]. The space is divided into four sectors by symmetry planes A and B, and further into eight sectors by two additional planes C and D perpendicular to A and passing through both oxygen atoms. The preferred conformation of the benzoyloxy group is assumed to be one in which it lies staggered between the carbinyl hydrogen and the smaller substituent [161].

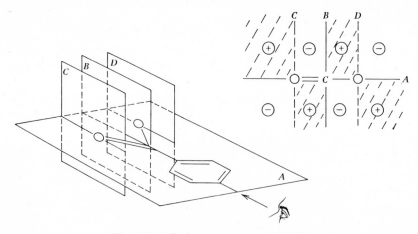

Fig. 3.21. The benzoate sector rule [181].

The benzoate is viewed from the *para*-position and the rotatory contributions of α,β- and β,γ-bonds are considered: the sector rule states that bonds falling in the shaded and unshaded sectors in Fig. 3.21 make positive and negative contributions, respectively, to the 230 nm Cotton effect. The contribution of a double bond would be larger than that of a single bond because of the larger polarizability, and the sector that carries a β,γ-double bond, if any, will make the dominant contribution. Similarly, the sector carrying a γ,δ-double bond will define the sign of the Cotton effect when this is the unsaturation closest to the carbinyl carbon atom [161]. Theoretical considerations [162] support the benzoate sector rule mentioned above. Several applications of the rule have been reported [161] and recently an extension called the dibenzoate chirality rule has been proposed for cyclic α-glycols [163].

In the case of α-glycol dibenzoates, the 230 nm absorption band is associated with two very strong Cotton effects of opposite signs at about 233 and 219 nm. This seems to indicate that both Cotton effects of dibenzoates are mainly due to a dipole-dipole interaction between electric transition moments of the two benzoate chromophores. If the chiralities of dibenzoates are defined as being positive or negative, respectively, according to whether the rotation is in the sense of a right- or left-hand screw, then the sign of the first Cotton effect is in agreement with the chirality, dibenzoate chirality rule [163]. The dibenzoate chirality rule can also be applied to other aromatic chromophores [163].

Episulfides, Thiocarbonates, Thionocarbonates

Spectroscopic studies of simple episulfides have shown the presence of a low-intensity absorption maximum in the 260 nm region, which is similar to the n-π^* absorption of the carbonyl chromophore. Several groups of workers [164, 165] have investigated the optical properties of various active episulfides.

From the data listed in Table 3.3, it is apparent that either the sign of the Cotton effect or the rotational strength, or both parameters, can be used for differentiating between the position and/or configuration of the episulfide function in the steroid and triterpene molecule.

In the case of the episulfide, as in carbonyl containing molecules, the experimental Cotton effect results only from the asymmetry induced in the episulfide group by the rest of the molecule. Moreover, whereas the magnetic dipole moment of an n-π^* transition is directed along the internuclear axis, it may be at right angles in an n-σ^* transition, as in a sulfide. Cookson et al. [165] have discussed the shape of episulfide orbitals and the conformational factors responsible for the Cotton effect.

A sector rule has been proposed for the episulfide chromophore [164]. This rule explains both the sign and magnitude of the n-σ^* Cotton effect associated with episulfides. This rule, which is illustrated in Fig. 3.22, results from theoretical calculations. In order to deduce the Cotton effect we must consider two views of each molecule, that is (a) the view along the bisectrix of the C-S-C angle; and (b) the view of the molecule from above, projected on the plane of the episulfide ring.

Table 3.3.   Cotton Effects of Some Episulfides

Compound	Circular Dichroism Molecular Ellipticity
17β-Hydroxyandrostan-2α,3α-episulfide	$[\theta]_{269} = -5990$
Cholestan-2α,3α-episulfide	$[\theta]_{268} = -3840$
Cholestan-2β,3β-episulfide	$[\theta]_{264} = +\ 630$
Lanostan-2α,3α-episulfide	$[\theta]_{267} = -6530$
Lanostan-2β,3β-episulfide	$[\theta]_{265} = +1910$
Lanost-8-en-2α,3α-episulfide	$[\theta]_{265} = -5250$
Cholestan-3α,4α-episulfide	$[\theta]_{267} = +4600$
17β-Acetoxyandrostan-3β,4β-episulfide	$[\theta]_{266} = +\ 325$
3β-Hydroxycholestan-5α,6α-episulfide	$[\theta]_{272} = +6680$
3β-Hydroxycholestan-5β,6β-episulfide	$[\theta]_{268} = -4460$
Cholestan-5β-6β-episulfide	$[\theta]_{267} = -4850$
3β,20β-Dihydroxy-5α-pregnan-11α,12α-episulfide	$[\theta]_{267} = +4060$
3β-20β-Dihydroxy-5α-pregnan-11β,12β,episulfide	$[\theta]_{261} = +3460$
3β-Acetoxy-5α-androstan-16α,17α-episulfide	$[\theta]_{264} = +\ 700$
3β-Acetoxy-5α-androstan-16β,17β,episulfide	$[\theta]_{268} = +5370$

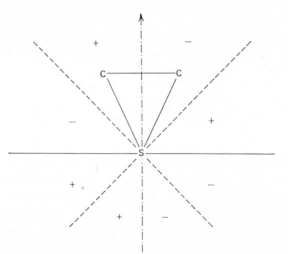

Fig. 3.22.   The sector rule proposed for the *n*-σ* transition of episulfides [184*d*].

In cyclic dithio- and trithiocarbonates, if the plane formed by the two ring heteroatoms and the carbon atom of the thiocarbonyl group is looked at from the thiocarbonyl sulfur atom through its carbon atom, the sign of the *n*-π* and π-π* Cotton effects will result from the chirality of the chromophoric system, as shown in Table 3.4.

Table 3.4. Sign of the Cotton Effects
in Dithio- and Trithiocarbonates

	Cotton Effect	
Chirality	*n-π* Transition*	*π-π* Transition*
	Negative	Positive

	Positive	Negative

Thionocarbonates exhibit two U.V. bands at approximately 315 and 230 nm, which are assigned to n-π^* and π-π^* transitions, respectively.

During a recent examination of the CD properties of the n-π^* band in cyclic thionocarbonates, Haines and Jenkins [165a] have shown that the sign and magnitude of the Cotton effect is related to the chirality of the ring.

Amines and Derivatives

Amines present a complex absorption pattern, usually at low wavelengths. Ultraviolet absorptions of simple aliphatic amines in the vapor state show several transitions below 240 nm, two of which appear between 190 and 240 nm. Hence, in the case of aliphatic and alicyclic amines, devoid of chromophores absorbing above 240 nm, the correct stereochemistry cannot easily be ascertained from their optical properties, since no Cotton effect appears in the spectral region investigated with the usual RD and CD instruments. In such cases, we may refer to derivatives presenting favorable spectroscopic properties in the wavelength range easily accessible for RD and CD. This is why many derivatives of amines have been examined and their optical properties thoroughly discussed. Indeed, the RD and CD data of a number of chromophoric derivatives of optically active amines and amino acids (*vide supra*) have been examined both for their intrinsic spectroscopic interest and to test their usefulness for the correlation of configuration. Among the most commonly used derivatives of the amino function are the isothiocyanate [90], the nitroso amines [166], nitroso amides [166], nitrosites [167], alkyl nitrites [168], Schiff bases [169, 170] (such as *N*-benzylidene [171], *N*-isopropylidene [172], *N*-salicylidene [173]), phthalimides [174, 175], maleyl [175, 176], phthaloyl [176], and itaconyl [176] derivatives, as well as *N*-phenylthioacetyl [177], *N*-thiobenzoyl [177], and sulphonamide [178] (see also next sections).

The Cotton effects associated with these chromophores and other derivatives of amines have been discussed in detail [7, 151, 186–176]. Unfortunately, many of these chromophoric compounds are hard to prepare and some are unsuitable because they exhibit undesirable optical properties. Among the various derivatives examined so far, the salicylidene chromophore, formed by condensation of salicylaldehydes with amines, is one of the most commonly used derivatives for the assignment of relative and/or absolute stereochemistry to optically active amines and amino acids [173]. It has been shown that most amines with the (S)-configuration exhibit positive Cotton effect curves, and (R)-derivatives negative RD and CD curves.

Recently, the Cotton effect of dimedone condensation products of several optically active aliphatic, alicyclic, and aromatic amines was examined by RD and CD [63, 179]. A correlation has been established between the sign of the Cotton effect and the absolute configuration of the asymmetric center. All dimedonyl derivatives of aliphatic and alicyclic amines investigated having the (R)-configuration exhibited a positive Cotton effect in the 280 nm region. A negative Cotton effect was observed for compounds presenting the (S)-configuration [179]. Moreover, the intensity of the 280 nm Cotton effect varies with the kind of amine under investigation. For example, the molecular amplitude of aliphatic amines is rather weak. In the case of saturated alicyclic amines the intensity of the Cotton effect is a function of the conformational rigidity of the system. In olefinic alicyclic amines and in aralkylamines, the intensity of the Cotton effects also depends on the proximity of the double bond or the aromatic system to the vinylogous amide chromophore [179].

Sometimes the sign of the Cotton effect presented by the dimedone condensation compound of aromatic amines can be used to establish the absolute configuration of the asymmetric center in the vicinity of the vinylogous amide chromophore. This is illustrated in Fig. 3.23, which reproduces the U.V., RD, and CD curves of the dimedonyl derivative of (R)- and (S)-α-phenylethylamine (31) [179]. The RD and CD curves of (R)-(31) and (S)-(31) shown in Fig. 3.23 clearly indicate that the Cotton effects of opposite sign are associated with the 279 nm transition in these stereoisomers. It should be emphasized that the intensity of these Cotton effects is very high. This seems to indicate that the chromophore is not merely the vinylogous amide grouping, but the homoconjugated system formed by the vinylogous amide and the aromatic ring [179].

In the case of active aralkylamines, the Cotton effects of the aromatic transitions will sometimes give useful stereochemical information [140, 155, 180] that can be used for configurational assignments. Very often one is, however, dealing with multiple Cotton effect RD and CD curves. Moreover, a change in the nature of the substituent on the aromatic ring, or in its position

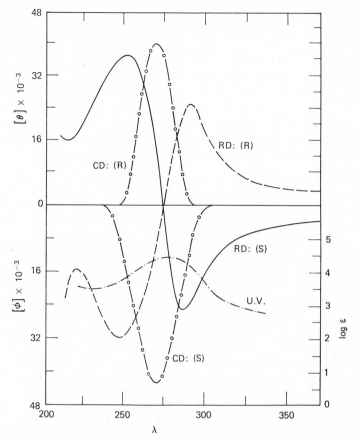

Fig. 3.23. RD, CD, and U.V. curves of the dimedone condensation compound (**31**) of (R)- and (S)-α-phenylethylamine [199].

will affect the sign and the intensity of the Cotton effects, sometimes rendering correlation of configurations extremely difficult.

In a recent study of the Cotton effects of tertiary amines, two optically active transitions have been observed, that is, a weak absorption at approximately 220–230 nm and a more intense transition around 195–205 nm [201].

Dithiocarbamates and Dithiourethanes

The dithiocarbamate [182], salicylidene [173], and *N*-phthaloyl [174, 175] derivatives are among the most commonly used derivatives for the determination of configuration of optically active amines and α-amino acids.

Recently, Ripperger [183] has prepared several asymmetric dithiocarbamates and dithiourethanes. The 340 nm band has been assigned to an *n-π** transition, whereas the absorption at about 280 nm was attributed to a π-π^* transition. For the n-π^* Cotton effect of dithiocarbamates a simple quadrant rule has been proposed [183]. This rule, which allows to predict the sign of the Cotton effects, is illustrated in Fig. 3.24.

Substituents falling in the upper-right and lower-left quadrants make a positive contribution, whereas substituents in upper-left and lower-right quadrants will make negative contributions to the Cotton effect. Although the Cotton effect associated with the n-π^* transition of the NCS_2 chromophore is rather weak, the quadrant rule for dithiocarbamates has been successfully applied to various aliphatic and alicyclic amino-derivatives [183].

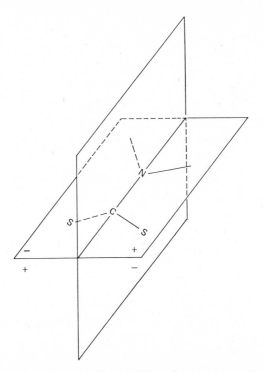

Fig. 3.24. The quadrant rule for the dithiocarbamate chromophore [203].

Azides

Alkyl azides show a weak transition around 280–290 nm, which is attributed [184] to the promotion of an electron from a nonbounding $2p_y$ orbital situated mainly on the nitrogen atom N_1 concerned with bonding to the alkyl group, to an antibonding π_x^* orbital associated principally with $2p_x$ atomic orbitals from the remaining two nitrogen atoms (N_2, N_3).

From the study of Djerassi, Moscowitz et al. [184], an octant rule for the azide chromophore has been proposed. As in the case of the saturated ketones, only two of the surfaces specifying the octants are well defined in terms of symmetry [184]. In order to determine the sign associated with a particular octant, it is convenient to look at the chromophore along the $N_3-N_2-N_1$ axis from N_3 towards N_1, with the bond specifying the lone pair of electrons on N_1 lying in a vertical plane, as indicated in Fig. 3.25. The signs of the azide octants are the same as those of the carbonyl octants. This is because the analogous viewpoint for $C=O$ would be along the $C-O$ bond from the carbon atom towards the oxygen atom, with the carbonyl group rotated through 90° about the $C-O$ axis from its normal orientation for application of the octant rule. This corresponds to a double change in sign, which is equivalent to no change at all [184].

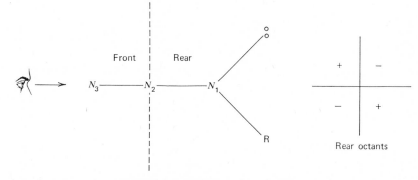

Fig. 3.25. The azide octant rule [204].

The azide octant rule which has been supported by experimental results [184], has been applied recently to numerous azido sugars and it has been shown that both the conformation of the ring system as well as the configuration of the substituents should be taken into consideration [185].

Azomethines, Nitrosamines, N-Chloro-amines, Nitro-, Nitryloxy-Derivatives, and Aziridines

Snatzke, Schreiber, and their co-workers have proposed several rules which allow us to predict the sign of the Cotton effect associated with cyclic azomethines [186], nitrosoamines, nitro-derivatives [187], as well as N-chloro-amines [188].

The $C=N$ chromophore of azomethines presents a weak absorption around 250 nm, which becomes optically active when this grouping is located in an asymmetric surrounding. A rule, based on numerous examples, states

[186] that cyclic azomethines of conformation **G** exhibit a positive Cotton effect, whereas a negative Cotton effect is associated with conformation **H**.

G   **H**

The n_N-π_3^* transition of N-nitroso-derivatives of optically active amines appears in the 370 nm region and presents a Cotton effect that is a function of the stereochemistry in the asymmetric vicinity [187]. A sector rule, illustrated in Fig. 3.26, has been proposed for this chromophore [187].

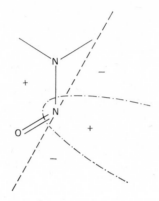

Fig. 3.26.   The sector rule proposed for the 370 nm transition of N-nitroso-amines [207].

Snatzke has also obtained the CD data for numerous nitro-derivatives [187, 189], thus completing the data obtained previously for this chromophore [7, 190]. Figure 3.27 shows the sector rule which has been suggested [189] for the nitro-chromophore.

The spiro-pyrazoline derivatives of camphor, borneol, and some steroids have been prepared by Snatzke and Himmelreich [191a]; the azo group of this heterocycle is optically active in an asymmetric vicinity. The Cotton effect, which appears at approximately 330 nm, is enhanced by the presence of a ketone in the neighborhood. In a recent study by Severn and Kosower [191b], the authors show that the long wavelength absorption of azolkanes (1,2-dialkyldiazenes) appearing at about 380 nm exhibits a CD maximum. According to the authors [191b] this observation supports the conclusion that the low energy n-π^* band of diazenes is a single n_+-π^* transition. More recently, Snatzke and collaborators [192] have obtained various nitryloxy-steroids; three weak Cotton effects are associated with this chromophore in

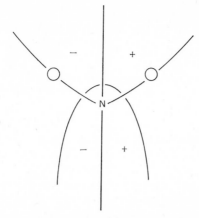

Fig. 3.27. The sector rule proposed for the nitro-chromophore [209].

the 270, 230, and 210 nm region. The authors emphasized that the CD bands can be used for the location of hydroxyl groups as well as for stereochemical assignments [192].

N-Chloro-amines and N-chloroamino-ketals exhibit a weak absorption band between 250 and 280 nm, which becomes optically active in an asymmetric surrounding [188]. Moreover, a correlation between the stereochemistry in the vicinity of the chromophore and the sign of the Cotton effect has been established [188].

Russian investigators have found recently [193] that an asymmetrical center in 2-alkylaziridine determines the formation of a stable asymmetric nitrogen atom in stereoselective N-halogenation reactions. The RD properties of such optically active halogenated aziridines have been reported [193].

The Thiocyanate Chromophore

RD and CD measurements of a number of steroidal thiocyanates have demonstrated that the 250 nm transition is optically active [194]. It has been assumed that this transition is qualitatively similar to the n-π^* transition of azides (see Section 5) so that an octant rule, illustrated in Fig. 3.28, has been proposed for the thiocyanate chromophore [194].

The discussion of the optical activity associated with the azide chromophore [184], which has been classified as "inherently symmetric," suggested to Djerassi et al. [194] that similar arguments might apply to the isoelectric thiocyanate chromophore ($-S-C\equiv N$). Hence, any optical activity associated with these chromophores is a result of their location in a dissymmetric environment. Accordingly, the sign and the magnitude of their Cotton effect depend on the nature and location of the atoms in the vicinity of the chromophore. The thiocyanate transition, at *ca.* 245 nm, may be attributed to

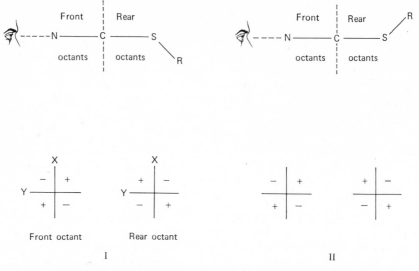

Fig. 3.28.   The octant rule for the thiocyanate chromophore [214].

the promotion of an electron from a nonbonding $3p_y$ orbital situated mainly on sulfur, to an antibonding π^* orbital determined largely by the carbon and nitrogen $2p_x$ atomic orbitals [194].

Looking along the axis from nitrogen through carbon to sulfur (see Fig. 3.28), the symmetry planes are: (a) XZ-plane containing the S, C, and N atoms and the carbon of group R attached to sulfur; (b) the YZ-plane, which is orthogonal to the XZ-plane, and contains the S, C, and N atoms; (c) a so far poorly defined surface approximated by a third plane (XY), orthogonal to the other planes and passing through the carbon atom of the S—C≡N grouping [194].

The rotational strength of the thiocyanate chromophore, as in the case of the azide, is weaker than that of the carbonyl group. Nevertheless, CD data of various thiosteroids have been obtained and the potential utility of the new octant rule has been demonstrated by analyzing the rotameric contributions of various steroidal thiocyanates [194]. Temperature-dependent CD curves of such derivatives have been obtained in order to study the effect of free rotation.

Sulfur Derivatives, Ethers, Ozonides, Nitrones, Nitroxides, and Diketopiperazines

Besides the sulfur derivatives mentioned previously, several other sulfur-containing compounds have been prepared and submitted to RD and CD techniques. Among the various functions investigated, are xanthates [195, 196], dithiocarbamates [196], thionocarbalkoxy derivatives [197], thiohydantoines [197], acylthioureas [198], sulfides [199], disulfides [200], dithianes [200], dithiolans [201], sulfoxides and related compounds [202–205].

Of particular importance in chemistry and biochemistry is the disulfide chromophore (—S—S—) [70e, 165, 200, 206–211]. The disulfide group presents various absorption bands in the U.V. Consequently, three Cotton effects, attributed to the disulfide chromophore, have been found in the metabolite acetylaranotin (**32**) at 345, 310, and 268 nm [212]. Desulfurization of acetylaranotin (**32**), which incidently is closely related to gliotoxin (**33**) [213], followed by reduction and deacetylation gave the diol (**34**) [212]. Its CD spectrum has a negative maximum at 222, a positive maximum at 210, and a negative maximum below 200 nm. In this wavelength region the diol (**34**) contains only the diketopiperazine chromophore [212], thus these three Cotton effects are optically active transitions of the cyclic diketopiperazine moiety [214].

Recently, the optical properties of active organic sulfites [215] and alkyl-sulfinyl-steroids [216] have been reported. Furthermore, the influence of ethylene acetal, monothioacetal, and dithioacetal functions on the Cotton effect associated with a vicinal ketone has been discussed [217].

While the RD and CD of various phosphine oxides and phosphine sulfides have been reported [218], Mislow et al. [219] have proposed a direct configurational correlation of sulfoxides and phosphine oxides by intersystem matching of the Cotton effects. The similarity in the dichroism of the two systems suggested to Mislow the possibility of a displacement rule embracing sulfoxides and phosphine oxides. At a wavelength λ, removed from the center of the optically active transitions, the molecular rotation $[\Phi]$ is proportional to the sum over all transitions of $\lambda_i^2 R_i/(\lambda^2 - \lambda_i^2)$, where λ_i and R_i are the position of the band center and the rotational strength of the ith transition, respectively [219].

Overberger and Weise have prepared and investigated the optical properties of asymmetric thiepan-2-ones [220] and poly(thiol esters) [221]. Two Cotton effects centered around 298 nm (n-π* transition) and 234 nm (π-π* band) are associated with the thiolactone chromophore.

The optical properties of some dialkyl ethers and cyclic ethers (tetra-hydrofuran, tetrahydropyran) have been reported [222, 223], as well as of some of their complexes. Important stereochemical conclusions could be reached from these measurements.

The correct configuration of some ozonides could be assigned from their RD curves [224].

Rassat and his collaborators [225] have prepared a number of nitroxides and shown that camphenyl t-butyl nitroxide is a stable, optically active, free radical, of which the CD has been studied [225].

Parello and Lusinchi [226] have analyzed the nitrone chromophore. At least four transitions have been found to be associated with the

$$\begin{array}{c}\diagdown\\ \diagup\end{array}\!\!C=\overset{+}{N}-\bar{O}$$

chromophore, that is, a weak n-π^* band at approximately 295 nm, a strong π-π^* transition around 230–250 nm, a n-σ^* band at about 207 nm, and a fourth transition presumably of π-σ^* type. In an asymmetric vicinity these transitions are active; a Cotton effect has been shown to be associated with each of them.

Parello et al. [227] have also shown that photochemical treatment of nitrones yields oxazirans. The oxaziran group is an optically active chromo-

phore, which exhibits at least two Cotton effects (i.e., around 225 and 195 nm).

Some optically active phosphorous organic compounds have also been reported and their optical properties thoroughly discussed [218, 219, 228].

The synthesis and specific rotation of (+)-bromochlorofluoromethane (**35**), one of the simplest organic compounds capable of optical activity, have been described recently [229]. Studies have been made of the optical properties of several compounds in which the asymmetric center is different from carbon, sulfur, or nitrogen (e.g., Si, Ge, Se, Te, etc.) [7d, 230].

Poly-amino Acids

The wealth of information and experimental data on the optical properties of poly-α-amino acids, the related polypeptides and proteins, and of nucleic acids is increasing at an exponential rate. Since the classical chapter by Blout [231], more recent detailed chapters and review articles [7, 210, 232–239] have been published on various aspects of this topic.

The peptide moiety common to numerous substances of these series has only one plane of symmetry [94], and the rotatory strength changes sign when a given perturbant is moved from one side of the peptide plane to the other. The electronic orbitals involved in the n-π^* transition have a higher symmetry than the peptide group itself. The n orbital of the C—O group has a nodal surface perpendicular to the plane of the peptide group. This surface is planar in symmetrical ketones (see Section 5) [93] but somewhat distorted in peptides. According to Schellman [94] calculations and experiments show that this surface plays the same role as a true symmetry plane, so that the resultant rule for the peptide group is a quadrant rule. Figure 3.29 shows the quadrant rule for the peptide function [94]. The vertical surface is not planar because of the horizontal distortion of the nonbonding electrons of the carbonyl group of the peptide moiety (see above) [94].

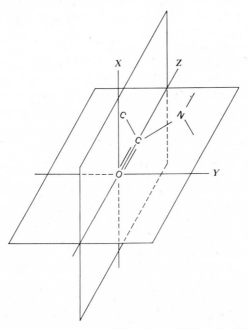

Fig. 3.29. The quadrant rule for the peptide function [114].

The discovery of the anomalous dispersion of α-helical polypeptides establishes conformational rotations due to the helical structure [238]. Moreover, it has been known that native proteins, in general, show much less negative rotations than denatured proteins, and the suggestion was made that such changes in rotation are associated with loss of helical structures. Moffitt [240] proposed an equation to describe the optical rotatory dispersion of α-helical polypeptides. This equation has been used as the basis for the estimation of the α-helix content of many types of synthetic polypeptides and many types of proteins.

In his chapter on CD of poly-α-amino acids and proteins, Beychok [239a] discusses the optical properties associated with the active disulfide transitions (*vide supra*), as well as the CD of some specific proteins such as myoglobin and hemoglobin, insulin, ribonuclease, serumalbumin, and lysozyme.

In an analysis of the visible and near U.V.-RD of synthetic polypeptides and proteins, Blout et al. [237] were able to determine the α-helix content of proteins *in solution*. Their analysis provides a basis for the differentiation of α-helix containing proteins from proteins involving other structures. The study of the conformation of biopolymers in solution can be performed by RD and CD techniques. Moreover, the optical methods have been used to provide information about the structure of nucleohistones and their addition complexes with acridines, stabilization of ribonucleic acids with natural or synthetic polybases, and action of urea and sodium dodecylsulfate on the structure of ovalbumine [241].

The RD and CD techniques have also been used to study charge-transfer absorption bands associated with intra- and intermolecular electron-donor-acceptor complexes in chiral molecules [242].

More recent papers have discussed the RD and CD properties associated with polyamino acids, polypeptides, proteins, and nucleic acids [243–250], thus showing the fundamental contribution of these techniques to a better understanding of the nature and origin of life.

Synthetic High Polymers

Although a rather large number of optically active synthetic polymers have been prepared, mainly by the schools of Pino [251] and Schultz [252], little attention has been paid to the optical activity of these compounds. Nevertheless, it has been shown that a relationship exists between the sign of the Cotton effects and the helical conformations of poly-α-olefins [253, 254].

The optical properties (mainly RD) associated with some polyhydrocarbons, polyalkenylethers, polyacrylic derivatives, and polyaldehydes have been discussed [253]. Interesting but sometimes not well understood optical phenomena have been observed with synthetic polymers such as polyacrylates and polylactides. Anomalies in RD curves have been noted, which are not present in the curves of low molecular weight esters and lactones.

One of the main problems in synthetic polymers is the determination of their chemical structure, since it is only in relatively few cases that highly stereoregular polymers have been obtained. Only when the structural problem has been solved can optical methods be applied to the investigation of the stereochemistry. Synthetic high polymers constitute another area where RD and CD techniques can be expected in the future to clarify conformational features which are not easily accessible by other methods.

Inorganic Compounds, Ligands, Metal Chelates

Although only few purely inorganic, optically active molecular species are known, it is important to bear in mind that optical activity is not a unique feature of asymmetric carbon atoms or asymmetric organic molecules. The optically active *tris*[dihydroxo tetrammin cobalt (III)] cobalt (III) hexanitrate

(Fig. 3.30) resolved by Werner as the bromo-camphor-sulfonate, was the keystone in the arch of Werner's coordination theory at the beginning of this century. Figure 3.30 shows the structural formula of the cobalt complex. The bidentate ligand shown in I, symbolized by A-A is an octahedral complex, namely the species represented in II. The A's represent the hydroxyl groupings forming a total of six bridges between the central and the three coordinated cobalt atoms [256a].

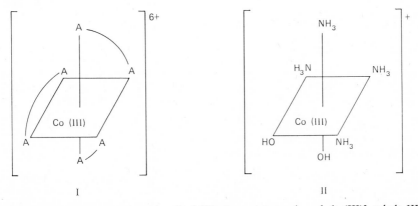

Fig. 3.30. The structural formula of *tris*[dihydroxo tetrammin cobalt (III)] cobalt III hexanitrate [276a].

Although most active inorganic compounds are metal-organic compounds, they differ substantially from the organic molecules reviewed above. In contrast to many optically active organic molecules, the active inorganic complexes have usually to be resolved; nature very seldom provides active metal-organic complexes, such as Vitamin B_{12} (for a discussion of the optical properties of cobalamines and related hemines, see [7c] and [262]). Nevertheless, the stereochemistry of numerous organometallic compounds has been investigated. For example, the phenomenon of isomerism due to nonplanarity of chelate rings, such as in complexes of amines, has been examined by RD and CD techniques. Moreover, since amino acid ions form stable chelate compounds with various metals, their stereochemistry can be deduced from optical properties. Similarly, the stereochemistry of metallocenes can be investigated through their optical properties.

In general, the long-wavelength absorption band of transition metal complexes is due to an electronic transition of the magnetic-dipole type. The optical properties of several metal complexes with symmetries lower than dihedral, as well as those with three bidendate ligands, have been measured in order to assign spectroscopic transitions and to establish absolute configurations. Moreover, the origin of the optical activity of metal complexes has been studied theoretically. The contribution played by RD and CD in this field has been fundamental, because the sign of the rotation at a particular wavelength cannot be used to relate the configurations of various substances, since they possess several absorption bands. The major role played by the RD and CD techniques for the assignment of configuration of ligands, metal complexes, chelates, and metallocenes has recently been reviewed [255–260], so that it will not be discussed in this section.

6 TABLE OF FUNCTIONAL GROUPS AND CHROMOPHORIC DERIVATIVES

Table 3.5 gives various functions and chromophoric derivatives that can easily be investigated by RD and CD. The position of the main optically active absorption bands and relevant references are also listed.

Table 3.5. Cotton effect of specific chromophores

Functional Groups	Wavelength Regions of Cotton Effects (in nm)	References
Acetates	210	137
Acids (organic)	210	124–126
Acylthioureas	340–345	151, 198, 284, 286
Adamantanones	295	281
Aldehydes		
Saturated	300	7, 93
α,β-Unsaturated	240–260. 340	7
β,γ-Unsaturated	300	7, 68c, 121
Alkyl dithiocarbamates	270, 330	7, 284
Alkyl nitrites	320–440	7, 168, 284
Alkylsulfinyl groups	210, 230	216
Allenes	220–250	85, 86, 87
Amides	210	138a
Amines (tertiary)	195–205, 220–230	191
Amino ketones	[a]	7, 287
α-Amino acids	210	124, 125, 139–149
Aromatic groups	200–380	7, 155
Anhydrides	210–220	7, 130
Arylketones	310	7, 155
Arylphosphoryl groups	[a]	264
Aryltetralines	230–245, 280–290	7, 155

[a] Depends on substituents and/or functional groups.

Table 3.5. *Continued*

Functional Groups	Wavelength Regions of Cotton Effects (in nm)	References
Azides	280–290	184, 185, 191
α-Azido acids	300–320	7
Aziridines	260 (290)	193
Azoalkanes	330–380	191
Azo derivatives	330–380	191
Azomethines	250	151, 186
Benzimidazoles	220–280	263
Benzoates	225–230	161
Benzquinolizidines	260, 280	7, 155, 271
Benzyl ketones	220, 300	7, 155
Benzyl olefins	265–290	7, 155
Benzyltetrahydroiso-quinolines	240, 270–290	7, 155, 270
Bianthryls	200–380	7, 155, 274
Biaryls	200, 260, 300, 340	7, 22, 155, 158
Binaphthyls	200–380	7, 155, 274
Biphenyls	200–380	7, 155, 158
Carboxylic acids	210	122–126
α,β-Unsaturated	250	7, 127
Dicarboxylic	200–210	125, 128
Cyclobutanones	300	7, 280
Cyclohexanones	300	7, 93, 94
Cycloheptanones	300	7
Cyclopentanones	300	7, 136
Dialkyldiazenes	330–380	191
Dialkyl ethers	low wavelength[a]	222, 223
Diazenes	330–380	191
Dibenzoates (cyclic α-glycols)	219, 233	163
Dienes (homoannular)	260–280	7, 76, 77
Dihydroindoles	275	7, 155
Dihydrophenanthrenes	200–380	7, 155
Dihydroresorcinol derivatives of amino acids	280–290	152

[a] Depends on substituents and/or functional groups.

Table 3.5. *Continued*

Functional Groups	Wavelength Regions of Cotton Effects (*in* nm)	References
Diketopiperazines	200, 210, 222	212, 214
Dimedonyl derivatives of amines	280–290	63, 179
Dimedonyl derivatives of amino acids	280–290	152
Dipyrrylmethanes	500	7
Diselenides	210	7
Disulfides	270, 310, 340[a]	7, 165, 200, 206–211, 286
Dithianes	280–350	7, 200
Dithiocarbamates	270, 330–350	7, 151, 182, 183, 195, 196, 284
Dithiocarbonates	280, 380	164, 286
Dithiolanes (dithioacetals)	245	164, 201, 286
Dithiourethanes	280, 340	7, 183, 195
Double bonds	200	7, 72, 73
Episulfides	260–265	164, 165, 286
Esters	210	137
Ethers	low wavelength[a]	222, 223, 261
Ethylenic compounds	200[a]	72, 73
Flavans	200–380	7, 155
Furans	220	7, 155
Glycol dibenzoates	219, 233	163
α-Halo-ketones	290–320	7
α-Hydroxy acids	210	124, 125, 141–150
Indanes	280	13, 22, 155, 273
Indolenines	240–290	155
Indoles	250–280	7, 155
Isothiocyanato derivatives of amino acids	250	90
Isothiocyanato derivatives of amino alcohols	200, 255, 340	90
Isoxazolines	[a]	191
Itaconyl derivatives of amines	[a]	176

[a] Depends on substituents and/or functional groups.

252

Table 3.5. Continued

Functional Groups	Wavelength Regions of Cotton Effects (*in* nm)	References
Ketones		
α-Cyclopropyl	300	101–104
Cyclopropenones	227, 260	261
α-Epoxy	300	101–104
Saturated	300	7, 93, 94
α-β-Unsaturated	240–260, 340	7
β,γ-Unsaturated	300	7, 68c, 121
Lactames	210	138c
Lactones		
Saturated	210	124, 132–135
α,β-Unsaturated	250	7, 269
Maleyl derivatives of amines		175, 176
Methyl xanthates	*a*	284
Morphinans	200–300	7, 155
N-Benzylidene derivatives of amines	250	171
N-Chloro amines	250–280	151, 188
N-Chloroamino ketals	250–280	188
N-Dithiocarbamates	270, 330–350	7, 151, 284, 286
N-Isopropylidene derivatives of amines	220	151, 172, 285
N-Nitroso derivatives of amines	370	151, 166, 187
N-Phenylthioacetyl	330	70d, 151, 177, 265
N-Phthaloyl derivatives of amines and amino acids	320	7, 151, 174, 175, 176, 284
N-(2-Pyridyl-N-oxide) amino acids	330	266
N-Salicylidene derivatives of amines	280, 315, 400	7, 151, 173
N-Thiobenzoyl derivatives	365	70d, 151, 177, 265

a Depends on substituents and/or functional groups.

Table 3.5. *Continued*

Functional Groups	Wavelength Regions of Cotton Effects (in nm)	References
N-Thionocarbethoxy derivatives	290	7
N-Thionocarbethoxy derivatives of amino acids	280	286
Nitrite esters	320–440	7, 275, 284
Nitrites	330–420	7
Nitroalkanes	270–280, 370	7, 284
Nitro derivatives	280, 370	7, 187, 189, 190, 284
Nitrones	207, 230–250, 295	226
Nitrosites (nitroso-nitrites)	680	7, 167, 276
Nitroso amides	300–450	7, 166
Nitroso amines	350–450	7, 151, 166, 187
Nitrous esters	320–440	7, 151, 284
Nitroxides	240, 275, 290	225
Nitryloxy groups	210, 230, 270	192
Nucleosides	230–350	155
Nucleotides	230–350	155
Olefins	200	7, 72, 73
Osmic esters	450, 550	74
Oxathianes	300	164
Oxathiolanes	240–250	164, 286
Oxazirans	195, 225	227
Oxazolines	[a]	191
Oximes		
Saturated	195–215	7, 153, 261
α,β-Unsaturated	240	7, 153, 261
Ozonides	low wavelength[a]	224
Peptides	190	7, 94, 231
Phenyls	200, 260	156, 157
Phenylosotriazole	225, 250–275	159
3-Phenyl-2-thiohydan-toines	310	286
Phosphinates	210, 260[a]	218

[a] Depends on substituents and/or functional groups.

254

Table 3.5. *Continued*

Functional Groups	Wavelength Regions of Cotton Effects (in nm)	References
Phosphine oxides	220, 230–240[a]	218, 219
Phosphine sulfides	260–280[a]	218
Phthalimides	320	174, 175, 284
Phthaloyl derivatives of amines and amino acids	320	174, 175, 176, 284
Poly-amino acids	[a]	7, 231–239
Polycarbonyl compounds	[a]	7, 290
Purines	[a]	160
Pyrazoles	260	7, 272
Pyrazolines	330	191
Quinoxalines	220–250, 320	7, 155, 263
Selenonaphthyl esters	[a]	268
Selenophenyl esters	[a]	268
Spiro pyrazolines	330	191
Styrenes	260–270	155, 160
Sulfinates	[a]	286
Sulfides	193–198, 200–215, 230–260[a]	199
Sulfites	[a]	215
Sulfonamides	[a]	178
Sulfoxides	220, 260–280[a]	202–205, 219, 286
Tetrahydrofurans	low wavelength[a]	222, 223, 283
Tetrahydroisoquinolines	270–290	7, 155
Tetrahydropyrans	low wavelength[a]	222, 223, 261
Thiocarbonates	235, 305, 430	7, 164, 286
Thiocyanates	245–250	194, 286
Thiohydantoines	265, 310	7, 197, 286
Thiolacetates	235, 270	282, 286
Thiolactones	235, 280–300	220, 221, 286
Thionamides	340–360	7, 138c, 151, 267, 277, 286, 288
Thiones (thioketones)	240, 490	7, 278, 286
Thionocarbalkoxy derivatives	270–370	151, 197

[a] Depends on substituents and/or functional groups.

Table 3.5. *Continued*

Functional Groups	Wavelength Regions of Cotton Effects (*in* nm)	References
Thioureas	340	151, 286
Thiourethanes	250–500	7, 195
Trithiocarbonates	235, 305, 430	7, 164, 286
Trithiones	200–550	7, 279, 286
Tropones	220, 320	7, 155
Xanthates	280, 350–355	7, 151, 195, 196, 284, 286

[a] Depends on substituents and/or functional groups.

7 CONCLUSION

This chapter does not intend to cover the whole field of applications of optical rotatory dispersion and circular dichroism in chemistry. Rather, its purpose is to emphasize the great utility and potential of these optical methods in stereochemistry. Thanks to the work of numerous schools, with those of Djerassi, Klyne, Legrand, Mislow, and Snatzke the most prolific, a large array of information has become available. The experimental data allow correlations to be made, and they constitute the basis from which general rules could be drawn, such as the octant rule for the carbonyl chromophore. At present, the techniques discussed provide a simple and reliable way to assign absolute configurations and conformations to optically active molecules, either by correlation—that is, comparison of the signs of Cotton effects with those of molecules of known stereochemistry—or by making use of the octant, sector, and quadrant rules.

Optical rotatory dispersion and circular dichroism have been applied to all fields of chemistry, from rather simple organic molecules, like terpenes, alkaloids, steroids, flavones, antibiotics, and porphyrins to more complex entities of biological importance (such as nucleic acids, polyamino acids, and proteins), as well as to organometallic compounds (ligands or metallocenes), thus showing the great potential of the method.

In view of the importance of the optical rotatory dispersion and circular dichroism techniques, illustrated in this chapter by the very broad range of applications, we can anticipate future advances in the three major directions outlined in the Introduction, namely in the field of instrumentation, applications to stereochemical problems and, last but not the least, in theoretical interpretations of experimentally observed phenomena.

References

1. A. M. Cotton, *Ann. Chim. Phys.*, Sec. 8, **8**, 347 (1896).
2. W. Kuhn and E. Braun, *Z. Physik. Chem.*, **B8**, 445 (1930).
3. S. Mitchell and S. Cormack, *J. Chem. Soc.*, 415 (1932).
4. M. Scherer, *Compt. Rend.*, **195**, 950 (1932).
5. H. Rudolph, *J. Opt. Soc. Am.*, **45**, 50 (1955).
6. C. Djerassi et al., *J. Am. Chem. Soc.*, **77**, 4354 (1955).
7. *a.* C. Djerassi, *Optical Rotatory Dispersion: Applications to Organic Chemistry*, McGraw-Hill, New York, 1960; *b.* P. Crabbé, *Optical Rotatory Dispersion and Circular Dichroism in Organic Chemistry*, Holden-Day, Inc., San Francisco, 1965; *c.* L. Velluz, M. Legrand, and M. Grosjean, *Optical Circular Dichroism. Principles, Measurements, and Applications*, Verlag Chemie, Weinheim, 1965; *d.* P. Crabbé, *Applications de la Dispersion Rotatoire Optique et du Dichroisme Circulaire Optique en Chimie Organique*, Gauthier-Villars, Paris, 1968; *e.* *Optical Rotatory Dispersion and Circular Dichroism in Organic Chemistry*, G. Snatzke, Ed., Heyden and Sons, Ltd., London, 1967.
8. P. M. Jones and W. Klyne, *J. Chem. Soc.*, 871 (1960).
9*a.* T. Bürer, M. Kohler, and H. H. Günthard, *Helv. Chim. Acta*, **41**, 2216 (1958).
9*b.* M. Billardon and J. Badoz, *Compt. Rend.*, **248**, 2466 (1959).
9*c.* M. Grosjean, A. Lacam, and M. Legrand, *Bull. Soc. Chim. France*, 1495 (1959).
10. F. Woldbye, in *Optical Rotatory Dispersion and Circular Dichroism in Organic Chemistry*, G. Snatzke, Ed., Heyden, London, 1967.
11. R. Newman and R. S. Halford, *Rev. Sci. Instr.*, **19**, 270 (1948).
12. G. R. Bird and W. A. Shurcliffe, *J. Opt. Soc. Am.*, **49**, 239 (1952).
13. N. J. Harrick, *J. Opt. Soc. Am.*, **54**, 1281 (1964).
14. W. C. Walker, *J. Opt. Soc. Am.*, **54**, 569 (1964).
15. W. A. Shurcliffe, *Polarized Light*, Harvard Univ. Press, Cambridge, 1962.
16. A. S. Makas, *J. Opt. Soc. Am.*, **52**, 43 (1962).
17. M. N. McDermott and R. Novick, *J. Opt. Soc. Am.*, **51**, 1008 (1961).
18. J. F. Archard and A. M. Taylor, *J. Sci. Instr.*, **25**, 407 (1948).
19. E. J. Gillam, *J. Sci. Instr.*, **34**, 435 (1957).
20. W. B. Fordyce, J. Green, and A. C. Parker, *Biochem. J.*, **68**, 33P (1958).
21. E. F. Daly, *Proc. 4th Intern. Meeting Mol. Spectr., Bologna, 1959*, Pergamon, New York, 1962, p. 1281.
22. B. R. Malcolm and A. Elliott, *J. Sci. Instr.*, **34**, 48 (1957).
23. E. J. Gillam, *Nature*, **178**, 1412 (1956).
24. B. H. Billings, *J. Opt. Soc. Am.*, **39**, 797 (1949).
25. A. S. Keston and J. Lospalluto, *Federation Proc.*, **12**, 229 (1953).
26. A. Savitsky, W. Slavin, and R. E. Salinger, *Proc. 4th Intern. Meeting Mol. Spectr., Bologna, 1959*, Pergamon, New York, 1962, p. 1360.
27. H. Tagasaki, *J. Opt. Soc. Am.*, **51**, 462 (1961).

28. R. J. King, *J. Sci. Instr.*, **43**, 617 (1966).
29. R. W. Ditchburn and G. A. J. Orchard, *Proc. Phys. Soc.*, B, **67**, 608 (1954).
30. V. A. Kizel et al., *Opt. Spectr.* (*USSR*), **17**, 248 (1964).
31. A. Abu-Shumays and J. J. Duffield, *Anal. Chem.*, **38**, 29A (1966).
32. H. Cary et al., *Appl. Opt.*, **3**, 329 (1964).
33. W. Budde, *J. Opt. Soc. Am.*, **52**, 343 (1962).
34. G. S. Wasserman, *J. Opt. Soc. Am.*, **54**, 1492 (1964).
35. E. J. Gillam and R. J. King, *J. Sci. Instr.*, **38**, 21 (1961).
36. B. Jirgensons, *Biochemistry*, **1**, 917 (1962).
37. R. J. King, *J. Sci. Instr.*, **43**, 924 (1966).
38. B. Carroll, M. B. Tillem, and E. S. Freeman, *Anal. Chem.*, **30**, 1099 (1958).
39. F. Woldbye, *Acta Chem. Scand.*, **13**, 2137 (1959).
40. J. P. Dirkx, P. J. van der Haak, and F. L. J. Sixma, *Anal. Chem.*, **36**, 1988 (1964); *Chem. Weekblad*, **56**, 151 (1960).
41. G. Nebbia and L. Notarnicola, *Boll. Sci. Fac. Chim. Ind. Bologna*, **21**, 92 (1963).
42. A. L. Rouy and B. Carroll, *Anal. Chem.*, **36**, 2501 (1964); **37**, 96 (1965).
43. F. Woldbye and S. Bagger, *Acta Chem. Scand.*, **20**, 1145 (1966).
44. R. J. King and M. J. Downs, *Photoelec. Spectrometry Group Bull.*, **19**, 579 (July 1970).
45. G. Holzwarth, W. B. Gratzer, and P. Doty, *J. Am. Chem. Soc.*, **84**, 3194 (1962); G. Holzwarth, *Rev. Sci. Instr.*, **36**, 59 (1965).
46. A. L. Rouy and B. Carroll, *Anal. Chem.*, **38**, 1367 (1966); **36**, 2501 (1964).
47. W. Heller, in *Technique of Organic Chemistry*, 3rd ed., A. Weissberger, Ed., Vol. I, Part III, Interscience, New York, 1960, p. 2292.
48. U. Meyer, *Ann. Physik*, **30**, 607 (1909).
49. T. M. Lowry and W. R. C. Coode-Adams, *Proc. Roy. Soc.* (*London*), **A226**, 391 (1927).
50. T. M. Lowry and C. Snow, *Proc. Roy. Soc.* (*London*), **A127**, 271 (1930).
51. H. S. Gutowski, *J. Chem. Phys.*, **19**, 438 (1951).
52. H. J. Hediger and H. H. Günthard, *Helv. Chim. Acta*, **37**, 1125 (1954).
53. H. R. Wyss and H. H. Günthard, *Appl. Opt.*, **5**, 1736 (1966).
54. C. C. Robinson, *Appl. Opt.*, **3**, 1163 (1964).
55. J. P. Craig, R. F. Gribbly, and A. A. Dougal, *Rev. Sci. Instr.*, **35**, 1501 (1964).
56. C. R. Pidgeon and S. D. Smith, *Infrared Phys.*, **4**, 13 (1964).
57. H. Rudolph, *J. Opt. Soc. Am.*, **49**, 1127 (1959).
58. J. G. Foss, *Anal. Chem.*, **35**, 1329 (1959).
59. G. Kemmner, *Reports on Analysis Techniques*, No. 5, Bodenseewerk Perkin-Elmer & Co. Uberlingen/Bodensee, 1966.
60. M. Grosjean and M. Legrand, *Compt. Rend.*, **251**, 2150 (1960).
61. P. Drude, *Lekrbuch der Optik*, S. Hirzel, Leipzig, 1906.
62. A. Cotton, *Compt. Rend.*, **120**, 989, 1044 (1895).
63. G. Aguilar Santos, E. Santos, and P. Crabbé, *J. Org. Chem.*, **32**, 2642 (1967).
64. *a.* A. Rassat, in *Optical Rotatory Dispersion and Circular Dichroism in Organic Chemistry*, G. Snatzke, Ed., Chap. 16, p. 314, Heyden and Sons, Ltd., London, 1967; *b.* A. Moscowitz, *op. cit.*, Chap. 17, p. 329.

65. L. H. Zalkow, R. Hale, K. French, and P. Crabbé, *Tetrahedron*, **26**, 4947 (1970).
66. A. Moscowitz, K. M. Wellman, and C. Djerassi, *Proc. Natl. Acad. Sci. U.S.*, **50**, 799 (1963).
67. *a.* O. E. Weigang, *J. Chem. Phys.*, **41**, 1435 (1964); *b.* E. A. Boudreaux, O. E. Weigang, and J. A. Turner, *Chem. Comm.*, **12**, 378 (1966).
68. *a.* G. Snatzke, *Proc. Roy. Soc. A.*, **297**, 43 (1967); *b.* see also Chap. 18 in [12]; *c.* G. Snatzke and K. Schaffner, *Helv. Chim. Acta*, **51**, 986 (1968); *d.* G. Snatzke, D. M. Piatah, and E. Caspi, *Tetrahedron*, **24**, 2899 (1968).
69. L. Velluz and M. Legrand, *Compt. Rend.*, **265**, 663 (1967).
70. *Inter alia: a.* J. Buchingham and R. D. Guthrie, *Chem. Comm.*, **12**, 570 (1967); *b.* T. Suga, T. Shishibori, and T. Matzuura, *Bull. Chem. Soc. Japan*, **41**, 1175 (1968); *c.* L. Verbit and J. W. Clark-Lewis, *Tetrahedron*, **24**, 5519 (1968); *d.* E. Bach, A. Kjaer, R. Dahlbom, T. Walle, B. Sjöberg, E. Bunnenberg, C. Djerassi, and R. Records, *Acta Chem. Scand.*, **20**, 2781 (1966); see also [300], [302], [304], [309]; *e.* G. Claeson, *Acta Chem. Scand.*, **22**, 2429 (1968).
71. N. L. Allinger and J. Ch. Tai, *J. Am. Chem. Soc.*, **87**, 2081 (1965).
72. *a.* A. Yogev and Y. Mazur, *Chem. Comm.*, **21**, 552 (1965); *b.* A. Yogev and Y. Mazur, *Tetrahedron*, **22**, 1317 (1966); *c.* A. Yogev, D. Amar, and Y. Mazur, *Chem. Comm.*, **7**, 339 (1967).
73. M. Legrand and R. Viennet, *Compt. Rend.*, **262**, 1290 (1966).
74. E. Bunnenberg and C. Djerassi, *J. Am. Chem. Soc.*, **82**, 5953 (1960).
75. *a.* E. Premuzic and A. I. Scott, *Chem. Comm.*, **21**, 1078 (1967); *b.* A. D. Wrixon, E. Premuzic, and A. I. Scott, *Chem. Comm.*, **11**, 639 (1968).
76. A. Moscowitz, E. Charney, U. Weiss, and H. Ziffer, *J. Am. Chem. Soc.*, **83**, 4661 (1961).
77. *a.* U. Weiss, H. Ziffer, and E. Charney, *Tetrahedron*, **21**, 3105 (1965); *b.* E. Charney, *Tetrahedron*, **21**, 3127 (1965).
78. H. Ziffer and U. Weiss, *J. Org. Chem.*, **31**, 2691 (1966).
79. H. J. C. Jacobs and E. Havinga, *Rec. Trav. Chim.*, **84**, 932 (1965).
80. H. Ziffer, U. Weiss, and E. Charney, *Tetrahedron*, **23**, 3881 (1967).
81. E. J. Corey and A. G. Hortmann, *J. Am. Chem. Soc.*, **85**, 4033 (1963).
82. T. Asao, S. Ibe, K. Takase, Y. S. Cheng, and T. Nozoe, *Tetrahedron Letters*, **33**, 3639 (1968).
83. E. Charney, H. Ziffer, and U. Weiss, *Tetrahedron*, **21**, 3121 (1965).
84. *a.* K. Mislow, *Ann. N.Y. Acad. Sci.*, **93**, 459 (1962); *b.* D. J. Sandman and K. Mislow, *J. Am. Chem. Soc.*, **91**, 645 (1969).
85. *a.* P. Rona and P. Crabbé, *J. Am. Chem. Soc.*, **91**, 3289 (1969); *b.* W. R. Benn, *J. Org. Chem.*, **33**, 3113 (1968).
86. *a.* P. Crabbé, M. Biollaz, H. Carpio, A. Failli, P. Rona, E. Velarde, and J. H. Fried, *J. Org. Chem.*, in press; *b.* E. Velarde and P. Crabbé, unpublished results.
87. S. F. Mason and G. W. Vane, *Tetrahedron Letters*, **21**, 1593 (1965).
88. P. M. Jones and W. Klyne, *J. Chem. Soc.*, 871 (1960); see also [4], [6], and [9].
89. L. Mamlok, A. M. Giroud, and J. Jacques, *Bull. Soc. Chim. France*, 1806 (1961).
90. B. Halpern, W. Patton, and P. Crabbé, *J. Chem. Soc.* (B), 1143 (1969).

91. S. T. K. Bukhari, R. D. Tuthrie, A. I. Scott, and A. D. Wrixon, *Chem. Comm.*, **24**, 1580 (1968).

92. *a.* For a review see [22]; *b.* R. C. Schulz, R. Wolf, and H. Mayerhöfer, *Kolloid-Z. & Z. Polymere*, **227**, 65 (1968); *c.* K. Heyns, K. W. Pflughaupt, and H. Paulsen, *Chem. Ber.*, **101**, 2800 (1968); *d.* W. S. Chilton, *J. Org. Chem.*, **33**, 4459 (1968) and references therein; *e.* T. Sticzay, C. Peciar, and S. Bauer, *Tetrahedron Letters*, **20**, 2407 (1968); *f.* M. Maeda, T. Kinoshita, and A. Tsuji, *Tetrahedron Letters*, **30**, 3407 (1968); *g.* I. Listowsky and S. England, *Biochem. Biophys. Res. Comm.*, **30**, 329 (1968); *h.* G. Lyle and M. J. Piazza, *J. Org. Chem.*, **33**, 2478 (1968); *i.* S. Inouye, T. Tsuruoka, and T. Niida, *J. Antibiotics*, **19**, 288 (1966); *j.* S. Inouye, *Chem. Pharm. Bull. Japan*, **15**, 1609 (1967); *k.* S. Inouye, T. Tsuruoka, T. Ito, and T. Niida, *Tetrahedron*, **23**, 2125 (1968); *l.* W. Voelter, E. Bayer, R. Records, E. Bunnenberg, and C. Djerassi, *Ann. Chem.*, **718**, 238 (1968); *m.* H. Paulsen, J. Brüning, and K. Heyns, *Chem. Ber.*, **102**, 459 (1969); *n.* H. Paulsen, K. Propp, and J. Brüning, *Chem. Ber.*, **102**, 469 (1969); *o.* F. Cramer, G. Mackensen, and K. Sensse, *Chem. Ber.*, **102**, 494 (1969); *p.* K. Sensse and F. Cramer, *Chem. Ber.*, **102**, 509 (1969); *q.* R. J. Ferrier, N. Prasad, and G. H. Sankey, *J. Chem. Soc.* (C), **587** (1969) and earlier papers by these authors; *r.* H. Iwamura and T. Hashizume, *J. Org. Chem.*, **33**, 1796 (1968).

93. W. Moffitt, R. B. Woodward, A. Moscowitz, W. Klyne, and C. Djerassi, *J. Am. Chem. Soc.*, **83**, 4013 (1961).

94. J. A. Schellman, *Accounts Chem. Res.*, **1**, 144 (1968).

95. J. A. Schellman, *J. Chem. Phys.*, **44**, 55 (1966).

96. G. Wagnière, *J. Am. Chem. Soc.*, **88**, 3937 (1966).

97. Y. H. Pas and D. P. Santry, *J. Am. Chem. Soc.*, **88**, 4157 (1966).

98. J. S. E. Holker, W. R. Jones, M. G. R. Leeming, G. M. Holder, and W. B. Whalley, *Chem. Comm.*, **2**, 90 (1967).

99. J. Hudec, *Chem. Comm.*, **11**, 539 (1967).

100. *a.* R. de L. Kronig, *J. Opt. Soc. Am.*, **12**, 547 (1926); *b.* H. A. Kramers, *Atti. Congr. Intern. Fisici. Como*, **2**, 545 (1927).

101. *a.* M. Legrand, R. Viennet, and J. Caumartin, *Compt. Rend.*, **253**, 2378 (1961); *b.* C. Djerassi, W. Klyne, T. Norin, G. Ohloff, and E. Klein, *Tetrahedron*, **21**, 163 (1965).

102. H. C. Brown and A. Suzuki, *J. Am. Chem. Soc.*, **89**, 1933 (1967).

103. S. B. Laing and P. J. Sykes, *J. Chem. Soc.* (C), **937** (1968).

104. W. Reusch and P. Mattison, *Tetrahedron*, **24**, 4933 (1968).

105. P. Crabbé, A. Cruz, and J. Iriarte, *Can. J. Chem.*, **46**, 349 (1968).

106. W. B. Whalley, *Chem. Ind.*, 1024 (1962).

107. R. E. Ballard, S. F. Mason, and G. W. Vane, *Disc. Faraday Soc.*, **35**, 43 (1963).

108. G. Snatzke, *Tetrahedron*, **21**, 413, 421, 439 (1965).

109. A. Moscowitz, communication at the NATO Summer School, Bonn, September 1965.

110. K. Kuriyama, M. Moriyama, T. Iwata, and K. Tori, *Tetrahedron Letters*, **13**, 1661 (1968).

111. R. Bucourt, D. Hainaut, J. C. Gasc, and G. Nominé, *Tetrahedron Letters*, **49**, 5093 (1968).

112. H. Ziffer and C. H. Robinson, *Tetrahedron*, **24**, 5803 (1968).

113. See also: *a.* G. Snatzke, B. Zeeh, and E. Müller, *Tetrahedron*, **20**, 2937 (1964); *b.* C. Djerassi and J. E. Gurst, *J. Am. Chem. Soc.*, **86**, 1755 (1964).

114. J. C. Bloch and S. R. Wallis, *J. Chem. Soc.* (B), 1177 (1966).

115. *a.* P. Crabbé, R. Grezemkovsky, and L. H. Knox, *Bull. Soc. Chim. France*, 789 (1968); *b.* P. Crabbé, P. Anderson, and E. Velarde, *J. Am. Chem. Soc.*, **90**, 2998 (1968); *c.* P. Anderson, P. Crabbé, A. D. Cross, J. H. Fried, L. H. Knox, J. Murphy, and E. Velarde, *J. Am. Chem. Soc.*, **90**, 3888 (1968).

116. H. Labhart and G. Wagnière, *Helv. Chim. Acta*, **42**, 2219 (1959).

117. *a.* R. B. Woodward and E. G. Kovach, *J. Am. Chem. Soc.*, **72**, 1009 (1950); *b.* R. B. Woodward and P. Yates, *Chem. Ind.*, 1391 (1954).

118. *a.* R. C. Cookson and J. Hudec, *J. Chem. Soc.*, 429 (1962); *b.* D. E. Bays, R. C. Cookson, and S. MacKenzie, *J. Chem. Soc.* (B), 215 (1967).

119. S. F. Mason, *Proc. Chem. Soc.*, **61** (1964) and earlier papers.

120. A. Moscowitz, *Proc. Chem. Soc.*, **60** (1964).

121. *a.* K. Mislow, M. A. W. Glass, A. Moscowitz, and C. Djerassi, *J. Am. Chem. Soc.*, **83**, 2771 (1961); *b.* A. Moscowitz, K. Mislow, M. A. W. Glass, and C. Djerassi, *J. Am. Chem. Soc.*, **84**, 1945 (1962); *c.* E. Bunnenberg, C. Djerassi, K. Mislow, and A. Moscowitz, *J. Am. Chem. Soc.*, **84**, 2823 (1962).

122. A. Moscowitz, in *Advances in Chemical Physics*, Prigogine, Ed., Vol. 4, I, Interscience, New York, 1962, p. 67.

123. *a.* P. Crabbé, A. Cruz, and J. Iriarte, *Photochem. and Photobiol.*, **7**, 829 (1968); *b.* P. Sunder-Plassman, P. H. Nelson, P. H. Boyle, A. Cruz, J. Iriarte, P. Crabbé, J. A. Edwards, and J. H. Fried, *J. Org. Chem.*, **34**, 3779 (1969).

124. *a.* J. P. Jennings, W. Klyne, and P. M. Scopes, *Proc. Chem. Soc.*, 412 (1964); *J. Chem. Soc.*, 7211, 7229 (1965); *b.* W. Klyne, P. M. Scopes, and A. Williams, *J. Chem. Soc.*, 7237 (1965); *c.* J. P. Jennings, W. Klyne, and P. M. Scopes, *J. Chem. Soc.*, 294 (1965); *d.* W. Klyne, *Proc. Roy. Soc.*, A, **297**, 66 (1967); *e.* J. D. Renwick and P. M. Scopes, *J. Chem. Soc.* (C), 1949 (1968), and references therein.

125. *a.* A. Fredga, J. P. Jennings, W. Klyne, P. M. Scopes, B. Sjöberg, and S. Sjöberg, *J. Chem. Soc.*, 3928 (1965); *b.* I. P. Dirkx and F. L. J. Sixma, *Rec. Trav. Chim.*, **83**, 522 (1964); *c.* G. Gottarelli, W. Klyne, and P. M. Scopes, *J. Chem. Soc.* (C), 1366 (1967), and references therein; *d.* G. Gottarelli and P. M. Scopes, *J. Chem. Soc.* (C), 1370 (1967); *e.* Y. Inouye, S. Sawada, M. Ohno, and H. M. Walborsky, *Tetrahedron*, **23**, 3237 (1967); *f.* L. Verbit and Y. Inouye, *J. Am. Chem. Soc.*, **89**, 5717 (1967).

126. G. Snatzke, H. Ripperger, C. Horstmann, and K. Schreiber, *Tetrahedron*, **22**, 3103 (1966).

127. U. Weiss and H. Ziffer, *J. Org. Chem.*, **28**, 1248 (1963).

128. G. Montaudo and C. G. Overberger, *J. Am. Chem. Soc.*, **91**, 753 (1969).

129. M. Kasha, *J. Radiation Res.*, **20**, 55 (1963).

130. F. A. Mikulski, Ph.D. Thesis, Princeton University, 1965; see also [7*d*].

131. *a.* W. Klyne, P. M. Scopes, R. C. Sheppard, and S. Turner, *J. Chem. Soc.* (C), 1954 (1968); *b.* T. G. Waddell, W. Stöcklin, and T. A. Geissman, *Tetrahedron Letters*, **17**, 1313 (1969); *c.* A. K. Banerjee and M. Gut, *J. Org. Chem.*, **34**, 1614 (1969).

132. H. Wolf, *Tetrahedron Letters*, **42**, 5151 (1966).

133. *a.* T. Okuda, S. Harigaya, and A. Kiyomoto, *Chem. Pharm. Bull. Japan*, **12**, 504 (1964); *b.* M. Gorodetsky, N. Danieli, and Y. Mazur, *J. Org Chem.*, **32**, 760 (1967); *c.* O. Cervinka and L. Hub, *Collect. Czech. Chem. Comm.*, **33**, 2927 (1968).

134. M. Legrand and R. Bucourt, *Bull. Soc. Chim. France*, 2241 (1967).

135. A. F. Beecham, *Tetrahedron Letters*, **32**, 3591 (1968).

135*a.* G. Snatzke and E. Otto, *Tetrahedron*, **25**, 2041 (1969).

136. W. Klyne, *Tetrahedron*, **13**, 29 (1961).

137. *a.* J. Cymerman Craig and S. K. Roy, *Tetrahedron*, **21**, 1847 (1965); J. Cymerman Craig, D. P. G. Hamon, K. K. Purushothaman, and S. K. Roy, *Tetrahedron*, **22**, 175 (1966); *b.* J. P. Jennings, W. Klyne, W. P. Mose, and P. M. Scopes, *Chem. Comm.*, **15**, 553 (1966); *c.* J. P. Jennings, W. P. Mose, and P. M. Scopes, *J. Chem. Soc.* (C), 1102 (1967); *d.* K. Kuriyama and K. Igaraski, Abstracts of Papers, **34**, 127, 20th Ann. Meeting, Chem. Soc. Japan, April 1967.

138. *a.* J. A. Schellman and P. Oriel, *J. Chem. Phys.*, **37**, 2114 (1962); *b.* See [125*e*]; *c.* H. Wolf, *Tetrahedron Letters*, **16**, 1075 (1965).

139. W. Gaffield, *Chem. Ind.*, 1460 (1964).

140. *a.* A. Moscowitz, A. Rosenberg, and A. E. Hansen, *J. Am. Chem. Soc.*, **87**, 1813 (1965); *b.* M. Legrand and R. Viennet, *Bull. Soc. Chim. France*, 679 (1965), 2798 (1966); *c.* J. Cymerman Craig and S. K. Roy, *Tetrahedron*, **21**, 391 (1965); *d.* R. D. Anand and M. K. Hargreaves, *Chem. Ind.*, 880 (1968); *e.* J. Horwitz, E. H. Strickland, and C. Billups, *J. Am. Chem. Soc.*, **91**, 184 (1969); *f.* D. G. Neilson, I. A. Khan, and R. S. Whitehead, *J. Chem. Soc.* (C), 1853 (1968).

141. *a.* J. M. Calvo, C. M. Stevens, M. G. Kalyanpur, and H. E. Umbarger, *Biochemistry*, **3**, 2024 (1964); *b.* E. Jonsson, *Acta Chem. Scand.*, **19**, 2247 (1965); *c.* J. Cymerman Craig and S. K. Roy, *Tetrahedron*, **21**, 1847 (1965); *d.* J. Cymerman Craig, R. J. Dummel, and S. K. Roy, *Biochemistry*, **4**, 2547 (1965).

142. *a.* Y. Tsuzuki, K. Tanabe, K. Okamoto, and M. Fukubayashi, *Bull. Chem. Soc. Japan*, **39**, 1387 (1966); *b.* J. Rétey, A. Umani-Ronchi, and D. Arigoni, *Experientia*, **22**, 72 (1966).

143. T. H. Applewhite, R. G. Binder, and W. Gaffield, *J. Org. Chem.*, **32**, 1173 (1967).

144. D. L. Dull and H. S. Mosher, *J. Am. Chem. Soc.*, **89**, 4230 (1967), and references cited.

145. R. D. Anand and M. K. Hargreaves, *Chem. Comm.*, **9**, 421 (1967).

146. J. Bolard, *Compt. Rend.*, **264**, 73 (1967).

147. H. C. Beyerman, L. Maat, D. de Ryke, and J. P. Visser, *Rec. Trav. Chim.*, **86**, 1057 (1967).

148. L. I. Katzin and E. Gulyas, *J. Am. Chem. Soc.*, **90**, 247 (1968), and references therein.

149. J. M. Tsangaris, J. Wen Chang, and R. B. Martin, *J. Am. Chem. Soc.*, **91**, 726 (1969).

150. *a.* I. Frič, V. Špirko, and K. Bláha, *Collection Czech. Chem. Comm.*, **33**, 4008 (1968); *b.* See also [140*a*] and [140*e*].

151. B. Sjöberg, in *Optical Rotatory Dispersion and Circular Dichroism in Organic Chemistry*, G. Snatzke, Ed., 1967, Chap. 11, p. 173.

152. P. Crabbé, B. Halpern, and E. Santos, *Tetrahedron*, **24**, 4315 (1968); see also [179*c*].

153. P. Crabbé and L. Pinelo, *Chem. Ind.*, 158 (1966).

154. G. G. Lyle and R. Mestrallet Barrera, *J. Org. Chem.*, **29**, 3311 (1964).

155. P. Crabbé and W. Klyne, *Tetrahedron*, **23**, 3449 (1967).

156. *a.* J. H. Brewster and J. G. Buta, *J. Am. Chem. Soc.*, **88**, 2233 (1966); *b.* See also [140*a*] and [140*e*]; *c.* G. De Angelis, Ph.D. Thesis, Iowa State University, 1966, and references cited therein; *d.* P. M. Scopes, *Ann. Reports*, 47 (1967); *e.* L. Verbit, A. S. Rao, and J. W. Clark-Lewis, *Tetrahedron*, **24**, 5839 (1968).

157. K. Kuriyama, T. Iwata, M. Moriyama, K. Kotera, Y. Hameda, R. Mitsui, and T. Takeda, *J. Chem. Soc.* (B), 46 (1967).

158. K. Mislow, M. A. W. Glass, R. E. O'Brien, P. Rutkin, D. H. Steinberg, J. Weiss, and C. Djerassi, *J. Am. Chem. Soc.*, **84**, 1455 (1962).

159. *a.* G. G. Lyle and M. J. Piazza, *J. Org. Chem.*, **33**, 2478 (1968); *b.* D. W. Miles, R. K. Robins, and H. Eyring, *J. Phys. Chem.*, **71**, 3931 (1967).

160. P. Crabbé, *Chem. Ind.*, 917 (1969).

161. N. Harada, M. Ohashi, and K. Nakanishi, *J. Am. Chem. Soc.*, **90**, 7349 (1968).

162. N. Harada and K. Nakanishi, *J. Am. Chem. Soc.*, **90**, 7351 (1968).

163. *a.* N. Harada and K. Nakanishi, *J. Am. Chem. Soc.*, **91**, 3989 (1969); *b.* M. Koreeda, N. Harada, and K. Nakanishi, *Chem. Comm.*, **10**, 548 (1969); *c.* N. Harada, K. Nakanishi, and S. Tatsuoka, *J. Am. Chem. Soc.*, **91**, 5896 (1969). We want to express our gratitude to Professor Nakanishi for providing us preprints of his publication.

164. *a.* C. Djerassi, H. Wolf, D. A. Lightner, E. Bunnenberg, K. Takeda, T. Komeno, and K. Kuriyama, *Tetrahedron*, **19**, 1547 (1963); *b.* D. A. Lightner and C. Djerassi, *Tetrahedron*, **21**, 583 (1965); *c.* D. A. Lightner, C. Djerassi, K. Takeda, K. Kuriyama, and T. Komeno, *Tetrahedron*, **21**, 1581 (1965); *d.* K. Kuriyama, T. Komeno, and K. Takeda, *Tetrahedron*, **22**, 1039 (1966).

165. D. E. Bays, R. C. Cookson, R. R. Hill, J. F. McGhie, and G. E. Usher, *J. Chem. Soc.*, 1563 (1964).

165*a*. A. H. Haines and C. S. P. Jenkins, *Chem. Comm.*, **7**, 350 (1969).

166. *a.* C. Djerassi, E. Lund, E. Bunnenberg, and B. Sjöberg, *J. Am. Chem. Soc.*, **83**, 2307 (1961); *b.* G. Snatzke, H. Ripperger, C. Horstmann, and K. Schreiber, *Tetrahedron*, **22**, 3103 (1966); *c.* A. La Manna and V. Ghislandi, *Il Farmaco*, **17**, 355 (1962).

167. *a.* S. Mitchell, *J. Chem. Soc.*, 3258 (1928); *ibid.*, 1829 (1930); *b.* S. Mitchell and S. B. Cormack, *J. Chem. Soc.*, 415 (1932).

168. *a.* W. Kuhn and H. L. Lehmann, *Z. Physik. Chem.*, **B18**, 32 (1932); *b.* W. Kuhn and H. Biller, *Z. Physik. Chem.*, **B19**, 1 (1935); *c.* H. B. Elkins and W. Kuhn, *J. Am. Chem. Soc.*, **57**, 296 (1935); *d.* W. Kuhn, *Ann. Rev. Phys. Chem.*, **9**, 417 (1958).

169. *a.* F. Nerdel, K. Becker, and G. Kresze, *Chem. Ber.*, **89**, 2862 (1956); *b.* G. Dudek, *J. Org. Chem.*, **32**, 2016 (1967).

170. *a.* A. P. Terentev and V. M. Potapov, *Zh. Obshch. Khim.*, **28**, 1161, 3323 (1958); *b.* V. M. Potapov, A. P. Terentev, and R. I. Sarylaeva, *Zh. Obshch. Khim.*, **29**, 3139 (1959); *c.* V. M. Potapov and A. P. Terentev, *Zh. Obshch. Khim.*, **30**, 666 (1960); *d.* V. M. Potapov, A. P. Terentev, and S. P. Spivak, *Zh. Obshch. Khim.*, **31**, 2415 (1961).

171. H. E. Smith, S. L. Cook, and M. E. Warren, *J. Org. Chem.*, **29**, 2265 (1964).

172. H. E. Smith, M. E. Warren, and A. W. Ingersoll, *J. Am. Chem. Soc.*, **84** 1513 (1962).

173. *a.* D. Bertin and M. Legrand, *Compt. Rend.*, **256**, 960 (1963); *b.* M. E. Warren and H. E. Smith, *J. Am. Chem. Soc.*, **87**, 1757 (1965); *c.* H. E. Smith and T. Ch. Willis, *J. Org. Chem.*, **30**, 2654 (1965); *d.* H. Ripperger, K. Schreiber, G. Snatzke, and K. Heller, *Z. Chem.*, **5**, 62 (1965); *e.* H. E. Smith and R. Records, *Tetrahedron,* **22**, 813 (1966); *f.* H. Ripperger, K. Schreiber, G. Snatzke, and K. Ponsold, *Tetrahedron*, **25**, 827 (1969).

174. *a.* J. H. Brewster and S. F. Osman, *J. Am. Chem. Soc.*, **82**, 5754 (1960); *b.* H. Wolf, E. Bunnenberg and C. Djerassi, *Chem. Ber.*, **97**, 533 (1964).

175. A. La Manna and V. Ghislandi, *Il Farmaco*, **19**, 480 (1964).

176. A. La Manna, V. Ghislandi, P. M. Scopes, and R. H. Swan, *Il Farmaco*, **20**, 842 (1965).

177. B. Sjöberg, B. Karlen, and R. Dahlbom, *Acta Chem. Scand.*, **16**, 1071 (1962).

178. V. M. Potapov, V. N. Demyanovich, and A. P. Terentev, *Vestn. Mosk. Univ. Ser. II, Khim.*, **20**, 56 (1965); *Chem. Abstr.*, **62**, 14461 (1965).

179. *a.* E. Santos, J. Padilla, and P. Crabbé, *Can. J. Chem.*, **45**, 2275 (1967); *b.* P. Crabbé, B. Halpern, and E. Santos, *Tetrahedron*, **24**, 4299 (1968); *c.* V. Tortorella, B. Halpern, and and P. Crabbé, in preparation.

180. *a.* A. H. Beckett and L. G. Brookes, *Tetrahedron*, **24**, 1283 (1968); *b.* H. E. Smith and M. E. Warren, *Tetrahedron*, **24**, 1327 (1968), and references therein; *c.* A. La Manna, V. Ghislandi, P. B. Hulbert, and P. M. Scopes, *Il Farmaco*, **23**, 1161 (1968).

181. J. Parello and F. Picot, *Tetrahedron Letters*, **49**, 5083 (1968).

182. *a.* C. Djerassi, H. Wolf, and E. Bunnenberg, *J. Am. Chem. Soc.*, **84**, 4552 (1962); *b.* W. S. Briggs and C. Djerassi, *Tetrahedron*, **21**, 3455 (1965); *c.* I. P. Dirkx and T. J. de Boer, *Rec. Trav. Chim.*, **83**, 535 (1964).

183. *a.* H. Ripperger, *Tetrahedron*, **25**, 725 (1969); *b.* H. Ripperger, *Angew. Chem. Intern. Ed.*, **6**, 704 (1967).

184. *a.* C. Djerassi, A. Moscowitz, K. Ponsold, and G. Steiner, *J. Am. Chem. Soc.*, **89**, 347 (1967); *b.* W. D. Closson and H. B. Gray, *J. Am. Chem. Soc.*, **85**, 290 (1963); *c.* E. Lieber, J. S. Curtice, and C. N. R. Rao, *Chem. Ind.*, 586 (1966).

185. H. Paulsen, *Chem. Ber.*, **101**, 1571 (1968).

186. *a.* H. Ripperger, K. Schreiber, and G. Snatzke, *Tetrahedron*, **21**, 1027 (1965), and references cited therein; *b.* R. Bonnett and T. R. Emerson, *J. Chem. Soc.*, 4508 (1965); *c.* Z. Badr, R. Bonnett, W. Klyne, R. J. Swan, and J. Wood, *J. Chem. Soc.* (C), 2047 (1966).

187. *a.* G. Snatzke, H. Ripperger, C. Horstman, and K. Schreiber, *Tetrahedron*, **22**, 3103 (1966); *b.* H. Ripperger and K. Schreiber, *Tetrahedron*, **23**, 1841 (1967); *c.* H. Ripperger and H. Pracejus, *Tetrahedron*, **24**, 99 (1968).

188. H. Ripperger, K. Schreiber, and G. Snatzke, *Tetrahedron*, **21**, 727 (1965), and references therein.

189. *a.* G. Snatzke, D. Becher, and J. R. Bull, *Tetrahedron*, **20**, 2443 (1964); *b.* G Snatzke, *J. Chem. Soc.*, 5002 (1965); *c.* J. R. Bull, J. P. Jennings, W. Klyne G. D. Meakins, P. M. Scopes, and G. Snatzke, *J. Chem. Soc.*, 3152 (1965)

190. *a.* T. M. Lowry, *Optical Rotatory Power*, Dover, New York, 1964; *b.* S Mitchell and R. R. Gordon, *J. Chem. Soc.*, 853 (1936).

191. *a.* G. Snatzke and J. Himmelreich, *Tetrahedron*, **23**, 4337 (1967); *b.* D. J. Severn and E. M. Kosower, *J. Am. Chem. Soc.*, **91**, 1710 (1969), and references therein.

192. G. Snatzke, H. Laurent, and R. Wiechert, *Tetrahedron*, **25**, 761 (1969).

193. R. G. Kostyanovsky, Z. E. Samojlova, and I. I. Tchervin, *Tetrahedron Letters*, **9**, 719 (1969).

194. C. Djerassi, D. A. Lightner, D. A. Schooley, K. Takeda, T. Komeno, and K. Kuriyama, *Tetrahedron*, **24**, 6913 (1968).

195. *a.* T. M. Lowry and H. Hudson, *Phil. Trans. Roy. Soc. London*, **232A**, 117 (1933); *b.* C. Djerassi, H. Wolf, and E. Bunnenberg, *J. Am. Chem. Soc.*, **84**, 4552 (1962); *c.* B. Sjöberg, D. J. Cram, L. Wolf, and C. Djerassi, *Acta Chem. Scand.*, **16**, 1079 (1962); *d.* Y. Tsuzuki, K. Tanabe, M. Akagi, and S. Tejima, *Bull. Chem. Soc. Japan*, **40**, 628 (1967).

196. B. Sjöberg, A. Fredga, and C. Djerassi, *J. Am. Chem. Soc.*, **81**, 5002 (1959).

197. C. Djerassi, K. Undheim, R. C. Sheppard, W. G. Terry, and B. Sjöberg, *Acta Chem. Scand.*, **15**, 903 (1961).

198. *a.* C. Djerassi and K. Undheim, *J. Am. Chem. Soc.*, **82**, 5755 (1960); *b.* C. Djerassi, K. Undheim, and A. M. Weidler, *Acta Chem. Scand.*, **16**, 1147 (1962).

199. *a.* P. Laur, H. Häuser, J. E. Gurst, and K. Mislow, *J. Org. Chem.*, **32**, 498 (1967); *b.* P. Salvadori, *Chem. Comm.*, **20**, 1203 (1968).

200. *a.* A. Fredga, *Acta. Chem. Scand.*, **4**, 1307 (1950); *b.* C. Djerassi, A. Fredga, and B. Sjöberg, *Acta Chem. Scand.*, **15**, 417 (1961); *c.* C. Djerassi, H. Wolf, and E. Bunnenberg, *J. Am. Chem. Soc.*, **84**, 4552 (1962); *d.* A. F. Beecham and A. McL. Mathieson, *Tetrahedron Letters*, **27**, 3139 (1966); *e.* see also [70e].

201. R. C. Cookson, G. H. Cooper, and J. Hudec, *J. Chem. Soc.* (B), 1004 (1967).

202. *a.* K. Mislow, *Angew. Chem. Intern. Ed.*, **4**, 717 (1965); *b.* K. Mislow, M . M Green and M. Raban, *J. Am. Chem. Soc.*, **87**, 2761 (1965) and earlier papers by the Princeton group; *c.* K. Mislow, *Record Chem. Progr.*, **28**, 217 (1967), and references therein; *d.* P. D. Henson and K. Mislow, *Chem. Comm.*, **8**, 413 (1969).

203. *a*. K. K. Andersen, W. Gaffield, N. E. Papanikolaou, J. W. Foley, and R. I. Perkins, *J. Am. Chem. Soc.*, **86**, 5637 (1964); *b*. C. R. Johnson and D. McCants, *J. Am. Chem. Soc.*, **87**, 5404 (1965); *c*. For early references see [7*d*].

204. *a*. S. I. Goldberg and M. S. Sahli, *J. Org. Chem.*, **32**, 2059 (1967); *b*. M. Axelrod, P. Bickart, M. L. Goldstein, M. M. Green, A. Kjaer, and K. Mislow, *Tetrahedron Letters*, **29**, 3249 (1968); *c*. R. Nagarajan, B. H. Chollar, and R. M. Dodson, *Chem. Comm.*, **11**, 550 (1967); *d*. P. B. Sollman, R. Nagarajan, and R. M. Dodson, *Chem. Comm.*, **11**, 552 (1967).

205. *a*. D. N. Jones, M. J. Green, M. A. Saeed, and R. D. Whitehouse, *Chem. Comm.*, **19**, 1003 (1967); *b*. D. N. Jones and M. J. Green, *J. Chem. Soc.* (C), 532 (1967); *c*. D. N. Jones, M. J. Green, and R. D. Whitehouse, *Chem. Comm.*, **24**, 1634 (1968); *d*. D. N. Jones, M. J. Green, M. A. Saeed, and R. D. Whitehouse, *J. Chem. Soc.* (C), 1362 (1968).

206. H. Herrmann, R. Hodges, and A. Taylor, *J. Chem. Soc.*, 4315 (1964).

207. M. Carmack and L. A. Neubert, *J. Am. Chem. Soc.*, **89**, 7134 (1967).

208. A. F. Beecham, J. W. Loder, and G. B. Russell, *Tetrahedron Letters*, **15**, 1785 (1968).

209. J. A. Barltrop, P. M. Hayes, and M. Calvin, *J. Am. Chem. Soc.*, **76**, 4348 (1954).

210. D. L. Coleman and E. R. Blout, in *Conformation of Biopolymers*, G. N. Ramachandran, Ed., Vol. I, Academic Press, New York, 1967, p. 123.

211. *a*. R. M. Dodson and V. C. Nelson, *J. Org. Chem.*, **33**, 3966 (1968); *b*. G. Claeson, *Acta Chem. Scand.*, **22**, 2429 (1968); *c*. P. C. Kahn and S. Beychok, *J. Am. Chem. Soc.*, **90**, 4168 (1968).

212. R. Nagarajan, N. Neuss, and M. M. Marsch, *J. Am. Chem. Soc.*, **90**, 6518 (1968).

213. *a*. See [80] and [206]; *b*. A. F. Beecham, J. Fridrichsons, and A. McL. Mathieson, *Tetrahedron Letters*, **27**, 3131 (1966); *c*. A. F. Beecham and A. McL. Mathieson, *Tetrahedron Letters*, **27**, 3139 (1966).

214. *a*. C. G. Overberger, G. Montaudo, J. Sebenda, and R. A. Veneski, *J. Am. Chem. Soc.*, **91**, 1256 (1969); *b*. H. Edelhoch and R. E. Lippoldt, *J. Biol. Chem.*, **243**, 4799 (1968).

215. M. K. Hargreaves, P. G. Modi, and J. G. Pritchard, *Chem. Comm.*, **21**, 1306 (1968) and references therein.

216. D. N. Jones, D. Mundy, and R. D. Whitehouse, *Chem. Comm.*, **24**, 1636 (1968).

217. C. H. Robinson, L. Milewich, G. Snatzke, W. Klyne, and S. R. Wallis, *J. Chem. Soc.* (C), 1245 (1968).

218. *a*. O. Korpium and K. Mislow, *J. Am. Chem. Soc.*, **89**, 4784 (1967); *b*. R. A. Lewis, O. Korpium, and K. Mislow, *J. Am. Chem. Soc.*, **89**, 4786 (1967); *c*. W. D. Balzer, *Tetrahedron Letters*, **10**, 1189 (1968).

219. F. D. Saeva, D. R. Rayner, and K. Mislow, *J. Am. Chem. Soc.*, **90**, 4176 (1968).

220. C. G. Overberger and J. K. Weise, *J. Am. Chem. Soc.*, **90**, 3525 (1968).

221. C. G. Overberger and J. K. Weise, *J. Am. Chem. Soc.*, **90**, 3538 (1968).

222. *a*. P. Salvadori, L. Lardicci, and P. Pino, *Tetrahedron Letters*, **22**, 1641 (1965); *b*. P. Salvadori, L. Lardicci, P. Pino, and G. Consiglio, *Tetrahedron Letters*, **44**, 5343 (1966).

223. *a*. J. S. Baran, *J. Med. Chem.*, **10**, 1039 (1967); *b*. W. Klyne, W. P. Mose, P. M. Scopes, G. M. Holder, and W. B. Whalley, *J. Chem. Soc.* (C), 1273 (1967); *c*. S. F. Mason and G. W. Vane, *Chem. Comm.*, **12**, 598 (1967); *d*. P. Vink, C. Blomberg, A. D. Vreugdenhil, and F. Bickelhaupt, *Tetrahedron Letters*, **52**, 6419 (1966).

224. R. W. Murray, R. D. Youssefyeh, and P. R. Story, *J. Am. Chem. Soc.*, **88**, 3655 (1966).

225. Y. Brunel, H. Lemaire, and A. Rassat, *Bull. Soc. Chim. France*, 1895 (1964).

226. J. Parello and X. Lusinchi, *Tetrahedron*, **24**, 6747 (1968).

227. J. Parello, R. Beugelmans, P. Millet, and X. Lusinchi, *Tetrahedron Letters*, **49**, 5087 (1968).

228. *a*. J. Riess, *Bull. Soc. Chim. France*, **18**, 29, 3552 (1965); *b*. J. Riess and G. Ourisson, *Bull. Soc. Chim. France*, 933 (1965); *c*. L. Horner, J. P. Berecz, and C. V. Bercz, *Tetrahedron Letters*, **46**, 5783 (1966); *d*. R. A. Lewis, O. Korpiun, and K. Mislow, *J. Am. Chem. Soc.*, **89**, 4786 (1967); *e*. C. Donninger and D. H. Huston, *Tetrahedron Letters*, **47**, 4871 (1968); *f*. J. N. Seiber and H. Tolkmith, *Tetrahedron*, **25**, 381 (1969).

229. M. Hargreaves and B. Modarai, *Chem. Comm.*, **1**, 16 (1969).

230. *a*. For leading references see L. H. Sommer, *Stereochemistry, Mechanism and Silicon*, McGraw-Hill, New York, 1965; *b*. L. H. Sommer and R. Mason, *J. Am. Chem. Soc.*, **87**, 1619 (1965); *c*. L. H. Sommer and J. McLick, *J. Am. Chem. Soc.*, **89**, 5806 (1967), *ibid.*, **91**, 2001 (1969) and references therein; *d*. L. Spialter and D. H. O'Brien, *J. Org. Chem.*, **31**, 3048 (1966); *e*. R. E. Hill and P. Simpson, *Chem. Comm.*, **18**, 1077 (1968); *f*. R. Corriu, J. Massé, and G. Royo, *Compt. Rend.*, **264**, 987 (1964); *g*. R. Corriu and G. Lanneau, *Compt. Rend.*, **267**, 782 (1968); *h*. See also: P. A. Hart and M. P. Tripp, *Chem. Comm.*, **4**, 174 (1969); *i*. R. J. P. Corriu and J. P. Massé, *Chem. Comm.*, **11**, 589 (1969).

231. E. R. Blout, in *Optical Rotatory Dispersion: Applications to Organic Chemistry*, C. Djerassi, Ed., McGraw-Hill, New York, 1960, Chap. 17.

232. P. Urnes and P. Doty, *Advan. Protein Chem.*, **16**, 401 (1961).

233. G. D. Fashman, in *Methods in Enzymology*, S. P. Colowick and N. O. Kaplan, Eds., Vol. VI, Academic Press, New York, 1962, p. 928.

234. D. Ridgeway, *Advan. Biol. Med. Phys.*, **9**, 271 (1963).

235. J. T. Yang, in *Newer Methods of Polymer Characterization*, B. Ke, Ed., Wiley, New York, 1964, Chap. III.

236. J. Schellman and C. Schellman, in *The Proteins*, Vol. II, H. Neurath, Ed., Academic Press, New York, 1964.

237. *a*. E. Shechter and E. R. Blout *Proc. Natl. Acad. Sci.*, **51**, 695 (1964); *b*. E. R. Blout, J. P. Carver, and E. Shechter, in *Optical Rotatory Dispersion and Circular Dichroism in Organic Chemistry*, G. Snatzke, Ed., Heyden and Sons, Ltd., London, 1967, Chap. 14, p. 224.

238. J. T. Yang, in *Poly-α-Amino Acids*, G. D. Fasman, Ed., Dekker, New York, 1967, Chap. 6, p. 239.

239. *a*. Sh. Beychok, in *Poly-α-Amino Acids*, G. D. Fasman, Ed., Dekker, New York, 1967, Chap. 7, p. 293; *b*. B. Jirgensons, *Optical Rotatory Dispersion of Proteins and Other Macromolecules*, Springer-Verlag, New York, 1969.

240. *a.* W. Moffitt, *J. Chem. Phys.*, **25**, 467 (1956); *b.* W. Moffitt and J. T. Yang, *Proc. Natl. Acad. Sci.*, **42**, 596 (1956); *c.* W. Moffitt, *Proc. Natl. Acad. Sci.*, **42**, 736 (1956).

241. M. E. Fredericq, *Bull. Soc. Chim. France*, Résumé Comm., **2b** (1969), p. 3.

242. P. Moser, *Helv. Chim. Acta*, **51**, 1831 (1968).

243. *a.* A. Bodansky, M. A. Ondetti, V. Mutt, and M. Bodansky, *J. Am. Chem. Soc.*, **91**, 944 (1969); *b.* E. Peggion, L. Strasorier, and A. Cosani, *Chem. Comm.*, **3**, 97 (1969); *c.* C. C. Yang, C. C. Chang, K. Hayashi, T. Suzuki, K. Ikeda, and K. Hamaguchi, *Biochim. Biophys. Acta*, **168**, 373 (1968); *d.* D. R. Dunstan and P. M. Scopes, *J. Chem. Soc.* (C), 1585 (1968).

244. *a.* R. Rocchi, A. Scatturin, L. Moroder, F. Marchiori, A. M. Tamburro, and E. Scoffone, *J. Am. Chem. Soc.*, **91**, 492 (1969); *b.* D. D. Kasarda, J. E. Bernardin, and W. Gaffield, *Biochemistry*, **7**, 3950 (1968); *c.* K. O. Lloyd, S. Beychok, and E. A. Kabat, *Biochemistry*, **7**, 3762 (1968); *d.* J. A. Gordon, *J. Bio. Chem.*, **243**, 4615 (1968).

245. *a.* D. W. Urry, *Proc. Natl. Acad. Sci.*, **60**, 1114 (1968); *b.* D. W. Urry, *J. Phys. Chem.*, **72**, 3035 (1968); *c.* P. Y. Cheng, *Biochemistry*, **7**, 3367 (1968).

246. *a.* K. J. Dorrington and Ch. Tanford, *J. Biol. Chem.*, **243**, 4745 (1968); *b.* K. McCarthy and W. Lovenberg, *J. Biol. Chem.*, **243**, 6436 (1968); *c.* L. Stevens, R. Townend, S. N. Timasheff, G. D. Fashman, and J. Potter, *Biochemistry*, **7**, 3717 (1968); *d.* A. J. Adler, R. Hoving, J. Potter, M. Wells, and G. D. Fashman, *J. Am. Chem. Soc.*, **90**, 4736 (1968).

247. D. L. Coleman and E. R. Blout, *J. Am. Chem. Soc.*, **90**, 2405 (1968).

248. S. P. W. Tang, J. E. Coleman, and Y. P. Myer, *J. Biol. Chem.*, **243**, 4286 (1968).

249. *a.* T. Nishimura, B. Shimizu, and I. Twai, *Biochim. Biophys. Acta*, **157**, 221 (1968); *b.* A. J. Adler, L. Grossman, and G. D. Fashman, *Biochemistry*, **7**, 3836 (1968); *c.* D. W. Miles and D. W. Urry, *J. Biol. Chem.*, **243**, 4181 (1968); *d.* M. Ikehara, M. Kaneko, and M. Sagai, *Chem. Pharm. Bull. Japan*, **16**, 1151 (1968); *e.* M. Ikehara, M. Kaneko, and Y. Nakahara, *Tetrahedron Letters*, **45**, 4707 (1968); *f.* T. Y. Tsong and J. M. Sturtevant, *J. Am. Chem. Soc.*, **91**, 2382 (1969); *g.* Y. Courtois, P. Fromageot, and W. Guschlbauer, *European J. Biochem.*, **6**, 493 (1968); *h.* J. E. Coleman, *J. Biol. Chem.*, **243**, 4574 (1968).

250. D. W. Miles, M. J. Robins, R. K. Robins, M. W. Winkley, and H. Eyring, *J. Am. Chem. Soc.*, **91**, 831 (1969) and earlier papers by these investigators.

251. *a.* P. Pino, *Advan. Polymer Sci.*, **4**, 393 (1965); *b.* P. Pino, G. P. Lorenzi, and O. Bonsignori, *Chim. Ind.* (*Milan*), **48**, 760 (1966); *c.* P. Salvadori, L. Lardicci, G. Consiglio, and P. Pino, *Tetrahedron Letters*, **22**, 1641 (1965); *ibid.*, **44**, 5343 (1966); *d.* P. Pino, C. Carlini, E. Chiellini, F. Ciardelli, and P. Salvadori, *J. Am. Chem. Soc.*, **90**, 5025 (1968) and references cited therein.

252. *a.* R. C. Schutz and E. Kaiser, *Advan. Polymer Sci.*, **4**, 236 (1965); *b.* R. C. Schultz and R. H. Jung, *Makromol. Chem.*, **96**, 295 (1966), and references therein.

253. M. Goodman, A. Abe, and Y. L. Fan, in *Polymer Handbook*, Interscience, New York, 1966.

254. P. Pino, in *Optical Rotatory Dispersion and Circular Dichroism in Organic Chemistry*, G. Snatzke, Ed., Heyden, London, 1967, Chap. 19, p. 341.

255. A. M. Sargeson, in *Chelating Agents and Metal Complexes*, F. D. Dwyer and D. P. Mellor, Eds., Academic, New York, 1964, Chap. 5, p. 183.

256. *a.* F. Woldbye, in *Optical Rotatory Dispersion and Circular Dichroism in Organic Chemistry*, G. Snatzke, Ed., Heyden, London, 1967, Chap. 6, p. 101; *b.* S. F. Mason, in *Optical Rotatory Dispersion and Circular Dichroism in Organic Chemistry*, G. Snatzke, Ed., Heyden, London, 1967, Chap. 7, p. 116; *c.* R. D. Gillard, in *Physical Methods in Advanced Inorganic Chemistry*, H. A. O. Hill and P. Day, Eds., Interscience, New York, 1968, p. 167.

257. K. Schlögl, in *Topics in Stereochemistry*, N. L. Allinger and E. L. Eliel, Eds., Vol. 1, Interscience, New York, 1967, p. 77.

258. J. Fujita and Y. Shimura, in *Spectroscopy and Structure of Metal Chelate Compounds*, K. Nakamoto and P. J. McCarthy, Eds., Wiley, New York, 1968, Chap. 3, p. 156.

259. *a.* B. Bosnich and A. T. Phillip, *J. Am. Chem. Soc.*, **90**, 6352 (1968); *b.* B. Bosnich and D. W. Watts, *J. Am. Chem. Soc.*, **90**, 6228 (1968); *c.* D. A. Buckingham, I. I. Olsen, and A. M. Sargeson, *J. Am. Chem. Soc.*, **90**, 6654 (1968); *d.* Y. T. Chen and G. W. Everett, *J. Am. Chem. Soc*, **90**, 6660 (1968); *e.* R. E. Ernst, M. J. O'Connor, and R. H. Holm, *J. Am. Chem. Soc.*, **90**, 5735 (1968).

260. *a.* S. F. Mason and J. W. Wood, *Chem. Comm.*, **23**, 1512 (1968); *b.* K. Yamasaki, J. Hidaka, and Y. Shimura, *Bull. Chem. Soc. Japan*, **42**, 45 (1969); *c.* T. Komorita, J. Hidaka, and Y. Shimura, *Bull. Chem. Soc. Japan*, **42**, 168 (1969); *d.* R. D. Gillard and M. G. Price, *Chem. Comm.*, **2**, 67 (1969); *e.* K. Yamasaki, J. Hidaka, and Y. Shimura, *Bull. Chem. Soc. Japan*, **42**, 119 (1969); *f.* Y. Sasaki, J. Fujita, and K. Saito, *Bull. Chem. Soc. Japan*, **42**, 146 (1969); *g.* S. Yamada and Y. Kuge, *Bull. Chem. Soc. Japan*, **42**, 152 (1969); *h.* K. Bauer, H. Falk, and K. Schlögl, *Angew. Chem. Intern. Ed.*, **8**, 135 (1969); *i.* C. J. Hawkins, *Chem. Comm.*, **14**, 777 (1969).

261. Unpublished observations.

262. See: H. Wolf, H. Brockmann, I. Richter, C. D. Mengler, and H. H. Inhoffen, *Annalen*, **718**, 162 (1968) and references therein.

263. W. S. Chilton and R. C. Krahn, *J. Am. Chem. Soc.*, **89**, 4129 (1967).

264. R. A. Lewis, O. Korpium, and K. Mislow, *J. Am. Chem. Soc.*, **89**, 4786 (1967).

265. G. C. Barrett, *J. Chem. Soc.* (C), 1 (1967).

266. V. Tortorella and G. Bettoni, *Chem. Comm.*, **7**, 321 (1967).

267. Y. Inouye, S. Sawada, M. Ohno, and H. M. Walborsky, *Tetrahedron*, **23**, 3237 (1967).

268. K. Blaha, I. Fric, and H. D. Jakubke, *Collection Czech. Chem. Comm.*, **32**, 558 (1967).

269. F. Burkhardt, W. Meier, A. Fürst, and T. Reichstein, *Helv. Chim. Acta*, **50**, 607 (1967).

270. *a.* A. R. Battersby, I. R. C. Bick, W. Klyne, J. P. Jennings, P. M. Scopes, and M. J. Vernengo, *J. Chem. Soc.*, 2239 (1965); *b.* J. Cymerman Craig, M. Martin-Smith, S. K. Roy, and J. B. Stenlake, *Tetrahedron*, **22**, 1335 (1966), and references therein.

271. Z. Horii, M. Ikeda, Y. Yamawaki, T. Tamura, S. Saito, and K. Kotera, *Tetrahedron*, **19**, 2101 (1964); *ibid.*, **20**, 1106 (1965).

272. H. Carpio, A. Cervantes, and P. Crabbé, *Bull. Soc. Chim. France*, 1256 (1969).

273. J. M. Brewster and J. G. Buta, *J. Am. Chem. Soc.*, **88**, 2233 (1966).

274. *a.* R. Grinter and S. F. Mason, *Trans. Faraday Soc.*, **60**, 274 (1964); *b.* G. M. Badger, R. J. Drewer, and G. E. Lewis, *J. Chem. Soc.*, 4268 (1962).

275. *a.* C. Djerassi, I. T. Harrison, O. Zagneetko, and A. L. Nussbaum, *J. Org. Chem.*, **27**, 1173 (1962); *b.* M. Legrand and R. Viennet, *Compt. Rend.*, **255**, 2985 (1962); *c.* C. Djerassi, H. Wolf, and E. Bunnenberg, *J. Am. Chem. Soc.*, **85**, 2835 (1963).

276. *a.* S. Mitchell, *J. Chem. Soc.*, 3258 (1928); *b.* S. Mitchell and S. B. Carmack, *J. Chem. Soc.*, 415 (1932).

277. J. V. Burakevich and C. Djerassi, *J. Am. Chem. Soc.*, **87**, 51 (1965).

278. *a.* C. Djerassi and D. Herbst, *J. Org. Chem.*, **26**, 4675 (1961); *b.* R. E. Ballard and S. F. Mason, *J. Chem. Soc.*, 1624 (1963).

279. H. Wolf, E. Bunnenberg, C. Djerassi, A. Lüttringhans, and A. Stockhausen, *Ann. Chem.*, **674**, 62 (1964).

280. *a.* J. M. Conia and J. Goré, *Bull. Soc. Chim. France*, 1968 (1964); *b.* J. Goré, C. Djerassi, and J. M. Conia, *Bull. Soc. Chim. France*, 950 (1967).

281. W. Scott Briggs, M. Šucký, and C. Djerassi, *Tetrahedron Letters*, **9**, 1097 (1968).

282. K. Takeda, K. Kuriyama, T. Komeno, D. A. Lightner, R. Records, and C. Djerassi, *Tetrahedron*, **21**, 1203 (1965).

283. J. S. Baran, *J. Med. Chem.*, **10**, 1039 (1967).

284. W. Scott Briggs and C. Djerassi, *Tetrahedron*, **21**, 3455 (1965).

285. Z. Badr, R. Bonnett, T. R. Emerson, and W. Klyne, *J. Chem. Soc.*, 4503 (1965).

286. K. Kuriyama and T. Komeno, in *Optical Rotatory Dispersion and Circular Dichroism in Organic Chemistry*, G. Snatzke, Ed., 1967, Chap. 21, p. 366.

287. N. J. Leonard, J. A. Adamcik, C. Djerassi, and O. Halpern, *J. Am. Chem. Soc.*, **80**, 4858 (1958).

288. J. V. Burakevich and C. Djerassi, *J. Am. Chem. Soc.*, **87**, 51 (1965).

289. *a.* K. Wellman and C. Djerassi, *J. Am. Chem. Soc.*, **87**, 60 (1965); *b.* K. Wellman, P. H. A. Laur, W. S. Briggs, A. Moscowitz, and C. Djerassi, *J. Am. Chem. Soc.*, **87**, 66 (1965); *c.* K. M. Wellman, W. S. Briggs, and C. Djerassi, *J. Am. Chem. Soc.*, **87**, 73 (1965).

290. J. J. Schneider, P. Crabbé, and N. S. Bhacca, *J. Org. Chem.*, **33**, 3118 (1968).

Chapter **IV**

STREAMING BIREFRINGENCE

Anton Peterlin and Petr Munk*

* Present address: Department of Chemistry, University of Texas, Austin, Texas. On leave of absence from Institute of Macromolecular Chemistry, Czechoslovak Academy of Sciences, Prague, Czechoslovakia.

I   INTRODUCTION

Streaming birefringence or Maxwell effect is the double refraction that is produced where pure liquids, solutions, or suspensions are subjected to shearing forces. The optical anisotropy of the sheared fluid is a consequence of orientation and deformation of molecules or particles in the flow field. The single kinetic unit—molecule or particle—must exhibit geometrical and optical anisotropy in the liquid at rest or must become anisometric and birefringent in flow if birefringence is to occur. Spherical and rigid units do not show streaming birefringence. Therefore, the effect is uniquely important if we are interested in the shape of molecules and particles, and it is indeed quite often used in biophysical investigations. On the other hand, it can be used for the study of flow fields, particularly of two-dimensional fields where the distribution of the velocity gradient can be easily mapped [1].

The orientation in the generally used shearing flow field with *transverse* gradient is kinematic and not static as is the case with electric [1a], magnetic [2], and inertial [3] birefringence. It is a consequence of nonuniform rotation of the kinetic units in the flow which has a nonvanishing rotational component.

Closely related to streaming birefringence is the double refraction observed in a purely *dilatational* flow field as occurring in a jet and in an acoustical wave. In this case the velocity gradient is *longitudinal*, the flow has no rotational component, and hence the orientation of kinetic units is static. But the orientation again occurs only with anisometric particles and the birefringence with optically anisotropic particles, the anisotropy being either intrinsic with the kinetic units at rest or caused by deformation in flow.

Until now the theoretical treatment is available only for isolated particles of simplest shape (spheroid, ellipsoid, random coil) with no hydrodynamic or optical interaction with other particles. The data so obtained have to be averaged over the actual particle size and shape distribution in order to be applicable to the polydisperse solutions employed experimentally.

The effect was discovered in 1873 by Maxwell [4], who observed that the extremely viscous Canada balsam becomes birefringent on flowing, the birefringence disappearing rapidly when the balsam ceases to flow. The effect was soon observed and investigated on many viscous liquids as resins and oils [5–10] in the concentric cylinder apparatus [11] which produces a very nearly constant velocity gradient and hence a nearly uniform birefringence in the flowing liquid. A renewed interest in streaming birefringence on pure liquids started with the work of Vorländer [12], Sadron [13], and Buchheim, Stuart, and Mentz [14].

Colloidal suspensions exhibit a much larger effect and were therefore investigated rather early. Systematic studies were initiated by Freundlich and his collaborators [15] who soon recognized that the effect is caused by the orientation of nonspherical particles in flow. This fact made streaming birefringence a valuable tool in biophysics [16]. The first investigation of polymer solutions was performed by Signer [17].

Streaming birefringence in a periodically oscillating flow with transverse gradient was first investigated by Adler, Sawyer, and Ferry [18]. Acoustic birefringence (Lucas effect) was discovered by Lucas [19] in castor oil and by Kawamura [20] in vanadium pentoxide sols.

Streaming birefringence depends on a great number of properties of solvent, and solute, and upon conditions of the experiment. While some of these properties may be considered as independent variables (viscosity and refractive index of solvent, concentration, molecular weight of solute and its refractive index, velocity gradient and/or shearing stress, temperature), other properties can be learned from streaming birefringence measurements. Let us enumerate some of them: size, shape, and optical anisotropy of the solute particles, their polydispersity, the deformability of particles under the influence of external forces (including internal viscosity and limitations due to the length of polymer chain), the oriented adsorption of solvent molecules on the solute, and the components of stress tensor. In such a situation it is sometimes difficult to separate the different influences and hence to find the right interpretation.

2 GENERAL CONSIDERATIONS

The Flow Field

FLOW WITH TRANSVERSE GRADIENT

The ideal linear flow field with constant transverse gradient

$$\vec{v} = (0, \, Gx, \, 0) \tag{4.1}$$

is used exclusively in all theoretical considerations (Fig. 4.1). The flow field is actually two-dimensional. The variation in the velocity between the yz-layers is described by the velocity gradient G, which is defined as dv_y/dx. Its dimension is $\sec^{-1}$. The absolute value may vary within wide limits, from 10^{-2} to

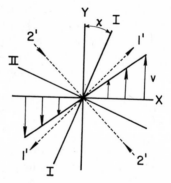

Fig. 4.1. Stationary state of linear laminar flow according to Eq. 4.1. The z-axis is perpendicular to and points out of the paper plane. The length of the arrows gives the velocity as function of x. The directions of maximum dilatation and compression are marked $1'$ and $2'$. The directions of the main refractive indices are marked I and II.

$10^6 \sec^{-1}$. The center of hydrodynamic resistance, coinciding in a great many cases, although not always, with the center of mass of the particle, is positioned at the origin. That means that the particle is participating fully in the purely translational component of flow. The latter does not affect its orientation or deformation and therefore is dropped in all theoretical considerations.

The flow field of (4.1) contains a rotational and a dilatational component:

$$\vec{v} = (0, Gx, 0) = \left(\frac{G}{2}\right)(-y, x, 0) + \left(\frac{G}{2}\right)(y, x, 0) = \vec{v}_{rot} + \vec{v}_{dil}; \quad (4.2)$$

that is, the volume rotates with a constant angular velocity $G/2$ and is subject to a linear dilatation $G/2$ and compression $G/2$ in the diagonals of the first and second quadrant, respectively. The angular velocity of the volume element is at the same time the circular frequency $2\pi\nu$ of volume element rotation. As a rule, a nonspherical particle in such a field rotates with a nonuniform angular velocity and during each rotation is twice compressed and twice extended. The nonuniform rotation produces a particle concentration in the direction of minimum angular velocity and a depletion in the direction of maximum angular velocity. The orientation is hence kinematic and not static.

The ideal realization of such a simple linear flow field would be achieved with two parallel plates of infinite extent, one at rest and the other moving with a velocity $v = G \cdot d$, where d is the distance between the plates. In stationary state, the velocity of the liquid is constant in any layer parallel to the plates. The best approximation to such a flow field can be obtained by rotating one or two concentric cylinders of Couette-type apparatus which contains the liquid in the annular gap between the cylinders. The velocity of the flowing liquid is that of a rotation with constant angular velocity on any given concentric cylindrical layer, with a radial variation characterized by the varying length of the arrows in Fig. 4.2. If the intercylinder gap is small compared with the cylinder radius, the gradient approaches that for the case of infinite plates, where it is constant. With finite gap width, we have

$$G = \frac{dv}{dr} - \frac{v}{r} = \frac{\omega r_a r_i}{r(r_a - r_i)} \sim \frac{\omega(r_a + r_i)}{2(r_a - r_i)} \quad (4.3)$$

if the inner (r_i) or the outer (r_a) cylinder rotates with the angular velocity $\omega = 2\pi n/60$, where n is number of revolutions per minute, and the other cylinder is fixed.

Laminar flow in capillary tubes is sometimes employed for qualitative observations [21]. In this case, the velocity is parallel to the tube axis and constant in any concentric layer. It increases from the wall toward the center of the tube as a quadratic function of the radius (Fig. 4.3). The gradient is zero at the axis and maximum at the wall. In any such nonlinear flow field, we assume that the particle is so small that its movement is completely determined by the local flow field which, in first approximation, is always linear with $G = dv/dr$ at the center of the particle. The nonlinear terms are completely neglected.

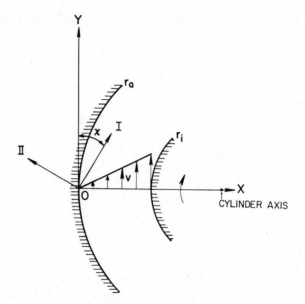

Fig. 4.2. Velocity distribution in the concentric cylinder instrument.

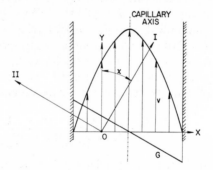

Fig. 4.3. Velocity and gradient distribution in a circular capillary or in the narrow cross section of of a quadrangular capillary.

The geometry of the tube makes the observation of birefringence difficult and a quantitative evaluation nearly impossible. The uniaxial symmetry of the flow field is automatically imparted on the observed optical properties even if the individual volume element has a lower (i.e., biaxial) optical symmetry.

Quadrangular capillaries [15] are better suited for quantitative work. If one side of the cross section is much larger than the other side, we may neglect the end effect and view the flow as that between two infinite fixed walls. The velocity profile is quadratic as in the case of circular tubes, the gradient being proportional to the distance from the center plane, that is, zero in the center and maximum at the wall.

The above considerations and the resulting velocity profiles in Figs. 4.2 and 4.3 apply only for Newtonian liquids. Most polymer solutions, however, exhibit a gradient dependent viscosity. As a rule, the shear stress increases less than linearly with the gradient; the viscosity decreases with increasing gradient. This effect distorts the velocity profile as soon as the gradient is not constant in the whole cross section. It enhances the gradient extremes by reducing the smallest and increasing the largest gradient values. The velocity profile becomes flatter in the flat and steeper in the steep sections. By choosing a very narrow gap in the Couette, one can so far reduce the gradient inhomogeneity that one may neglect the distortion of velocity profile caused by the non-Newtonian viscosity of the liquid investigated.

In recent years extensive experiments were performed in an oscillating laminar flow field [18, 22], that is, a field as shown in (4.1) and Fig. 4.1 but with a periodic gradient

$$\tilde{G} = G_0 \cos \omega t = G_o e^{i\omega t}. \qquad (4.4)$$

In the last notation we agree to take only the real part of the complex exponential function without marking that explicitly. Such a notation turns out to ease considerably the calculation of the amplitude and phase shift of orientation and birefringence as a function of frequency of applied flow field. As a rule the gradient of the oscillating flow is so small that we may neglect all but the linear effects.

FLOW WITH LONGITUDINAL GRADIENT

The linear flow field in a *circular jet* oriented in the z-axis reads

$$\vec{v} = \left(-\frac{Gx}{2}, -\frac{Gy}{2}, Gz \right). \qquad (4.5)$$

It has no rotational component. The volume element is radially compressed and axially extended so that the volume remains unchanged. Such a field describes fairly well the velocity distribution in a laterally contracting free jet or fiber after leaving the orifice or the spinneret, respectively. As a rule, the attainable values of longitudinal gradient are many orders of magnitude smaller than those of transverse gradient.

An interesting arrangement for very nearly linear flow field with longitudinal gradient [23] is shown in Fig. 4.4. Four cylinders rotating with constant angular velocity in the indicated directions produce in the centre a two-dimensional flow

$$\vec{v} = (-Gx, Gy, 0). \qquad (4.6)$$

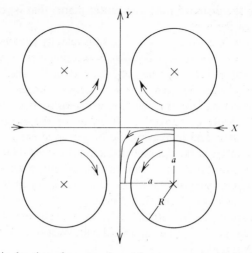

Fig. 4.4. Schematic drawing of an experimental arrangement with four parallel cylinders producing in the centre the two-dimensional laminar flow with longitudinal gradient $G = \omega R/2(a - R)$ according to Eq. (4.6).

Such an arrangement permits the observation of the optical birefringence with a light beam parallel to the z-axis in exactly the same manner as the Couette-type concentric cylinder instrument. The flow field may be either stationary or oscillating, depending on the rotation of the cylinders.

In the *acoustical wave* propagating in the z-axis direction, the velocity field reads

$$\vec{v} = (0, 0, v_o e^{i(\omega t - kz)}). \tag{4.7}$$

The linear approximation of the local field with the center of the particle at $z = 0$ turns out to be

$$v_z = v_o\left(1 - \frac{i\omega z}{c_a}\right)e^{i\omega t}. \tag{4.8}$$

Here, $k = 2\pi/\lambda = \omega/c_a$, and c_a is the velocity of sound in the medium. Again as in the case of streaming birefringence in an oscillating flow field, the amplitude of the gradient $\omega v_o/c_a$ is nearly always so small that only the linear effects in orientation and birefringence are observable. The flow field of the acoustical wave does not conserve the volume or the density of the liquid.

Optical Effects

The optical properties of the liquid can be studied by a polarized optical beam passing through the liquid. In the concentric cylinder instrument, the beam is parallel with the cylinder axis and hence perpendicular to the flow plane xy of Figs. 4.1 or 4.2. Upon emerging from solution, the light beam enters the analyzer crossed with respect to the polarizer. The field appears dark; no light passes through the analyzer if the cylinders are at rest. If one of the cylinders rotates and the liquid in the gap is sheared, the field brightens as a consequence of the resulting streaming birefringence of the liquid. By simultaneous rotation of the crossed polarizer and analyzer, we find the extinction position, that is, the direction of the main axes of refractive index of the birefringent liquid in the flow plane. The difference of the corresponding refractive indices n_1 and n_2 determines the birefringence Δn; the orientation in the first quadrant of the flow plane (between the positive x- and y-axes) determines the extinction angle χ (Fig. 4.1).

In the xz-plane, perpendicular to the flow vector, the refractive index is independent of orientation of the electric vector [23a]. Together with the non-vanishing extinction angle χ in the xy-plane that leads to the conclusion that the optical properties of the linearly sheared liquid have biaxial symmetry. Uniaxial symmetry is only achieved in the limiting case $\chi = 0$.

With low molecular weight liquids (e.g., benzene, alcohol, and castor oil), the birefringence $\Delta n = n_1 - n_2$ is proportional to the gradient or shearing stress $s = \eta G$, and the extinction angle is equal to 45° at all velocity gradients in the region of laminar flow. For suspensions of larger particles and polymer solutions, the proportionality of birefringence with the gradient and the 45° orientation of the main axes of refractive index are limiting cases for low gradient or shearing stress. At higher G or s the extinction angle diminishes and eventually approaches 0° as shown in Figs. 4.5 [24] and 4.6 [25]. The

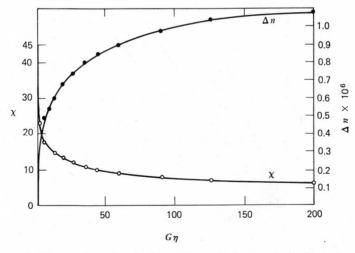

Fig. 4.5. Birefringence Δn and extinction angle χ of tobacco mosaic virus suspension in water as function of gradient [24]. Note the downward curvature of the Δn curve (negative departure from proportionality) and the approach to a saturation value.

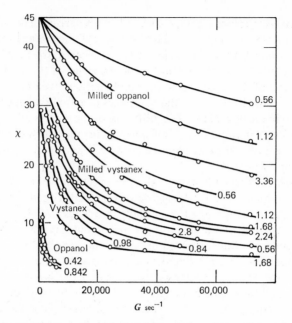

Fig. 4.6. Extinction angle χ of polyisobutylene solutions in gasoline as function of gradient [25]. The numbers indicate the volume concentration as percentages.

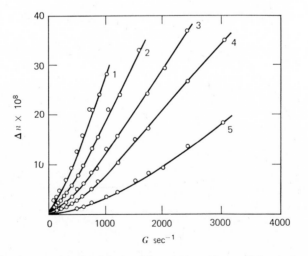

Fig. 4.7. Birefringence Δn of polyisobutylene solutions (Oppanol) in gasoline as function of gradient [25]. $c = 84$ (1), 67 (2), 56 (3), 42 (4), and 28 (5) $\times 10^{-4}$. Note the upward curvature of all curves (positive departure from proportionality).

behavior of birefringence depends on the type of solute particle or macromolecule. For rigid particles (Fig. 4.5, suspension of tobacco mosaic virus), Δn approaches a limiting saturation value (negative departure from proportionality, downward curvature of the Δn-G or Δn-s curve); for deformable particles (Fig. 4.7, polyisobutylene solutions [25]), it may increase more rapidly (positive departure from proportionality, upward curvature of the Δn curve).

At sufficiently low concentration c, the extinction angle χ and the specific birefringence $\Delta n/c$ become independent of concentration. At higher concentration, one achieves the superposition of experimental data on a single master curve (Fig. 4.8 [26]) if one plots $\Delta n/c$ over $(\eta - \eta_s)G/c$, where η and η_s

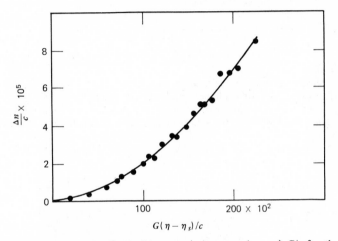

Fig. 4.8. Master curve of specific birefringence $\Delta n/c$ versus $(\eta - \eta_s)\, G/c$ for the polyisobutylene solutions of Fig. 4.7 [26].

are the viscosity of solution and solvent, respectively, at the gradient G. The superposition of the extinction angle is less perfect, although it may be very helpful in the extrapolation to zero concentration (Fig. 4.9 [27]). The limiting values, intrinsic birefringence $(\Delta n/nc\eta_s)_{c=0}$ and $\chi_{c=0}$, correspond to the isolated particle or molecule. They may be compared with the theory which, to date, has been able to treat successfully for such a case only the hydrodynamics and the optical contribution as a function of gradient.

The frequency dependence of streaming birefringence was measured mainly at small gradients. Hence the extinction angle is 45° independent of frequency. The birefringence drops continuously from the steady-state value at $\omega = 0$ and vanishes at high ω (Fig. 4.10 [28]). The phase angle δ between the applied stress and the observed birefringence is 0 at $\omega = 0$, increases to a slightly sloped plateau at 45°, and finally rises to 90° at very high ω.

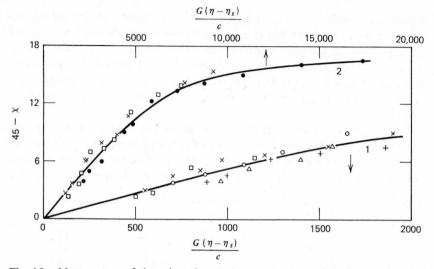

Fig. 4.9. Master curve of the orientation angle χ versus $(\eta - \eta_s)G/c$ for a polymethyl methacrylate ($M = 2.3 \times 10^6$) solution in (1) acetone lower scale and (2) a mixture of bromoform and tetrabromoethane upper scale [27]. The concentration range is between 0.28 and 0.98 $\times 10^{-2}$ g/cm³ in the former and between 0.12 and 0.23 $\times 10^{-2}$ g/cm³ in the latter case.

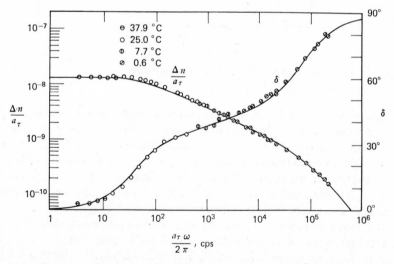

Fig. 4.10. Frequency dependence of birefringence and phase angle δ of polystyrene S111 ($M = 239,000$) in Aroclor 1248 [28]. Solid lines are theoretical functions for $Z = 200$, $h^* = 0.1$, $\varphi/\eta_s W_o = 2.0$ [(see Eq. 4.69)].

The birefringence in a flow field with longitudinal gradient was mainly observed in the jet after the liquid leaves the spinneret. The lowering of temperature and evaporation of solvent influence the viscosity and the bire-, fringence. A typical curve of the gradient dependence of birefringence of polycapronamide melt [29] is shown in Fig. 4.11 together with the information on the temperature gradient $-d \log T/dz = Z$ in the jet. The birefringence is very small up to $G = 0.5$ sec^{-1} and later increases rather rapidly up to a saturation value reached at $G = 10$ sec^{-1}.

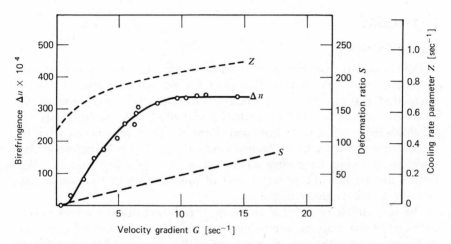

Fig. 4.11. Birefringence of polycapronamide melt in the jet as function of longitudinal gradient [29]. The temperature gradient along the jet is $Z = -d \log T/dz$.

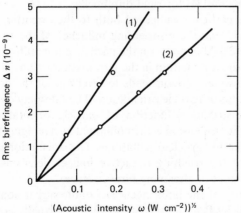

Fig. 4.12. Acoustic birefringence of polystyrene ($[\eta] = 100$ cm^3/g, $c = 8 \times 10^{-2}$g/cm^3) in toluene for two frequencies (1) 3.24 Mc/s and (2) 0.99 Mc/s as function of the square root of acoustic intensity in (W cm^{-2})$^{1/2}$ [32].

Acoustic birefringence is proportional to the intensity of the acoustic wave for a suspension of large particles comparable with the acoustical wavelength [20] and proportional to the square root for much smaller particles and macromolecules [19, 30–32] (Fig. 4.12). In the latter case, with increasing frequency the birefringence approaches a limiting value $(\Delta n)_{ac,\infty}$. The extinction directions are parallel and perpendicular to the propagation vector of the acoustic wave. No observations are yet available of the phase angle between birefringence and the acoustic wave.

3 THEORY

Rigid Particles

The anisometric particles are assumed to be identical in shape and present in such a small concentration that there are no hydrodynamic or optical interactions between the solute particles. Exact solutions of the hydrodynamic equations (viscous flow with neglected inertial effects) are available only for spheroids (ellipsoid of revolution) and ellipsoids. The former have an axis of rotational symmetry of length $2a_1$ and identical axes $(2a_2)$ in the perpendicular direction; the latter have three different axes of lengths $2a_1$, $2a_2$, $2a_3$. We further assume that the principal axes of polarizability of the particle are coincident with its geometrical axes.

The very simple geometrical forms (e.g., rods and disks) are very nearly untractable and must be approximated by very long prolate or very thin oblate spheroids with the axial ratio $p = a_1/a_2$ close to ∞ and 0, respectively. Approximate or partial solutions were given by Boeder [33] for needles, $(p = \infty)$ and by Kuhn [34] for rigid dumbbells.

In the flow the particle is subjected both to the orienting influence of the hydrodynamic field and the disorienting influence of the Brownian motion. The hydrodynamic field imposes on the particle a nonuniform rotation which results in a kinematic orientation in the $y(x)$-axis direction for prolate (oblate) particles. The Brownian motion tends to level out all the deviations from a uniform distribution of particle orientation. The interplay of both tendencies determines the distribution function of particle orientation. The optical polarizability of the suspension is then obtained by averaging the contribution of the particle over all particle orientations. The orientation of the main axes and the values of the resulting refractive indices determine the extinction angle and the amount of streaming birefringence.

Today the theory of the birefringence of dilute suspensions of rigid spheroids as developed by Peterlin and Stuart [35] is virtually complete and has been tested experimentally on many systems. An extension to ellipsoidal particles is straightforward but requires an improportionately larger computing effort. The first step, that is, calculation of the corresponding distribution function, has recently been made by Workman and Hollingsworth [35a].

Distribution Function of Spheroids

The orientation of the spheroid is specified by the polar angles θ and φ. The symmetry axis is at an angle θ with the z-axis, which is perpendicular to the flow plane. The plane through z and the symmetry axis is rotated by an angle φ from the xz-plane.

According to Jeffery [36] the laminar flow (4.1) imposes on the particle the angular velocity

$$\dot{\theta} = (\tfrac{1}{4})bG \sin 2\theta \sin 2\varphi,$$

$$\dot{\varphi} = (\tfrac{1}{2})G(1 + b \cos 2\varphi). \tag{4.9}$$

The anisometry parameter $b = (p^2 - 1)/(p^2 + 1)$ is positive for prolate, zero for spherical, and negative for oblate particles. The particles in the flow plane $(\theta = 90°)$ and at the pole $(\theta = 0)$ remain in that plane and at the pole, respectively. All other particles oscillate between a minimum and maximum θ value, the limits depending on the initial θ and φ.

The nonuniform rotation produces a nonuniform distribution of particle axes in the flow as shown in Fig. 4.13 [37]. If we start at $t = 0$ with a uniform distribution, the variation of $\dot{\varphi}$ with the angle produces after one-quarter of the rotational period $T = (2\pi/G)(p + 1/p)$ an extreme concentration at $\varphi = 90°$

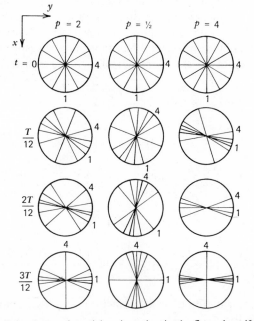

Fig. 4.13. Time dependence of particle orientation in the flow plane if at time 0 they are uniformly distributed. Axial ratio $p = 2, \tfrac{1}{2}, 4$ [37]. The period of rotation $T = 2\pi(p + 1/p)/G$.

(0°) and depletion at $\varphi = 0$ (90°) for prolate (oblate) particles. The initial distribution is restored at $T/2$. The effect increases with increasing $|b|$, is maximum at $b = 1$ ($p = \infty$) and $b = -1$ ($p = 0$), and disappears at $b = 0$ ($p = 1$), that is, with spherical particles.

If the distribution of particle orientation is $\phi(\theta, \varphi)$, the current density j_G caused by the liquid flow field $\bar{v}$ (4.1) is

$$j_G = (\dot{\theta}\phi, \dot{\varphi}\phi) = \phi \bar{v}_{\theta,\varphi}.$$

Superimposed on it is the current j_D caused by the Brownian motion

$$j_D = -D \operatorname{grad} \phi = -\left(\frac{D}{\sin \theta}\right)\left[\frac{\partial(\phi \sin \theta)}{\partial \theta}, \frac{\partial \phi}{\partial \varphi}\right] \qquad (4.11)$$

which tends to level out the existing nonuniformity of particle orientation. The gradient induced current has the same rotational direction in all four quadrants. In the case of prolate spheroids, the Brownian current has the same direction as the former in the second and fourth quadrants, but the opposite one in the first and third quadrants. As a consequence, the maximum concentration of particle axis is first at $\varphi = 45°$, which coincides with the direction of maximum dilatation. With prevailing orientation by flow, the maximum concentration shifts to the flow direction $\varphi = 90°$. The situation with oblate particles reverses the role of the quadrants; that is, the main axes are first oriented at $\varphi = -45°$ and approach $\varphi = 0°$ with increasing gradient.

The rotational diffusion constant D of the spheroid suspended in the liquid of viscosity η_s about an axis perpendicular to the main axis reads [38, 39]

$$D = \frac{kT}{\eta_s W}$$

$$W = Vv(p)$$

$$\frac{1}{v(p)} = \frac{p^2}{p^4 - 1}\left[-1 + \frac{2p^2 - 1}{2p(p^2 - 1)^{1/2}} \ln \frac{p + (p^2 - 1)^{1/2}}{p - (p^2 - 1)^{1/2}}\right] \qquad p > 1 \quad (4.12)$$

$$= 6 \qquad\qquad\qquad p = 1$$

$$= \frac{p^2}{1 - p^4}\left[1 + \frac{1 - 2p^2}{2p(1 - p^2)^{1/2}} \tan^{-1} \frac{(1 - p^2)^{1/2}}{p}\right] \qquad p < 1$$

A graphical representation of D as a function of particle volume V and axial ratio is given in Fig. 4.14.

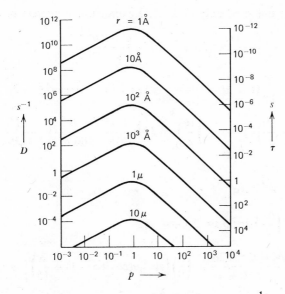

Fig. 4.14. Rotational diffusion constant D and relaxation time $\tau = \dfrac{1}{6D}$ for rigid spheroids as function of p at constant volume $4\pi a_1 a_2^{\,2}/3 = 4\pi r^3/3$ [37]. Parameter is the radius $r = (a_1\, a_2^{\,2})^{1/3}$ of the sphere. The viscosity of the medium is assumed 0.01 P, $T = 20°C$.

The continuity requirement of particle rotation in the flow field yields

$$\mathrm{div}(\mathbf{j}_G + \mathbf{j}_D) = - \frac{\partial \phi}{\partial t}, \tag{4.13}$$

the differential equation for the distribution function. In the steady flow, ϕ is independent of time so that (4.13) reads

$$\Delta\phi - \sigma\,\mathrm{div}\!\left(\frac{b\sin 2\theta \sin 2\varphi}{4}, \frac{1 + b\cos 2\varphi}{2}\right) = 0 \tag{4.14}$$

with the parameter

$$\sigma = \frac{G}{D} = \eta_s \frac{WG}{kT} \tag{4.15}$$

measuring the ratio between the orienting effect of flow and the disorienting effect of thermal motion. As Peterlin [37] has shown, the solution of (4.14) in the form of a series expansion in σ cannot be extended beyond $\sigma = 6$. A general solution for the whole range from incipient (predominating Brownian motion) to maximum possible orientation (vanishing influence of Brownian motion) can be obtained only by a series expansion in powers of b:

$$\phi = \left(\frac{1}{4\pi}\right)\left\{1 + b\left(-\frac{1}{2}\cos 2\varphi + \frac{3}{\sigma}\sin 2\varphi\right)\frac{3\sin^2\theta}{1 + 36/\sigma^2}\right.$$

$$+ b^2\left[-\frac{3}{14}(\cos^2\theta - 1) + \frac{9}{560}(35\cos^4\theta - 30\cos^2\theta + 3)\right.$$

$$\left.\left.+ \left\{\left(1 - \frac{60}{\sigma^2}\right)\cos 4\varphi - \left(\frac{16}{\sigma}\right)\sin 4\varphi\right\}\frac{15\sin^4\theta}{16(1 + 100/\sigma^2)}\right]\frac{1}{1 + 36/\sigma^2} + \cdots\right\}$$

$$(4.16)$$

For small σ, we have

$$\phi = \left(\frac{1}{4\pi}\right)\left\{1 + \left(\frac{\sigma b}{4}\right)\sin^2\theta\sin 2\varphi + \right.$$

$$\left.+ \left(\frac{\sigma^2 b}{8}\right)\left[\frac{\sin^2\theta\cos 2\varphi}{3} + \left(\frac{b}{8}\right)\left(\sin^4\theta(\cos 4\varphi - 1) + \frac{8}{15}\right)\right] + \cdots\right\} \quad (4.17)$$

which for $b = 1$ coincides with the distribution function of Boeder [33]. The linear approximation of (4.17) was used by Sadron [40] for the calculation of streaming birefringence at zero gradient. The ϕ so obtained was applied not only to the calculation of streaming birefringence but also to that of streaming dichroism [41] and non-Newtonian viscosity [37, 42]. The series expansion (4.16) converges extremely slowly if $|b|$ is close to unity, which is the case with all p above 3 or below $\frac{1}{3}$. Therefore, a full evaluation of birefringence based on (4.16) for more anisometric particles was only possible after the availability of sufficiently rapid computers [43].

The distribution function assumes a minimum and a maximum value in the flow plane and an intermediate value normal to this plane. The orientation of the directions of minimum and maximum ϕ is $+45$ and $-45°$, respectively, with respect to the direction of the velocity vector as σ approaches zero but shifts to $0°$ and $90°$, respectively, as σ increases.

Optical Anisotropy of the Spheroid

If the axes of the principal indices of refraction of the particle, n_1 and n_2, have the same orientation as the geometrical axes of the spheroid, the optical polarizabilities of the particle are [35]

$$Vg_1 = \frac{V}{4\pi}\frac{n_1^2 - n_s^2}{1 + (1 - 2e)(n_1^2 - n_s^2)/3n_s^2},$$

$$Vg_2 = \frac{V}{4\pi}\frac{n_2^2 - n_s^2}{1 + (1 + e)(n_2^2 - n_s^2)/3n_s^2}. \quad (4.18)$$

Here, n_s is the refractive index of the liquid and e is a shape factor dependent on p (Fig. 4.15). The latter is 0.5 for rods ($p = \infty$), 0 for $p = 1$, and -1 for $p = 0$. It varies rapidly at $p = 1$ but soon approaches saturation values 0.5 and -1 with large p and $1/p$, respectively.

The optical anisotropy factor of the spheroid

$$g_1 - g_2 = \frac{(3n_s^2)^2}{4\pi} \frac{(n_1^2 - n_2^2) + e(n_1^2 - n_s^2)(n_2^2 - n_s^2)/n_s^2}{[(n_1^2 + 2n_s^2) - 2e(n_1^2 - n_s^2)][(n_2^2 + 2n_s^2) + e(n_2^2 - n_s^2)]}$$

(4.19)

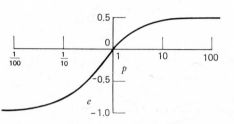

Fig. 4.15. Shape factor e of the anisotropy of the optical field in the spheroid as function of p [35].

is the sum of two terms: intrinsic and shape anisotropy. The *intrinsic anisotropy* is proportional to $(n_1^2 - n_2^2)$. It is a property of the material and vanishes if the medium is isotropic ($n_1 = n_2$), for instance, glass or crystals with cubic symmetry. The *shape anisotropy* is proportional to the shape factor e and hence disappears with spherical particles ($e = 0$). With rods and disks, it rapidly approaches saturation values with increasing p and $1/p$, respectively. The shape anisotropy also disappears if the refractive index of the solvent equals n_1 or n_2, although some modification of the intrinsic anisotropy remains, caused by the terms with e in the denominator. But it remains finite even with optically isotropic particle material ($n_1 = n_2$).

The interplay of intrinsic and shape anisotropy and their dependence on particle shape and refractive index of the solvent are shown in Fig. 4.16. We see that by variation of solvent, the optical anisotropy varies in wide limits and may eventually change the sign if $n_1 < n_2$ with a prolate spheroid, and $n_1 > n_2$ with an oblate spheroid, that is, if the optical and geometrical spheroids have an inverse axial ratio.

Birefringence

Using the distribution function (4.16) and the optical anisotropy factor (4.19), we calculate the main refractive indices of the streaming solution in the preferred directions I, II, and z (Fig. 4.1)

$$n_I^2 = \frac{n_{xx}^2 + n_{yy}^2}{2} + \left[\frac{(n_{yy}^2 - n_{xx}^2)^2}{4} + n_{xy}^4\right]^{1/2}$$

$$n_{II}^2 = \frac{n_{xx}^2 + n_{yy}^2}{2} - \left[\frac{(n_{yy}^2 - n_{xx}^2)^2}{4} + n_{xy}^4\right]^{1/2}$$

$$n_z^2 = n_{zz}^2 \tag{4.20}$$

with the extinction angle

$$\cot 2\chi = \frac{n_{yy}^2 - n_{xx}^2}{2n_{xy}^2} \tag{4.21}$$

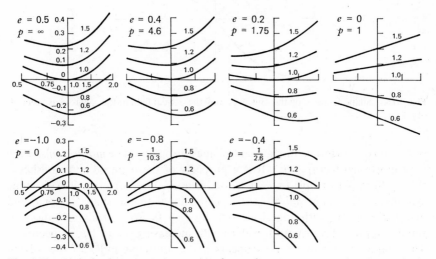

Fig. 4.16. Optical anisotropy $6\pi(g_1 - g_2)/(n_1^2 + 2n_2^2)$ as function of $3n_s^2/(n_1^2 + 2n_2^2)$ for different spheroids; prolate: $p = \infty$, 4.6, 1.75, sphere: $p = 1$, oblate $p = 0$, 0.097, 0.385 with n_1^2/n_2^2 as parameter [35].

and the refractive indices

$$n_{jk}^2 = n_s^2\,\delta_{jk} + 4\pi\left(\frac{c}{\rho}\right)\int[3g_2 + \phi(g_1 - g_2)]\cos\alpha_j \cdot \cos\alpha_k\,d\Omega \tag{4.22}$$

$$j, k = x, y, z \qquad \delta_{jk} = 1 \quad \text{if} \quad j = k$$

$$0 \quad \text{if} \quad j \neq k$$

where α_x, α_y, α_z are the angles between the particle and the coordinate axes

x, y, z, respectively, c is the concentration, ρ, the density, c/ρ the total volume fraction of the particles in solution. The three main values of refractive index (4.20) are all different; the streaming solution has the properties of an optically biaxial crystal.

The normal mode of observation is along the z-axis. For that case, we obtain the birefringence [35]

$$\Delta n = n_{\mathrm{I}} - n_{\mathrm{II}} = \left(\frac{c}{\rho}\right)(g_1 - g_2)\left(\frac{2\pi}{n_s}\right)f(\sigma, p)$$

$$\cot 2\chi = h(\sigma, p). \qquad (4.23)$$

The birefringence is proportional to the product of the volume concentration c/ρ, the optical anisotropy, and the orientation factor of the particle. The extinction angle, however, is independent of the optical anisotropy and concentration. For small σ, we have the power series

$$f(\sigma, p) = \frac{\sigma b}{15}\left[1 - \frac{\sigma^2}{72}\left(1 + \frac{6b^2}{35}\right) + \cdots\right],$$

$$\chi = \frac{\pi}{4} - \frac{\sigma}{12}\left[1 - \frac{\sigma^2}{108}\left(1 + \frac{24b^2}{35}\right) + \cdots\right]. \qquad (4.24)$$

The gradient dependence of f and χ for $p = 3$ and ∞ is plotted in Figs. 4.17 and 4.18, respectively, and for a wide range of $p > 1$ (prolate spheroids) is tabulated in Tables 4.1 and 4.2, respectively [43]. In a plot of Δn and χ versus σ, the initial slope of birefringence is proportional to b and thus depends on the shape of the particle, but that of the extinction angle is independent of it. With increasing σ, the birefringence increases more slowly than σ and the corresponding curve bends towards the σ-axis (negative departure from proportionality), and so more rapidly approaches saturation the smaller p or $1/p$. The saturation values of the orientation factor f are plotted in Fig. 4.19. The extinction angle at higher σ decreases more rapidly for particles with smaller p or $1/p$ but in any case approaches the common limit $\chi = 0$ for $\sigma \to \infty$.

The initial slopes of Δn and χ in a plot versus shear stress $G\eta_s$ yield two intrinsic constants of the suspension, the intrinsic birefringence (Maxwell constant) at zero gradient

$$[\Delta n]_o = \lim_{\substack{c=0 \\ G=0}} \frac{\Delta n}{n_s cG\eta_s} = \frac{2\pi}{15} \frac{g_1 - g_2}{\rho n_s^2} \cdot \frac{b}{D\eta_s} \qquad (4.25)$$

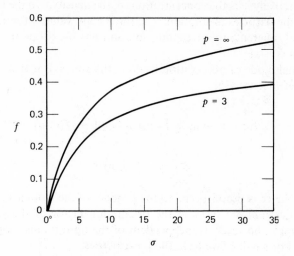

Fig. 4.17. Orientation factor $f(\sigma, p)$ of birefringence for prolate spheroids ($p = 3, \infty$) as function of σ [35, 43].

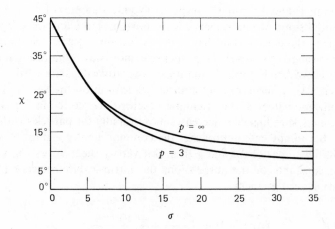

Fig. 4.18. Extinction angle χ for prolate spheroids ($p = 3, \infty$) as function of σ [35, 43].

292

and the intrinsic orientation

$$[\omega]_o = \lim_{\substack{c=0 \\ G=0}} \frac{\pi/4 - \chi}{G} = \frac{1}{12D} = \frac{W\eta_s}{12kT} \tag{4.26}$$

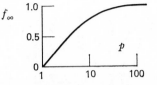

Fig. 4.19. Saturation values $f_\infty(p)$ of the orientation factor of birefringence according to Eq. (4.24) [35].

The latter quantity immediately yields the rotational resistance coefficient $W = kT/D\eta_s$, which is a function of particle size and shape only ($8\pi a^3$ for spheres), that is, of volume V and axial ratio p. With known particle volume, it yields the axial ratio of the hydrodynamically equivalent spheroid. Since the intrinsic viscosity at zero gradient [44] depends exclusively on p and not on V, we may indeed, by combination of $[\omega]_o$ and $[\eta]_o$, calculate p or $1/p$ and V [45, 46]. A combination of viscosity, sedimentation, and diffusion [47] also yields p or $1/p$ and V. Hence, these data can be used as a check of the values derived from viscosity and birefringence. They also permit us to distinguish between prolate ($p > 1$) and oblate ($p < 1$) spheroids [48]. With known V and p, we can derive the optical anisotropy factor $g_1 - g_2$ from $[\Delta n]_o$ and hence the main refractive indices n_1 and n_2 if we know the sum $n_1 + 2n_2$ from the refractive index of the suspension at rest. All these calculations have never yet been fully applied to experimental data because the poorly defined polydispersity of particle shape and size of most suspensions prohibits the straightforward application of the above-mentioned particle functions, which are all derived for strictly monodisperse systems. Biological particles (for instance, proteins), however, are often monodisperse and hence invite such an analysis.

The theory of streaming birefringence of rigid spheroids was extended to the case of absorbent and optically active particles and solvent by Snellman and Björnstahl [49].

For the static flow field with longitudinal gradient according to (4.5), Takserman-Krozer and Ziabicki [50] have calculated the particle distribution

Table 4.1. Orientation Factor, f, of Birefringence as Function of σ for Various Axial Ratios p (Prolate Spheroids) [28]

σ	$p = 2.00$	3.00	4.00	5.00	7.00	10.00	16.00	25.00	50.00	∞
0.00	0.0000	0.0000	0.0000	0.0000	0.0000	0.0000	0.0000	0.0000	0.0000	0.0000
0.25	0.0100	0.0133	0.0147	0.0154	0.0160	0.0163	0.0165	0.0166	0.0166	0.0167
0.50	0.0199	0.0266	0.0293	0.0307	0.0319	0.0325	0.0329	0.0331	0.0332	0.0332
0.75	0.0298	0.0397	0.0437	0.0458	0.0476	0.0486	0.0492	0.0494	0.0495	0.0496
1.00	0.0394	0.0525	0.0579	0.0606	0.0630	0.0643	0.0651	0.0654	0.0656	0.0656
1.25	0.0489	0.0651	0.0718	0.0751	0.0781	0.0797	0.0807	0.0811	0.0813	0.0813
1.50	0.0581	0.0774	0.0853	0.0892	0.0927	0.0947	0.0958	0.0963	0.0965	0.0966
1.75	0.0671	0.0893	0.0984	0.1029	0.1069	0.1092	0.1105	0.1110	0.1113	0.1114
2.00	0.0757	0.1007	0.1110	0.1161	0.1206	0.1231	0.1246	0.1252	0.1255	0.1256
2.25	0.0840	0.1118	0.1231	0.1287	0.1338	0.1366	0.1382	0.1388	0.1392	0.1393
2.50	0.0920	0.1223	0.1348	0.1409	0.1464	0.1494	0.1512	0.1519	0.1523	0.1524
3.00	0.1069	0.1421	0.1565	0.1636	0.1700	0.1735	0.1756	0.1764	0.1768	0.1769
3.50	0.1204	0.1601	0.1762	0.1842	0.1914	0.1954	0.1977	0.1986	0.1991	0.1992
4.00	0.1326	0.1763	0.1941	0.2029	0.2108	0.2152	0.2177	0.2187	0.2192	0.2194
4.50	0.1436	0.1909	0.2103	0.2198	0.2284	0.2331	0.2359	0.2370	0.2376	0.2377
5.00	0.1534	0.2041	0.2249	0.2351	0.2444	0.2494	0.2524	0.2536	0.2542	0.2544

6.00	0.1700	0.2268	0.2502	0.2617	0.2721	0.2778	0.2812	0.2825	0.2832	0.2834
7.00	0.1835	0.2456	0.2712	0.2839	0.2954	0.3017	0.3054	0.3069	0.3076	0.3079
8.00	0.1945	0.2613	0.2891	0.3028	0.3153	0.3222	0.3262	0.3278	0.3286	0.3289
9.00	0.2035	0.2746	0.3044	0.3191	0.3326	0.3399	0.3443	0.3460	0.3469	0.3472
10.00	0.2111	0.2860	0.3176	0.3334	0.3477	0.3556	0.3603	0.3621	0.3630	0.3633
12.50	0.2253	0.3086	0.3444	0.3624	0.3788	0.3879	0.3933	0.3953	0.3964	0.3968
15.00	0.2351	0.3254	0.3649	0.3848	0.4032	0.4133	0.4193	0.4216	0.4228	0.4232
17.50	0.2421	0.3383	0.3810	0.4028	0.4229	0.4340	0.4406	0.4431	0.4444	0.4449
20.00	0.2473	0.3485	0.3942	0.4177	0.4393	0.4513	0.4585	0.4612	0.4626	0.4631
22.50	0.2513	0.3568	0.4052	0.4302	0.4533	0.4661	0.4737	0.4766	0.4782	0.4787
25.00	0.2544	0.3637	0.4147	0.4412	0.4659	0.4796	0.4878	0.4910	0.4926	0.4932
30.00	0.2589	0.3744	0.4299	0.4592	0.4867	0.5020	0.5113	0.5148	0.5166	0.5173
35.00	0.2619	0.3823	0.4418	0.4736	0.5037	0.5206	0.5308	0.5347	0.5367	0.5374
40.00	0.2640	0.3883	0.4513	0.4854	0.5181	0.5366	0.5478	0.5524	0.5543	0.5551
45.00	0.2656	0.3930	0.4589	0.4952	0.5304	0.5505	0.5626	0.5673	0.5698	0.5706
50.00	0.2667	0.3967	0.4653	0.5037	0.5413	0.5630	0.5763	0.5814	0.5841	0.5850
60.00	0.2683	0.4021	0.4750	0.5169	0.5590	0.5838	0.5991	0.6051	0.6083	0.6091
80.00	0.2699	0.4082	0.4868	0.5338	0.5826	0.6125	0.6314	0.6389	0.6429	0.6442
100.00	0.2707	0.4114	0.4933	0.5434	0.5966	0.6298	0.6511	0.6596	0.6642	0.6657
200.00	0.2718	0.4161	0.5034	0.5588	0.6199	0.6592	0.6850	0.6954	0.7010	0.7029

Table 4.2. Extinction Angle, χ, as Function of σ for Various Axial Ratios p (Prolate Spheroids) [28]

σ	$p = 2.00$	3.00	4.00	5.00	7.00	10.00	16.00	25.00	50.00	∞
0.00	45.00	45.00	45.00	45.00	45.00	45.00	45.00	45.00	45.00	45.00
0.25	43.81	43.81	43.81	43.81	43.81	43.81	43.81	43.81	43.81	43.81
0.50	42.62	42.62	42.62	42.62	42.62	42.62	42.62	42.62	42.62	42.62
0.75	41.44	41.44	41.44	41.45	41.45	41.45	41.45	41.45	41.45	41.45
1.00	40.28	40.29	40.29	40.29	40.30	40.30	40.30	40.30	40.30	40.30
1.25	39.14	39.15	39.16	39.16	39.17	39.17	39.17	39.17	39.17	39.17
1.50	38.02	38.04	38.05	38.06	38.07	38.07	38.07	38.07	38.07	38.08
1.75	36.92	36.96	36.98	36.99	37.00	37.01	37.01	37.01	37.01	37.01
2.00	35.86	35.91	35.94	35.96	35.97	35.98	35.98	35.99	35.99	35.99
2.25	34.82	34.90	34.94	34.96	34.98	34.99	35.00	35.00	35.00	35.00
2.50	33.82	33.93	33.98	34.01	34.03	34.04	34.05	34.05	34.06	34.06
3.00	31.93	32.09	32.17	32.21	32.25	32.27	32.28	32.29	32.29	32.29
3.50	30.18	30.41	30.52	30.58	30.63	30.66	30.68	30.69	30.69	30.69
4.00	28.56	28.87	29.02	29.09	29.17	29.21	29.23	29.24	29.25	29.25
4.50	27.08	27.47	27.66	27.75	27.85	27.89	27.93	27.94	27.95	27.95
5.00	25.73	26.20	26.42	26.54	26.65	26.71	26.75	26.77	26.77	26.78

6.00	23.35	23.98	24.29	24.45	24.60	24.68	24.73	24.75	24.76	24.76
7.00	21.35	22.13	22.51	22.71	22.90	23.00	23.06	23.09	23.10	23.11
8.00	19.65	20.57	21.02	21.26	21.48	21.60	21.68	21.70	21.72	21.73
9.00	18.20	19.25	19.75	20.02	20.27	20.41	20.50	20.53	20.55	20.55
10.00	16.95	18.09	18.66	18.96	19.24	19.39	19.49	19.53	19.54	19.55
12.00	14.45	15.80	16.48	16.84	17.18	17.37	17.49	17.54	17.56	17.56
15.00	12.60	14.07	14.84	15.25	15.64	15.86	16.00	16.05	16.08	16.09
17.50	11.16	12.72	13.55	14.00	14.44	14.68	14.84	14.89	14.92	14.93
20.00	10.02	11.62	12.51	13.00	13.47	13.74	13.90	13.97	14.00	14.01
22.50	9.08	10.71	11.64	12.16	12.67	12.97	13.14	13.21	13.24	13.26
25.00	8.30	9.95	10.91	11.46	12.00	12.31	12.50	12.58	12.62	12.63
30.00	7.08	8.71	9.72	10.32	10.91	11.26	11.48	11.57	11.61	11.62
35.00	6.17	7.76	8.80	9.43	10.07	10.45	10.69	10.78	10.83	10.85
40.00	5.46	7.00	8.04	8.70	9.38	9.79	10.06	10.16	10.21	10.23
45.00	4.90	6.37	7.41	8.08	8.79	9.23	9.51	9.62	9.68	9.69
50.00	4.43	5.84	6.87	7.54	8.28	8.74	9.03	9.15	9.22	9.23
60.00	3.73	5.00	5.99	6.66	7.41	7.89	8.20	8.33	8.40	8.42
80.00	2.82	3.88	4.75	5.36	6.08	6.55	6.86	6.98	7.05	7.08
100.00	2.27	3.15	3.90	4.45	5.09	5.52	5.80	5.92	5.98	6.00
200.00	1.14	1.62	2.04	2.35	2.74	2.99	3.16	3.23	3.27	3.28

function and the resulting birefringence (Fig. 4.20)

$$\Delta n = \left(\frac{c}{\rho}\right)(g_1 - g_2)\left(\frac{2\pi}{n_s}\right) f_l(\sigma, p) \qquad (4.27)$$

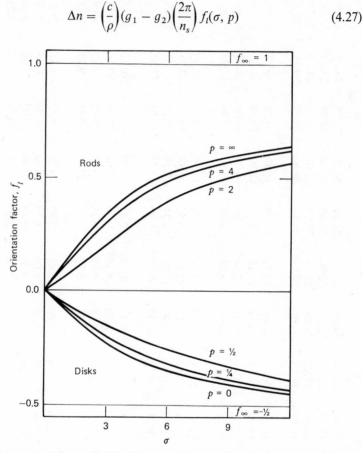

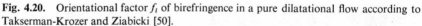

Fig. 4.20. Orientational factor f_l of birefringence in a pure dilatational flow according to Takserman-Krozer and Ziabicki [50].

The orientation of the main axes of refractive index is parallel and perpendicular to the velocity. The extinction angle is $0°$, independent of the gradient. Hence the information about the particles obtained by this method is less complete than in the case of transverse gradient.

Birefringence in an Oscillating Flow Field

The gradient is a periodic function of time [4.4]. The amplitude G_o is usually so small that nonlinear effects are not observable. Thus we obtain the extinction angle $\chi = 45°$, the amplitude of intrinsic birefringence [51]

$$[\Delta n]_\omega = \frac{\Delta n}{n_s c G_o \eta_s} = \frac{2\pi}{15} \frac{g_1 - g_2}{\rho n_s^2} \frac{b}{D\eta_s} (1 + \omega^2\tau^2)^{-1/2} = [\Delta n]_{G=0}(1 + \omega^2\tau^2)^{-1/2}$$

(4.28)

with the relaxation time

$$\tau = \frac{1}{6D},$$

(4.29)

and the phase angle δ between the maximum of birefringence and gradient

$$\tan \delta = \omega\tau = \frac{\omega}{6D}.$$

(4.30)

The relaxation times τ as a function of the spheroid dimensions are plotted in Fig. 4.14. The birefringence in the oscillating flow field equals that in the static flow at zero gradient divided by the relaxation factor $(1 + \omega^2\tau^2)^{1/2}$, which yields the experimentally observed decrease of birefringence with increasing frequency (Fig. 4.10).

A very similar approach applies to the acoustic birefringence. Since the flow field [(2.7) and (2.8)] has no rotational component, the orientation of particles is quasi-static as in a potential field and not the consequence of non-uniform particle rotation. According to Peterlin [52, 53], we have for the amplitude

$$\left(\frac{\Delta n}{n_s}\right)_{ac} = c \cdot \frac{2\pi}{15} \cdot \frac{g_1 - g_2}{\rho n_s^2} \cdot \frac{b}{D} \cdot \left(\frac{2I_a}{\rho c_a^3}\right)^{1/2} \frac{\omega}{(1 + \omega^2\tau^2)^{1/2}}$$

(4.31)

with the same relaxation time and phase angle as in the case of streaming birefringence in an oscillating flow field. The birefringence is proportional to the square root of the intensity of the acoustic wave $I_a = \frac{1}{2}\rho v_o^2 c_a$. Equation (4.31) contains the same factors as the intrinsic Maxwell constant [(4.25)] so that we can write for the intrinsic acoustic birefringence (Lucas constant)

$$[\Delta n]_{ac} = \frac{\Delta n}{n_s c I_a^{1/2}} = [\Delta n]_{G=0}\left(\frac{2}{\rho c_a^3}\right)^{1/2} \frac{\omega\eta_s}{(1 + \omega^2\tau^2)^{1/2}}.$$

(4.32)

Equations (4.27), (4.28), and (4.32), relating the intrinsic birefringences, are general and apply not only to suspensions of rigid particles but also to solutions and pure, low-molecular-weight liquids.

In contrast with streaming birefringence in an oscillating flow field, the acoustic birefringence increases first linearly with the frequency and reaches a limiting value $[\Delta n]_{G=0} (2/\rho c_a^3)^{1/2}(\eta_s/\tau)$. This difference results from the fact that at constant intensity of the acoustic wave, the gradient is proportional to the frequency, that is, to the inverse wavelength $1/\lambda = \omega/2\pi c_a$.

With very large particles comparable in their dimensions to the wavelength ($\geq \lambda/10$), the orientation seems to be due primarily to the wave pressure that tends to orient the particles perpendicular to the wave propagation vector [54]. According to Oka [55] the resulting birefringence amplitude is proportional to the intensity of the acoustical field

$$\left(\frac{\Delta n}{n}\right)_{ac} = \frac{16\pi}{45} \cdot \frac{c}{\rho} \cdot \frac{g_1 - g_2}{n^2} \cdot \frac{r^3}{kT} \cdot \frac{I}{c_a} \frac{m - m_s}{m - m_s + m_1}, \tag{4.33}$$

in agreement with birefringence data of Kawamura [20] and orientational experiments of Pohlmann [56] on suspensions of aluminum particles in xylene. Here r is the radius of the equivalent Raleigh disk, m is the mass of the particle, $m_1 = 8\rho_s r^3/3$, $m_s = m\rho_s/\rho$, and ρ and ρ_s are the densities of the particle and liquid, respectively.

Pure Low Molecular Weight Liquids

From a strictly formal point of view the orientation of every single molecule in the flow field can be treated in exactly the same way as the isolated rigid particle in suspension [53]. The smallness of the molecules and the ensuing large rotational diffusion constants (Fig. 4.14) yield such a small value of the parameter σ that only the initial linear range of birefringence and $\chi = 45°$ are observable. If the molecule is imagined as an ellipsoid with three main axes a_1, a_2, a_3 and main polarizabilities $\alpha_1, \alpha_2, \alpha_3$, then we derive from (4.25) the Maxwell constant of such a liquid

$$M' = \frac{\Delta n}{nG\eta} = \frac{\pi}{15} \left(\frac{n^2 + 2}{3n}\right)^2 N \left[\frac{(\alpha_1 - \alpha_2)b_{12}}{\eta D_3} + \frac{(\alpha_2 - \alpha_3)b_{23}}{\eta D_1} + \frac{(\alpha_3 - \alpha_1)b_{31}}{\eta D_2}\right],$$
$$\tag{4.34}$$

where N is Avogadro number, $b_{jk} = (a_j^2 - a_k^2)/(a_j^2 + a_k^2)$ with j, $k = 1, 2, 3$, and D_l with $l = 1, 2, 3$ is the diffusion constant for rotation about the lth axis [57]. However, we must not forget that the molecule is not isolated but is imbedded in a homogeneous matrix, and the internal field is not the isotropic Lorentz-Lorentz field. Therefore, a disagreement between the geometrical (particle dimensions) and optical (optical polarizability) data obtained from experiments in gases and from x-ray investigation and the values obtained from streaming birefringence of the liquid must not be considered as an argument against (4.34) [57a].

Raman and Krishnan [58] developed a theory of streaming birefringence of pure liquids based on the orientation of molecules in the stress field of the dilatational component of the laminar flow [(4.2)]. In contrast with the kinematic orientation theory based on nonuniform rotation, their theory supposes a static orientation as in the case of electric or magnetic birefringence. We so obtain $\chi = 45°$ and proportionality between Δn and G in agreement with experiment. The molecular parameters involved are subjected to the same criticism as those in (4.34). But the static orientation theory fails to explain the gradual rotation with increasing gradient of the extinction direction from 45° to 0° (flow direction) generally observed with larger particles.

Viscoelastic Sphere

In the flowing liquid, tensions and compressions occur in the flow plane at right angles to each other and at 45° with respect to the velocity [(4.2)]. As a result of these forces a deformation effect will be superimposed on the orientation if the particles are deformable, even if they are spherical [59]. The behavior of such a birefringence is different from that observed in solutions of rigid particles.

According to Cerf [60], who employed the model for randomly coiled linear macromolecules in solution, the elastic sphere with Lamé's shearing elasticity coefficient μ and viscosity η_i is deformed into an ellipsoid by the hydrodynamic field. An orientation effect like that of rigid ellipsoids occurs if η_i is much larger than η_s and a deformation effect if η_i is much smaller than η_s. The birefringence at small gradients is given approximately by

$$\frac{\Delta n}{n_s} = c \frac{G\eta_s}{\rho} \cdot \frac{F}{\mu} \cdot \frac{\eta_s/\eta_i + 4\,\delta_o^2/5}{\eta_s/\eta_i + 4\,\delta_o^2/3}$$

$$= \sim c\,\frac{G\eta_s}{\rho}\frac{F}{\mu} \qquad \eta_i \ll \eta_s$$

$$F = \left(\frac{1}{n_o}\right)\left[5\gamma + 2\left(1 - \frac{n_s}{n_o}\right)^2\right]$$

$$\delta_o^2 = \frac{kT}{2\mu V} \tag{4.35}$$

where n_o is the refractive index of the particle material, and γ is an elasto-optic coefficient that characterizes the anisotropy acquired by the sphere when it is deformed. In a small interval of variation of n_s, (4.35) predicts a parabolic dependence of the birefringence on n_s, the minimum occurring at $n_o = n_s$. If $\gamma \geq 0$, the birefringence is always positive but may change its sign if $\gamma < 0$. The extinction angle at small gradient reads (Fig. 4.21)

$$\frac{\pi}{4} - \chi = \frac{G\eta_i}{2\mu} \frac{1 + 5\eta_s/2\eta_i}{1 + (4\eta_i/3\eta_s)\delta_o^2(1 + 5\eta_s/2\eta_i)} = \frac{G\eta_i}{\mu}\,\psi\!\left(\frac{\eta_s}{\eta_i}, \delta_o\right)$$

$$\psi = -\left(\frac{\mu}{\eta_i}\right)\left(\frac{d\chi}{dG}\right)_{G=0} = \left(\frac{3}{8\delta_o^2}\right)\frac{\eta_s}{\eta_i} \qquad \eta_i \gg \eta_s$$

$$= 0.5 + 1.25\,\frac{\eta_s}{\eta_i} \qquad \eta_i \ll \eta_s. \tag{4.36}$$

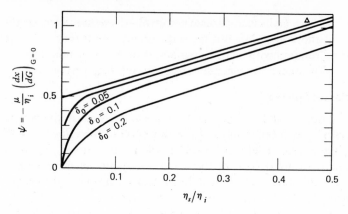

Fig. 4.21. Intrinsic Orientation ψ [Eq. (4.36)] versus η_s/η_i for several values of $\delta_o = (kT/2V\mu)^{1/2}$ [60].

For small viscosity η_s of the solvent, the variation of extinction angle with the gradient is proportional to η_s, the behavior characteristic of rigid particles. When η_s is large and η_i small, the phenomenon becomes a deformation effect. The initial variation of χ with the gradient ceases to be proportional to η_s. Such an effect indeed occurs in polymer solutions [see (4.65)].

The viscoelastic sphere with small internal viscosity may exhibit a non-vanishing limiting value for χ at high gradient [61], if the particle possesses not only a "static birefringence" due to its deformation in the hydrodynamic field, but also a "dynamic birefringence" due to the velocity gradient within the particle. With the static birefringence, the main axes of refractive index are parallel to the main axes of deformation. The dynamic birefringence has the main optical axes at 45° to the streaming direction inside the particle. The superposition of both yields at high gradient

$$\Delta n = AG$$

$$\sin 2\chi = B\left(1 + \frac{\eta_i}{2\eta_s}\right)^{-1}. \tag{4.37}$$

The constants A and B depend on the mass concentration and elasticity of the sphere.

Badoz [31] applied the elastic sphere model to the acoustic birefringence and also considered the internal optical field according to the theory of Böttcher [62]. His results, as far as frequency and intensity dependence of the effect are concerned, agree with those of Peterlin [52, 53] [see (4.31) and (4.32)].

The viscoelastic sphere seems to describe very well some subtle effects which are not so easily incorporated in other models, for example, the nonproportionality of intrinsic orientation at zero gradient with the viscosity of the medium and the nonvanishing extinction angle at very high gradient. But it is very difficult to associate the viscoelastic and optical parameters of the model with the known parameters of the molecule or the particle and particularly of the randomly coiled macromolecule, which this model was mainly intended to describe.

Linear Macromolecules

The randomly coiled macromolecule is usually described by a model consisting of Z links of length b_o with $Z - 1$ joints [63]. At the two free ends and at the joints, one concentrates the hydrodynamic resistance $W_o = 6\pi a_h$ in $Z + 1$ beads with a hydrodynamic radius a_h. The average square of the end-to-end distance is $\langle r^2 \rangle = h_o^2 = Zb_o^2$. The distribution function of end-to-end distances r

$$\phi_o = \left(\frac{\mu}{\pi}\right)^{3/2} \exp(-\mu r^2) \qquad \mu = \tfrac{3}{2}h_o^2 = \tfrac{3}{2}Zb_o^2 \qquad (4.38)$$

is obtainable from the steady-state diffusion equation if a purely entropic restoring force

$$\vec{F} = -2\mu kT\vec{r} \qquad (4.39)$$

is assumed to act on each free end.

The hydrodynamic and optical properties of the macromolecular coil were calculated either by considering explicitly only the positions of the free ends and deriving from them rather schematically the position of the remaining $Z - 1$ beads (*elastic dumbbell model*) or by taking into account explicitly the positions of all the beads (*elastic necklace model*). In the former case the links have a constant length b_o; in the latter, however, they are elastic springs with the root mean square length b_o. Indeed, with the necklace model every link is a subchain consisting of a finite number of sublinks with fixed length. The properties of such a link are identical with those of the elastic dumbbell model.

The general dependence of hydrodynamical and optical properties on gradient or frequency is the same for both models. The values of numerical factors and the finer details are, of course, different. The more elaborate necklace model is expected to be more realistic and has to be employed for checking the theory with experimental data. The much simpler dumbbell model, however, is very useful for obtaining the first estimate of an effect, for instance, of the influence of coil rigidity, of the dependence of hydrodynamic interaction on coil deformation and orientation in flow. Mathematically both models are described by formally identical equations, with scalar (1 dimension) for the dumbbell and tensorial ($Z + 1$ dimensions) coefficients for the necklace model.

Both models assume proportionality between molecular weight and average square of the end-to-end distance and, hence, apply to the macromolecule in an ideal solvent. No theoretical treatment of birefringence is available for polymers in a good solvent with $h_o^2 \sim M^{1+\epsilon}$, $\epsilon > 0$.

Optical Anisotropy

According to Kuhn and Grün [64], the ensemble of all conformations of a model with Z links of fixed length b_o yielding the same end-to-end vector $\vec{r}$ exhibits a uniaxial anisotropy of optical polarizability with

$$\gamma_1 = Z\frac{\alpha_1 + 2\alpha_2}{3} + 4\mu r^2 \frac{\alpha_1 - \alpha_2}{15} \qquad (4.40a)$$

in the direction of $\vec{r}$ and

$$\gamma_2 = Z\frac{\alpha_1 + 2\alpha_2}{3} - 2\mu r^2 \frac{\alpha_1 - \alpha_2}{15} \qquad (4.40b)$$

in the direction perpendicular to $\vec{r}$. Here, α_1 and α_2 are the polarizability of the link in the link direction and perpendicular to it, respectively. The difference

$$(\gamma_1 - \gamma_2)_i = 6\mu r^2 \frac{\alpha_1 - \alpha_2}{15} = \frac{3r^2}{5h_o^2}(\alpha_1 - \alpha_2) = \left(\frac{r^2}{h_o^2}\right)\theta_i \qquad (4.41)$$

is the *intrinsic optical anisotropy* of the ensemble. It is proportional to the square of the end-to-end distance r and inversely proportional to the mean square end-to-end distance, that is, to the number Z of links. Equation (4.41) needs corrective terms if r approaches the extended length Zb_o of the model

$$\gamma_1 - \gamma_2 = \frac{3r^2}{5h_o^2}(\alpha_1 - \alpha_2)\left(1 + \frac{8\mu r^2}{35Z} + \cdots\right) \rightarrow \frac{r^2}{h_o^2}(\alpha_1 - \alpha_2) \qquad \text{for} \quad r \rightarrow Zb_o.$$

$$(4.42)$$

The limiting anisotropy for an extended chain is by a factor $\frac{5}{3}$ larger than the value derived from (4.41).

From the approximate expression (4.41), we immediately derive the excess optical properties of the sheared solution, that is, the contribution of the randomly coiled macromolecules with molecular weight M and concentration c

$$n_{xx}^2 - n_s^2 = C\left[Z\frac{(\alpha_1 + 2\alpha_2)}{3} + (\tfrac{3}{5})(\alpha_1 - \alpha_2)\frac{\langle x^2 \rangle}{h_o^2}\right] \qquad (4.43)$$

$$2n_s n_{xy} = C(\tfrac{3}{5})(\alpha_1 - \alpha_2)\frac{\langle xy \rangle}{h_o^2}$$

$$n_{yy}^2 - n_s^2 = C\left[Z\frac{(\alpha_1 + 2\alpha_2)}{3} + (\tfrac{3}{5})(\alpha_1 - \alpha_2)\frac{\langle y^2 \rangle}{h_o^2}\right]$$

$$n_{zz}^2 - n_s^2 = C\left[Z\frac{(\alpha_1 + 2\alpha_2)}{3} + (\tfrac{3}{5})(\alpha_1 - \alpha_2)\frac{\langle z^2 \rangle}{h_o^2}\right]$$

$$C = \left(\frac{4\pi cN}{M}\right)\frac{(n_s^2 + 2)^2}{9}.$$

Here $\langle x^2 \rangle$ means the average of x^2 over all positions of the end-to-end vector of the model in flow. The anisotropic part of the main excess refractive indices of the solution in flow hence is

$$n_I^2 - n_s^2 = \left(\frac{3C}{10h_o^2}\right)(\alpha_1 - \alpha_2)[\langle x^2 + y^2 \rangle + (\langle y^2 - x^2 \rangle^2 + 4\langle xy \rangle^2)^{1/2}] \quad (4.44)$$

$$n_{II}^2 - n_s^2 = \left(\frac{3C}{10h_o^2}\right)(\alpha_1 - \alpha_2)[\langle x^2 + y^2 \rangle - (\langle y^2 - x^2 \rangle^2 + 4\langle xy \rangle^2)^{1/2}]$$

$$n_{zz}^2 - n_s^2 = \left(\frac{3C}{5h_o^2}\right)(\alpha_1 - \alpha_2)\langle z^2 \rangle$$

yielding the birefringence $\Delta n = n_I - n_{II}$ as

$$\frac{\Delta n}{n_s} = \frac{cN}{M}\cdot\frac{6\pi}{5h_o^2}\left(\frac{n_s^2 + 2}{3n_s}\right)^2(\alpha_1 - \alpha_2)(\langle y^2 - x^2 \rangle^2 + 4\langle xy \rangle^2)^{1/2} \quad (4.45)$$

and the extinction angle

$$\cot 2\chi = \frac{\langle y^2 - x^2 \rangle}{2\langle xy \rangle}. \quad (4.46)$$

As a rule the sheared solution has three different main indices and hence has the optical properties of a biaxial crystal.

This derivation immediately applies to the contribution to optical anisotropy of the elastic link of the necklace model. In order to obtain the excess refractive indices of solution, we simply replace h_o^2 by b_o^2, $\langle x^2 \rangle$ by the corresponding sum over all links $\langle \mathbf{x}^T \mathbf{A} \mathbf{x} \rangle$, with the $(Z + 1)$ dimensional vector $\mathbf{x}$ and tensor $\mathbf{A}$

$$\mathbf{x} = \begin{vmatrix} x_0 \\ x_1 \\ x_2 \\ \vdots \\ x_Z \end{vmatrix} \quad \mathbf{A} = \begin{vmatrix} 1 & -1 & 0 & 0 & \cdots & \cdot & \cdot \\ -1 & 2 & -1 & 0 & \cdots & \cdot & \cdot \\ 0 & -1 & 2 & -1 & \cdots & \cdot & \cdot \\ \cdot & \cdot & \cdot & \cdot & \cdots & \cdot & \cdot \\ \cdot & \cdot & \cdot & \cdot & \cdots & 2 & -1 \\ \cdot & \cdot & \cdot & \cdot & \cdots & -1 & 1 \end{vmatrix}, \quad (4.47)$$

and correspondingly $\langle y^2 \rangle$ and $\langle xy \rangle$ by $\langle \mathbf{y}^T \mathbf{A} \mathbf{y} \rangle$ and $\langle \mathbf{x}^T \mathbf{A} \mathbf{y} \rangle$, respectively.

The optical tensor so derived completely neglects the modification of the optical field inside the model caused by the presence of macromolecule which usually has a refractive index different from that of the solvent. In the first approximation according to Čopič [65], we may obtain the contribution caused by the modified optical field inside the coiled macromolecule by replacing the ensemble of all conformations with the same end-to-end vector by an optically equivalent ellipsoid with an average polarizability of the macromolecule and solvent involved. We hence obtain the so-called *macroform anisotropy* [66]

$$(\gamma_1 - \gamma_2)_f = \left(\frac{n_s^2 + 2}{3}\right)^2 \left(\frac{n_o^2 - n_s^2}{4\pi n_s \rho N}\right)^2 \frac{M^2}{v} \, 4\pi e = 4\pi e \theta_f, \qquad (4.48)$$

which is always positive. Here, n_o is the refractive index of the dissolved polymer, N is the Avogadro number, $v = 0.36 \, h_o^3$ is the volume of the molecular coil in solution (including the solvent in the coil). The optical shape factor e is a function of the axial ratio p of the optically equivalent spheroids (Fig. 4.15). It can also be expressed [67] as a function of the parameter $\xi = r/h_o$ (Fig. 4.22). The factor M^2/v can be replaced by $\phi M/0.36 \, [\eta]$ with $\phi = 2.4 \times 10^{23}$. The macroform anisotropy is proportional to $M/[\eta]$, whereas θ_i and θ_{fs} are independent of molecular weight.

The linear regularity of neighboring elements in the chain causes an additional axially symmetric optical interaction. It manifests itself in a local anisotropy of the internal optical field which is in many respects analogous to the average macroform anisotropy. The local anisotropy increases with chain extension, that is, with the length of the statistical segment. According

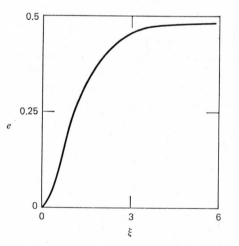

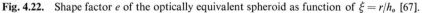

Fig. 4.22.   Shape factor e of the optically equivalent spheroid as function of $\xi = r/h_o$ [67].

to Kuhn and Kuhn [68] and Tsvetkov [69], the *microform anisotropy* reads

$$(\gamma_1 - \gamma_2)_{fs} = \frac{3}{5} \left(\frac{n_s^2 + 2}{3} \right)^2 \left(\frac{n_o^2 - n_s^2}{4\pi n_s} \right)^2 \frac{M_o \zeta}{N} 4\pi e_s \frac{r^2}{h_o^2} = \frac{r^2}{h_o^2} \theta_{fs}. \quad (4.49)$$

Here ζ is the number of monomeric units per segment, M_o is molecular weight of the monomer, and e_s is the optical shape factor of the segment, which is independent of the gradient. As a rule, e_s and hence θ_{fs} are positive. The microform anisotropy may be quite large, that is, of the same order of magnitude as θ_i or even larger. It is also closely related to the optical anisotropy of solvent molecules, which may be appreciably oriented in the immediate neighborhood of the solute chain, hence either adding to or subtracting from its optical anisotropy [70].

The total optical anisotropy of the macromolecule is the sum of the three effects [(4.41), (4.48), and (4.49)].

$$\gamma_1 - \gamma_2 = \frac{r^2}{h_o^2} (\theta_i + \theta_{fs}) + 4\pi e \theta_f. \quad (4.50)$$

The intrinsic anisotropy is independent of the refractive index of the solvent, the microform and macroform anisotropy are proportional to the square of the difference of refractive indices of macromolecule and solvent eventually modified by the Frisman-Dadivanyan effect. If $n_o = n_s$ (matching solvent), the last two contributions vanish and the birefringence of the solution is exclusively caused by the intrinsic anisotropy. The microform term increases and the macroform factor stays nearly constant with expanding coil, that is, with increasing end-to-end distance r. Experimental values of the ratio θ_{fs}/θ_f as a function of the gradient are a measure of the coil deformability in flow.

Hydrodynamics of Chain Molecule in Theta Solvent

The forces acting on a single bead are the force of the flow field $\bar{v}$, the entropy force $\vec{F}$ of the chain, and the diffusion force. The last two forces yield the equilibrium distribution of beads or of the chain ends [(4.38)] in solution at rest. The flow of the beads hence reads in tensor notation [(4.47)]

$$\mathbf{j} = \bar{v}\phi - 2\mu_o D_o \mathbf{H} \mathbf{A} \bar{r} \phi - D_o \mathbf{H} \nabla \phi \quad (4.51)$$

where $\mu_o = \frac{3}{2} b_o^2$, $D_o = kT/\eta_s W_o$, and $\mathbf{H}$ is the tensor of hydrodynamic interaction

$$\mathbf{H} = \mathbf{1} + 0.0733 \, W_o \begin{vmatrix} 0 & 1/h_{12} & 1/h_{13} & \cdots & & \cdot \\ 1/h_{21} & 0 & 1/h_{23} & \cdots & & \cdot \\ \vdots & \vdots & \vdots & \cdots & & \cdot \\ \cdot & \cdot & \cdot & & \cdots & 1/h_{z,z+1} \end{vmatrix} \quad (4.52)$$

h_{jk} is the root mean square distance between the jth and kth bead. The coefficient 0.0733 is derived under the assumption that the distribution of intrachain distances is Gaussian. The continuity condition yields the differential equation for the distribution function

$$\mathbf{V} \mathbf{j} = -\frac{\partial \phi}{\partial t}. \tag{4.53}$$

This linear equation with $3(Z + 1)$ independent variables becomes separable in $(Z + 1)$ three-dimensional equations by the introduction of normal coordinates [71]

$$\bar{\mathbf{u}} = (\xi, \eta, \zeta) = \mathbf{Q}^{-1} \bar{\mathbf{r}}. \tag{4.54}$$

The solution turns out to be the product of all Z distribution functions ϕ_p (Peterlin as quoted in [71])

$$\phi = \Pi \phi_p \qquad p = 1, 2, 3 \ldots Z$$

$$\phi_p = \left[\frac{\mu_o \mu_p}{\pi(1 + \beta_p^2)}\right]^{3/2} \exp\left\{-\frac{\mu_o \mu_p}{1 + \beta_p^2}\right.$$

$$\times \left[(1 + 2\beta_p^2)\xi_p^2 - 2\beta_p \xi_p \eta_p + \eta_p^2 + (1 + \beta_p^2)\zeta_p^2\right]\right\} \tag{4.55}$$

$$\beta_p = \frac{G}{4\mu_o D_o \lambda_p} = \frac{\beta_o}{\lambda_p}$$

The λ_p and μ_p are eigenvalues of the tensor $\mathbf{\Lambda} = \mathbf{Q}^{-1}\mathbf{H}\mathbf{A}\mathbf{Q}$ and $\mathbf{M} = \mathbf{Q}^T \mathbf{A} \mathbf{Q}$, respectively; and $\mathbf{Q}^{-1}$ and $\mathbf{Q}^T$ are the inverse and transposed $\mathbf{Q}$ matrix, respectively. The general form of distribution function for the pth eigenmode is the same as that found for the elastic dumbbell model [72], if we replace $\mu_o \mu_p$ by $\mu = \frac{3}{2} h_o^2$ and β_p by

$$\beta = \Sigma \beta_p = M[\eta]\eta_s \frac{G}{RT} = \frac{G}{4\mu D}, \tag{4.56}$$

where D is the translational diffusion constant of the free end of the model and R is the gas constant.

The linear entropy force $\vec{F} \sim \vec{r}$ permits an unlimited coil deformation in flow. The real macromolecule, however, cannot be extended beyond its full length $L = Zb_o$. The appropriate modification of F was calculated by Kuhn and Grün [64] and applied by Peterlin to the problem of streaming birefringence in a flow field with transverse [73] and longitudinal [74] gradient. The formal procedure consists in introducing a nonlinearity factor $\mathbf{E}$ in the $2\mu_o D_o \mathbf{H}\mathbf{A}\vec{r}\phi$ term of (4.51). The eigenvalues E_p were calculated by Reinhold and Peterlin [75] for the free-draining coil and applied to the gradient dependence of intrinsic viscosity. The birefringence, instead of growing without limits, exhibits saturation effects which occur at so much lower β values the shorter the macromolecule. For actual observation of saturation, however, a high M is required because $\beta_{\text{sat}} \sim M^{1/2}$ and $\beta \sim M^{3/2} G$ yielding $G_{\text{sat}} \sim M^{-1}$.

Cerf [76] refined the necklace model by considering the final resistance of the molecule to rapid shape changes. He introduced an *internal viscosity* force

$$\left(\frac{p\varphi}{Z}\right)\vec{u}_{p,\,\text{def}} = \varphi_p \vec{u}_{p,\,\text{def}} = \varphi_p (\dot{\vec{u}} - \vec{\Omega} \times \vec{u})_p \tag{4.57}$$

opposing the deformational velocity component of the pth eigenmode, that is, the difference between the total velocity and the velocity of rotation. The internal viscosity factor is φ/Z for the deformational motion of the free ends and φ for that of adjacent beads. The latter is independent of molecular weight; the former decreases as $1/M$. Kuhn and Kuhn [77], who first introduced the internal viscosity for the explanation of the gradient dependence of intrinsic viscosity of linear macromolecules, derived the $1/M$ dependence from a consideration of conformational changes necessary for a change of the end-to-end distance of the dumbbell model. With inclusion of the internal viscosity force, we have one additional current term in (4.51), and (4.53).

With small internal viscosity, the molecule rotates with the same angular velocity $G/2$ as the volume element. The distribution function reads like that in (4.55),

$$\phi_p = \left[\frac{\mu_o \mu_p (1 + \beta_p'^2 - \beta_p^2)}{\pi(1 + \beta_p'^2)}\right]^{3/2} \exp\left(-\frac{\mu_o \mu_p}{1 + \beta_p'^2}\left[(1 + \beta_p'(\beta_p' + \beta_p))\xi_p^2\right.\right.$$

$$\left.\left. - 2\beta_p \xi_p \eta_p + (1 + \beta_p'(\beta_p' - \beta_p))\eta_p^2 + (1 + \beta_p'^2)\zeta_p^2\right]\right) \tag{4.58}$$

$$\beta_p' = \beta_p \left(1 + \frac{v_p \varphi_p}{\eta_s W_o}\right).$$

The main difference between (4.55) and (4.58) is the partial replacement of β_p by β_p', which contains the internal viscosity terms multiplied by the eigenvalues v_p of the hydrodynamic interaction tensor $\mathbf{N} = \mathbf{Q}^{-1}\mathbf{H}\mathbf{Q}^{-1T}$. With large internal viscosity, the molecule is nearly rigid and hence rotates in the flow with an orientation-dependent angular velocity. The corresponding distribution function for small values of the gradient was obtained as a power series in D_o^{-1} [78].

The distribution functions for the coil in laminar flow depend on the eigenvalues λ_p, μ_p, and $v_p = \lambda_p/\mu_p$ of the three tensors $\mathbf{HA}$, $\mathbf{A}$, and $\mathbf{H}$. The second tensor, A, has constant components, thus yielding flow-independent μ_p. The hydrodynamic interaction, however, is proportional to the reciprocal intramolecular distances, which increase in flow as a consequence of coil deformation. The increase is larger for distant and smaller for close chain elements. Such a nonuniform deformation hence changes the eigenvalues λ_p and ν_p, the relative changes for small p being larger than those for large p. In all theoretical studies of streaming birefringence based on the necklace model, these changes were neglected. For the elastic dumbbell, Peterlin [73, 79] has considered the effect under the assumption that the relative increase of the mean square distance between the jth and kth bead is a linear function of $|j - k|$, being 0 for $|j - k| = 1$ and maximum for $|j - k| = Z$.

The change of eigenvalues with the gradient becomes irrelevant for the experiment in an oscillating flow field. This is because the applied gradient in such a case is always so small that for all practical purposes the macromolecular coil has the same dimensions as it has in solution at rest.

The hydrodynamics of the ideally flexible linear macromolecule in a flow field with longitudinal gradient [(4.5)] was treated by Takserman-Krozer [80] (necklace model) and Peterlin [74] (elastic dumbbell with finite chain length). The distribution function has a singularity at $\beta = 0.5$ and -1.0 if the molecule is not limited in extension by a nonlinear entropy force F which becomes infinite at full chain extension $r = Zb_o$.

Birefringence in Steady Flow

In a matching solvent the birefringence is caused exclusively by the intrinsic anisotropy of the molecule [(4.41)]. The shape anisotropy vanishes according to (4.48) and (4.49) because $n_o = n_s$. In order to obtain the birefringence and extinction angle, we must calculate with the aid of the distribution function the averages $\langle y^2 - x^2 \rangle$ and $\langle xy \rangle$ (dumbbell model) or $\langle \mathbf{y}^T A\mathbf{y} - \mathbf{x}^T A\mathbf{x} \rangle$ and $\langle \mathbf{x}^T A\mathbf{y} \rangle$ (necklace model) and insert them into (4.45) and (4.46), respectively. We obtain for a coil with moderate stiffness (small $\varphi/\eta_s W_o$)

$$\langle \mathbf{x}^T A\mathbf{x} \rangle = \Sigma\mu_p\langle \xi_p^2 \rangle = \frac{1}{2\mu_o}\left[Z - \Sigma\frac{2\beta_p(\beta_p' - \beta_p)}{4 + (\beta_p' - \beta_p)(\beta_p' + 3\beta_o)}\right]$$

$$\langle \mathbf{x}^T A\mathbf{y} \rangle = \Sigma\mu_p\langle \xi_p\eta_p \rangle = \frac{1}{2\mu_o}\Sigma\frac{4\beta_p}{4 + (\beta_p' - \beta_p)(\beta_p' + 3\beta_p)}$$

$$\langle \mathbf{y}^T A\mathbf{y} \rangle = \Sigma\mu_p\langle \eta_p^2 \rangle = \frac{1}{2\mu_o}\left[Z + \Sigma\frac{2\beta_p(\beta_p' + 3\beta_p)}{4 + (\beta_p' - \beta_p)(\beta_p' + 3\beta_p)}\right]. \quad (4.59)$$

These expressions substantially simplify if the internal viscosity is neglected (*ideally flexible coil*):

$$\langle \mathbf{x}^T \mathbf{A} \mathbf{x} \rangle = \frac{Z}{2\mu_o}$$

$$\langle \mathbf{x}^T \mathbf{A} \mathbf{y} \rangle = \frac{1}{2\mu_o} \Sigma \beta_p$$

$$\langle \mathbf{y}^T \mathbf{A} \mathbf{y} \rangle = \frac{Z}{2\mu_o} + \frac{1}{\mu_o} \Sigma \beta_p^2 \tag{4.60}$$

yielding

$$\frac{\Delta n}{n_s} = \frac{4\pi}{5} \left(\frac{n_s^2 + 2}{3n_s} \right)^2 \frac{cN}{M} (\alpha_1 - \alpha_2) \beta_o (\Sigma \lambda_p^{-1}) \left[1 + \beta_o^2 \frac{(\Sigma \lambda_p^{-2})^2}{(\Sigma \lambda_p^{-1})^2} \right]^{1/2}$$

$$\cot 2\chi = \beta_o \frac{\Sigma \lambda_p^{-2}}{\Sigma \lambda_p^{-1}}. \tag{4.61}$$

The extinction angle is independent of the optical factors. The intrinsic birefringence and orientation at zero gradient read

$$[\Delta n]_{G=0} = \frac{2\pi}{15} \left(\frac{n_s^2 + 2}{3n_s} \right)^2 \cdot \frac{N b_o^2 W_o}{MkT} (\alpha_1 - \alpha_2) \Sigma \lambda_p^{-1} = \frac{2K}{n} [\eta]_{G=0} \tag{4.62}$$

$$[\omega]_{G=0} = \eta_s \frac{b_o^2 W_o}{12kT} \frac{\Sigma \lambda_p^{-2}}{\Sigma \lambda_p^{-1}} = B \frac{M}{2RT} [\eta]_{G=0} \eta_s$$

$$K = \frac{2\pi}{5} \frac{(n_s^2 + 2)^2}{9n_s} \frac{(\alpha_1 - \alpha_2)}{kT}$$

$$B = \frac{\Sigma \lambda_p^{-2}}{(\Sigma \lambda_p^{-1})^2}.$$

The proportionality factor K between intrinsic birefringence and viscosity is the so-called *stress-optical coefficient* [see (4.77) and (4.78)].

It is worthwhile to mention that the model yields $\langle \mathbf{x}^T \mathbf{A} \mathbf{x} \rangle = \langle \mathbf{z}^T \mathbf{A} \mathbf{z} \rangle$ and $\langle \mathbf{x}^T \mathbf{A} \mathbf{z} \rangle = 0$, that is, the refractive index is independent of orientation of the electric vector in the xz-plane in perfect agreement with experiments [23a]. Such a result is also obtained from the simple elastic dumbbell model [68, 72].

In the validity range of the model, K is a constant independent of Z, that is, of molecular weight. But it depends on M at such a small molecular weight that the random coil model with statistical subchain is not more appropriate. In such a case, we must introduce either a model with a finite valency angle between subsequent chain atoms and limited rotation about the valency bond or the wormlike model of Kratky and Porod [81], which reduces the complexity of a short chain to one single parameter, the persistence length a [82]. According to Gotlib and Svetlov [83], the constant K has to be multiplied by a factor $0.3 \, \phi_1(x)/\phi_2(x)$ plotted in Fig. 4.23 versus $x = L/a$. It has the value $0.6x$ at small x (rodlike molecule) and approaches 1 at large x (random coil).

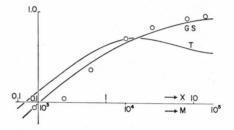

Fig. 4.23. The relative stress-optical coefficient $K_x/K_\infty = 0.3 \times \phi_1 (x)/\phi_2 (x)$ according to Gotlib and Svetlov [83] and that according to Tsvetkov [82] for short wormlike chains as functions of $x = L/a$. Circles are experimental data of polystyrene in Aroclor [22].

By using the relationship between β and β_p [(4.56)], we can transcribe (4.61) to

$$\frac{\Delta n}{c} = 2K \frac{RT}{M} \beta(1 + B^2\beta^2)^{1/2} = 2K \frac{RT}{M} \frac{\beta}{\sin 2\chi}$$

$$\cot 2\chi = B\beta. \tag{4.63}$$

But for the factor B, which is 1 in that case, these equations are identical with the ones obtained with the elastic dumbbell model [72, 68]. By plotting $\cot 2\chi$ versus β, we obtain B as a function of the gradient or cf the generalized, dimensionless parameter β. A straight line means constant B. A plot of Δn versus $\beta/\sin 2\chi$ has to be a straight line independent of whether B is constant or not. Some interesting combinations derivable from (4.63) were suggested by Munk and Peterlin [84]:

$$\frac{\Delta n}{c} \frac{\sin 2\chi}{\beta} = 2K \frac{RT}{M}, \tag{4.64a}$$

$$\frac{\Delta n}{c} \frac{\cos 2\chi}{\beta^2} = 2K \frac{RT}{M} B, \tag{4.64b}$$

$$\frac{\Delta n}{c} \sin 2\chi \tan 2\chi = 2K \frac{RT/M}{B}, \tag{4.64c}$$

$$\frac{\cot 2\chi}{\beta} = B. \tag{4.64d}$$

All four expressions on the left side are invariants independent of concentration and gradient, if K is a constant and the measurements are performed in the linear concentration range at sufficiently low β values such that no saturation effects are playing a role. As a consequence of nonuniform coil expansion [85] in flow, we expect a slight decrease of B with the gradient in good agreement with recent experiments of Munk and Peterlin [84] on polystyrene solutions in Aroclor and with older data of Philippoff on polystyrene [86], polyisobutylene [87], and nitrocellulose [88].

In the limit of *small β*, we know λ_p for the free-draining [89] and impermeable [71] coil and also for the intermediate cases [90] for the coil in theta ($\epsilon = 0$) and good ($\epsilon > 0$) solvent under the assumption that the same ϵ applies to all intramolecular distances [91]. The constant $1/B$ for these cases is shown in Fig. 4.24. The B is markedly below the value 1 of the dumbbell model.

Finite coil rigidity affects neither the initial intrinsic birefringence nor its relationship with the initial intrinsic viscosity [(4.62)]. But it significantly modifies the initial intrinsic orientation [(4.62)], which, according to Kuhn and Kuhn [77], is three times as large with rigid molecules as with ideally

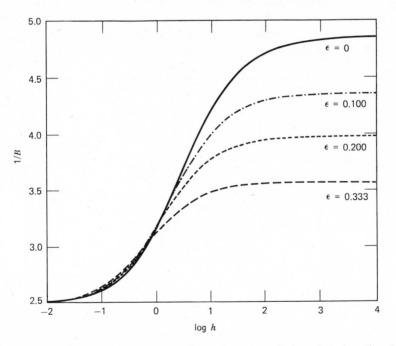

Fig. 4.24. The ratio $1/B = (\Sigma\lambda_p^{-1})^2/(\Sigma\lambda_p^{-2})$ at $\beta = 0$ for coils in a theta ($\varepsilon = 0$) and in good solvents ($\varepsilon = 0.1, 0.2, 0.333$) as function of coil permeability $h = 0.0519\ Z^{1/2}\ W_o/b_o$ [90].

flexible molecules (dumbbell model). For the necklace model, Cerf [76] obtains

$$[\omega]_{G=0} = 0.9M[\eta]\frac{\eta_s}{RT} \qquad \text{small } \eta_s$$

$$= 0.1M[\eta]\frac{\eta_s}{RT} + 0.0062N\frac{h^2\varphi}{RT} \qquad \text{large } \eta_s \qquad (4.65a)$$

for small hydrodynamic interaction (nearly free draining coil) and

$$[\omega]_{G=0} = 0.7M[\eta]\frac{\eta_s}{RT} \qquad \text{small } \eta_s$$

$$= 0.1M[\eta]\frac{\eta_s}{RT} + 0.0045N\frac{h^2\varphi}{RT} \qquad \text{large } \eta_s \qquad (4.65b)$$

for large hydrodynamic interaction. In a plot of $[\omega]_{G=0}$ versus η_s, we have first a linear increase that soon bends down into a new straight line with a finite ordinate intercept [92, 27] (Fig. 4.25), an effect which is very similar to that obtained for a deformable sphere [(4.36)], (Fig. 4.21). At low solvent viscosity the coil gets oriented like a rigid particle. At high η_s the coil defor-

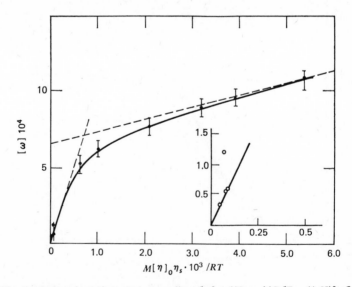

Fig. 4.25. Intrinsic orientation at zero gradient $[\omega] = (45 - \chi)/G$ [Eq. (4.65)] of a polymethyl methacrylate fraction ($M = 2.3 \times 10^6$) exhibiting the transformation from an orientational to deformational birefringence with increasing solvent viscosity [27].

mation prevails yielding a nonvanishing ordinate intercept. A more precise calculation of the initial slope in a low viscosity solvent in the whole range of hydrodynamic interaction was performed by Chaffey [78], who finds by a factor of about 0.7 smaller values than those given in (4.65). His data are plotted in Fig. 4.26. The ordinate intercept of the asymptote (Fig. 4.25) yields the coefficient φ of internal viscosity. According to (4.65) it turns out to be 1.6×10^{-5} g/sec for polystyrene investigated by Leray [92] in a series of solvents in a wide range of viscosity.

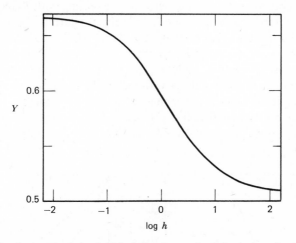

Y

Fig. 4.26. The factor Y of the initial slope at $\eta_s \to 0$ of intrinsic orientation at zero gradient $[\omega]_o = Y M [\eta] \eta_s / RT$ as function of the hydrodynamic interaction parameter $h = 0.0519 \, Z^{1/2} \, W_o / b_o$ [78].

It must be stressed that this orientational effect at zero gradient is independent of the actual composition of the observed birefringence (intrinsic birefringence, microform and macroform birefringence) as long as the shape of the macromolecule and the refractive index differences between the solute and solvent remain unchanged. The situation is particularly favorable if either intrinsic (polystyrene [92]) or macroform (polymethyl methacrylate [27]) birefringence is so prevalent that other contributions may be safely neglected. In such a case even the change of solute refractive index does not affect the results noticeably.

With increasing gradient the coil rigidity reduces coil deformation and hence the birefringence. Instead of the upward curvature of the Δn versus $G\eta$ curve for ideally flexible macromolecule, we obtain eventually a straight line (e.g., cellulose acetate, Fig. 4.40) and a downward curvature for completely rigid molecules (micellar solution of acetyl cellulose, Fig. 4.40). The extinction angle drops faster with increasing coil rigidity but later approaches more slowly the limiting value $\chi = 0$. The extreme cases, perfectly flexible and rigid dumbbell models, yield the curves shown in Fig. 4.27 [77].

No calculations are available of Δn and χ based on the necklace model considering the change of hydrodynamic interaction with coil deformation. A good feeling for the expected effects is obtained from the data calculated by Peterlin [73] for the elastic dumbbell model with finite chain length (Fig. 4.28). The steepest increase of Δn occurs for the model with constant hydrodynamic interaction, that is, with $B = 1$ in (4.63). The nonuniform decrease in hydrodynamic interaction with coil deformation, increasing with β, reduces B and hence the birefringence. The saturation effects start to be noticeable at $\beta \sim Z^{1/2}$. The relative effect of B at large β on extinction angle (Fig. 4.29) is the same as on Δn, as suggested by the same factor B in Δn and cot 2χ in (4.63). The absolute values, however, are small and hence more difficult to observe because χ rapidly vanishes with increasing β. On the whole, however, except for saturation effects the changes of B with β are small, so that neither the birefringence nor the extinction angle differ drastically from the values obtained from (4.63) with constant B, which is given as a function of ϵ and h in Fig. 4.24.

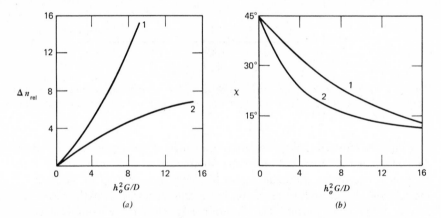

Fig. 4.27.   Birefringence and extinction angle for a soft (1) and rigid (2) dumbbell as function of $h_o{}^2 G/D$ [77].

In a *nonmatching* solvent the modification of optical anisotropy by the microform and macroform factor and their different dependence on β very strongly affect the birefringence and extinction angle. The effects are particularly spectacular if the intrinsic anisotropy, that is, the birefringence of the monomer, is negative. The always positive macroform anisotropy may cause a reversal of sign of birefringence coupled with a retrograde shift and inflection of χ at the gradient value which makes the total birefringence disappear.

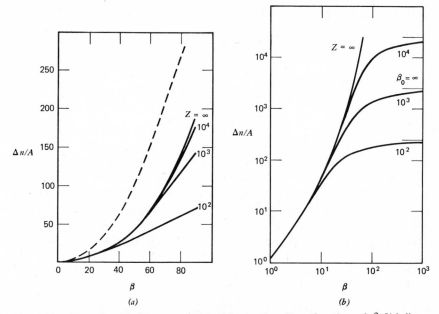

Fig. 4.28. Streaming birefringence $\Delta n/A$ with $A = 2\ ncK$ as function of β [(a) linear scale; (b) logarithmic scale] for $Z = 10^2$, 10^3, 10^4, and ∞ according to the dumbbell model with hydrodynamic interaction affected by coil deformation in flow [73]. The broken line in (a) corresponds to gradient independent, constant hydrodynamic interaction ($B = 1$) and $Z = \infty$.

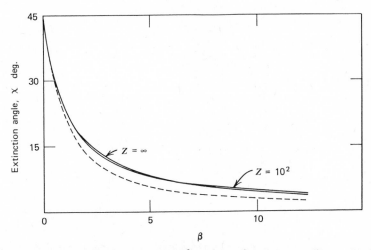

Fig. 4.29. Extinction angle χ as function of β for $Z = 10^2$ and ∞ according to the dumbbell model with hydrodynamic interaction affected by coil deformation in flow [73]. The broken line corresponds to gradient independent, constant hydrodynamic interaction ($B = 1$) and $Z = \infty$.

According to (4.50) and (4.63), we have for the specific birefringence $(\Delta n/c)_{c=0}$

$$\lim_{c=0} \left(\frac{\Delta n}{n_s c}\right) \cdot \frac{3M}{4\pi N} \left(\frac{3n_s}{n_s^2 + 2}\right)^2 = (\theta_i + \theta_{fs})\beta(1 + B^2\beta^2)^{1/2} + \tfrac{9}{4}\theta_f\,\psi(\beta). \quad (4.66)$$

The function $\psi(\beta)$ has limiting values β for $\beta \to 0$, and $4\pi/3$ for $\beta \to \infty$ [65, 67, 69]. The anisotropy factor $\theta_i + \theta_{fs}$ in the first term can be either positive or negative. The corresponding birefringence as function of β is plotted as curves 2 and 3 in Fig. 4.30 ($B = 1$) [93]. The second term is always positive but after the initial linear increase with β rapidly approaches the constant limit $3\pi\theta_f$ (curve 1). The resulting streaming birefringence [the right side of (4.66)] is plotted in Fig. 4.30 for cases where $\theta_i + \theta_{fs}$ are positive (curve 4) and negative (curve 5). In the limit of $\beta \to 0$ the optical factor K

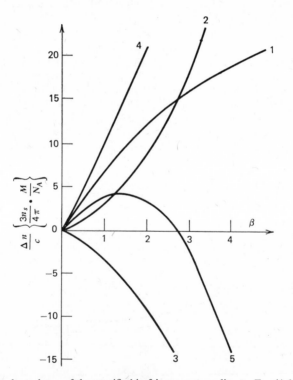

Fig. 4.30. The dependence of the specific birefringence according to Eq. (4.66) for ideally flexible linear macromolecule ($B = 1$; dumbbell model with constant hydrodynamic interaction). (1) The macroform anisotropy ($\theta_f = 3$).; (2) The intrinsic and microform anisotropy for $\theta_i + \theta_{fs} = 2$; (3) $\theta_i + \theta_{fs} = -2$; (4) The summed effect of 1 and 2; (5) The summed effect of 1 and 3 [93].

from (4.62) contains the sum of all three anisotropy factors of the macromolecule

$$(\alpha_1 - \alpha_2)_{tot} = (\alpha_1 - \alpha_2) + \frac{(n_o^2 - n_s^2)^2}{4\pi n_s \rho N} \left(\frac{e_s}{Z} + 0.83 \frac{M^2}{\rho N h_o^3}\right) \tag{4.67}$$

corresponding to θ_i, θ_{fs}, and θ_f.

The extinction angle depends on the relative contributions to the observed birefringence of the macroform (θ_f) and the combined microform and intrinsic ($\theta_i + \theta_{fs}$) anisotropy. If the former term is negligible, the extinction angle depends only on the hydrodynamic properties of the macromolecule, as in the case of matching solvent [(4.63)]. Otherwise we have for χ the more complicated expression [65, 67, 69]

$$\tan 2(\chi - \chi_o) = \frac{1 - x}{1 + x} \tan 2\delta$$

$$x = \frac{\theta_i + \theta_{fs}}{9\theta_f/4} \frac{\beta(1 + \beta^2)^{1/2}}{\psi(\beta)}$$

$$\chi_o = \frac{\chi_f + \chi_{i,fs}}{2}$$

$$\delta = \frac{\chi_f - \chi_{i,fs}}{2}. \tag{4.68}$$

χ_f and $\chi_{i,fs}$ are the extinction angles corresponding to the birefringence component caused by the anisotropy contributions θ_f and ($\theta_i + \theta_{fs}$), respectively. According to (4.66) and (4.67), the birefringence changes sign at $x = -1$, yielding anomalous behavior of χ in this region (Fig. 4.31) in good agreement with experimental data [27].

AVERAGE SHAPE OF MACROMOLECULE IN FLOW

The general expressions for Δn and $\cot 2\chi$ [(4.45) and (4.46)] of a macromolecule in a matching solvent contain the average squares of the projections of the molecule (dumbbell) or of the links (subchains of the necklace) on the coordinate axes. For the dumbbell model, we immediately derive [94]

$$\langle y^2 - x^2 \rangle_\beta = \frac{2h_o^2}{3} \beta \frac{[\Delta n]_\beta}{[\Delta n]_o} \cos 2\chi \tag{4.69}$$

and under the assumption that $\langle x^2 \rangle_\beta$ in flow is practically identical with the value $\langle x^2 \rangle$ at rest

$$h_\beta^2 = \langle y^2 - x^2 \rangle_\beta + 2\langle x^2 \rangle_o + \langle z^2 \rangle_o = \langle y^2 - x^2 \rangle_\beta + h_o^2. \tag{4.70}$$

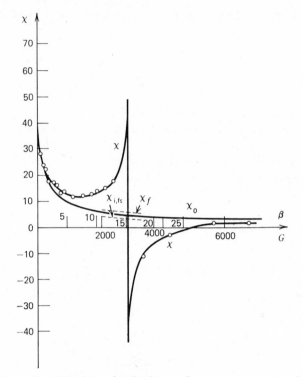

Fig. 4.31. Anomalous behaviour of extinction angle χ, χ_o, χ_f, $\chi_{i,fs}$ as function of β according to Eq. (4.68). Experimental points are from a solution of poly(p-t-butylphenyl methacrylate) fraction ($M = 7.4 \times 10^6$) in tetrachloromethane [27].

The average square chain deformation hence reads

$$\frac{h_\beta^2}{h_o^2} = 1 + \frac{2\beta}{3}\frac{[\Delta n]_\beta}{[\Delta n]_o}\cos 2\chi = \frac{b_\beta^2}{b_o^2}. \tag{4.71}$$

dumbbell   necklace

In the case of the necklace model, $\langle x^2 \rangle$ is replaced by $\langle \mathbf{x}^T \mathbf{A}\mathbf{x} \rangle$ and h_β^2 by the sum over all square link length. The experimental data on birefringence hence determine the ratio b_β^2/b_o^2 as already written in (4.71). An ideally flexible coil expands in flow as

$$\left(\frac{h_\beta^2}{h_o^2}\right)_{\text{id.fl}} = 1 + \frac{2\beta^2}{3} \tag{4.72a}$$

and a subchain, that is, elastic link of the necklace model as

$$\left(\frac{b_\beta^2}{b_o^2}\right)_{\text{id.fl}} = 1 + \frac{2B\beta^2}{3}. \tag{4.72b}$$

The deformation of the elastic links is maximum in the center and minimum at the ends of the necklace [75]. From the analogy of the necklace and dumbbell model, we may conclude that (4.72b) in first approximation also represents the average square deformation of the whole molecule.

The experimental data on polyisobutylene [25, 87, 95, 96] and nitrocellulose [97, 98, 99], permit a calculation of coil expansion according to (4.71). The results are plotted in Fig. 4.32. PIB expands linearly with β^2, the proportionality factor being 0.55, which is less than $\frac{2}{3}$ of (4.72a) but more than $2B/3 = 0.267$ (Rouse) or 0.136 (Zimm) of (4.72b). A polydispersity factor [(4.83), Table 4.3] $\kappa = 2.1$ or 4.0 corresponding to a M_w/M_n ratio 1.4 or 2.0 yields agreement with the free-draining or the impermeable, ideally flexible coil, respectively. Very recent investigation by Cottrell [100] of coil deformation in laminar flow by light scattering, however, leads to the conclusion that PIB in decaline is very nearly rigid. According to his data, at $\beta = 10$, the h_β/h_o ratio is about 1.2 instead of 3.8 as predicted for an ideally flexible impermeable coil in striking contrast to streaming birefringence data. In the case of nitrocellulose the average square of coil deformation is very nearly proportional to β instead to β^2, thus indicating a very limited deformability of the macromolecule. The mean square of coil expansion at $\beta = 21$ is less than 20% of that expected for an impermeable coil.

The same considerations can be applied to the birefringence in a flow field with longitudinal gradient if cos 2χ in (4.71) is replaced by $1(\chi = 0)$.

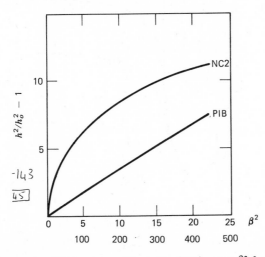

Fig. 4.32. Coil expansion $h^2/h_o^2 - 1$ according to Eq. (4.71) versus β^2 for polyisobutylene (PIB) and nitrocellulose (NC-2). The upper scale of β^2 relates to PIB, the lower to NC-2 [94].

FLOW FIELD WITH LONGITUDINAL GRADIENT

With a flow field of (4.5) the birefringence turns out to be [80]:

$$\left(\frac{\Delta n}{c}\right)_l = 2K \frac{RT}{M} \sum_p \frac{\beta_p}{(1 + \beta_p)(0.5 - \beta_p)}. \tag{4.73}$$

The calculation was performed for the ideally elastic necklace with no or constant hydrodynamic interaction. This expression goes to infinity if β_p approaches 0.5 or -1. Such an unrealistic behavior is the consequence of the linear entropy force F which does not limit the link expansion. By introducing the nonlinearity factor E in the restoring force, which prevents expansion beyond the extended length of the link or molecule, Peterlin [74] obtained a finite value for the birefringence (Fig. 4.33) and pronounced saturation effects. The theoretical data seem to be in fairly good qualitative agreement with experiments of Ziabicki and Kedzierska [29], which show first a relatively

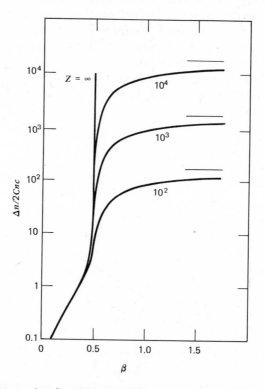

Fig. 4.33. Birefringence in a flow field according to Eq. (4.5) (jet flow) with longitudinal gradient for molecules with $Z = 10^2$, 10^3, 10^4, ∞. No or constant hydrodynamic interaction. Elastic dumbbell model [74]; $C = \eta_s KRT/nM$.

long, nearly linear increase of Δn with the gradient and at $G = 10$ sec^{-1}, a rapid saturation. This saturation may be primarily caused by solidification of the thread.

Birefringence in an Oscillating Flow Field

STREAMING BIREFRINGENCE IN AN OSCILLATING FLOW FIELD

As a consequence of the small amplitude of the gradient G_o and β_o, the macromolecule is practically undeformed so that the correct λ_p are those of the molecule in solution at rest; the birefringence is that of (4.66) extrapolated to $G = 0$ or $\beta = 0$ (intrinsic birefringence) modified by the relaxation terms, and the extinction angle is 45°. The complex intrinsic birefringence is [101]:

$$[\Delta n]_\omega^* = \frac{BRT}{M\eta_s}(\mathrm{I} - i\mathrm{II}) = [\Delta n]_\omega e^{-i\delta}$$

$$\tan \delta = \mathrm{II}/\mathrm{I}$$

$$\mathrm{I} = \Sigma \frac{\tau_p}{1 + \omega^2 \tau_p'^2}$$

$$\mathrm{II} = \Sigma \frac{\omega \tau_p \tau_p'}{1 + \omega^2 \tau_p'^2}. \tag{4.74}$$

The relative amplitude $[\Delta n]_\omega/[\Delta n]_{\omega = G = 0} = [\Delta n]_\omega/[\Delta n]_o$ is plotted in Fig. 4.34 (free-draining and impermeable coil, $Z = 100$ and 1) as function of $\omega\tau_1$ for different values of internal viscosity parameter φ. The corresponding phase angle δ is plotted in Fig. 4.35 [102]. The calculations for a set of Z values were performed by Thurston and Peterlin [103]. According to (4.74) the birefringence steadily drops with increasing frequency. The asymptote has a slope corresponding to ω^{-1}. At lower frequencies there is an intermediate straight section with a slope corresponding to $\omega^{-1/2}$ (free-draining coil) and $\omega^{-1/3}$ (impermeable coil). The length of this section increases with Z and decreases with hydrodynamic interaction and internal viscosity. The phase angle δ first increases rather rapidly, but at $\omega\tau_1 = 1$ it begins to go up at a lower rate, thus producing a more or less inclined plateau centering about $\delta = 45°$; and later it rapidly reaches again the asymptotic value 90°. The length of the plateau increases with Z and decreases with hydrodynamic interaction and internal viscosity. The dumbbell model exhibits neither the transition region in birefringence nor the plateau in the phase angle. The theoretical results are in good agreement with experimental data, as shown in Fig. 4.10 [28], and permit the calculation of Z, h, and φ.

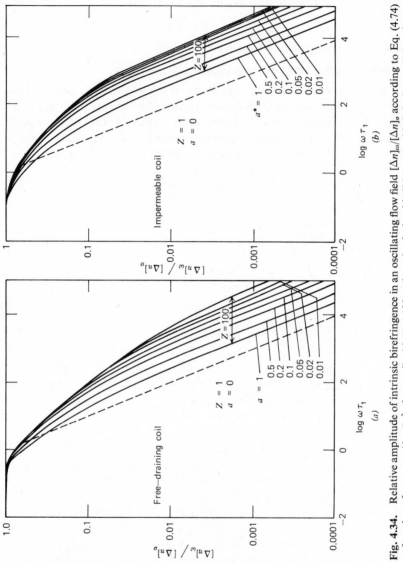

Fig. 4.34. Relative amplitude of intrinsic birefringence in an oscillating flow field $[\Delta n]_{\omega}/[\Delta n]_{o}$ according to Eq. (4.74) as function of $\omega\tau_1 = \omega/4\mu_o D_o\lambda_1$ for a coil with $Z = 100$ and an elastic dumbbell ($Z = 1$) with finite internal viscosity φ: (a) free-draining coil, parameter $a = \varphi/Z\eta_s W_o$; (b) impermeable coil, parameter $a^* = h\varphi/Z\eta_s W_o = 0.0519\varphi/h_o n_s$ [101].

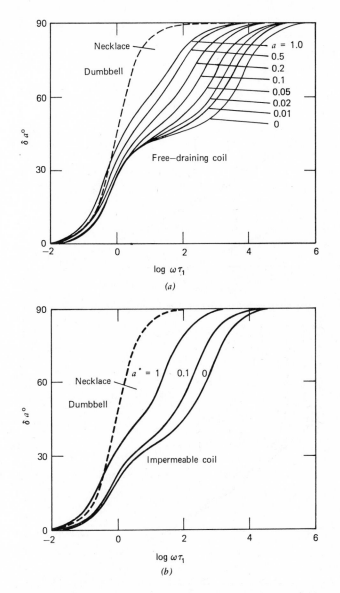

Fig. 4.35. Phase angle δ in an oscillating flow field or in an acoustic wave field according to Eq. (4.74) for the same models as in Fig. 4.34 [102].

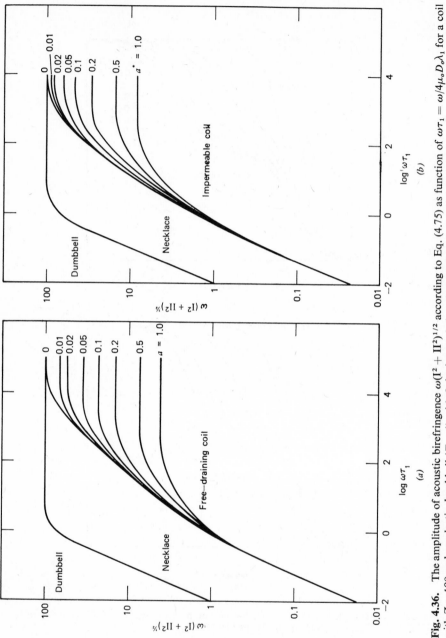

Fig. 4.36. The amplitude of acoustic birefringence $\omega(I^2 + II^2)^{1/2}$ according to Eq. (4.75) as function of $\omega\tau_1 = \omega/4\mu_o D_o\lambda_1$ for a coil with $Z = 100$ and an elastic dumbbell ($Z = 1$) with finite internal viscosity: (a) free-draining coil, parameter $a = \varphi/Z\eta_o W_o$; (b) imper-

ACOUSTIC BIREFRINGENCE

The problem was treated on the basis of the elastic dumbbell [104], elastic sphere [31], and elastic necklace model [102]. The last treatment yields the complex intrinsic Lucas constant

$$[L]^* = \lim_{c=0}\left[\frac{\Delta n}{ncI_a^{1/2}}\right]^* = \frac{4\pi}{5}\left(\frac{n_s^2 + 2}{3n_s}\right)^2\left(\frac{2}{\rho c_a^3}\right)^{1/2}\frac{N\omega}{M}(\alpha_1 - \alpha_2)(I - iII)$$

$$= [L]e^{-i\delta}$$

$$\tan \delta = II/I \qquad\qquad (4.75)$$

with the values I and II from (4.74). The amplitude $\omega(I^2 + II^2)^{1/2}$ is plotted in Fig. 4.36 as function of $\omega\tau_1$ for the dumbbell $(Z = 1)$ and necklace $(Z = 100)$ with finite internal viscosity for the free-draining and impermeable coil. The phase angle δ is identical with that in an oscillating flow field with transversal gradient (Fig. 4.35). The birefringence is proportional to the square root of the wave intensity and to the frequency, as long as $\omega\tau_1$ is far below 1. Later it rapidly approaches a saturation value which, in turn, diminishes with increasing internal viscosity The saturation birefringence as a function of φ is plotted in Fig 4.37 for the free-draining and impermeable coil. In the bilogarithmic plot of Fig. 4.36, the birefringence has a nearly straight section with a slope $\frac{1}{2}$ (free-draining) and $\frac{2}{3}$ (impermeable coil) between the original linear and final saturation region. The length of this transition increases with Z and decreases with hydrodynamic interaction and internal viscosity. It is

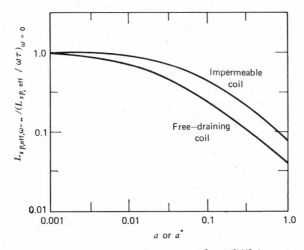

Fig. 4.37. Limiting values of acoustic birefringence $\omega(I^2 + II^2)^{1/2}$ for $\omega = \infty$ as function of parameter a (free-draining coil) and a^* (impermeable coil) as defined in Fig. 4.36. Necklace model with $Z = 100$ [102].

exactly the same effect as with streaming birefringence in an oscillating flow field ($\Delta n_\omega \sim \omega^{-1} \Delta n_a$) and with the dynamic shear modulus of a coiled macromolecule of finite Z [105].

Concentration Dependence (Stress-Optical Law)

All theoretical treatment applies to an isolated particle. Every type of hydrodynamic or optical interaction among the solute is completely neglected. In this range the birefringence is proportional to the number of particles (i.e., to the concentration) and the extinction angle is independent of it. The deviations from linearity in the former quantity and the dependence of extinction angle on concentration are indications of particle interaction. There are only a few rather unsuccessful attempts to treat theoretically the concentration dependence of birefringence for rigid [106] and soft [107] particles based on analogy with the polarization of a homogeneous liquid in an electric field. The situation is much better in the case of solution of randomly coiled linear macromolecules where one has developed a random network theory [108] and also discovered a far-going parallelism between optical and rheological properties (*rheo-optical law*).

At finite concentration the randomly coiled macromolecules partially overlap and interpenetrate, thus forming a great many intermolecular mechanical cross-links or entanglements of finite lifetime. A network theory of such a solution, which differs only in detail from a molecular theory of an imperfectly elastic rubber-like solid [109], was developed by Yamamoto and Inagaki [110], Lodge [108], and Yamamoto [111].

In its simplest form the theory for polymer solutions rests on certain assumptions, as follows. In the solution at rest, enough long-chain molecules are linked at a few points along their length by temporary physical entanglements so as to form a homogeneous network extending throughout the solution. Due to the effect of thermal motion, entanglement junctions are continuously being lost and new ones formed. In flow the network will deform and the associated change of configurational entropy will give rise to a nonisotropic contribution to the stress and optical polarizability. The results so obtained depend on a memory function of the mechanical cross-links which has not been interpreted in simple molecular properties.

A more straightforward approach to the concentration dependence of birefringence is based on the fact that the anisotropic part of the excess shear stress tensor p_{jk} and the excess birefringence tensor in a matching solvent, that is, the component derived from θ_i, depend on the same molecular parameters (necklace model). From Eq. 4.39, 4.43–47 one derives

$$p_{jk} = \frac{cN}{M} \langle \mathbf{x}_j \mathbf{F}_k \rangle = \frac{3cRT}{Mb_o^2} \langle \mathbf{x}_j \mathbf{A} \mathbf{x}_k \rangle$$

$$n_{jk} = \frac{6\pi cN}{5Mb_o{}^2} \frac{(n_s{}^2 + 2)^2}{9n_s} (\alpha_1 - \alpha_2) \langle \mathbf{x}_j \mathbf{A} \mathbf{x}_k \rangle$$

$$n_{jk} = K p_{jk}$$

$$K = \frac{2\pi}{5} \frac{(n_s^2 + 2)^2}{9n_s} \frac{\alpha_1 - \alpha_2}{kT}. \tag{4.76}$$

In particular, one has for the normal stress difference

$$p_{yy} - p_{xx} = \frac{n_{yy} - n_{xx}}{K}, \tag{4.77a}$$

the excess shear stress

$$p_{xy} = (\eta - \eta_s)G = \frac{n_{xy}}{K}, \tag{4.77b}$$

and the orientation angle $\chi_{s.t.}$ of stress tensor in the flow plane (*recoverable strain s*)

$$s = \cot 2\chi_{s.t.} = \frac{p_{yy} - p_{xx}}{2p_{xy}} = \frac{n_{yy} - n_{xx}}{2n_{xy}} \tag{4.78}$$

The orientation angle $\chi_{s.t.}$ of the stress tensor is identical with the extinction angle of streaming birefringence. This *stress-optical* law first formulated by Lodge [108] with the stress-optical coefficient K [see also (4.62)] applies not only to dilute solutions but also to higher concentrations, because we can assume that the mechanical interactions do not affect the optical factor θ_i. Moreover, we can expect the approximate validity of (4.77) and (4.78) at higher concentrations even for nonmatching solvents, because with increasing concentration the average refractive index of the environment of every single molecule will rapidly approach the average value inside the molecule and thus eliminate the term θ_f, which is proportional to the square of the difference of the respective indices. The term θ_{fs}, having the same dependence on molecular dimensions as θ_i, only changes the factor $\alpha_1 - \alpha_2$ in K and hence does not interfere with the stress-optical law. At very high concentration, however, it has to decrease for the same reason as does θ_f. This effect hence begins to invalidate the stress-optical law.

The first rather empirical approach to the proportionality of the optical and rheological properties of streaming solutions was made by Peterlin et al. [26, 112, 113], who plotted Δn over $G(\eta - \eta_s)$ and χ over $G(\eta - \eta_s)/c$. They obtained a perfect master curve for solutions in matching solvents in the former case (see Fig. 4.8), and either a master curve (see Fig. 4.9) or a set of slightly displaced curves in the latter case. The main consequence of this type of plotting is the use of the modified parameter

$$\beta_c = M(\eta - \eta_s)\frac{G}{cRT} \tag{4.79}$$

instead of the value from (4.56), which is valid only in the limit of zero concentration. The use of the parameter β_c is now a generally accepted practice for plotting experimental data and extrapolating to zero concentration.

The stress-optical law yields a straight line if the birefringence Δn is plotted versus the excess stress anisotropy $p_{xy}/\sin 2\chi$, or $\beta_c/\sin 2\chi$, or the xy component of birefringence $\Delta n \cdot \sin 2\chi$ versus the excess shearing stress $(\eta - \eta_s)G$ or $c\beta_c$ (Fig. 4.38). This was amply demonstrated on polystyrene [108, 114, 86, 115], polyisobutylene [96, 87, 116], polysilicones [117], and nitrocellulose [88, 98].

The theoretical and experimental limitations of the stress-optical law are as follows:

1. Saturation effects at very high β—if the coil is nearly completely extended the birefringence approaches a saturation value and the excess shearing stress

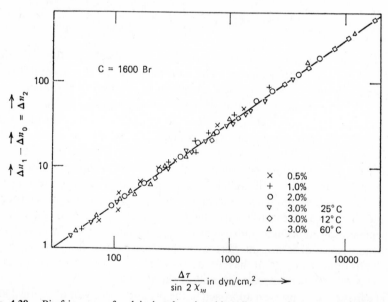

Fig. 4.38. Birefringence of polyisobutylene in white oil versus $\Delta\tau/\sin 2\chi$ demonstrating the constancy of the stress-optical coefficient K over five decades of principal stress difference [87]: Δn_1, Δn_o, and Δn_2 are the birefringence of solution, solvent and solute. (See Polar Diagrams, p. 335.)

increases more than linearly with the mean square of coil dimensions; as a consequence, the ratio K between birefringence and shearing stress decreases; no such effect was yet unambiguously reported;

2. Marked decrease with very short chains of the stress-optical coefficient K, a "constant" of the polymer-solvent system, with decreasing molecular weight (Fig. 4.23);

3. Decrease of K with increasing concentration of polymer in nonmatching solvents—decrease of θ_f (Fig. 4.58).

The polydispersity has no effect so long as K is constant for all macromolecules present (see limitation 2) and the interaction among the coils is independent of M. The latter effect becomes important as soon as coil entanglement starts to influence the viscosity leading to the well-known $M^{3.4}$ dependence.

Combining (4.78) with (4.64d) yields

$$s = \cot 2\chi = B\frac{M}{cRT}p_{xy} = B\beta_c. \tag{4.78a}$$

The coefficient B has the value derived from (4.63) (Fig. 4.24). The proportionality between $\cot 2\chi$ and the excess shearing stress p_{xy} or β_c is correct only for small gradients, exactly speaking for the limit of $G = 0$. At larger gradients, $\cot 2\chi$ increases more slowly than p_{xy} or β, as demonstrated by Philippoff on polystyrene [86], polyisobutylene [87], and particularly on nitrocellulose [88, 98]. It is also dependent on concentration as shown by Peterlin and coworkers on the same three polymers [26, 112, 113]. A plot of $\cot 2\chi$ over β_c or p_{xy}/c does not yield a master curve independent of concentration, but a set of slightly different curves which at sufficiently small concentration can be employed for the extrapolation to $c = 0$.

These effects are a consequence of the specific dynamics and hydrodynamics of polymer molecules: eventual internal viscosity opposing rapid change of coil shape, change of hydrodynamic interaction in the macromolecule caused by concentration (interaction between partially interpenetrating coils), and coil deformation in flow.

Polydispersity

The theories presented for rigid and deformable particles and for linear macromolecules apply only to a monodisperse system. If the solute particles are heterogeneous with respect to size, shape, optical, or mechanical properties, then the birefringence phenomena are considerably different from those for a monodisperse system. The main complication arises because not only the amount of birefringence but also the orientation of the main axes vary with particle properties.

If the solution is infinitely dilute so that no interaction occurs between the particles, the excess refractive index is simply the sum of contributions of all particles present. The summation is carried out over all species of particles which, if present alone in solution, would give the observed value of Δn_i and χ_i of birefringence and extinction angle, respectively, at the same velocity gradient. We hence obtain for the birefringence of the polydisperse system [118]

$$\Delta n = \left[\left(\sum_i \Delta n_i \sin 2\chi_i \right)^2 + \left(\sum_i \Delta n_i \cos 2\chi_i \right)^2 \right]^{1/2} \qquad (4.80)$$

and for the extinction angle

$$\cot 2\chi = \frac{\sum_i \Delta n_i \cos 2\chi_i}{\sum_i \Delta n_i \sin 2\chi_i}. \qquad (4.81)$$

If the components differ appreciably in Δn_i and χ_i, the observed gradient dependence may be rather unusual, as seen in Fig. 4.39 for a mixture of two

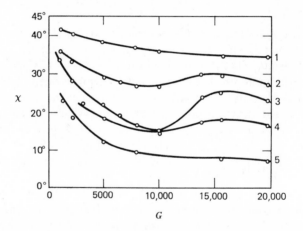

Fig. 4.39. Extinction angle curves for nitrocellulose fractions ($M = 72{,}000$ and $255{,}000$) in cyclohexane as solvent mixed in the ratio (1) 1 : 0, (2) 0.8 : 0.2, (3) 0.5 : 0.5, (4) 0.2 : 0.8, (5) 0 : 1 [119].

nitrocellulose fractions ($M = 72{,}000$ and $255{,}000$) in cyclohexanone investigated by Signer and Liechti [119]. The individual fractions give the usual extinction angle curves, whereas the mixtures show the apparently anomalous behavior with a minimum and a maximum. More complicated effects occur in mixtures of particles exhibiting birefringence of opposite sign and widely differing extinction angles.

Microgel or micellar particles are so large that they become oriented extremely rapidly. The extinction angle drops very nearly to zero and the birefringence soon reaches saturation and becomes constant. The presence of such particles in a molecular solution affects the observed birefringence and extinction angle significantly, as shown in Fig. 4.40 for a mixture of molecular and micellar acetyl cellulose in cyclohexanone [120]. After a very steep initial increase, the birefringence shows, for a while, a less than linear dependence on the gradient, somewhat similar to that for rigid particles, and finally becomes proportional to the gradient. The experimental data correspond well to values calculated according to (4.80) and (4.81).

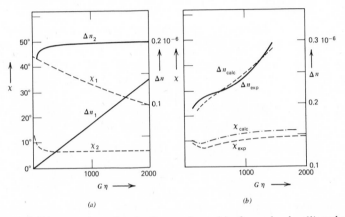

Fig. 4.40. Birefringence Δn and extinction angle χ: (a) of a molecular (1) and micellar (2) solution of cellulose acetate in cyclohexane; (b) of a 1:1 mixture of both components as measured (Δn_{exp}, χ_{exp}) and as calculated (Δn_{calc}, χ_{calc}) according to Eqs. (4.80) and (4.81) [120].

As a rule, the largest particles present are responsible for the effects at small gradient, that is, for the intrinsic Maxwell constant and orientation at $G = 0$, the smallest particles for the effects at very large gradient. The presence of very small particles that yield a finite extinction angle even at the highest gradient applied provides a nonvanishing limiting value χ, which is, therefore, a consequence of polydispersity of the sample and not of some particular property of the particles. An important case is that of a birefringent solvent, which invariably has $\Delta n_s \sim \eta G$ and $\chi_s = 45°$, thus modifying to some extent the initial values of birefringence and extinction angle of solution, but yielding a nonvanishing χ_∞.

The polydispersity significantly affects the parameter β and its square occurring in the expressions for birefringence and extinction angle. If the molecular weight dependence of intrinsic viscosity can be expressed by the Kuhn-Houwink-Mark power law $[\eta] \sim M^a$, $0.5 \le a \le 1$ and the molecular weight distribution is of the Schulz-Zimm type

$$dn = \left(\frac{z+1}{M_w}\right)^{z+1} \frac{M^z}{z!} \exp\left[-(z+1)\frac{M}{M_w}\right] dM, \qquad (4.82)$$

we obtain

$$\langle\beta\rangle_n = \frac{M_n[\eta]_n \eta_s G}{RT} \cdot \frac{(z+1+a)!}{z!(z+1)^{1+a}} \qquad (4.83)$$

$$\langle\beta^2\rangle_n = \langle\beta\rangle_n^2 \cdot \frac{z!(z+2+2a)!}{(z+1+a)!^2} = \kappa_n \langle\beta\rangle_n^2$$

$$\cos 2\chi = \kappa_n B \langle\beta\rangle_n.$$

The factor κ_n as a function of the polydispersity $M_w/M_n = (z+2)/(z+1) = w$ is given in Table 4.3 for $a = 0.5$, 0.75, and 1.0 [121, 122]. If instead of $\langle\beta\rangle_n$

Table 4.3. Polydispersity Factor κ_n of Eq.(4.83) as Function of w and a [121]

$w = \dfrac{M_w}{M_n}$	$a = 0.5$	0.75	1.00
1	1.000	1.000	1.000
1.05	1.113	1.155	1.204
1.10	1.227	1.314	1.418
1.20	1.459	1.641	1.867
1.33	1.774	2.093	2.499
1.50	2.174	2.676	3.333
2.00	3.396	4.497	6.000

we use $\langle\beta\rangle_w$, we obtain [123]

$$\kappa_w = \frac{z!(z+1+2a)!}{(z+1)(z+a)!^2} \qquad (4.84)$$

$$= \kappa_n \frac{(z+1+a)^2}{(z+1)(z+2+2a)}.$$

The factors κ_n and κ_w rapidly increase with increasing polydispersity and deviation from theta condition, thus yielding an increase of the experimental value of $\kappa_n B$ or $\kappa_w B$ above that of the monodisperse solution [(4.62) and (4.63)].

4 TREATMENT AND INTERPRETATION OF EXPERIMENTAL DATA

Polar Diagrams

When examining streaming birefringence data, the first aim is to find what type or types of particles are present. To do this the method of polar diagrams [124] is often useful. This method is based on the often neglected fact that the two basic quantities in streaming birefringence measurements, the extinction angle and the magnitude of birefringence, are closely related to one another. Therefore, the interdependence of these two quantities may provide additional information as to the conventional dependences of them on the velocity gradient. It is easy to show [87, 125] that the polar plot of Δn against 2χ, which is identical with the Cartesian plot of $\Delta n \sin 2\chi = 2n_{xy}$ against $\Delta n \cos 2\chi = n_{yy} - n_{xx}$, can be treated as a vector diagram in two-dimensional space. The resulting curve is situated in the first quadrant for positive and in the third one for negative optical anisotropy of the particles. Cartesian components of this diagram are just the two quantities to be summed when adding birefringences of more components, according to Sadron's rule [(4.80) and (4.81)] [118]. Thus, the combination of birefringences is reduced to a simple combination of vectors. It is obvious that this property will be found with all plots of the type $(\Delta n \sin 2\chi) F_1(G)$ against $(\Delta n \cos 2\chi) F_2(G)$ where $F_1(G)$ and $F_2(G)$ are any arbitrary functions of the velocity gradient. It appears that the most useful of such diagrams is that where $F_1(G) = F_2(G) = 1/G$.

For a pure solvent whose orientation angle is 45° at all experimentally possible velocity gradients and the magnitude of birefringence is proportional to the gradient, the plot reduces to a single point on the axis $\Delta n \sin 2\chi/G$. The same is true for solutions of polymers whose determining parameter $\beta = M[\eta]\eta_s G/RT$ is sufficiently small. At higher values of this parameter the shape of the plot depends on the type of particles present.

For rigid particles the plot tends toward origin, because there always exists a region of velocity gradients, where kinematic orientation of all particles is complete and Δn approaches its saturation value. With ellipsoids, the plot is close to a semicircle with the center on the axis $\Delta n \sin 2\chi/G$, its exact shape depending on the axial ratio (Fig. 4.41).

For deformable particles, the shape of the plot depends on the extent of possible maximum deformation and how easily the particles deform. Let us recall that, for deformable particles, the excess stress and optical tensors are proportional if we take into account the intrinsic part of optical anisotropy only. Accordingly, the quantity $\Delta n \sin 2\chi/G$ is proportional to excess viscosity $(\eta - \eta_s)$. Therefore, Hermans' [72] dumbbell model and Zimm's [71] segment model, both predicting Newtonian viscosity, are represented as straight

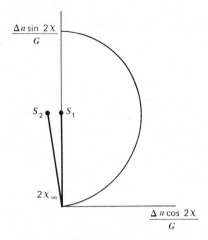

Fig. 4.41. Polar diagram for ellipsoids of axial ratio, $p = 5$; S_1, centre of the quarter circle at low gradients; S_2, center of the quarter circle at high gradients [124].

horizontal lines. The plot for particles that exhibit a non-Newtonian viscosity approaches slowly the $\Delta n \cos 2\chi/G$ axis. The extension in the horizontal direction depends on the particle deformability: highly deformable particles (high molecular weight polymers with low chain rigidity) exhibit plots [115] extensively departing from coordinate origin (Fig. 4.42), while plots for particles with lower deformability do not change their distance from origin appreciably or even return to the origin [124] (Fig. 4.43).

While a polar diagram may offer a valuable guide for finding the type of particles in one-component systems, its main value is in interpreting bire-

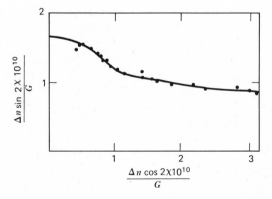

Fig. 4.42. Polar diagram after correction for solvent birefringence. Polystyrene ($M_w = 2.40 \times 10^6$) in dimethylbenzophenone, $c = 0.00306$ g/ml [115].

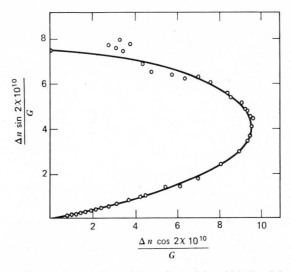

Fig. 4.43. Polar diagram for solution of deoxyribonucleic acid isolated from *Escherichia coli*; $c = 2.2 \times 10^{-5}$ g/ml [124].

fringent behavior of multicomponent systems and in the ease of subtraction of the solvent contribution from the observed birefringence. Let us consider a system that contains two sorts of particles widely differing in size. At low velocity gradients, the decisive parameter β for *smaller* particles is so small that they contribute to the plot by a constant vertical vector. The shape of the plot for this gradient range is determined entirely by the larger particles. Their characteristic plot is shifted only vertically. Hence we are able to find the characteristics of the larger particles regardless of the presence of the smaller particles. On the other hand, some solutions may contain rigid particles (sometimes microgels) so large that their orientation is virtually complete even for the smallest experimentally accessible gradients. In such a case their birefringence is constant, the extinction angle is zero, and the contribution to the plot is situated in the abscissa axis. The more or less constant initial value of $\Delta n \cos 2\chi$ of the observed birefringence is caused by their contribution and may be easily subtracted from the overall dependence. The remainder is the contribution of other particles present. A number of possible combinations of two types of particles exists, each of them easily recognizable in the polar plot, although their gradient dependences of Δn and χ may seem rather anomalous (Fig. 4.44). Even if the two types of particles do not differ so much in size, it is sometimes possible to distinguish their birefringences by successive approximations, especially when their anisotropies have opposite signs.

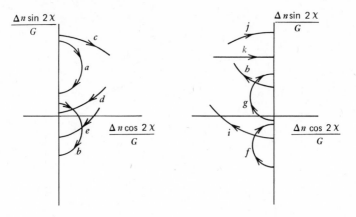

Fig. 4.44.   Scheme of a polar diagram for various mixtures of high molecular weight component A and a low molecular weight component B. The arrows indicate the direction of the increasing velocity gradient $(\Delta n_A/G)_{G \to 0} = a$; $(\Delta n_B/G)_{G \to 0} = b$. β_A is the determining parameter of the component A. (a) sg $a =$ sg $b = +1$, $a \gtrless b$; (b) sg $a = +1$, sg $b = -1$, $a > b$; (c) sg $a = +1$, sg $b = \pm 1$, $a > b$, $\beta_A < 1$; (d) sg $a = +1$, sg $b = +1$, $a > b$, $\beta_A > 1$; (e) sg $a = +1$, sg $b = -1$, $a > b$, $\beta_A \gg 1$; (f) sg $a =$ sg $b = -1$, $a \gtrless b$; (g) sg $a = -1$, sg $b = +1$, $a > b$; (h) sg $a = -1$, sg $b = +1$, $a < b$, $\beta_A < 1$; (i) sg $a = -1$, sg $b = +1$, $a > b$, $\beta_A < 1$; (j) sg $a = -1$, sg $b = +1$, $a \gtrless b$, $\beta_A \gg 1$; (k) sg $a = -1$, sg $b = +1$, $a \gg b$, $\beta_A \to \infty$ [124].

On the other hand, polydispersity of the sample of the same polymer species does not affect significantly the shape of the polar plot. It only slightly compresses it in the horizontal direction. This holds if the distribution function has not more maxima at very much different molecular weights.

Chain Molecules

When the type of particles present has been found and the birefringence separated, if necessary and possible, into the contributions of the components, it is usually worthwhile to confirm the "diagnosis" by means of other relationships.

For linear macromolecules in matching solvent, the stress-optical law [(4.76)] holds. The constancy of K evaluated for various gradients and concentrations according to (4.64a) really proves the chain character of the sample. The concentration and gradient variation of the B-value evaluated from expressions (4.64b,c,d) is usually small for linear molecules. For experiments at nonzero concentrations and gradients, the parameter β in (4.64) should be, of course, replaced by β_c defined by (4.79) where η is the actual viscosity at given gradient and concentration.

Providing that the stress-optical law holds, (4.64b), (4.64c), and (4.64d) are all equivalent. They have, of course, various advantages and disadvantages in evaluation of experimental data. Equation (4.64d) does not include Δn; the extinction angle influences (4.64b) very little in the region of high deformations (χ close to zero); only optical quantities are included in (4.64c), which is to be used when viscosity data are not available or are too imprecise.

The literature contains insufficient data for testing these relations in a broad region of β-values. According to Janeschitz-Kriegl [126], the B-values from (4.64) are nearly constant for polystyrene up to $\beta = 0.5$, but afterwards decrease to about half the initial value and then remain more or less constant. Munk and Peterlin [84] confirmed the constancy of the B-value in the high β-value region up to $\beta = 30$ for high molecular weight polystyrene in highly viscous solvents (Fig. 4.45). Similar results have been obtained also for deoxyribonucleic acid in spite of its obviously restricted deformability [84] (Fig. 4.46).

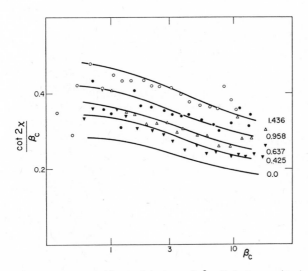

Fig. 4.45. Constant $B = \cot 2\chi/\beta_c$ as function of β_c. Polystyrene in Aroclor 1248, $M = 984,000$, numbers at curves mean concentration in g/100 ml. Extrapolated curve to $c = 0$ is marked 0.0 [84].

The interpretation of the above-mentioned experimental data on the basis of existing theories must be considered with caution because only a few data were obtained on polymers with narrow molecular weight distribution. Even with a narrow distribution (e.g., $M_w/M_n = 1.05$), the sample comprises a rather wide range of molecular weights which deeply influence the birefringent behavior. We indeed do not yet know with sufficient precision the actual gradient dependence of birefringence of a monodisperse polymer sample, which could be used for critical check of the theory. Hence for a particular sample, we

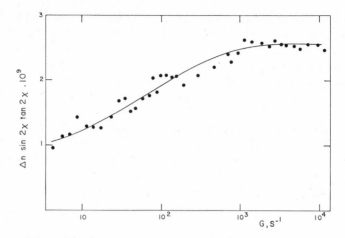

Fig. 4.46. Product $\Delta n \sin 2\chi \tan 2\chi$ as a function of velocity gradient, desoxyribonucleic acid, same sample as in Fig. 4.43 [84].

usually cannot estimate the influence of polydispersity without knowing the behavior of monodisperse equivalent. One semiquantitative rule applies: the more non-Newtonian the sample is, the stronger is the influence of poly-dispersity.

While the shape of birefringent dependences may provide information about the type and deformability of the particles, many molecular quantities may be deduced from values extrapolated to zero velocity gradient and zero concentration. The best way to obtain these extrapolated values is probably to extrapolate the slowly varying expressions $\Delta n \sin 2\chi/G(\eta - \eta_o)$, $\cot 2\chi/\beta$, $\Delta n \cos 2\chi/\beta G(\eta - \eta_o)$, and $\Delta n \sin 2\chi \tan 2\chi(M/cRT)$ to both limits. The extrapolated values K and B may then be calculated. In the literature, however, the independent extrapolation of optical and mechanical variables is used more frequently, giving expressions $[\Delta n]_o$, $[\omega_o]$, and $[\eta]_o$ [(4.62)]. Obviously,

$$\frac{[\Delta n]_o}{[\eta]_o} = \frac{2K}{n},$$

$$\frac{[\omega]_o}{[\eta]_o} = B \frac{M}{2RT} \eta_s.$$

$$(4.85)$$

With B known, (4.85) seems to offer a means for the calculation of molecular weight. The B-value is, however, dependent on the type of particle [(4.65ab), Fig. 4.24] and polydispersity [(4.83) and (4.84)]. For monodisperse linear particles in a theta solvent, it depends on the hydrodynamic interaction and kinematic rigidity. For particles with high hydrodynamic interaction (which is the most usual case) $B = 0.2$ for soft and 1.8 for rigid particles.

The separation of the contributions θ_i, θ_{fs}, and θ_f can be obtained by plotting the zero gradient values of the anisotropy $(\alpha_1 - \alpha_2)$ versus $M/[\eta]$ or n_s as described in connection with (4.67) and demonstrated in Figs. 4.55 and 4.56.

5 EXPERIMENTAL RESULTS

Solvents

In agreement with all theories of birefringence of low molecular weight liquids [(4.23) and (4.24) for $\sigma \to o$], we observe that the magnitude of birefringence is proportional to the product of viscosity and velocity gradient divided by absolute temperature, and the extinction angle is 45° at all experimentally obtainable velocity gradients. In mixtures of chlorinated diphenyls (Aroclor, a Monsanto trademark), the viscosity at constant temperature decreases with increasing gradient. The ratio $\Delta n/G\eta$, however, remains constant [84].

The same reduction $(T\,\Delta n/G\eta = \text{const})$ is supposed to hold for different temperatures, too. However, some tacit assumptions are included in this postulate, namely, that the structure of the liquid is not changing with temperature. If we suppose that solvent molecules form clusters that are oriented by the flow, then temperature-dependent changes in the size and structure of these clusters may simultaneously influence the quantity $T\,\Delta n/G\eta$. Some measurements of Munk [127] and Janeschitz-Kriegl [128] indicate that such changes really can play a role (Fig. 4.47).

Generally, we may expect that the flow birefringence of a solvent will increase with its viscosity and with anisotropy of its molecules. As the anisotropy of single bonds is usually low, the birefringence of water, aliphatic

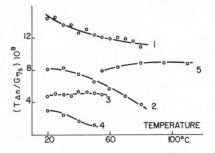

Fig. 4.47. Dependence of $T\Delta n/G\eta_s$ on temperature for pure solvents (1) nitrobenzene, (2) ethyl benzoate, (3) dibutyl phthalate, (4) benzene, (5) benzophenone. Data for benzophenone from Janeschitz-Kriegl and Burchard [128] [127].

Table 4.4. Flow Birefringence of Pure Solvents

Solvent	$t,°C$	η,cP	$\dfrac{\Delta n}{G} \times 10^{12}$	$\dfrac{\Delta n}{G\eta} \times 10^{10}$	$\dfrac{\Delta nT}{G\eta} \times 10^{8}$	Reference
Cyclohexanone	25	1.987	0.36	0.18	0.54	151
Cyclohexanone	25	2.06	0.39	0.19	0.57	183a
Tetrahydrofuran	25	0.473	0.16	0.34	1.0	151
Octyl alcohol	25			0.37	1.10	88
Tetrahydrofuryl alcohol	25	5.28	0.80	0.15	0.45	151
Butyl acetate	25	0.688	0.35	0.51	1.52	160
Dimethylformamide	25	0.812	0.44	0.54	1.6	151
Dimethyl sulfoxide	25	1.987	0.053	0.027	0.08	151
Bromoform	25	1.89	1.27	0.67	2.00	115
Benzene	25	0.606	0.62	1.02	3.05	160
Toluene	25	0.551	0.59	1.07	3.19	160
Bromobenzene	25	1.056	2.27	2.15	6.40	151
Bromobenzene	25	1.038	2.06	1.99	5.92	126
Nitrobenzene	25	1.82	8.90	4.89	16.4	151
Ethyl benzoate	20	2.19	6.12	2.80	8.19	160
Benzyl benzoate	25			2.48	7.40	88
Ethyl cinnamate	20	6.00	44.3	7.38	21.6	183b
Dibutyl phthalate	25	16.3	26.0	1.59	4.75	151
Tricresyl phosphate	25	55.8	117.6	2.11	6.30	151
Decalin	25			0.38	1.13	183c
Butylnaphthalene	25	5.42	18.7	3.45	10.3	115
Benzophenone	55	4.70	11.4	2.43	7.95	128
2,5-Dimethyl benzophenone	40	12.04	18.7	1.55	4.87	115
Aroclor 1248	25	275	665	2.42	7.11	84

hydrocarbons, alcohols, ethers, and even chlorinated compounds is very often close to the error of measurement. Birefringence of aliphatic ketones, esters, and amides is usually small, too. Higher birefringence is exhibited by aromatic solvents, especially when there is extended conjugation as in naphthalene derivatives, ethyl cinnamate, or nitrobenzene. Bromine increases birefringence also. A few examples of solvent birefringences are in Table 4.4.

Rigid Particles

The term "rigid particles" is used rather vaguely in the literature. It includes sometimes not only really rigid (i.e., completely undeformable) particles but also particles with high "thermodynamic" rigidity (i.e., long segment length) or particles with high kinematic rigidity (i.e., high internal viscosity). The expected behavior characteristics of all these systems have one thing in common: the magnitude of birefringence approaches some saturation value at high velocity gradients. Historically, the rheo-optic behavior of rigid particles (spheroids) was quantitatively predicted much earlier than that of other models. As a result most older, and even many quite recent, papers discuss the birefringence phenomena in terms of rigid models, even if the system under examination clearly does not belong to this category. It is rather surprising that the theory for rigid particles has not yet been rigorously tested, apparently because of the lack of a proper system composed of reasonably rigid monodisperse particles. Some inorganic sols apparently exhibit rigidity, vanadium pentoxide V_2O_5, for example, but their polydispersity is completely unknown. Suspensions of proteins or viruses may be the best systems for testing but the proteins are usually not large enough for measurement covering the whole region of interest. Consequently, only measurements on tobacco mosaic virus (TMV) meet the rigorous criteria.

Actually, repeated measurements [129–132] on TMV solutions have shown that the general behavior of the birefringence corresponds to the theory. However, from the quantitative point of view, only the dependence of extinction angle on shear stress was analyzed, and even its fit with the theoretical function for rigid ellipsoids was not too good. The apparent length of the particle calculated from experimental data according to formulas for ellipsoids decreases with increasing velocity gradient (Fig. 4.48), even for TMV preparation with rather narrow length distribution [130]. The evaluation according to Fig. 4.48 actually diminishes the discrepancy because the apparent length is proportional to the cubic root of the apparent rotation diffusion constant. All authors claim that the discrepancy is due to the residual poly-

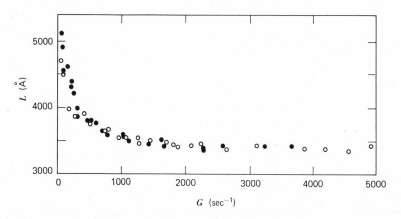

Fig. 4.48. Flow birefringence length of two samples of TMV in phosphate buffer [130].

dispersity of the sample and to the presence of aggregates being heavily accentuated at low shear stresses. However, the data of Leray [131] suggest that even at very high shear stresses the extinction angle does not approach the 0° limit but some limit in the vicinity of 7° (Fig. 4.49). That may mean that the discrepancy is of a more basic nature. Similar deviations (explained again as due to the polydispersity) were also observed on less precise data for fibrinogen and other proteins [133–136].

A very important difference in behavior of rigid and soft particles was observed by Philippoff [137]. Rigid particles [TMV, V_2O_5 sols, cellulose crystallites, poly(benzyl glutamate)] exhibit flow birefringence in the 1,3 plane while soft linear polymers do not—both in accord with theory.

Highly branched biopolymers, such as amylopectins and limit dextrins [138] or dextrans [139], exhibit a behavior similar to rigid particles. The magnitude of birefringence exhibits slight saturation while the extinction angle with increasing product $G\eta/T$ approaches rather rapidly a saturation value very different from 0°. This value is independent of stress in very broad limits (Fig. 4.50). This anomaly cannot be explained by polydispersity. It is remarkable that the value of limiting extinction angle decreases with increasing intrinsic viscosity of the sample (Fig. 4.51). Similar anomalies were observed on amylose solutions [140, 141]. In this case, the saturation of Δn was less pronounced, probably because of the deformable character of the amylose molecules.

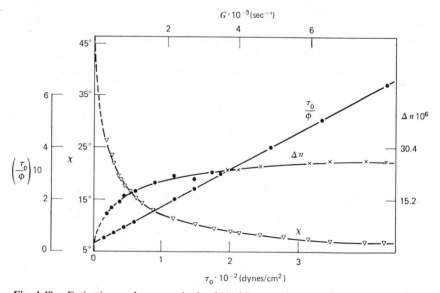

Fig. 4.49. Extinction angle χ, magnitude of birefringence Δn and τ_o/φ ($\varphi = 45° - \chi$) as a function of shearing stress τ_o. TMV in glycerin solution 2.5×10^{-5} g/ml [131].

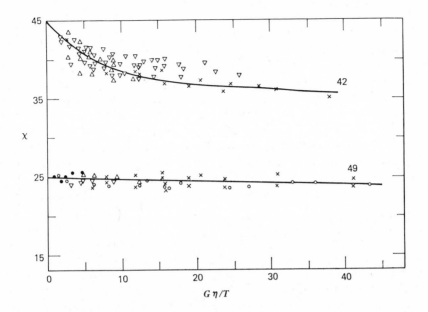

Fig. 4.50. The extinction angle of dextrans APC 42 and 49 in different solvents— mixtures of water, NaOH, KOH, glycerol, formamide, and ethylenediamine. Each experiment was done at two or more temperatures and at two or three concentrations [139].

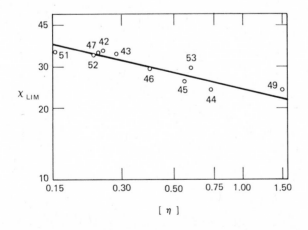

Fig. 4.51. Empirical relation between intrinsic viscosity in dl/g and the limiting extinction angle (at $G\eta_s/T = 45$) of dextran fractions [139].

Crystallites and Micelles

Some materials may be prepared in the form of sols composed from submicroscopic crystals. As the birefringence of crystals is usually very high and the size of these particles is large compared with those of true macromolecules, these sols exhibit very strong birefringence. A well-known example is the sol of vanadium pentoxide [142]. Unfortunately, strong optical absorption in the visual spectral region prevents precise measurements of V_2O_5 birefringence.

The suspension of the organic dye milling yellow [143] is also strongly birefringent. The unusually marked dependence on concentration and temperature of its suspension is probably due to the limited solubility of the dye (Fig. 4.52). The extinction angle approaches the nonzero high-shear limit in this case too.

Detergent molecules in moderately concentrated solution form micelles. These micelles, having a highly regular structure, behave hydrodynamically as do rigid particles exhibiting fairly high birefringence. Their overall birefringent behavior corresponds to that of other rigid particles. Moreover, streaming birefringence proves to be a valuable method for studying the change of size and structure of micelles with the composition of the solution. Thus, soaps change their positive anisotropy in alkaline solution to negative anisotropy in acid solution [144]. For n-alkyl trimethylammonium bromides the length of micelles increases with the length of the alkyl group, detergent concentration, and KBr molarity, but markedly decreases with the addition of glycerine. Thus the addition of glycerine actually diminishes the micelle

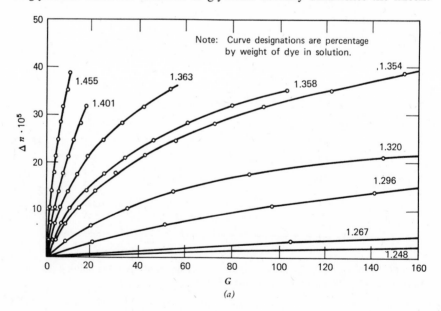

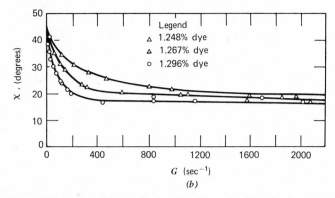

Fig. 4.52. Milling yellow (*a*) magnitude of birefringence, (*b*) extinction angle [143].

orientation. Simultaneously, the anisotropy decreases and eventually changes its sign from positive to negative, probably because of decrease of the form anisotropy both with the decreasing size of the micelles and the decreasing refractive index difference [145, 146].

Soft Particles

Most linear polymers have considerable freedom of internal rotation and therefore can have a large number of energetically equivalent conformations. When the transition among different conformations does not need high activation energy, the particles can change their conformation easily under the influence of external forces; they are soft. High activation energy causes high internal viscosity. Lack of internal freedom of rotation leads to only a few possible conformations; the particles are semirigid (rigid particles have no possibility of changing conformation). Most linear single-chain polymers form in solution soft random coils and may be well treated by the theory of soft particles as developed for elastic dumbbell and necklace models.

Stress-Optical Coefficient

One of the most important predictions of the theory of soft particle in matching solvent is the proportionality between stress and optical tensors [(4.76) and (4.77)]. This prediction is based on the very general assumption that both optical anisotropy and elastic entropic force for the whole macromolecule and all its segments depend on the conformation of the chain in the same way. As a result, the stress-optical coefficient for a given polymer-solvent pair should not depend on such factors as molecular weight, concentration, velocity gradient, or polydispersity. The test for the shear (1,2) component of both tensors is relatively easy, as the shear stress may be measured with good precision. On polyisobutylene, Philippoff [87, 23*a*], finds excellent agreement with theory over a very broad concentration and stress range (Fig. 4.38).

Munk et al. [115] and Munk and Peterlin [84] performed this test on polystyrene in different solvents and with different molecular weights. They were not able to find any systematic deviation of measured stress-optical coefficient with respect to any of the basic variables.

On the other hand, the measurement of the difference of normal stress p_{yy}-p_{xx} is experimentally very difficult. Nevertheless, Philippoff [147] was able to show that, for polyisobutylene in white oil and polystyrene in Aroclor, the stress-optical law holds with reasonable accuracy also for the difference of normal components of both tensors, or in other words, that the axes of both tensors coincide.

The anisotropy of the polymer segment may be deduced from the stress-optical coefficient [(4.62)]. Tsvetkov's group found segmental anisotropies for many polymer systems; a comprehensive table is given in Tsvetkov's review [93] in Ke's monograph. We shall not repeat those data here.

Segmental Anisotropy

The anisotropy of a polymer molecule depends primarily on the anisotropy of polarizability of a statistical segment of the polymer chain and this, in turn, depends on the anisotropy of the polarizability and the average position of all atomic groups in the segment. Calculations [148] are usually based on the anisotropy of the polarizability of chemical bonds (Denbigh [149] is most frequently quoted in the literature; Le Fevre and Le Fevre [1a] have probably the best data available). Once all principal tensor components of bond polarizabilities are known, we can compute the anisotropy of any molecule or group in any conformation and hence, by comparing measured and calculated values of anisotropy, decide what conformation is prevalent in the given molecules or polymer segment. However, the anisotropy of polymer segment measured by streaming birefringence is composed from the intrinsic anisotropy of the segment [(4.41)] and from the contributions of inhomogeneity in the refractive index [effect of microform (4.49)] and of adsorbed molecules of the solvent. So before drawing conclusions about polymer conformation, we must take all these effects into consideration.

Unlike the macroform effect [(4.48)], the microform effect decreases with increasing concentration so slowly that it is virtually independent of concentration in the usual concentration range; hence it adds algebraically to the intrinsic anisotropy [(4.50)]. It may thus be measured only in highly concentrated systems as swollen gels [150] and even then with low accuracy. The microform anisotropy is proportional to the molecular weight of a statistical segment. Consequently, its contribution to the segmental anisotropy is proportional to the contribution of intrinsic anisotropy. For high refractive index differences, the microform effect may equal or even outweigh the intrinsic anisotropy.

As Frisman and Dadivanyan have shown [70], the segmental anisotropy is strongly dependent on the solvent even for solvents with the same refractive index (Fig. 4.53). It is caused by the polymer-solvent interaction leading to adsorption of solvent molecules in at least partly oriented position along the chain. As the aromatic solvents are usually more anisotropic, the segmental anisotropy is more positive in aromatic than in aliphatic solvents [122]. Munk [151] has shown that for polyvinyl chloride the segmental anisotropy may be correlated to the anisotropy of the solvent expressed as its stress-optical coefficient (Fig. 4.54). A large effect of this type, however, may make rather difficult the determination of segmental anisotropy and eliminate the possibility of calculating it from chemical bond data.

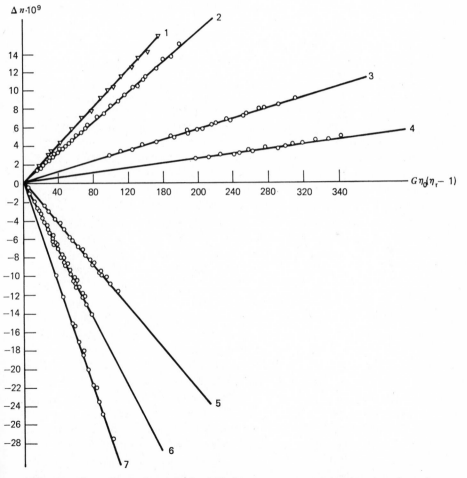

Fig. 4.53. Dependence of magnitude of birefringence on excess shearing stress for poly-vinyl acetate in various solvents: (1) chlorobenzene; (2) toluene; (3) benzene; (4) o-xylene; (5) tetrachloromethane; (6) chloroform and cyclohexanone; (7) dichloroethane [70].

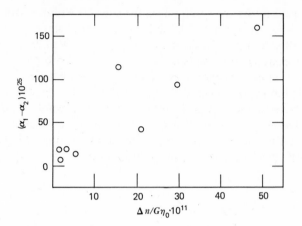

Fig. 4.54. Correlation of the segmental anisotropy of polyvinyl chloride to the anisotropy of solvent expressed as $\Delta n/G\eta_s$ [151].

Relation Between Shear and Normal Components of Optical Tensor

The relation between shear and normal components of optical and stress excess tensors is expressed by the coefficient B in (4.62) and (4.63). Its experimental value in the limit of low concentration and low β is very sensitive to the polydispersity and to experimental errors. The most recent experimental results [84, 126] show that for polystyrenes with narrow molecular weight distribution the B-value is in the vicinity of 0.3, that is, quite close to the theoretical value for the impermeable elastic necklace with ideal flexibility.

The coefficient B depends on concentration and on β (Fig. 4.45). According to Daum [152] it exhibits a maximum in the concentration dependence. The position of the maximum shifts to lower concentrations with increasing molecular weight. From the definition of B in (4.62) it is obvious that the uniform changes in eigenvalues λ_p caused by increased viscosity of the medium cancel out. With increasing polymer concentration, however, the coils start to interpenetrate. The first mode corresponding to the movement of the whole chain is influenced first. The relaxation time τ_1 increases and the eigenvalue λ_1 decreases more than other eigenvalues [85], leading to an increase in B. At higher concentrations the other modes are influenced too, and B decreases again. Similarly, with an increasing β, the deformation of the whole coil is larger than that of adjacent segments [84, 85]. The hydrodynamic interaction decreases nonuniformly and the same applies to the increase of eigenvalues λ_p. Again, the first eigenvalue is influenced most, thus causing the decrease of B. Characteristically this decrease of B manifests itself in the same region of β as the non-Newtonian decrease of viscosity, which is also caused by the nonuniform decrease of hydrodynamic interaction.

Macroform Anisotropy

The macroform anisotropy [(4.48)] of a polymer dissolved in a solvent with a different refractive index is proportional to the square of refractive index difference between the polymer and the solvent and to $M/[\eta]$. While intrinsic anisotropy may be either positive or negative, the form anisotropy is always positive. As both anisotropies depend on molecular weight in different ways, they may be separated by investigating the molecular weight dependence of molecular anisotropy $[\Delta n]/[\eta]$. This separation was successfully performed for many systems. A typical example [153] is given for polystyrene in Fig. 4.55. The Flory constant Φ is calculable from the slope of experimental curves giving $\Phi = 2.40 \times 10^{23}$, in good agreement with direct measurement. The parabolic dependence of constant $[\Delta n]$ on the refractive index of solvent was also tested successfully [153] on many systems (Fig. 4.56).

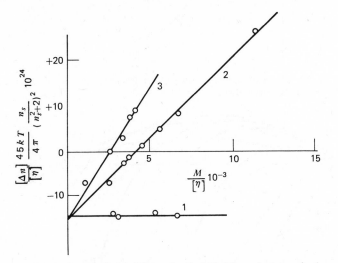

Fig. 4.55. Relation between ratio $[\Delta n]/[\eta]$ and ratio $M/[\eta]$ for polystyrene in three solvents: (1) bromoform: $\alpha_1 - \alpha_2 = -144 \times 10^{-25}$ cm³; (2) dioxane $\alpha_1 - \alpha_2 = -148 \times 10^{-25}$ cm³; (3) butanone: $\alpha_1 - \alpha_2 = -152 \times 10^{-25}$ cm³ [153].

The macroform effect may cause many apparent anomalies, especially if the intrinsic anisotropy is negative (Figs. 4.30, 4.31, 4.55). Not only the magnitude but also the sign of molecular anisotropy for a given polymer may be changed with change in molecular weight of the polymer or refractive index of the solvent. Moreover, the macroform birefringence increases with β more slowly than the intrinsic birefringence [(4.66)]. That, in some cases, may lead to the change of sign of birefringence with increasing gradient [153] (Fig. 4.57). In spite of the instructive value of Fig. 4.57, it must be emphasized that it is oversimplified. The inversion of sign considered in terms of the polar diagram

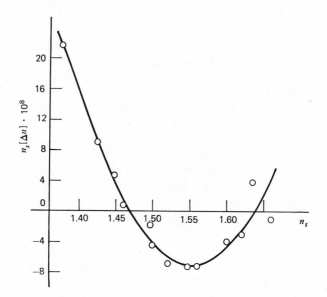

Fig. 4.56. Relation between constant [Δn] and refractive index of solvents for poly-*p*-tert-butylphenylmethacrylate [153].

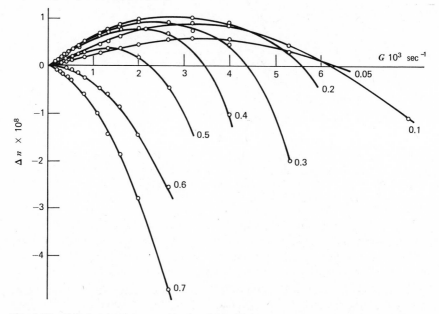

Fig. 4.57. Relation between magnitude of birefringence and velocity gradient for poly-styrene ($M = 3.3 \times 10^6$) in dioxane. The numerals at the curves indicate the concentration in g/100 ml [153].

means that the curve goes from the first to the third quadrant. If there is no polydispersity of molecular weight and chain conformation and if the orientation angles of intrinsic and form birefringence are the same, the curve in the polar diagram would pass from the first quadrant to the third at the origin. As none of these conditions is fulfilled, the curve should pass through either the second or the fourth quadrant, which means that the extinction angle must behave anomalously. Such anomalies were indeed observed [27, 153] (Fig. 4.31). Since the magnitude of birefringence never equals zero, the curves in Fig. 4.57 should exhibit discontinuities.

As the polymer concentration increases, the polymer coils approach each other, the refractive index inhomogeneity of solution decreases, and so does the form birefringence. The ratio of form anisotropy at given concentration θ_f^* to that of zero concentration θ_f depends on the homogeneity of segment distribution through the solution, that is, on $[\eta]c$ (Fig. 4.58). However, a very high concentration of polymer is necessary for complete suppression of the form effect [114] (Fig. 4.59). Thus, the change of concentration may invert the sign of birefringence, too (Fig. 4.57).

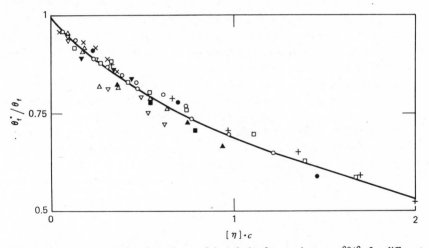

Fig. 4.58. The concentration dependence of the relative form anisotropy θ_f^*/θ_f for different fractions of polyisobutylene in hexane and poly(p-carbethoxyphenylmethacrylamide) in ethyl acetate [93].

Internal Viscosity

The crucial result of Cerf's [76] theory of internal viscosity is the form of dependence of $[\omega]_o$ on the solvent viscosity, as shown in Fig. 4.25 and (4.65ab). Experimentally, Leray [92] studied this dependence on polystyrene in mixed solvents and on deoxyribonucleic acid solutions with varying amounts of glycerine. Polystyrene solutions really fit on the asymptotic part of the curve.

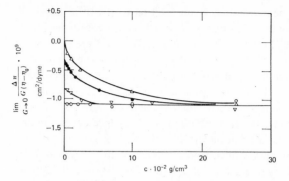

Fig. 4.59. The stress-optical coefficient as a function of concentration for polystyrene ($M_w = 3 \times 10^5$) $\triangle$ dioxane; ● cyclohexanone, $\bigtriangledown$ chlorobenzene, $\bigcirc$ bromobenzene [114].

The effect of internal viscosity is proportional to the molecular weight [(4.65), $h^2 \sim M$]. Data for DNA followed the whole theoretical curve. Tsvetkov and Budtov [27] explored polymethyl-methacrylate in different solvents. The data seem to follow theoretical predictions (Fig. 4.25). The extinction angle of cellulose tricarbanilate oligomers [128] ($M = 27, 38, 57, 90,$ and 152×10^3) yields a higher B-value than expected for perfectly flexible coils (see Fig. 4.60), which can be interpreted in terms of partial coil rigidity or as a transition from coil to ellipsoid.

However, Janeschitz-Kriegl [154] expressed serious doubts about the experimental reliability of the above data. DNA, with its properties varying from preparation to preparation, does not seem to be a good material for such an investigation. In other studies on polystyrene [84, 115, 126] no dependence of the parameter B on solvent viscosity was observed; such dependence would be conspicuous if the internal viscosity played any role. Thus, the experimental evidence for the internal viscosity is contradictory. It cannot yet be considered as experimentally proved.

Short Chain Particles

By short chain particles, we understand linear chain particles composed from a small number of statistical segments. The theory for wormlike particles is probably more suitable for them than is that for gaussian chains. Short polymers (oligomers) and polymers with very extended structure or secondary structure (α-helix, double helix, ladder polymers, cellulose derivatives with bulky side groups) form this group.

According to the theory of Gotlib and Svetlov [83], the ratio $[\Delta n]/[\eta]$ should increase with the ratio $x = L/a$, where L is the extended length of the polymer molecule and a its persistence length. With increasing x the ratio should approach a limiting value, holding for a gaussian chain. Experimental tests of this theory yielded dependences close to theoretical ones and reasonable values for persistence lengths. The test was performed on oligomers of poly(oxypropylene-glycol) [155], polystyrene [22, 155] (Fig. 4.23) and polyisobutylene [156], on many synthetic polypeptides in helical conformation [157], on cyclolinear polyphenylsiloxane [158], on cellulose tricarbanilates [128], and on deoxyribonucleic acid [159, 159a]. So, for instance, from the data of Tsvetkov et al. [155] we derive $M_a = 477 = 4.6 \times 10^4$, that is, 4.6 monomer unit per persistence length of polystyrene in benzene at 20°C, which well agrees with the literature value 4 [159b] and with a little larger value 6 derived by Thurston [28] from the frequency dependence of streaming bire-

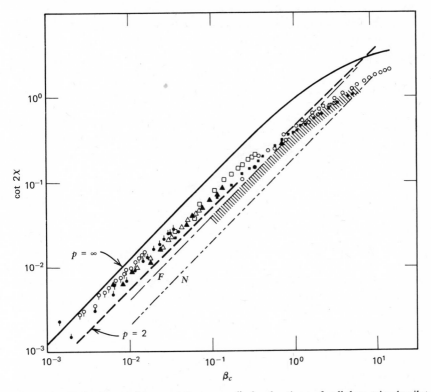

Fig. 4.60. Extinction angles as cot 2χ versus β_c for fractions of cellulose tricarbanilate ($\stackrel{\circ}{|}$) 1%, $M = 27 \times 10^3$: ($\stackrel{\bullet}{|}$) 1.7%, $M = 38 \times 10^3$; (Δ) 1.8%, $M = 57 \times 10^3$, ($\blacktriangle$) 1%, $M = 9 \times 10^4$, ($\square$) 0.8%, $M = 15.2 \times 10^4$, ($\blacksquare$) 0.74%, $M = 28 \times 10^4$, ($\bigcirc$) 0.6%, $M = 5 \times 10^5$, ($\bullet$) 0.4%, $M = 72 \times 10^4$, ($\diamond$) 0.2%, $M = 1.2 \times 10^6$; hatched area denotes location of the measured points for anionic polystyrenes in bromobenzene solutions at 25°C; (– –) prolate rotational ellipsoids with the indicated axial ratios; (...) according to the theory for perfectly flexible gaussian chains; F = free-draining approximation; N = strong hydrodynamic interaction (nondraining) [128].

fringence of polystyrene in Aroclor 1248 ($\eta_s = 2.237\ P$). Janeschitz-Kriegl and Burchard [128] obtain 18.3 monomer units per persistence length of cellulose tricarbanilate in benzophenone yielding $a = 76.8$Å. By small angle x-ray scattering in acetone solution a value of 89.1Å has been found [159c]. This slightly higher value is in accordance with the fact that, at corresponding molecular weights, higher intrinsic viscosities are also found in acetone [159d]. The high segmental anisotropy for these systems is in accord with their segmental length.

The extinction angle in higher molecular weight fractions of cellulose tricarbanilate depends on β virtually in the same way as for higher molecular weight polystyrene. In the range of lower molecular weights a transition toward a behavior of ellipsoids or of partially rigid chains is observed, that is, an increase of parameter B (Fig. 4.60). The reduction of experimental data to one master curve using parameter β is rather good even for the temperature dependences.

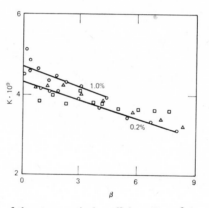

Fig. 4.61. Dependence of the stress-optical coefficient K on β for cellulose tricarbanilate in benzophenone at 55°C ($M = 1.2 \times 10^6$) [128].

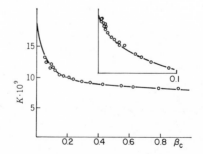

Fig. 4.62. The dependence of stress-optical coefficient on β for cyclolinear polyphenylsiloxane ($M = 5 \times 10^5$) $c = 7.35 \times 10^{-3}$ g/ml in bromoform [160].

Some unusual phenomena were observed with solutions of short-chain polymers. The stress-optical coefficient for cellulose tricarbanilate decreases with increasing β (Fig. 4.61). A similar decrease at much lower β was observed for polyvinyl chloride [151] and cyclolinear polyphenylsiloxane [160] (Fig. 4.62). In all these cases the stress-optical coefficient decreases with increasing β at low β-values and approaches a limiting value at higher shear stresses. The explanation of this behavior is not known, but undoubtedly it is connected with short-chain polymers.

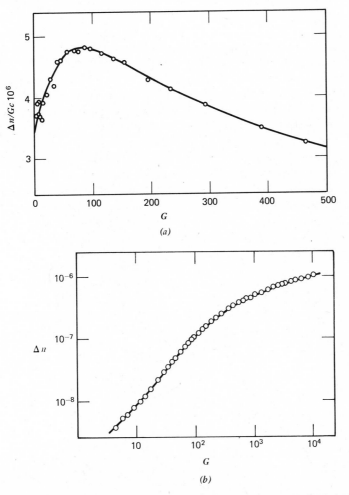

Fig. 4.63. Deoxyribonucleic acid from *Escherichia coli* $c = 2.2 \times 10^{-5}$ g/ml; (*a*) $\Delta n/Gc$ versus gradient [161]; (*b*) bilogarithmic plot of magnitude of birefringence versus gradient.

Deoxyribonucleic acids have such long statistical segments that they may be considered as short chains even with molecular weights in the range of millions. This leads to one phenomenon that has not yet been observed with any other system: the magnitude of birefringence first increases more than linearly with the gradient, as it is the case with soft molecules, but later slowly approaches saturation at high velocity gradients [161] (Fig. 4.63). The probable reason is that due to the small number of segments, the coil deformation soon approaches saturation (Fig. 4.28b: elastic dumbbell model with finite chain length).

Polyelectrolytes

The average size of polyelectrolyte molecules increases rapidly with a high degree of ionization, low ionic strength, and low concentration. The counterions shield the repulsion and diminish the expansion. The local concentration of counter-ions may be increased either by higher concentration of polyelectrolyte or by the presence of low molecular electrolyte. As the macroform birefringence depends on molecular weight and on the reciprocal coil volume, its relative importance decreases with increasing expansion. However, with the expansion of the coil, more extended conformations are favored leading to an increased intrinsic and microform anisotropy.

The enhanced coil expansion at decreasing concentration of polyelectrolyte reverses the concentration dependence of extinction angle [162] (Fig. 4.64) and reduces that of viscosity and birefringence. The validity of the rheo-optical law, however, is not affected, as may be seen from the master curve of χ versus $(\eta - \eta_o)G/c$ (Fig. 4.65).

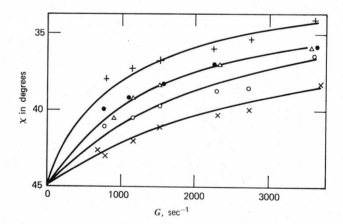

Fig. 4.64. Orientation angles for poly-(4-vinyl-N-n-butylpyridinium bromide) in aqueous solution: $\times$ $c = 1.885\%$; $\bigcirc$ 1.520%, $\triangle$ 0.506%; $\bullet$ 0.326%, $+$ 0.164% [162].

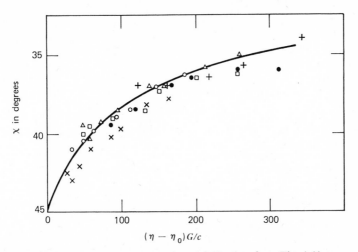

Fig. 4.65.   Master curve χ versus $(\eta - \eta_s)G/c$, data from Fig. 4.64.

Poly(4-vinyl-*N*-*n*-butylpyridinium bromide) has a negative birefringence in aqueous solution [163], its orientation [162] is apparent from Figs. 4.64 and 4.65. Polyvinylpyridine [164] and its chloride [165] have positive anisotropy, which increases with the coil expansion. Polyvinylpyridine behaves like a typical soft macromolecule with normal concentration dependence, while its chloride exhibits reversed concentration dependence as a consequence of the change in ionic strength. Δn versus G curves are bent downward; that need not mean saturation and chain rigidity but only the influence of pronounced non-Newtonian viscosity.

The intrinsic anisotropy of polyacrylic acid [166] with 50% ionization is negative, as may be seen from the negative birefringence in high dilution, where the macroform effect is negligible. With decreasing ionization, higher concentration, or addition of NaCl, the macroion shrinks and the large macroform effect reverses the sign of the birefringence. On the other hand, in isoionic solution, according to Tsvetkov [93] the anisotropy decreases with increasing concentration (Fig. 4.66), as is usual for systems exhibiting a prevailing macroform effect. Surprisingly, the behavior of polymethacrylic acid, according to Tsvetkov, suggests absence of the form effect (Fig. 4.67). On the other hand the apparent anisotropy decreases as the viscosity increases with increasing degree of ionization [167] as would be expected for strong form effect. Reexamination of this discrepancy is highly advisable.

Deoxyribonucleic acid is a polyelectrolyte, but its double helix is so rigid that the electrical forces apparently cannot change its conformation. Thus the extinction angle of DNA is not changed by changing the ionic strength by four orders of magnitude [168]. Consequently, its birefringent behavior may

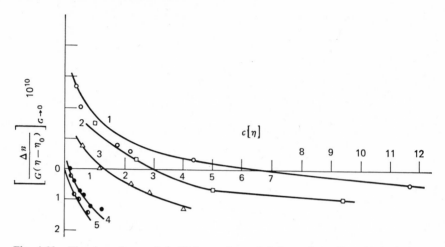

Fig. 4.66. The dependence of $\Delta n/G$ $(\eta - \eta_s)$ on $c[\eta]$ for polyacrylic acid in buffered solution. Ionic strength 0.012 mole/liter. (1) $M = 1.8 \times 10^6$; (2) 1.3×10^6; (3) 0.53×10^6; (4) 0.16×10^6; (5) 0.038×10^6 [93].

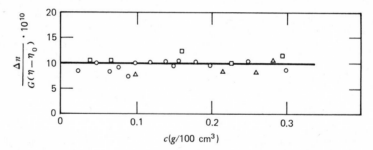

Fig. 4.67. The dependence of $\Delta n/G$ $(\eta - \eta_s)$ on concentration for polymethacrylic acid in buffered solution; Ionic strength 0.012 mole/liter; Molecular weights between 0.6 $\times 10^5 - 8.3 \times 10^5$ [93].

be successfully treated by methods used for soft or semirigid particles [92, 161, 169]. It is remarkable that the theory of the birefringence of a molecular coil is suitable [170], even for DNA with a molecular weight higher than 10^8. Also for synthetic polynucleotides, the existence of the helix has crucial importance; the complex of polyadenine and polyuracil has a much larger anisotropy than any of the components [171].

Copolymers

Streaming birefringence of copolymers is strongly dependent on their structure. For random and block copolymers, the apparent molecular or segmental anisotropy is virtually the linear combination of anisotropies of constituent homopolymers [172]. Anisotropy of graft copolymers depends on the length and frequency of the grafted chains. The apparent anisotropy is composed of the backbone anisotropy (the length and anisotropy of the backbone segment may be increased by grafting) and the side-chain contribution. When the side chains are few and long, they are oriented in the flow direction more or less independently of the backbone. Thus, a copolymer with few long grafted polystyrene (negative anisotropy) chains on a polydiphenylpropene (positive anisotropy) backbone exhibits negative anisotropy, which is larger than for homopolystyrene [173]. The increase in anisotropy is probably due to enlarged steric hindrances. On the other hand, a copolymer with many short polystyrene grafts on a polymethyl methacrylate or polybutyl methacrylate backbone has very strong positive anisotropy, one or even two orders of magnitude larger in absolute value than polystyrene [173, 174]. The anisotropy strongly increases with the frequency and length of the grafts. Apparently, the macromolecule behaves as if composed of rather stiff segments with side chains rigid because of the lack of space; even the largest grafts were relatively short in these experiments. The benzene rings, being perpendicular to the direction of the side chain, are parallel to the backbone, thus producing positive anisotropy.

Extremely Large Particles

Some polymer solutions exhibit birefringence which, in the range of very small gradients, does not seemingly decrease with decreasing velocity gradient, while the extinction angle approaches 0° or 90° instead of usual 45° or 135° [175–178]. These solutions contain very large particles, either incompletely dissolved polymer or crosslinked microgel particles or undissolved crystallites of crystalline polymers (usually cellulose derivatives), or products of polymer aging. The particles are sometimes so large that they are completely oriented at all experimental velocity gradients. As they are usually nondeformable, the angle and the magnitude of birefringence are gradient independent. Typical anomalous dependence is depicted in Fig. 4.68. If the viscosity of the solution is very high and the diffusion of the particles extremely slow, the birefringence may persist even after the flow is stopped, and decrease very slowly, as Philippoff [175] was able to observe (Fig. 4.69). In many cases these particles can be removed by filtration through fine filters; the rest of the solution then behaves normally. As Munk [178] has shown, the presence of such particles manifests itself clearly in their polar diagram (Fig. 4.70). It may serve even for their characterization if they are not large enough to be completely oriented at all velocity gradients (Fig. 4.71).

Time Dependence of Birefringence

It takes some time after one has started the Couette cell in motion before the steady distribution of suspended particles is achieved. This time is equal to a low multiple of the longest relaxation time of particles in the system and

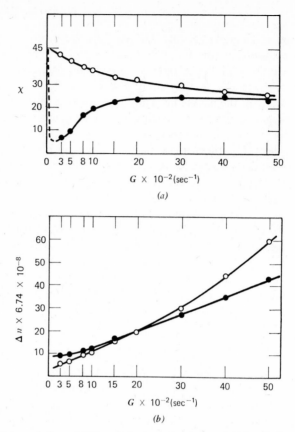

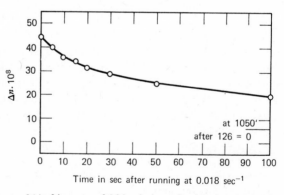

Fig. 4.68. Water solution of 7% polyvinyl alcohol before (–O–) and after (–●–) aging. Velocity gradient dependence of (a) extinction angle; (b) magnitude of birefringence [176].

Fig. 4.69. Decay of birefringence of 9% solution of polyisobutylene (Vistanex B-100) in decalin—unfiltered solution [175].

362

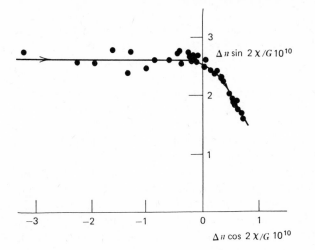

Fig. 4.70. Polar diagram for a solution of polyvinyl chloride in dibutyl phthalate, $M = 17.10^4$, $c = 0.95 \times 10^{-2}$ g/ml; the arrow designates the direction of the increasing velocity gradient [178],

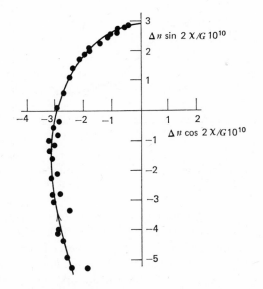

Fig. 4.71. Polar diagram for a solution of polyvinyl chloride in tricresyl phosphate, $M = 7.4 \times 10^4$, $c = 1.07 \times 10^{-2}$ g/ml [178].

is, in most cases, well under one second. Only for extremely viscous solutions of very large particles does the time increase to a few seconds. Such fast changes usually cannot be measured and only steady birefringence is observed. For long observation times the birefringence may change because of insufficient temperature control or degradation of polymer in shear. Usually these changes are slow.

For high molecular weight polystyrene in highly viscous solvents [179] and for high molecular weight deoxyribonucleic acid [180] a different process was observed at $\beta_c > 10$. The magnitude of birefringence decreases and the extinction angle increases in such a way that the stress-optical coefficient $\Delta n \sin 2\chi/2G(\eta - \eta_o)$ does not change with time (Fig. 4.72). However, the parameter B decreases two to ten times. For lower concentrations and gradients the decrease exhibits some sort of induction period; for higher concentrations and gradients this induction period cannot be observed. The process is relatively slow; the product of its characteristic time with velocity gradient is usually nearly constant for a given sample; its value is $10^4 - 10^5$. Sometimes $\beta_c > 10$ may be achieved at rather low velocity gradients; then the decrease is very slow. The phenomenon is completely reversible; after the rotor has been stopped, the solution slowly returns to its original state.

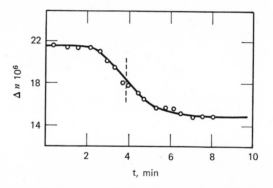

Fig. 4.72. Decrease in birefringence with the time of action of the hydrodynamic field. Deoxyribonucleic acid from rat spleen, $c = 2.2 \times 10^{-5}$ g/ml, $G = 390$ s^{-1} [180].

The suggested explanation is as follows [181]. Very long molecular chains assume rather extended conformation at sufficiently high β_c. For such elongated particles there is high probability that molecular collisions induced by flow will lead to permanent entanglements. Further flow produces still denser entanglement of the aggregates formed in the first phase. For viscoelastic behavior the number of effective particles is important. As each part of the chain between two entanglements acts as an independent unit, the number of effective particles *increases* during aggregation and subsequent entanglement:

the effective molecular weight and β decrease. Because the original β is used for calculation of B the latter value decreases.

By contrast, polymethyl methacrylate in poor solvents has a tendency toward aggregation and the temporary crosslinks are very firm. Under similar circumstances polymethyl methacrylate forms a continuous network and the viscosity, together with the birefringence, increases considerably [182]. Further shearing leads to the destruction of the network associated with the degradation of polymer molecules.

Oscillatory Birefringence

The birefringence in steady flow is influenced mainly by the longest relaxation times of the macromolecule and their dependence on concentration and velocity gradient. On the contrary in the oscillatory birefringence, the importance of the longest relaxation times decreases with increasing frequency and shorter

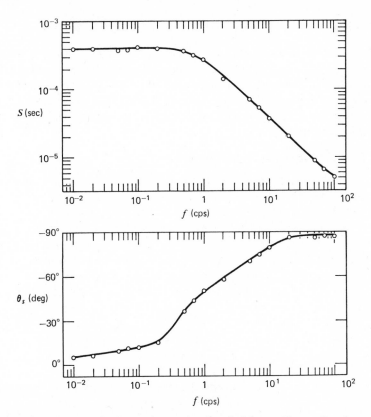

Fig. 4.73. Components of the complex mechano-optical factor S^* as a function of frequency f for 1.72% milling yellow solution at 23°C [183].

relaxation times are excited. For long chains with a large number of relaxation times, the decrease of the mechano-optical factor with frequency is more gradual and the phase angle exhibits a plateau. The steepest decrease is observed for rigid particles with only one relaxation time. The experimental dependences are clearly different for suspensions of milling yellow [183] with essentially rigid particles (Fig. 4.73), and for polystyrene [28] (Fig. 4.10).

From the shape of the experimental curves, we can estimate not only the number of relaxation times Z but also the hydrodynamic interaction factor h^* and ratio of internal viscosity to friction factors (φ/f). While the number of segments and the interaction factor are found from birefringence measurements with reasonable accuracy, the internal viscosity may be only guessed at and should be obtained from an independent measurement, usually from the oscillatory viscosity. Thus the oscillatory birefringence may supply complementary data to the steady-state streaming birefringence.

6 MEASUREMENT OF STREAMING BIREFRINGENCE

In this section we shall deal with the basic rules that govern the design of flow birefringence instruments and determine the experimental precision. Every instrument includes three basic parts: (1) a cell where the hydrodynamic field acts on the liquid; (2) drive mechanism of the cell allowing variation of the hydrodynamic field in broad limits; and (3) an optical device for measurement of birefringence.

Cell for Birefringence Measurements

Birefringence is measured not directly but by means of the phase difference introduced by the birefringent layer into the optical system. This phase difference is proportional to the thickness of the layer. Birefringences of streaming liquids are usually very small, so for sensitive measurement a thick layer of measured liquid is necessary. This condition, together with the requirement of a linear hydrodynamic field, virtually restricts the possible geometries of the cell to the concentric cylinder Couette cell.

If the gap between cylinders is narrow enough, then the magnitude of the inhomogeneity of the field is small [(4.3, Fig. 4.2)]. It is then justified to consider measured values (which are some averages of values for the whole gap) as corresponding to the average value of the velocity gradient in the gap. On the other hand, the influence of the local inhomogeneity on the motion of suspended particles has never been studied.

However, other factors causing deviations from this basic behavior are more important. They are end effects at cylinder bases, turbulence, and viscous heating of the liquid. End effects may be completely neglected in optical measurements unlike the case of viscosity measurements, because the optical path through the medium with another character of flow is usually negligible compared to the path in the medium studied.

The onset of turbulence for a liquid between two cylinders was studied by Taylor [184]. With inner cylinder rotating, the flow ceases to be laminar at critical gradient G_c defined by

$$G_c = \frac{\eta}{\rho} \frac{\pi^2}{2} \left[\frac{(r_i + r_a)^3}{0.114(r_a - r_i)^5 r_i^2} \right]^{1/2}, \qquad (4.86)$$

where ρ is the density of the liquid. The conditions for laminar flow are decidedly more favorable when the outer cylinder rotates, because the flow is stabilized by centrifugal forces. The factor describing the improvement decreases with the decreasing gap width; some of Taylor's values are given in Table 4.5.

Table 4.5.
Improvement Factor
for Outer Rotor[184]

$\dfrac{2(r_a - r_i)}{r_a + r_i}$	f
0.38	1000
0.10	50
0.017	6

In flowing liquid, the energy dissipates as heat. The evolved heat raises the temperature of the liquid and must be taken away by thermostating. The temperature rise may sometimes be appreciable. However, the secondary heat effect is usually even more important. If the heat is led away through cylinder walls, a temperature gradient is established in the liquid. It is connected with density and refractive index gradients. The light path in the medium with refractive index gradient is curved. The light beam spreads out and hits the wall in the narrow gap. The intensity of the beam is reduced and the polarization changes that are caused by reflection at the walls prevent measurement. If both walls are kept at the same temperature, then according to Björnståhl [185], the temperature maximum is close to the gap center and some finite thickness of light beam goes through without being reflected at the walls. This useful thickness depends on the geometry of the cell, the properties of the liquid, and the amount of heat dissipated. It sharply diminishes with increasing velocity gradients. With usual cell constructions, it is so thin that it prevents measurements when the amount of dissipated heat exceeds 10^6 erg/ml sec. When both walls are not thermally equivalent (as is the case with most constructions) the useful thickness may not exist at all. The practical limits of possible measurements are lower in this case but not appreciably.

The thermal effect may be partly avoided if the cylinder walls are made from thermally insulating material. The thermostating of the liquid must, of course, be achieved by another means. So Kuhn [166] replaced the stator by an islet in a larger space filled by liquid. In this case the liquid is in well-defined streaming only for a very short time, and the steady phenomena may be obscured by entrance effects. If the islet constitutes a larger part of the original stator, then the liquid sticks to the rotor and becomes overheated [186]. The most promising construction of this kind is that of Janeschitz-Kriegl [186]. His cylinders are made from black glass and another axial streaming is superposed on the liquid forcing it into another thermostated space. The velocity gradient tensor is slightly changed, but for a suitably chosen ratio of axial and tangential velocities the change may be neglected. However, all these constructions prevent measurement of contingent time effects.

The design of the cell is restricted by technological aspects, too. A rotor with a diameter of a few centimeters cannot be manufactured with smaller eccentricity than about 1μ; the imperfection of the best bearing is about the same. But 2μ represents 1% of the width of the 0.2-mm gap between cylinders. Thus it seems that 0.2 mm is the lowest reasonable limit for the gap width. The necessity to avoid any reflections of the light beam on cylinder walls also limits the degree of narrowness of the gap that can be designed.

Special attention must be paid to the construction of windows passing the light beam. Every stress—either external or internal—acting on the glass window induces an optical anisotropy and parasite birefringence. Even with slight stresses, these parasite effects are comparable to effects introduced in the liquid by shearing, when measured birefringences are close to the limit of measurability. Therefore, it is necessary to keep all stresses on the window as low as possible. The windows must be made from optical glass and very carefully thermally treated to reduce internal stresses. They must be attached to the body of the cell by some rather soft cement. The cement must, of course, be insoluble in all liquids that are to be measured. From our experience, it is advisable to screw the window between two polyethylene gaskets and two metal parts of the window holder, to heat the assembled piece above the melting temperature of polyethylene, and then to cool it slowly. Another rather sophisticated window holder was designed by Janeschitz-Kriegl [187].

When the whole annulus between cylinders is observed while measuring birefringence, large circular windows covering the whole annulus are necessary. The same applies for most constructions of cells with an outer rotor. When an inner rotor is used and birefringence is measured only at one point in the gap, the easier construction with only small windows is possible. A remarkable design due to Tsvetkov and Frisman [188] is to use a thin-wall cylinder as the rotor between two stators, thus forming two gaps, one with the inner and the other with the outer rotor; here small windows may be used also.

For evaluating streaming birefringence measurements, knowledge of the viscosity is usually necessary. Therefore, some flow birefringence apparatuses were designed allowing simultaneous measurements of viscosity and birefringence. The stator in these instruments is attached to a spring system, the torque of which is measured. These constructions are very convenient, but the alignment of a spring-supported stator is much more difficult than that of a firm one. On the other hand, the construction of the cell bottom and cover dictated by birefringence measurement interferes with the elimination of end effects in viscosity measurement. As a result, both measurements in these combined instruments are usually less precise than measurements in two separate apparatuses.

Driving Unit

It is desirable to perform streaming birefringence measurements in as broad a range of velocity gradients as possible. The low gradient limit is given by the least measurable birefringence. As some highly viscous solutions of large particles are strongly birefringent at very low velocity gradients, it is advisable to design the drive also for very low gradients, say 0.1–1.0 sec^{-1}. The upper gradient limit is governed by the heat effect for highly viscous liquids and by the onset of turbulence for liquids of low viscosity. Usually measurements above 10,000 sec^{-1} are not possible. It is difficult to regulate a motor in such broad limits of velocity; therefore, some combination of mechanical and electrical regulation is necessary. The rotor velocity should be stable within about 1 %; higher stability is not necessary because errors in measurement of birefringence usually exceed 1 % (3 % overall accuracy means good measurement). Stroboscopic measurement of motor velocity is easy and reasonably precise. Munk uses a stroboscopic disk with many rings having different numbers of white and black fields and a stroboscopic bulb fed by the public power net (50 or 60 sec^{-1}). The motor need not be very powerful. When 0.5 W is dissipated in each milliliter of sheared liquid, the heat effect prevents measurement. Most cell construction needs less than 50 ml of liquid, so a 100-W motor is large enough.

Measurement of Birefringences

For measurement of streaming birefringence, two optical arrangements are used. In one of them the whole annulus is observed. The principal optical directions oriented at a constant extinction angle to the flow direction rotate with the radius along the circumference of the annulus. When such a cell is placed between crossed polarizers, four points exist on the circumference where the principal optical directions coincide with the directions of the polarizers. Thus a dark cross is formed in the annulus and from its position the extinction angle may be found. Such an arrangement is used in the Edsall [189] design of a streaming birefringence apparatus and is realized in the Rao Flow Birefringence Viscometer. In the optical path of this instrument, there is a rotatable wire cross with a scale attached. It is rotated until it coincides with the dark cross. Its position describes the extinction angle. The magnitude of birefringence is measured by the Senarmont method, the principle of which will be described later. In this arrangement, the quarter-wave plate is inserted in front of the analyzer and the latter is rotated. The dark cross undergoes a scissorlike movement and the Senarmont position of analyzer is achieved when the cross degenerates into a line. Measurement by this method is very simple but not very sensitive. The sensitivity may be increased, according to Zimm [190], by adding another rotating plate with the hole which follows the annulus. The intensity of the transmitted light depends on the position of the hole and is measured by means of a photomultiplier. The photoelectric signal is analyzed. Its phase relation to the scanning hole gives the extinction angle.

In most apparatuses the narrow light beam passes through the annulus at just one point on the circumference. The streaming in the lighted area may be considered in good approximation as plane simple shear, and the optical properties may be measured just as in a birefringent crystal. The measurement of the extinction angle is based on the fact that the birefringent layer between two crossed polarizers brightens the optical field and the darkness can be restored by rotating the layer (or, in the case of streaming birefringence, by rotating the coupled polarizers) so that its principal optical directions coincide with the principal directions of polarizers. For higher sensitivity, half-shade methods were developed: the Bravais double plate for white-light measurements, or the Nakamura double plate from right and left rotating quartz. The optical arrangements for measuring the phase differences may be also used to find the extinction directions.

For designing and analyzing compensators for measurement of birefringence, it is very important to know the intensity of the light passing through the optical system. In most arrangements there are some birefringent layers situated between the polarizer and the analyzer. The ratio I/I_o of the light intensities emerging from the polarizer and the analyzer, respectively, depends on the phase differences of the birefringent layers Δ_i $(i = 1, 2, 3, \ldots)$, on the azimuths α_i of their planes of faster wave propagation with respect to the polarizer, and on the azimuth σ of the analyzer. For three birefringent layers, the relation reads [191]

$$I/I_o = \tfrac{1}{2}[1 + C \cos 2\sigma + S \sin 2\sigma] \tag{4.87}$$

$$C = \left[1 - 2\sin^2 2\alpha_1 \sin^2\left(\frac{\Delta_1}{2}\right)\right]\left[1 - 2\sin^2 2\alpha_2 \sin^2\left(\frac{\Delta_2}{2}\right)\right]$$

$$\times \left[1 - 2\sin^2 2\alpha_3 \sin^2\left(\frac{\Delta_3}{2}\right)\right]$$

$$+ \sin 4\alpha_1 \sin 4\alpha_2 \sin^2\left(\frac{\Delta_1}{2}\right)\sin^2\left(\frac{\Delta_2}{2}\right)\left[1 - 2\sin^2 2\alpha_3 \sin^2\left(\frac{\Delta_3}{2}\right)\right]$$

$$+ \sin 4\alpha_1 \sin 4\alpha_3 \sin^2\left(\frac{\Delta_1}{2}\right)\sin^2\left(\frac{\Delta_3}{2}\right)\left[1 - 2\cos^2 2\alpha_2 \sin^2\left(\frac{\Delta_2}{2}\right)\right]$$

$$+ \sin 4\alpha_2 \sin 4\alpha_3 \sin^2\left(\frac{\Delta_2}{2}\right)\sin^2\left(\frac{\Delta_3}{2}\right)\left[1 - 2\sin^2 2\alpha_1 \sin^2\left(\frac{\Delta_1}{2}\right)\right]$$

$$+ \sin 2\alpha_1\left\{\cos 2\alpha_2 \sin 4\alpha_3 \sin^2\left(\frac{\Delta_3}{2}\right)\right.$$

$$\left. - \sin 2\alpha_2\left[1 - 2\sin^2 2\alpha_3 \sin^2\left(\frac{\Delta_3}{2}\right)\right]\right\} \sin \Delta_1 \sin \Delta_2$$

$$+ \sin 2\alpha_3\left\{\cos 2\alpha_2 \sin 4\alpha_1 \sin^2\left(\frac{\Delta_1}{2}\right)\right.$$

$$\left. - \sin 2\alpha_2\left[1 - 2\sin^2 2\alpha_1 \sin^2\left(\frac{\Delta_1}{2}\right)\right]\right\}\sin \Delta_2 \sin \Delta_3$$

$$- \sin 2\alpha_1 \sin 2\alpha_3 \sin \Delta_1 \cos \Delta_2 \sin \Delta_3$$

$$S = \sin 4\alpha_1 \sin^2\left(\frac{\Delta_1}{2}\right)\left[1 - 2\cos^2 2\alpha_2 \sin^2\left(\frac{\Delta_2}{2}\right)\right]\left[1 - 2\cos^2 2\alpha_3 \sin^2\left(\frac{\Delta_3}{2}\right)\right]$$

$$+ \sin 4\alpha_2 \sin^2\left(\frac{\Delta_2}{2}\right)\left[1 - 2\sin^2 2\alpha_1 \sin^2\left(\frac{\Delta_1}{2}\right)\right]\left[1 - 2\cos^2 2\alpha_3 \sin^2\left(\frac{\Delta_3}{2}\right)\right]$$

$$+ \sin 4\alpha_3 \sin^2\left(\frac{\Delta_3}{2}\right)\left[1 - 2\sin^2 2\alpha_1 \sin^2\left(\frac{\Delta_1}{2}\right)\right]\left[1 - 2\sin^2 2\alpha_2 \sin^2\left(\frac{\Delta_2}{2}\right)\right]$$

$$+ \sin 4\alpha_1 \sin 4\alpha_2 \sin 4\alpha_3 \sin^2\left(\frac{\Delta_1}{2}\right)\sin^2\left(\frac{\Delta_2}{2}\right)\sin^2\left(\frac{\Delta_3}{2}\right)$$

$$+ \sin 2\alpha_1\left\{\cos 2\alpha_2\left[1 - 2\cos^2 2\alpha_3 \sin^2\left(\frac{\Delta_3}{2}\right)\right]\right.$$

$$\left. - \sin 2\alpha_2 \sin 4\alpha_3 \sin^2\left(\frac{\Delta_3}{2}\right)\right\}\sin \Delta_1 \sin \Delta_2$$

$$+ \cos 2\alpha_3 \left\{ \sin 2\alpha_2 \left[1 - 2 \sin^2 2\alpha_1 \sin^2 \left(\frac{\Delta_1}{2} \right) \right] \right.$$

$$\left. - \cos 2\alpha_2 \sin 4\alpha_1 \sin^2 \left(\frac{\Delta_1}{2} \right) \right\} \sin \Delta_2 \sin \Delta_3$$

$$+ \sin 2\alpha_1 \cos 2\alpha_3 \sin \Delta_1 \cos \Delta_2 \sin \Delta_3.$$

As the right hand side of (4.87) is not symmetrical in indices 1, 2, 3, the order of introducing the layers is important. However, the expression for C is symmetrical in indices 1 and 3, so for crossed polarizers ($\sin 2\sigma = 0$) the first and third layers may be interchanged without changing the light intensity. The middle layer cannot be interchanged.

Equation (4.87) is very flexible. We can deduce relations for two layers or one layer just by putting the redundant Δ_i equal to zero. For methods using the dark field, we simply put $I/I_o = 0$ and get the equation for the quantity which is to be adjusted. When this equation has no solution, we may seek the minimum of the intensity by setting the derivation of I/I_o, according to the variable quantity, equal to zero. For half-shade methods, the intensities of both halves of the field are equated, thus yielding one equation for the quantity of interest. In the following section, the two most frequently used compensators, Senarmont's and Brace's are described.

Senarmont's Compensator

Senarmont's compensator is based on the fact that elliptically polarized light created from linearly polarized light by passing through birefringent layer oriented in the azimut $\pi/4$ to its plane of polarization is converted to linearly polarized light by passing through a birefringent layer with phase difference $\pi/2$ (quarter-wave plate) oriented parallel to the original plane of polarization. This light is now polarized in a direction σ to the polarizer and may be completely extinguished by rotation of the analyzer by σ. Equation (4.87) for these conditions [$\Delta_1 = \Delta$, $\alpha_1 = \pi/4$, $\Delta_2 = \pi/2$, $\alpha_2 = 0$, $\Delta_3 = 0$, $I/I_o = 0$] yields

$$2\sigma = \Delta + \pi \tag{4.88}$$

where Δ is the phase difference of the unknown layer. As polarizers are not crossed for this method, the order of the layers cannot be changed; the quarter-wave plate must always be after the measured layer. When the quarter-wave plate is not exact and has the phase difference $\delta \neq \pi/2$, the method may still give reasonable precision. However, the light cannot be extinguished completely and its minimum intensity increases as $(1/4)\sin^2 \Delta \cos^2 \delta$, thus complicating finding the "extinction."

The human eye is more sensitive in evaluating small differences in brightness than in finding the extinction. This substantiates the use of half-shade methods where the field is divided in two halves and their brightness is matched. As the Senarmont method converts the elliptically polarized light into a linear one, the measurement of polarization is in principle the same task as measurement of optical rotation. Thus, all half-shade devices used for the latter measurement may be used also for the Senarmont method. The elliptic half-shade (i.e., birefringent plate covering only half of the field) is very simple, flexible in use, and reasonably precise.

The half-shade plate may be situated as the first, second, or third plate. Each position has advantages and disadvantages. If the plate is at the first or second position, it has to be rather thin; that is, its phase difference ϑ should be less than 0.1. The intensity of the light in the half-shade position and the sensitivity may be varied by changing the azimuth ω of the half-shade plate. Defining hypothetic phase difference θ by

$$\tan\left(\frac{\theta}{2}\right) = \sin 2\omega \tan\left(\frac{\vartheta}{2}\right),\tag{4.89}$$

we get for the half-shade position of the analyzer

$$\tan 2\sigma = \tan\left(\Delta + \frac{\theta}{2}\right)\tag{4.90}$$

or

$$\Delta = 2\sigma - 2\sigma_o\tag{4.91}$$

where σ_o is the half-shade position for the analyzer with the measured layer absent.

It may be shown that the error of the Senarmont method does not exceed 1% when the azimuth of the measured layer is $45° \pm 4°$. Similarly, the actual phase difference of the quarter-wave plate may differ by 10% from the ideal value.

The sensitivity of the method increases with decreasing θ; on the other hand, θ must be large enough to keep the intensity of light at a level sufficient for visual observation and higher than parasite light and background light caused by improper adjustment.

The Senarmont compensator may be used also for locating the principal optical directions: the half-fields are matched in absence of the measured layer, then the layer is introduced and rotated until the match is restored. That happens if the azimuth α of the layer is zero when the half-shade plate is in the second position. However, if it is in the first position, the half-shade azimuth α is given by

$$\sin 2\alpha = -\tan\left(\frac{\theta}{2}\right)\left[\tan\left(\frac{\Delta}{2}\right) + \tan\left(\frac{\vartheta}{2}\right)\cos 2\omega\right]. \qquad (4.92)$$

When θ and ϑ are sufficiently small, the error is negligible for small Δ and approaches 10-35' for the quarter-wave layer. The error increases close to the half-wave layer. On the other hand, the half-shade in the second position introduces a larger error in measuring phase difference when ω is far from $\pi/4$; thus the sensitivity of the compensator cannot be varied.

If the half-shade is in the third position, then plates with a phase difference close to the half-wave are most convenient and the half-shade has to rotate together with the analyzer. The difference of their azimuths determines the sensitivity.

The sensitivity of all three arrangements is virtually the same.

Brace's Compensator

When the measured phase differences are very small, then the accuracy of reading the azimuth σ may become the factor determining the overall precision. In that case Brace's compensator [191–193] is to be used because the measured azimuths in the absence and the presence of the measured layer are farther apart, just removing the above source of inaccuracy.

In Brace's arrangement the polarizers are crossed and the measured layer, half-shade plate, and compensator plate are situated between them. The half-shade position is achieved by rotating the compensator plate. For a measured layer at azimuth $\pi/4$, the matching condition reads

$$\tan\left(\Delta + \frac{\theta}{2}\right) = -\sin 2\beta \frac{\sin \delta}{[1 - 2\sin^2 2\beta \sin^2(\delta/2)]}, \qquad (4.93)$$

where β and δ are the azimuth and the phase difference of the compensator plate, respectively. Other symbols have the same meaning as previously. The θ value may be obtained by measurement outside the unknown layer. For small Δ's the approximate relation holds:

$$\Delta \doteq -\sin \delta(\sin 2\beta - \sin 2\beta_o). \qquad (4.94)$$

It is apparent that Δ should be smaller than δ. The analysis shows that the last relation is exact to within 1 % if $\delta < 0.25$ and $\vartheta < 0.06$. For small Δ's, the relative error caused by the limited sensitivity of the eye is about the same as with the Senarmont method; with an increasing Δ over 0.25, the errors increase, so that at higher Δ's the Senarmont method is preferable. As with the

Senarmont method, the sensitivity of the method can be adjusted by rotating the half-shade (changing θ) unless the half-shade is in the middle position. Also, the method can be used for finding extinction directions. Putting the measured layer in the middle position leads to an error α in this measurement given by

$$\sin 2\alpha = -\tan\left(\frac{\theta}{2}\right)\left[\tan\left(\frac{\delta}{2}\right) + \tan\left(\frac{\Delta}{2}\right) + \cos 2\omega \tan\left(\frac{\vartheta}{2}\right)\right]. \quad (4.95)$$

From the foregoing, it is apparent that for very small birefringences the Brace compensator is most useful, while for larger effects the Senarmont method is to be preferred. However, a more universal compensator can be designed easily, allowing measurements both by Senarmont's and Brace's method. The transition between both methods is achieved simply by changing the compensator plate and the azimuth which is read (analyzer for Senarmont's and compensator for Brace's method).

Photoelectric Measurements

In recent years some photoelectric systems for measurement of birefringence were designed. The common idea of these systems lies in introducing some periodically variable element in the path of the light. This variable element, together with the unknown plate, influences the intensity of the light in a complex manner. When the variable element changes sinusoidally, the photoelectric response may be analyzed into a steady component, a component with the basic frequency, and one with double frequency. By a selective amplifier only the basic frequency is retained; it can be extinguished by proper adjustment of other optical elements. The condition for extinction is equivalent to the half-shade condition with visual compensators and yields relations suitable for finding birefringence characteristics of the measured layer.

The number of possible designs is large. Wayland [194] uses a Faraday cell for periodic modulation, Penkov and Stepanenko [195] the photoelastic response of some periodically stressed material, and Leray and Scheibling [196] a rotating birefringent plate (note that in the last case the basic frequency is the double frequency of the rotation as 180° rotation brings the plate in the equivalent position). The arrangement of other elements may correspond to either Brace's or Senarmont's method. When measuring the principal optical direction, the polarizers should be crossed and the compensator plate in both arrangements should be parallel to the polarizers or removed completely. The reference signal disappears when the unknown layer is parallel to the polarizers. For phase difference measurements, the unknown layer is oriented in $\pi/4$ azimuth with respect to the polarizer. Then the compensator (Brace) or analyzer (Senarmont) is rotated until the reference signal disappears. The known

Brace's and Senarmont's formulas then apply for the measured layer ($\beta_o = 0$, $\sigma_o = \pi/2$ in these cases). It should be mentioned, however, that some arrangements of optical elements are not admissible. For example, in Brace's method with the modulator in the middle position, the photoelastic modulator may be used; the rotating plate will not serve. With the measured layer in the middle, both modulators may be used. Thus, for every design, (4.87) should be checked.

At the end of this section let us compare the dark-field methods (both visual and photoelectric), visual half-shade methods, and photoelectric modulation methods. The light intensity in the dark-field methods goes through a minimum at the extinction position (i.e., $\partial I/\partial x = 0$, x the variable of measurements); the position of the extreme is therefore difficult to find. On the other hand, in both half-shade methods and modulating methods, the difference in light intensities of both half-fields or the photoelectric signal vary linearly. That increases the sensitivity considerably. However, there is a basic difference in visual and photoelectric methods. The human eye is sensitive to the ratio of two light intensities (logarithmic sensitivity) while the photomultiplier is sensitive to the absolute intensity (linear sensitivity). Thus, the sensitivity of the eye is diminished by the presence of "background" intensity which, in turn, increases as the square of the "half-shade phase difference" or its equivalent. All extra sources of light, as parasite light or the light flux caused by improper optical elements (e.g., inexact quarter-wave plate), further diminish the sensitivity. On the other hand, in photoelectric measurement these extra signals are retained by electronics and large modulating effects may be used just for increasing the signal, that is, the sensitivity.

Most designers claim for photoelectric instruments a sensitivity at least five times better than for visual observation. However, photoelectric methods also have disadvantages. The photomultiplier cannot distinguish reflected light and light scattered by dirt from the desired light; the probability of artifacts is, therefore, higher in photoelectric measurements.

Apparatuses

All streaming birefringence apparatuses combine the above basic ideas in different ways. The actual designs vary widely, however. A description of them would be outside the scope of the present review; further details can be found in the original literature [188–190, 194–200].

Recently, some apparatuses have been designed that allow measurement of birefringence even along other directions or for other types of flow. Thus, Foreman [142] constructed a two-disk cell allowing measurement in the 1,3 plane. Oriel and Schellman [136] developed a Couette instrument for measurement both in axial (1,2 plane) and radial (1,3 plane) directions. For measurement in the 2,3 plane, as far as we know, only a rather primitive capillary instrument of Philippoff [137] was used. Wales and Janeschitz-Kriegl [201] constructed an instrument for streaming birefringence of polymer melts based on the plate-and-cone geometry of the cell. Thurston and Schrag [183] designed a slit instrument for measurement of oscillatory birefringence. Gill [202–203] used completely new geometry—a tube rotating in an elliptic sleeve—for straining viscoelastic liquids and measuring their birefringence.

References

1. J. W. Prados and F. N. Peebles, *A.I.Ch.E. J.*, **5**, 225 (1959); S. Sutera and H. Wayland, *J. Appl. Phys.*, **32**, 721 (1961).

1a. Kerr effect, see Chap. 5 F of Vol. 1, by Professor and Mrs. R. J. W. LeFevre.

2. Cotton-Mouton effect, A. Cotton, *Proc. 1930 Solvay Congr.*, p. 391; A. Cotton and H. Mouton, Conf. Report, "Les Progrès de la Physique Moléculaire," Gauthier Villars, Paris, 1917, pp. 127, 164.

3. A. Lösche, *Kolloid-Z.*, **122**, 94 (1941); *Ann. Physik*, **5**, 381 (1950).

4. J. C. Maxwell, *Collected Papers*, Cambridge Univ. Press, 1890, Vol. II, p. 379; *Proc. Roy. Soc. (London)*, **A22**, 46 (1873).

5. A. von Kundt, *Ann. Physik*, **13**, 110 (1881).

6. G. de Metz, *Ann. Physik*, **35**, 497 (1888).

7. K. Umlauf, *Ann. Physik*, **45**, 304 (1892).

8. J. E. Almy, *Phil. Mag.* (5), **44**, 499 (1897).

9. B. V. Hill, *Phil. Mag.* (5), **48**, 485 (1899).

10. C. Zakrzewski, *Ann. Akad. Krakau*, **50** (1904); C. Zakrzewski and C. Kraft, *ibid.* **506** (1905); V. Bernatzki, *Physik. Z.*, **6**, 730 (1905).

11. J. C. Maxwell, *Pogg. Ann.*, **151**, 151 (1874).

12. D. Vorländer and R. Walter, *Z. Physik. Chem.*, **118**, 1 (1925).

13. Ch. Sadron, *J. Phys. Radium* (7), **7**, 263 (1936).

14. W. Buchheim, H. A. Stuart, and H. Mentz, *Z. Physik*, **112**, 407 (1939).

15. H. Diesselhorst and H. Freundlich, *Physik. Z.*, **16**, 419 (1915); H. Freundlich, H. Diesselhorst, and K. Leonhardt, *Elster-Geitel Festschr.*, Braunschweig, 1915, p. 453; H. Freundlich, F. Stapelfeldt, and H. Zocher, *Z. Physik. Chem.*, **114**, 161, 190 (1925).

16. A. L. von Muralt and J. T. Edsall, *J. Biol. Chem.*, **89**, 315, 351 (1930); *Trans. Faraday Soc.*, **26**, 837 (1930).

17. R. Signer, *Z. Physik. Chem.* (A) **150**, 247, 257 (1930).

18. F. T. Adler, W. M. Sawyer, and J. D. Ferry, *J. Appl. Phys.*, **20**, 1036 (1949).

19. R. Lucas, *Compt. Rend.*, **206**, 827 (1938); *J. Phys. Radium* (7), **10**, 151 (1939); *Rev. Acoust.*, **8**, 121 (1939).

20. M. Kawamura, *Kagaku* (Nature), **7**, 6, 54, 139 (1938).

21. H. Zocher, *Z. Physik. Chem.*, **98**, 296 (1921); M. Aschenbrenner, *ibid.*, **127**, 415 (1927); W. Philippoff, *Kolloid-Z.*, **83**, 163 (1938); K. Hess, M. Kiessig, and W. Philippoff, *Naturwissenschaften*, **26**, 184 (1938).

22. G. B. Thurston and J. L. Schrag, *J. Chem. Phys.*, **45**, 3373 (1966); *J. Polymer Sci. A2*, **6**, 1331 (1968).
23. A. Peterlin, *Pure Appl. Chem.*, **12**, 563 (1966).
23a. W. Philippoff, *Trans. Soc. Rheol.*, **5**, 149 (1961), *J. Polymer Sci.*, **B8**, 107 (1970).
24. A. Wissler, Ph.D. Thesis, University Bern, 1940.
25. V. N. Tsvetkov and E. Frisman, *Acta Physicochim. U.R.S.S.*, **20**, 61 (1945).
26. A. Peterlin and M. Čopič, *Reports J. Stefan Inst.*, Ljubljana, **1**, 65 (1953).
27. V. N. Tsvetkov and V. P. Budtov, *Vysokomolekul. Soedin.*, **6**, 1209 (1964). (See also [93].)
28. G. B. Thurston, *J. Chem. Phys.*, **47**, 3582 (1967).
29. A. Ziabicki and K. Kedzierska, *J. Appl. Polymer Sci.*, **2**, 14 (1959).
30. V. Tsvetkov, A. Mindlina, and C. Makarov, *Acta Physicochim. U.R.S.S.*, **21**, 135 (1946); V. Tsvetkov and V. Marinin, *Dokl. Akad. Nauk SSSR*, **63**, 653 (1948); V. Tsvetkov and V. E. Eskin, *Zh. Eksp. Teor. Fiz.*, **18**, 614 (1948).
31. J. Badoz, *J. Phys. Radium*, **15**, 777 (1954).
32. N. C. Hilyard and M. G. Jerrard, *Nature*, **194**, 173 (1962), *J. Appl. Phys.*, **33**, 3470 (1962).
33. P. Boeder, *Z. Physik*, **75**, 258 (1932).
34. W. Kuhn, *Z. Physik. Chem.* (A) **161**, 1 (1932); *Angew. Chem.*, **49**, 858 (1936).
35. A. Peterlin and H. A. Stuart, *Z. Physik*, **112**, 1 (1939).
35a. H. J. Workman and C. A. Hollingsworth, *J. Colloid & Interface Sci.*, **29**, 664 (1969).
36. G. B. Jeffery, *Proc. Roy. Soc.* (*London*), **A102**, 161 (1922).
37. A. Peterlin, *Z. Physik*, **111**, 232 (1938).
38. R. Gans, *Ann. Physik*, **86**, 628 (1928).
39. F. Perrin, *J. Phys. Radium* (7), **5**, 497 (1934).
40. Ch. Sadron, *J. Phys. Radium* (7), **8**, 481 (1937).
41. G. R. Bird and E. R. Blout, *J. Chem. Phys.*, **25**, 798 (1956).
42. H. A. Scheraga, *J. Chem. Phys.*, **23**, 1526 (1955).
43. H. A. Scheraga, J. T. Edsall, and J. O. Gadd, Jr., *J. Chem. Phys.*, **19**, 1101 (1951).
44. R. Simha, *J. Phys Chem.*, **44**, 25 (1940).
45. A. Peterlin and H. A. Stuart, *Z. Physik.*, **112**, 129 (1939).
46. C. Sadron, *Progr. in Biophys. Chem.*, **3**, 237 (1953).
47. H. A. Scheraga and L. Mandelkern, *J. Am. Chem. Soc.*, **75**, 179 (1953).
48. A. Peterlin, *Makromol. Chem.*, **34**, 89 (1959).
49. O. Snellman and Y. Björnstahl, *Kolloid-Beih.*, **52**, 403 (1941).
50. R. Takserman-Krozer and A. Ziabicki, *J. Polymer Sci.*, **A1**, 491 (1963).
51. R. Cerf and G. B. Thurston, *J. Chem. Phys.*, **61**, 1457 (1964).
52. A. Peterlin, *Zbornik Prirodoslovnega Društva Ljubljana*, **2**, 24 (1941); *J. Phys. Radium*, **11**, 45 (1950).
53. A. Peterlin and H. A. Stuart, "Doppelbrechung, insbesondere künstliche Doppelbrechung," *Hand- und Jahrbuch der Chemischen Physik*, A. Eucken and K. L. Wolf, Eds., Vol. 8, IB, Akad. Verlagsges., Leipzig, 1943; Edward Bros., Inc., Ann Arbor, Mich., 1948.

54. L. V. King, *Proc. Roy. Soc. London* (A), **153**, 17 (1935).
55. S. Oka, *Kolloid-Z.*, **87**, 37 (1939); *Z. Physik*, **116**, 632 (1940).
56. R. Pohlmann, *Z. Physik*, **107**, 497 (1937); **113**, 697 (1939).
57. F. Perrin, *J. Phys. Radium* (7) **5**, 497 (1934).
57a. J. V. Champion and G. H. Meeten, *Proc. Phys. Soc.*, **88**, 1033 (1966); *Trans. Faraday Soc.*, **64**, 238 (1968); *J. Chim. Phys.*, **66**, 1049, 1565 (1969).
58. C. V. Raman and K. J. Krishnan, *Phil. Mag.* (7), **5**, 769 (1928).
59. W. Haller, *Kolloid-Z.*, **61**, 26 (1932).
60. R. Cerf, *J. Chim. Phys.*, **48**, 59, 85 (1951); **52**, 53 (1955); *J. Polymer Sci.*, **12**, 15, 35 (1954).
61. R. Cerf, *J. Phys. Radium*, **15**, 145 (1954).
62. C. J. F. Böttcher, *Theory of Dielectric Polarization*, Elsevier, New York, 1952.
63. W. Kuhn, *Kolloid-Z.*, **68**, 2 (1934).
64. W. Kuhn and F. Grün, *Kolloid-Z.*, **101**, 248 (1942).
65. M. Čopič, *Ricerca Sci.*, **A25**, 942 (1955); *J. Chim. Phys.*, **53**, 440 (1956); *J. Polymer Sci.*, **20**, 493 (1956); *J. Chem. Phys.*, **26**, 1382 (1957).
66. V. N. Tsvetkov and E. Frisman, *Acta Physicochim. U.R.S.S.*, **20**, 363 (1945); *Dokl. Akad. Nauk SSSR*, **97**, 647 (1954); V. N. Tsvetkov, *J. Polymer Sci.*, **23**, 151 (1957).
67. V. N. Tsvetkov, E. V. Frisman, O. B. Ptitsyn, and S. Ya. Kotlyar, *Zh. Tekhn. Fiz.*, **28**, 1428 (1958).
68. W. Kuhn and H. Kuhn, *Helv. Chim. Acta*, **26**, 1394 (1943).
69. V. N. Tsvetkov, *Vysokomolekul. Soedin.*, **5**, 740 (1963).
70. E. V. Frisman and A. K. Dadivanyan, *Vysokomolekul. Soedin.*, **8**, 1359 (1966); *J. Polymer Sci.*, **C16**, 1001 (1967).
71. B. H. Zimm, *J. Chem. Phys.*, **24**, 269 (1956).
72. J. J. Hermans, *Physica*, **10**, 777 (1943), *Rec. Trav. Chim.*, **63**, 219 (1944).
73. A. Peterlin, *Polymer*, **2**, 257 (1961).
74. A. Peterlin, *J. Polymer Sci.*, **B4**, 287 (1966).
75. Chr. Reinhold and A. Peterlin, *J. Chem. Phys.*, **44**, 4333 (1966).
76. R. Cerf, *J. Phys. Radium*, **19**, 122 (1959); *Advan. Polymer Sci.*, **1**, 382 (1959).
77. W. Kuhn and H. Kuhn, *Helv. Chim. Acta*, **28**, 1533 (1945); **29**, 72 (1946).
78. Ch. Chaffey, *J. Chim. Phys.*, **63**, 1385 (1966).
79. A. Peterlin, *J. Phys. Radium*, **22**, 407 (1961).
80. R. Takserman-Krozer, *J. Polymer Sci.*, **A1**, 2477 (1963); **C16**, 2855 (1967).
81. O. Kratky and G. Porod, *Rec. Trav. Chim.*, **68**, 1106 (1949).
82. V. N. Tsvetkov, *Vysokomolekul. Soedin.*, **4**, 894 (1962); **7**, 1468 (1965); *Dokl. Akad. Nauk SSSR.*, **165**, 360 (1965); *European Polymer J. Suppl.*, 1969, p. 237.
83. Yu. Ya. Gotlib and Yu. E. Svetlov, *Dokl. Akad. Nauk SSSR*, **168**, 621 (1966).
84. P. Munk and A. Peterlin, *Rheol. Acta*, **9**, 288, 294 (1970).
85. A. Peterlin, *Rheol. Acta*, in press.
86. W. Philippoff, *Proc. 4th Intern. Congr. Rheology*, Providence, 1963, Interscience, New York, 1965, Part 2, p. 343.
87. W. Philippoff, *Trans. Soc. Rheol.*, **4**, 159 (1960).
88. W. Philippoff, *Acta Rheol.*, **1**, 371 (1961).
89. P. E. Rouse, Jr., *J. Chem. Phys.*, **21**, 1272 (1953).

90. N. W. Tschoegl, *J. Chem. Phys.*, **44**, 4615 (1964).
91. A. Peterlin, *J. Chem. Phys.*, **23**, 2464 (1955).
92. J. Leray, *Compt. Rend.*, **241**, 1741 (1955), *J. Polymer Sci.*, **23**, 167 (1957), *J. Chim. Phys.*, **57**, 323 (1960).
93. V. N. Tsvetkov, in *Newer Methods of Polymer Characterization*, B. Ke, Ed., Interscience, New York, 1964, Chapter 14.
94. A. Peterlin, *Kolloid-Z. & Z. Polymere*, **187**, 68 (1963); *J. Chem. Phys.*, **39**, 224 (1963).
95. W. Philippoff, *Nature*, **178**, 811 (1956); *J. Appl. Phys.*, **27**, 989 (1956).
96. J. G. Brodnyan, F. H. Gaskins, and W. Philippoff, *Trans. Soc. Rheol.*, **1**, 109 (1957).
97. B. Schmidli, Ph.D. Thesis, ETH Zürich (1952).
98. W. Philippoff, F. M. Gaskins, and J. G. Brodnyan, *J. Appl. Phys.*, **28**, 1118 (1957).
99. W. Philippoff, *Ind. Eng. Chem.*, **51**, 883 (1959).
100. F. R. Cottrell, Thesis, MIT 1968, F. R. Cottrell, E. W. Merrill, and K. A. Smith, *J. Polymer Sci.*, **A2**, **7**, 1415 (1969).
101. A. Peterlin, *J. Polymer Sci.*, **B5**, 113 (1967).
102. A Peterlin, *Polymer*, **8**, 21 (1967).
103. G. B. Thurston and A. Peterlin, *J. Chem. Phys.*, **46**, 4881 (1967).
104. A. Peterlin, *Rec. Trav. Chim.*, **69**, 14 (1950).
105. A. Peterlin and Chr. Reinhold, *Trans. Soc. Rheol.*, **11**, 1, 15 (1967).
106. M. Joly, *J. Chim. Phys.*, **48**, 536 (1951); *Kolloid-Z.*, **126**, 77 (1952).
107. P. Munk, *J. Polymer Sci.* A2, **5**, 685 (1967).
108. A. S. Lodge, *Proc. 2nd Intern. Congr. Rheology*, Oxford 1953, Butterworths, London, 1954, p. 229; *Nature*, **176**, 838 (1955); *Trans. Faraday Soc.*, **52**, 127 (1956); *Elastic Liquids*, Academic Press, London–New York, 1964, p. 101.
109. L. R. G. Treloar, *The Physics of Rubber Elasticity*, 2nd ed., Clarendon, Oxford, 1958, p. 197.
110. M. Yamamoto and H. Inagaki, *Busseiron Kenkyu*, No. 55, 26 (1952).
111. M. Yamamoto, *J. Phys. Soc. Japan*, **11**, 413 (1956).
112. A. Peterlin, *Proc. 2nd Intern. Congr. Rheology*, Oxford 1953, Butterworths, London, 1954, p. 343; *Bull. Sci. Conseil Acad. RPF Yougoslavie*, **1**, 40 (1953); *J. Polymer Sci.*, **12**, 45 (1954).
113. A. Peterlin and R. Signer, *Helv. Chim. Acta*, **36**, 1575 (1953).
114. H. Janeschitz-Kriegl, *Makromol. Chem.*, **40**, 140 (1960); J. L. S. Wales and H. Janeschitz-Kriegl, *Rheol. Acta*, **7**, 19 (1968).
115. P. Munk, D. Poupětová, M. Bohdanecký, and M. Dillingerová, *J. Polymer Sci.*, **C16**, 1125 (1967).
116. W. Philippoff and R. S. Stratton, *Trans. Soc. Rheol.*, **10**, 467 (1966).
117. W. Philippoff, *Trans. Soc. Rheol.*, **4**, 169 (1960).
118. Ch. Sadron, *J. Phys. Radium* (7), **9**, 381 (1938).
119. R. Signer and H. Liechti, *J. Chim. Phys.*, **44**, 58 (1947).
120. Ch. Sadron and H. Mosimann, *J. Phys. Radium* (7), **9**, 384 (1938).
121. U. Daum, *J. Polymer Sci.* A2, **6**, 141 (1968).
122. H. Janeschitz-Kriegl and U. Daum, *Kolloid-Z.*, **210**, 112 (1966).

123. P. Munk, *Collection Czech. Chem. Commun.*, **32**, 2715 (1967).
124. P. Munk, *J. Polymer Sci.*, **C16**, 111 (1967).
125. W. Philippoff, *Trans. Soc. Rheol.*, **5**, 163 (1961).
126. H. Janeschitz-Kriegl, *Kolloid-Z.*, **203**, 119 (1965).
127. P. Munk, unpublished data.
128. H. Janeschitz-Kriegl and W. Burchard, *J. Polymer Sci.*, A2, **6**, 1953 (1968).
129. J. W. Rowen and W. Ginoza *Biochim. Biophys. Acta*, **21**, 416 (1956).
130. H. Boedtker and N. S. Simmons, *J. Am. Chem. Soc.*, **80**, 2550 (1958).
131. J. Leray, *J. Chim. Phys.*, **58**, 316 (1961).
132. J. T. Yang, *J. Am. Chem. Soc.*, **83**, 1316 (1961).
133. J. T. Edsall, J. F. Foster, and H. Scheinberg, *J. Am. Chem. Soc.*, **69**, 2731 (1947).
134. J. T. Edsall, G. A. Gilbert, and H. A. Scheraga, *J. Am. Chem. Soc.*, **77**, 157 (1955).
135. H. Boedtker and P. Doty, *J. Am. Chem. Soc.*, **78**, 4267 (1956).
136. P. J. Oriel and J. A. Schellman, *Biopolymers*, **4**, 469 (1966); **5**, 399 (1967).
137. W. Philippoff, *Trans. Soc. Rheol.*, **5**, 149 (1961).
138. C. J. Stacy and J. F. Foster, *J. Polymer Sci.*, **20**, 67 (1956).
139. J. T. Yang and J. F. Foster, *J. Polymer Sci.*, **18**, 1 (1955).
140. J. F. Foster and I. H. Lepow, *J. Am. Chem. Soc.*, **70**, 4169 (1948).
141. J. F. Foster and D. Zucker, *J. Phys. Chem.*, **56**, 174 (1952).
142. W. T. Foreman, *J. Chem. Phys.*, **32**, 277 (1960).
143. F. N. Peebles, J. W. Prados, and E. H. Honeycutt, *J. Polymer Sci.*, **C5**, 37 (1963).
144. H. Thiele, *Kolloid-Z.*, **112**, 73 (1949); **113**, 155 (1949).
145. J. K. Backus and H. A. Scheraga, *J. Colloid Sci.*, **6**, 508 (1951).
146. H. A. Scheraga and J. K. Backus, *J. Am. Chem. Soc.*, **73**, 5108 (1951).
147. W. Philippoff, *Trans. Soc. Rheol.*, **10**, Part 2, 467 (1966); **12**, Part 1, 85 (1968).
148. R. Kubo, *J. Phys. Soc. Japan*, **2**, 84 (1947), Yu. Ya. Gotlib, *Zh. Tekhn. Fiz.*, **27**, 707 (1957); K. Nagai, *J. Chem. Phys.*, **40**, 2818 (1964); R. L. Jernigan and P. J. Flory, *J. Chem. Phys.*, **47**, 1999 (1967).
149. K. D. Denbigh, *Trans. Faraday Soc.*, **36**, 936 (1940).
150. V. N. Tsvetkov and A. E. Grishchenko, *J. Polymer Sci. Part C16*, 3195 (1968).
151. P. Munk, *Collection Czech. Chem. Commun.*, **32**, 1541 (1967).
152. U. Daum, unpublished results, quoted in [154].
153. E. V. Frisman and V. N. Tsvetkov, *J. Polymer Sci.*, **30**, 297 (1958).
154. H. Janeschitz-Kriegl, *Advan. Polymer Sci.*, **6**, 170 (1969).
155. V. N. Tsvetkov, T. I. Garmonova, and R. P. Stankevich, *Vysokomolekul. Soedin.*, **8**, 980 (1966).
156. E. V. Frisman, M. A. Sibileva, and M. A. Chebishyan, *Vysokomolekul. Soedin.*, **9**, 1071 (1967).
157. V. N. Tsvetkov, I. N. Shtennikova, V. S. Skazka, and E. I. Rjumtsev, *J. Polymer Sci.*, Part C16, 3205 (1968).
158. V. N. Tsvetkov, K. A. Andrianov, E. L. Vinogradov, I. N. Shtennikova, S. E. Yakushkina and V. I. Pakhomov, *J. Polymer Sci.*, Part C23, 385 (1968).

159. E. V. Frisman, V. I. Vorob'ev, and L. V. Shchagina, *Vysokomolekul. Soedin.*, **6**, 884 (1964).

159a. V. N. Tsvetkov, L. N. Andrews, and L. N. Kvitchenko, *Vysokomolekul. Soedin.*, **7**, 2001 (1965).

159b. V. N. Tsvetkov, *Rubber Chem. Tech.*, **36**, 337 (1963).

159c. O. Kratky, I. Pilz, and W. Burchard, as quoted in [154], p. 275.

159d. W. Burchard and E. Husemann, *Makromol. Chem.*, **44/46**, 358 (1961).

160. P. Munk, unpublished results.

161. P. Munk, J. Šponar, and L. Pivec, *Collection Czech. Chem. Commun.*, **29**, 886 (1964).

162. D. O. Jordan, A. R. Mathieson, and M. R. Porter, *J. Polymer Sci.*, **21**, 463 (1956).

163. R. M. Fuoss and R. Signer, *J. Am. Chem. Soc.*, **73**, 5872 (1951).

164. D. O. Jordan, A. R. Mathieson, and M. R. Porter, *J. Polymer Sci.*, **21**, 473 (1956).

165. D. O. Jordan and T. Kurucsev, *Polymer*, **1**, 202 (1960).

166. W. Kuhn, H. Oswald, and H. Kuhn, *Helv. Chim. Acta*, **36**, 1209 (1953).

167. W. Kuhn, O. Künzle, and A. Katchalsky, *Helv. Chim. Acta*, **31**, 1994 (1948).

168. H. Schwander and R. Signer, *Helv. Chim. Acta*, **34**, 1344 (1951).

169. V. N. Tsvetkov, *Vysokomolekul. Soedin.*, **5**, 747 (1963).

170. R. E. Harrington, *Biopolymers*, **6**, 105 (1968).

171. S. Takashima, *Biopolymers*, **6**, 1437 (1968).

172. S. Ya. Magarik and V. N. Tsvetkov, *Zh. Fiz. Khim.*, **33**, 835 (1959).

173. P. Gramain, J. Leray, and H. Benoit, *J. Polymer Sci.*, Part C16, 3983 (1968).

174. V. N. Tsvetkov, G. A. Andreeva, I. A. Baranovskaya, V. E. Eskin, S. I. Klenin, and S. Ya. Magarik, *J. Polymer Sci.*, Part C16, 239 (1967).

175. W. Philippoff, *Trans. Soc. Rheol.*, **7**, 45 (1963).

176. S. Fujishige, *J. Colloid Sci.*, **13**, 193 (1958).

177. F. Debeauvais, P. Gramain, and J. Leray, *J. Polymer Sci.*, Part C16, 3993 (1968).

178. P. Munk, *Collection Czech. Chem. Commun.*, **32**, 787 (1967).

179. P. Munk and A. Peterlin, *Trans. Soc. Rheol.*, **14**:1, 65 (1970).

180. P. Munk, J. Šponar, and L. Pivec, *Collection Czech. Chem. Commun.*, **29**, 1222 (1964).

181. A. Peterlin, *J. Lubrication Technol.*, **90**, 571 (1968).

182. A. Peterlin, D. T. Turner, and W. Philippoff, *Kolloid-Z. & Z. Polymere*, **204**, 21 (1965).

183. G. B. Thurston and J. L. Schrag, *Trans. Soc. Rheol.*, **6**, 325 (1962).

183a. H. Janeschitz-Kriegl, G. Henrici-Olivé, and S. Olivé, *Kolloid-Z. & Z. Polymere*, **191**, 97 (1963).

183b. J. Leray, *J. Chim. Phys.*, **52**, 755 (1955).

183c. W. Philippoff, *Trans. Soc. Rheol.*, **3**, 153 (1959).

184. G. I. Taylor, *Proc. Roy. Soc.* (*London*), **A157**, 546, 565 (1936).

185. Y. Björnståhl, *Z. Physik*, **119**, 245 (1942).

186. H. Janeschitz-Kriegl, *J. Polymer Sci.*, **23**, 181 (1957).
187. E. Van Kuik-Van Meerten and H. Janeschitz-Kriegl, *J. Sci. Instr.*, **39**, 301 (1962).
188. E. V. Frisman and V. N. Tsvetkov, *Zh. Tekhn. Fiz.*, **25**, 447 (1955).
189. J. T. Edsall, A. Rich, and M. Goldstein, *Rev. Sci. Instr.*, **23**, 695 (1952).
190. B. H. Zimm, *Rev. Sci. Instr.*, **29**, 360 (1958).
191. P. Munk and P. Kratochvíl, *Collection Czech. Chem. Commun.*, **26**, 1591 (1961).
192. D. B. Brace, *Phys. Rev.*, **18**, 70 (1904); **19**, 218 (1904).
193. N. Wedeneewa, *Z. Instrumentenk.*, **43**, 17 (1923).
194. H. Wayland, *Compt. Rend.*, **249**, 1228 (1959).
195. S. N. Penkov and V. Z. Stepanenko, *Opt. Spectry (USSR)*, **14**, 156 (1963).
196. J. Leray and G. Scheibling, *Compt. Rend.*, **251**, 349 (1960).
197. V. Nováček and P. Munk, *Collection Czech. Chem. Commun.*, **28**, 1945 (1963).
198. H. Janeschitz-Kriegl, *Rev. Sci. Instr.*, **31**, 119 (1960).
199. H. Janeschitz-Kriegl and R. Nauta, *Rev. Sci. Instr.*, **42**, 880 (1965).
200. M. Intaglietta and H. Wayland, *Biorheology*, **2**, 195 (1965).
201. J. L. S. Wales and H. Janeschitz-Kriegl, *J. Polymer Sci.*, A2, **5**, 781 (1967).
202. S. J. Gill, *J. Appl. Polymer Sci.*, **1**, 17 (1959).
203. S. J. Gill and F. R. Dintris, *J. Polymer Sci.*, **57**, 251 (1962).

General References

Boehm, G., "Methodik der Untersuchung der Strömungsdoppelbrechung," in *Abderhalden's Handbuch Biol. Arbeitsmethoden*, Abt. II, Teil 3, Urban and Schwarzenberg, Berlin, Wien, 1939, pp. 3939–4004.

Cerf, R., "La Dynamique des Solutions de Macromolécules dans un Champ de Vitesse," *Advan. Polymer Sci.*, Vol. 1, Springer, Berlin, 1959, pp. 383–450.

Cerf, R., and H. A. Scheraga, "Flow Birefringence in Solutions of Macromolecules," *Chem. Rev.*, **51**, 185 (1952).

Edsall, J. T., "Streaming Birefringence and Its Relation to Particle Size and Shape," *Advan. Colloid Sci.*, Vol. I, Interscience, New York, 1942, pp. 269–316.

Janeschitz-Kriegl, H., "Flow Birefringence of Elastic Viscous Polymer Systems," *Advan. Polymer Sci.*, Vol. 6, Springer, Berlin, 1969, pp. 170–318.

Jerrard, H. G., "Theories of Streaming Double Refraction," *Chem. Rev.*, **59**, 345–428 (1959).

Kuhn, W., H. Kuhn, and P. Buchner, "Hydrodynamisches Verhalten von Makromolekulen in Lösung," *Ergeb. Exakt. Naturw.*, **25**, 1–108 (1951).

Peterlin, A., "Streaming and Stress Birefringence," in *Rheology*, F. R. Eirich, Ed., Vol. 1, Academic, New York, 1956, pp. 615–652.

Peterlin, A., and H. A. Stuart, "Doppelbrechung, insbesondere künstliche Doppelbrechung," in *Hand- und Jahrbuch der Chemischen Physik*, A. Eucken and K. L. Wolf, Eds., Vol. 8, IB, Akadem. Verlagsgesellschaft, Leipzig, 1943, pp. 1–115. Reprinted by Edward Bros., Inc., Ann Arbor, Michigan 1948.

Peterlin, A., and H. A. Stuart, "Künstliche Doppelbrechung," in *Das Makromolekul in Lösungen* (Vol. 2 of *Die Physik der Hochpolymeren*), H. A. Stuart, Ed., Springer, Berlin-Gottingen-Heidelberg, 1953, pp. 569–617.

Scheraga, H. A., and R. Signer, "Streaming Birefringence," in *Physical Methods of Organic Chemistry*, A. Weissberger, Ed., Vol. 1, 3rd ed., Interscience, New York, 1960, pp. 2388–2457.

Tsvetkov, V. N., "Flow Birefringence," in *Newer Methods of Polymer Characterization*, B. Ke, Ed., Interscience, New York, 1964, pp. 563–665.

THE FARADAY EFFECT

James M. Thorne

1 INTRODUCTION

The Faraday effect includes two phenomena that occur when light interacts with matter in a magnetic field, the theory of which is treated in the preceding chapter. The first phenomenon to be discovered was the rotation of plane polarized light. It will be referred to as magneto-optical rotatory dispersion, or MORD in this chapter. The second effect is magnetic circular dichroism, or MCD. This effect arises from the fact that in an external magnetic field the absorption of right circularly polarized light is not equal to the absorption of left circularly polarized light. The units and nomenclature of the Faraday effect are covered in the preceding chapter. However, since the macroscopic manifestations are the same for natural and magnetically induced optical activity, we may also refer to the literature of ORD and CD [1, 2] for units and nomenclature.

Commercial polarimeters and dichrographs are commonly used to measure MORD and MCD. It is necessary only to install a magnet in the instrument to convert from ORD to MORD, or from CD to MCD.

Several aspects of the Faraday effect which are important in instrumentation can be pointed out with the aid of Fig. 5.1. The wavelength where a given maximum or minimum in an MORD or MCD curve occurs depends only on the material in the sample cell; but the absolute magnitude of the Faraday

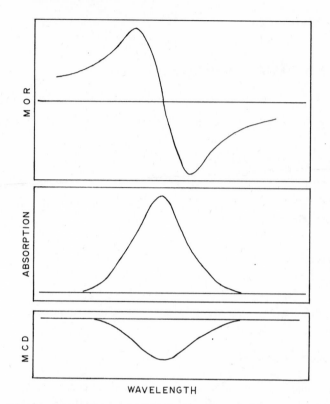

Fig. 5.1. Wavelength dependence of MORD, absorption, and MCD. Negative MORD and MCD bands are shown, but both positive and negative (and more complicated) bands occur naturally.

effect is directly proportional to the total amount of material in the light path (concentration times path length) and the magnetic field strength. (Actually it is the component of the magnetic field parallel to the light path that is important, but because instruments are built to keep the field and light parallel we may simply use the field strength.) If the magnetic field is reversed in direction, both MORD and MCD are reversed in sign: clockwise rotation of plane polarized light becomes anticlockwise rotation, and the relative absorbances of left and right circularly polarized light are reversed. This can be useful in determining the baseline (halfway between the curves of the forward and reversed fields) even when optically active compounds are present in solution.

Note that the MCD band has a shape similar to the absorption band, with MCD going to zero at wavelengths not far from the center of the band. But the MORD still has a significant magnitude at very distant wavelengths.

When there are several bands in the spectrum of a compound, the MORD tails overlap and this results in a spectrum that is more difficult to interpret than the corresponding MCD spectrum. For this reason, most Faraday effect measurements are MCD rather than MORD. It should be mentioned, however, that if an instrument's wavelength range does not include an MCD band of interest, it may still be possible to get information about that band from the tail of the MORD curve. But the main point to be made in comparing the two curves is that both rotation and dichroism occur simultaneously in magnetic fields. So when MCD is being measured, any plane polarized light in the light beam will be rotated. When MORD is being measured, MCD introduces ellipticity into the light beam so the light cannot be completely extinguished between crossed polarizers. Until now, these effects have not been serious problems, but they may become important as stronger magnets come into use and as measurements are made on highly optically active compounds, such as liquid crystals. Reference will be made later to the instrumental importance of these two effects.

The magnitude of the Faraday effect is small compared to natural optical activity (at least for the magnetic field strengths used to date). Very few compounds will yield more than $0.5°$ rotation in a 10 kG field when the absorbance of the sample is 1.0 or less. The difference in the absorbance of left and right circularly polarized light is seldom as great as 0.1. However, current instruments are accurate to about $10^{-3}°$ and 10^{-6} absorbance unit, and magnetic fields between 40 and 100 kG are being used [3, 4].

One useful feature of the Faraday effect is that rotation and dichroism are proportional to the total length of sample traversed by the light, even when a reflection has occurred during traversal. So reflection back through the sample cell cancels rotation and dichroism arising from natural optical activity, and doubles rotation and dichroism that are magnetically induced. This provides an instrumental method for separating the natural from the magnetically induced effects. The mirror should not be in a strong magnetic field, or unwanted rotation and ellipticity will arise from the Kerr magneto-optic effect [5]. But unwanted multiple reflections in the sample cell can result in spurious Faraday effect signals that would not occur if only natural optical activity were being measured. The solution for this is to mount the sample cell so that it is no longer exactly normal to the light beam. The reflected light will then miss the instrument's light-detection system.

2 OPTICAL INSTRUMENTATION

Most instruments used now for measuring the Faraday effect are commercial spectropolarimeters or dichrographs with magnets in the sample compartment. Since they have been described in detail in another chapter of this volume, and elsewhere [6, 7], only aspects peculiar to the Faraday effect will be discussed here.

The most obvious source of trouble is the interaction of the stray magnetic field with the components of the instrument. Fortunately, stray fields are small for iron-containing magnets because of the relatively low reluctance of the return path. Measurements made by the author on a permanent magnet with a 3-mm gap having a 9.0 kG field showed the stray field was down to 11 G 10 cm along the light beam, and down to 4 G at 10 cm perpendicular to the light beam. A superconducting magnet cannot be expected to have such a small percentage of stray field because its return path has a much higher reluctance.

According to the Lorentz relation, magnetic as well as electric fields can change the trajectories of moving electrons. This implies the stray magnetic field in a Faraday effect instrument may defocus the cascading electrons in the photomultiplier tube. In at least one case [8] the high stray field of a 50-kG superconducting magnet required the magnetic shielding and repositioning of the photomultiplier. When stray fields are small, the defocusing of photomultiplier electrons can be tolerated because no artifacts are produced—only the efficiency of the photomultiplier is reduced. Even this effect can be minimized by using an end-window, venetian-blind photomultiplier that keeps the electrons moving almost parallel to the magnetic field, thus reducing trajectory changes.

Electrons and ions also move in arc lamps which are often used as light sources for polarimeters and dichrographs. The highly intense portion of the arc has a tendency to wander from one point of contact on an electrode to another. In a properly functioning instrument this noise is likely to set the limit on the sensitivity of the measurement (at least in the ultraviolet). Stray magnetic fields can increase arc wander and noise, and stray fields have even been observed to extinguish the arc by bending it far out of its normal path [8]. Fortunately, arc lamps can be adequately shielded with iron and/or other magnetic shielding materials.

No other deleterious effects of stray fields have been reported, although instrumentalists would do well to be on guard against Hall effect disturbances in preamplifiers, and MORD arising from polarizers, lenses, and electro-optic light modulators.

If the magnet blocks any part of the light beam, it is likely that the instrument baseline will be changed, because of the changes in the distribution of the light falling on the photomultiplier. Light polarization, as well as intensity, affects the response of most photocathodes, and the response varies from one area of the photocathode to another. But this is not the only reason for the variation—the amount of light reflected from the photomultiplier envelope varies according to the angle of incidence and polarization (Fresnel laws of reflection), so changes in light distribution produce changes in photomultiplier current. Photomultipliers with flat windows do not entirely solve this problem because any strain in the windows will cause somewhat the same response to changes in light distribution. Selection of a good photomultiplier and careful alignment will usually bring the problem under control.

A similar problem concerns the light distribution over the magnet aperture. If the aperture is close to an image of a monochromator slit and the magnetic field is inhomogeneous, then the effective field (averaged over the light beam) will change with the slit width and/or wavelength. It is usually easier to change the position of the magnet than to make the field homogeneous.

Experimental evidence indicates that scattering and fluorescence from the sample are not particularly serious. They decrease the purity of the polarization of the light beam, but they do so symmetrically so that the noise increases but no artifacts appear. However, it is impossible to measure the Faraday effect accurately if the color purity (monochromaticity) of the light entering the sample is not adequate. The light from a poor or internally dusty monochromator will cover too broad a range of the spectrum, so that when the instrument is supposed to be measuring the Faraday effect in an absorption band, much of the light reaching the photomultiplier will be stray light of wavelengths outside the band. This stray light carries with it its own polarization, which is likely to be much different than the polarization of the light of interest. This, of course, gives spurious readings. Using samples of low absorbance helps eliminate this problem. In any case no sample with absorbance greater than about 1.0 should be used. (The signal-to-noise ratio is maximum at this absorbance for most instruments that are operating under optimum conditions.)

In some cases, magnetic field homogeneity can be important. Variations in field strength are not as significant as variations in the direction of the field. The Faraday effect varies linearly with the component of the magnetic field parallel to the light beam; the components perpendicular to the light beam result in the Voigt and/or Cotton-Mouton effects (birefringence of plane polarized light proportional to the square of the magnetic field strength) [5, 9]. No rotation is introduced into the beam; hence MORD measurements remain reliable except for the slight increase in noise caused by the ellipticity from these transverse magnetic field effects. With MCD, the problem is more serious. All of the circular dichrographs now in use depend on conversion of a beam of linearly polarized light alternately to left and then right circularly polarized light. The transverse magnetic field effects impose additional ellipticity of one sign on the beam so that one type of light (say, left circular) will leave the sample cell undermodulated (part way back to plane polarization) and the other type will leave the cell overmodulated (part way to plane polarized light that has been rotated 90°). The detection system will respond to the average of the two and will indicate less circular dichroism than is actually present. This effect has been minor in measurements made to date, but it may become important as very strong magnetic fields are used, or as the MCD is measured for materials showing large Cotton-Mouton effects (e.g., liquid crystals).

3 COMPARISON OF MAGNET TYPES

The ideal magnet for Faraday effect measurements would provide:

1. intense, stable magnetic field;
2. easy sample access;
3. ability to vary the sample temperature;
4. low initial cost;
5. low maintenance cost;

Needless to say, no magnet fulfills all of these requirements.

Superconducting and permanent magnets will now be discussed with respect to each of the criteria in the above list. Electromagnets will not be discussed because they are inferior to the other types in every category.

Both permanent and superconducting magnets (in the persistent mode) have fields that are extremely stable over long periods of time. Permanent magnets will decrease in field strength if exposed to other magnetic fields, or if the air gap is lengthened and then returned to the original position. Alnico V magnets are extremely insensitive to irreversible magnetization loss through heating (about 2% at 300°C), and through mechanical shock (cracking or breakage usually occurs before more than 0.5% of the magnetization is lost) [10]. Superconducting magnet field strengths are an order of magnitude greater than those achievable with permanent magnets. What is more, there appears to be no theoretical limit to the fields we can achieve with superconducting magnets if mechanical problems can be overcome.

The sample space in a permanent magnet is usually more readily accessible than that of a superconductor. Neither presents a very formidable problem if properly designed. To achieve usefully large fields with a permanent magnet, the gap must be less than 1 cm; but superconducting magnets can be made to accommodate much longer sample cells (but at greater cost). Actually a 1 mm path-length sample cell is adequate for measuring the Faraday effect in all but the most weakly absorbing compounds if they are moderately soluble and if high concentrations do not interfere with the measurements.

Superconducting magnets can be constructed to give a variety of sample temperatures, those most easily attainable being room temperature, liquid nitrogen temperature (77°K), and liquid helium temperature (4.2°K). Changing the temperature of the sample in a permanent magnet usually entails changing the temperature of the magnet also. Large temperature changes are possible only if the magnet is well insulated. The author has made MORD measurements with a permanent magnet at 160°K, but the accuracy was poor because the temperature was difficult to maintain. In addition, the field increased by about 10% and had to be recalibrated at this temperature.

As for cost, a superconducting magnet runs about a dollar per G—a permanent magnet costs about two cents per Gauss. Maintenance costs are even more lopsided, mainly because a superconducting magnet requires about $2\frac{1}{2}$ liters of liquid helium ($5–$10 per liter), and about the same amount of liquid nitrogen ($.20–$.90 per liter) for cooldown. The magnet will also require about 0.10 liter per hour of liquid helium and 0.25 liter of liquid nitrogen per hour while in operation [11].

4 MAGNET CALIBRATION

If a magnet is known to have good homogeneity over the volume of the sample, its field can be measured reliably with a fluxmeter or a gaussmeter. A magnet with an inhomogeneous field presents a much more difficult problem. The average field in the sample volume is not the value of interest, because each volume element contributes to the rotation or dichroism in proportion to the light intensity as well as the parallel component of the field. What is needed is the effective field, H_{eff}, which may be defined as

$$\frac{\int H_x I_x \, dV}{\int I_x \, dV} \tag{5.1}$$

where H_x and I_x are the components of the field and light intensity in the direction of the light beam. The easiest way to obtain H_{eff} is to measure the MORD of a sample whose Verdet constant is known very accurately (e.g., water or benzene) [12]. Some useful values for the MORD of 1 cm of water are $-2.1672°$ at 5890 Å and $-4.1937°$ at 4360 Å in a 10 kG field. The effective magnetic field can then be calculated from $H_{eff} = \alpha/Vl$ for the pure liquid, where α is the observed rotation, V the verdet constant, and l the length of the sample. Because the distribution of the light over the sample may change with wavelength, H_{eff} measured this way may not be accurate at other wavelengths.

If only MCD can be measured, the problem becomes more difficult because standard measurements of MCD are not so readily available. However, two alternative methods are available. First, we might insert a precalibrated magnet in the dichrograph and measure the MCD of an arbitrary sample. The MCD of the same sample can then be measured using the uncalibrated magnet. The second method of MCD calibration involves measuring the MCD of a suitable sample with the uncalibrated system, and transforming this MCD to MORD (Kronig-Kramers transform) [2]. The resulting value is then compared with the MORD actually observed for the sample in a calibrated magnet. Dratz [13] lists several criteria the sample must meet:

1. The MORD must be fairly large compared to the absorbance to insure precise data.

2. The rotation caused by other MORD bands must be small in the wavelength range where data are being taken. (That is, one must choose a relatively isolated MORD-MCD band.)

3. A dilute solution must be used, or the accuracy will be limited by the correction that must be made for the displacement of solvent by solute [14].

4. A solvent must be used which is transparent at least 500–1000 Å from the wavelengths of interest so refractive index corrections will not have to be made when using the Kronig-Kramers transform.

5. The compound must have sufficiently wide absorption bands to avoid errors arising from lack of resolution in either instrument.

Certain metal porphyrins appear to be ideal (e.g., nickel (II) deuteroporphyrin dimethyl ester, about $5 \times 10^{-6} M$ in chloroform, 1 cm path) [13]. If instrumental resolution is adequate, several easily obtainable aromatic hydrocarbons are also suitable (pyrene, acenaphthene, or phenanthrene in chloroform).

From the foregoing, it should be clear that the most accurate calibration can be obtained by electronically measuring the strength of a homogeneous magnetic field.

5 MAGNET DESIGN

In the author's opinion, the design of superconducting magnets should be left to the experts. The chemist's role is to specify: (1) the limits on the dimensions (both sample and external); (2) the sample temperatures required; (3) the homogeneity necessary; and (4) the maximum field (which is related to the cost of the magnet).

A permanent magnet system achieving up to 12 kG can be constructed rather inexpensively. The design considerations for such a magnet system will be discussed next.

The most important parameters for determining the magnitude of the magnetic field in the gap of a simple magnetic circuit (such as the one shown in Fig. 5.2) are the following:

1. l_m, length of magnet;
2. A_m, cross-sectional area of magnet;
3. l_g, length of gap;
4. A_g, area of gap;
5. B_m, magnetic flux density (induction) in the magnet;
6. H_m, magnetic field strength in the magnet.

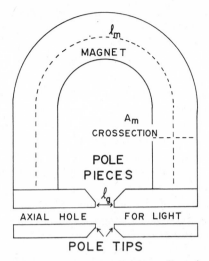

Fig. 5.2. Simplified permanent magnet system shown in cross section.

Equations (5.2) and (5.3) are the main design equations that relate these parameters to B_g, the field in the gap [15]:

$$FB_g A_g = B_m A_m \tag{5.2}$$

$$f B_g l_g = H_m l_m \tag{5.3}$$

where F and f are factors that correct for leakage (the magnetic flux that passes through the air in places other than the gap).

For Faraday-effect measurements, H in (5.3) should be as large as possible. H_g is very nearly equal to B_g since the magnetic permeability of the sample and cell are essentially the same as the permeability of a vacuum. The first step, then, in designing a permanent magnet system is to maximize B_g. Equation (5.3) indicates that l_m, the length of the magnet, must be as large as instrumental space limitations allow.

Combining (5.2) and (5.3) we obtain:

$$B_g^2 = \frac{B_m H_m}{F f} \frac{l_m A_m}{l_g A_g} \tag{5.4}$$

To maximize B_g, a magnetic material with a large energy product, $B_m H_m$, must be chosen.

Platinum cobalt (76.7% Pt, 23.3% Co) shows one of the highest energy products ($BH_{max} = 7.5 \times 10^6$ kG-Oe), but the price of a sufficiently large magnet would be prohibitive. Alnico VAB DG and Alnico VB DG have

$BH_{max} = 6.5 \times 10^6$ kG-Oe and the more common Alnico V has $BH_{max} = 5.5 \times 10^6$ kG-Oe. Ceramic magnets are down by an order of magnitude in BH_{max} from Alnico V.

Choosing a particular magnet from a catalog fixes BH_{max}, A_m, and l_m, as well as the overall geometry of the complete system. At this point it is possible to estimate the leakage factors, f and F. (It is impractical to calculate accurate values for these two factors.) A system of nomographs has been developed [15] to give an approximate value of F for any magnetic circuit of known geometry. The value of F for horseshoe magnets is given approximately by $F = 1.8\, l_g + 2$. The value of f can be safely taken as 1.35.

The next step is to find the value of A_g that will allow the magnet to work near its optimum point on the demagnetization curve (i.e., where $B_m H_m$ is maximum). If A_g is too small, the gap will have a large reluctance and impose a large demagnetizing force on the magnet, causing it to work too far left on the demagnetization curve (Fig. 5.3). If A_g is too large, the converse is true.

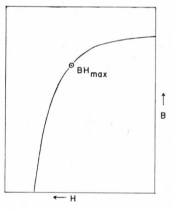

Fig. 5.3.   Typical demagnetization curve showing demagnetization force H, and magnetic induction, B, inside a permanent magnet.

The best value of A_g is found by eliminating B_g between (5.2) and (5.3). This gives:

$$A_g = \frac{B_m l_g f A_m}{H_m l_m F}$$

which can be calculated. This value is only as accurate as the combined estimations of l_m, f, and F.

The value of B_g calculated above is the field strength directly between solid portions of the pole pieces, not the effective field in the light beam passing through the sample (Fig. 5.4). This effective field originates primarily at the edges of the axial hole through which the light beam traverses the pole pieces.

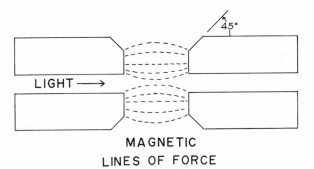

45°

LIGHT ⟶

MAGNETIC
LINES OF FORCE

Fig. 5.4. Cross section of magnet pole pieces showing magnetic lines of force in the sample area, light path through the pole pieces, and 45° pole taper.

The narrower the hole, the higher effective field achieved for a given B_g. But too narrow a hole blocks off so much light that the instrumental noise becomes intolerable. A good compromise for the Cary 60 Spectropolarimeter and Jasco ORD/UV dichrograph is a hole 1.5 × 15 mm.

Experiments performed by the author indicate that the taper of the tips of the pole pieces should be 45° ± 5° for best results.

6 SOURCES OF ERROR

In measuring MCD or MORD the same precautions must be taken as in measuring optical rotation. These include sample cell cleanliness, high sample purity, proper absorbance range, sufficiently narrow instrumental spectral band width, and control of solvent evaporation. In addition, Faraday effect measurements are influenced by the Cotton-Mouton and Voigt effects mentioned previously [5, 9]. Also, if the magnetic field being used is inhomogeneous, H_{eff} at one wavelength may not be the same as H_{eff} at another.

A number of additional sources of error in MORD measurements arise from the fact that everything in a magnetic field rotates plane polarized light. This includes the solvent and sample cell. It is not uncommon for the maximum rotation of the solute to be down at least an order of magnitude from that of the cell and solvent. Early values of MORD were obtained by subtracting the observed MORD of the cell and pure solvent from that of the cell and solution of interest. This procedure proved to be inaccurate for two reasons: (1) the value sought was the small difference between two large numbers, and (2) both MORD curves were steep functions of wavelength, so small errors in wavelength result in large errors in the MORD of the solute.

The obvious way to avoid this problem is to cancel (instrumentally) the rotation of the cell and solvent. One way to do this is to insert into the light path both the sample and another cell filled with pure solvent. If the magnetic field for each has the same magnitude, but opposite sign, the cell and solvent rotation will be canceled out of the light beam (except for the solvent molecules that have been displaced by the solute in the sample) [14]. Even with a pair of electromagnets wired in series electrically, it is difficult to maintain fields of the same magnitude because heating and slight dimension changes in the magnets cause the two magnetic fields to be different. It is even more difficult with permanent magnets because their field strength can be controlled only by placing pieces of soft iron across the gap.

Another approach is to place a sucrose solution of the proper concentration in the light path. The natural ORD of the sugar is very similar to the MORD of quartz cells and transparent solvents in the visible and near U.V. By choosing the right concentration of sugar, and the right direction of the magnetic field, the two rotations will be equal to within $\pm 1\%$ over a very wide wavelength range. Some instruments (e.g., the Cary 60) allow us to compensate electronically for the remaining mismatch by using the baseline multipots. This process makes it even more important to maintain color purity in the light beam and avoid excessively high sample absorbances because the multipot correction is far from the correct compensation for any stray light passing through the sample. If the proper precautions are not taken, a MORD band shaped like the absorption band will appear in the MORD spectrum wherever the absorbance is too high. The size of this band will decrease as the instrument slit width is decreased.

Whether the double magnet or sugar solution is used for compensation, the displaced solvent correction should be made (or else proved negligible). The true rotation of the solute, α, is given by:

$$\alpha = \theta_s - \left[\frac{(\rho_s - C)}{\rho_o}\right]\theta_o$$

where s refers to the sample, o to the pure solvent, θ is the rotation due to the liquid (excluding the rotation of the empty cell), ρ is density, and C is the concentration of solute in g/cc [14]. The displaced solvent correction does not apply to MCD measurements if the MCD of the solvent is small in the wavelength region where measurements are being taken, as is usually the case.

References

1. H. Eyring, H. Liu, and D. Caldwell, *Chem. Rev.*, **68**, 525 (1968).
2. A. Moscowitz, *Optical Rotatory Dispersion*, C. Djerassi, Ed., McGraw-Hill, New York, 1960.
3. D. A. Schooley, E. Burnenberg, and C. Djerassi, *Proc. Natl. Acad. Sci.*, **56**, 1377 (1966).

4. P. J. Stephens, W. Suetaak, and P. N. Schatz, *J. Chem. Phys.*, **44**, 4592 (1966).
5. F. A. Jenkins and H. E. White, *Fundamentals of Optics*, 3rd ed., McGraw-Hill, New York, 1957.
6. H. Cary, R. C. Hawes, P. B. Hooper, J. J. Duffield, and K. P. George, *Appl. Opt.*, **3**, 329 (1964).
7. F. Woldbye, *Record Chem. Progr.*, **24**, 197 (1963).
8. J. J. Duffield, A. Abu-Shumays, and S. Hedelman, Paper #232, 1969 Pittsburgh Conference on Analytical Chemistry and Applied Spectroscopy.
9. J. R. Partington, *An Advanced Treatise on Physical Chemistry*, Vols. 4 and 5, Longmans, London, 1954.
10. R. J. Parker and R. J. Studders, *Permanent Magnets and Their Application*, Wiley, New York, 1962.
11. *Westinghouse Application Guide for Superconducting Magnets*, Westinghouse Cryogenic Systems Dept., P.O. Box 8606, Pittsburgh, Pa.
12. C. E. Waring and R. L. Custer, *Technique of Organic Chemistry*, A. Weissberger, Ed., **I** (III), 3rd ed., Interscience, New York, 1960, pp. 2497–2552.
13. E. A. Dratz, Ph.D. Thesis, Univ. Calif., Berkeley, 1966.
14. V. E. Shashoua, *J. Am. Chem. Soc.*, **86**, 2109 (1964).
15. E. M. Underhill, Ed., *Permanent Magnet Handbook*, Crucible Steel Co. of America, Mellon Square, Pittsburgh, Pa., 1957.

Chapter **VI**

THE KERR EFFECT

Catherine G. Le Fèvre and R. J. W. Le Fèvre

I INTRODUCTION

Historical

Any account of this subject would be incomplete without reference to John Kerr (1824–1907), who, at the age of 51, announced his discovery of electric double refraction in the *Philosophical Magazine*. He had concluded his academic career at Glasgow University in 1849, where for three years he studied natural philosophy under William Thomson. He then taught for a time, was ordained to the Ministry of the Free Church of Scotland, and in 1857 was appointed as lecturer in mathematics and physical sciences in the Free Church Training College for Teachers in Glasgow. It was there in a cellar, and at his own expense, that he carried out the experiments which led to the observation [1] in 1875 of the effect that Faraday had sought but failed to find. Kerr noticed that when a block of glass was placed between crossed nicols, and the leads from the secondary of a Rumkorff machine attached to it in such a way that the applied electric field was perpendicular to the beam of plane polarized light passing through the glass, light was found to emerge through the analyzer. This illumination could not be extinguished by rotating the nicol. In other words, the glass had become doubly refracting. Kerr measured the ellipticity produced in the glass by means of a compensating device, which he introduced before the analyzer. He established the following relation, which now usually bears his name:

$$\delta = 2\pi Bl E^2 = \frac{2\pi l(n_p - n_s)}{\lambda} \qquad \text{(Kerr's law)} \qquad (6.1)$$

where δ is the magnitude of the double refraction in radians, B is the "Kerr constant," E the electric field strength in esu, l in cm the length of the light path through the field, n_p and n_s are the refractive indices for the components of the light parallel and perpendicular to the field, and λ the appropriate wave length. Kerr proved that this phenomenon was of electrical origin. He showed that his law held for isotropic nonviscous liquids, and that some of these were positively birefringent (e.g., carbon bisulphide, benzene, and toluene) and others (such as olive oil, colza, and seal oil) were negatively birefringent.

Following the pioneer work of Kerr, physicists in Britain, the United States, and Germany investigated the electric birefringence of many substances, and their results compose the approximately two hundred Kerr constants listed in Volume VII of the 1930 *International Critical Tables*. Different devices and techniques for measurement were adopted; these have been fully described in papers, references to which are given elsewhere [2]. The theory of the Kerr effect commonly accepted today began to evolve in

1910. It relates the intrinsic electro-optical polarizability properties of the molecule of a substance to its Kerr constant B. It is strictly applicable only to gases and vapors. From 1911 onward a few studies have been made of organic and inorganic substances in the gaseous state and, to date, there are about eighty such values for B on record.

The number of materials which can be vaporized without decomposition is, however, limited and accordingly, in order to be able to extend investigations beyond this restricted range, a technique has been developed recently to determine the Kerr effect of dissolved substances at infinite dilution. Good agreement is found for information obtained from the two sources.

The facts that the electro-optical polarizability properties of links are additive, and that they can be used successfully to compute that property of a molecule as a whole, have also been demonstrated.

Thus, through a study of the Kerr effect of gases or dilute solutions, not only can the electro-optical polarizabilities of both molecules and bonds be ascertained, but there also becomes available a new way of examining a wide variety of stereochemical problems [3].

An extension of this subject is the study of colloidal and other conducting systems, protein solutions, for example. Under certain experimental conditions, the Kerr effects in these systems can be determined, and deductions can be made concerning dimensions and shapes of dispersed particles.

In nonchemical fields, the Kerr effect has also been introduced successfully. The phenomenon played a part in the early development of television, and the employment of the Kerr cell as a light shutter is well known. However, such uses have no chemical applicability, and information concerning them is not included in this article.

2 THE MEASUREMENT OF ELECTRIC BIREFRINGENCE

Essential Requirements

A parallel beam of monochromatic light must pass successively through a polarizer N_1 (by which it is plane polarized), a cell K (in which the substance under examination is subjected to voltage gradients at 45° to the plane of polarization), a compensator C (with which ellipticity in the light from K may be estimated), an analyzer N_2 (crossed with reference to N_1), and a telescope T (focused on C). Appropriate components need to be mounted along a common axis on an optical bench; their details will depend on the measurement procedure that is adopted. If C is of the Szivessy-Dierkesmann type (see later) and dc voltages are used, electric birefringences can be determined visually [4]. Alternatively, if C is a "whole-plate" compensator, T may be replaced by a photomultiplier tube and observations made photometrically

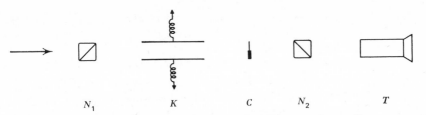

N_1 K C N_2 T

Fig. 6.1.   Essential components for visual measurements of Kerr effects.

[5–7]. Other arrangements, in which the voltages applied to K are alternating or pulsed, or in which a second cell—similar to K but containing a standard liquid—is used as a compensator, are also noted below.

Certain components are common requirements for all methods; directions for their construction will now be given.

Optical Benches

Optical benches, in various lengths, are conveniently made from H-section steel girders (height about 3 in., width about 5 in.), the top edges of which are accurately machined as inverted V's making straight lines along the girder length. Saddles (of various sizes according to the component to be carried) are cut from 1 in.-thick steel plate. Underneath each saddle are two parallel pads, one finished flat and the other grooved to marry the V-edge of the girder. On top of each a rectangular plate is held between parallel beveled guides and can be moved across the bench by screws acting through threaded blocks fixed at opposite sides of the saddle. During assembly, one of the guides must be removed temporarily. Each plate is attached to the base of a conical support (height *ca.* 4 in., diameter at top *ca.* 1 in., at bottom *ca.* 2 in.) bored vertically to take the 0.5 in.-diameter pillars on which optical and other components are mounted. These pillars are threaded to hold circular nuts which, by bearing onto the top of the conical support, enable the heights of the units above to be adjusted. Rotation of a pillar is prevented by a locking screw inserted horizontally through the conical support.

The Polarizing Unit

A brass tube about 6 cm long, with internal and external diameters of 2.5 and 3.5 cm, respectively, is mounted at 90° on a pillar. An inner tube, of corresponding dimensions 1.8 and 2.5 cm, painted matte-black within, carries a square-ended (Glan-Thompson) polarizing prism N_1 (cross section 10×10 mm), preceding and following which are knife-edged circular diaphragms with apertures of approximately 6 mm. The axes of the prism and its containing cylinders must be parallel. A new prism is first carefully measured in all directions, and a metal replica fabricated; in the end of the replica a short rod is inserted axially. A cylindrical length of 1.88-mm diameter balsa wood is then split and the two halves shaped to take the replica and hold it firmly in the inner tube. Axial location can be checked by observing the projecting rod when the assembly is spun in a lathe. The replica is then replaced by the prism, the inner tube inserted into the outer and turned 45° away from the setting giving minimum transmission of light reflected obliquely from a horizontal glass plate. Final adjustment is made by two small screws, mounted on the outer tube, between which is a flange attached to the inner tube and passing through a slot cut in the outer tube.

The Compensating Unit

Essentially, a birefringent plate has to be rotatable in a vertical plane normal to the light beam entering N_2; settings should be readable to 0.01°. A circular scale graduated in degrees and equipped with a main release, vernier, slow-motion device, and so on, is therefore necessary. A more easily constructed alternative is to mount a 360-toothed disk against a 360 : 1 worm drive, the head of which is itself divided into 100 divisions, and the spindle of which is geared 1 : 1 with a revolution counter. A clutch, whereby the worm may be turned in either direction by an electric motor, is also a convenience. The central bearing is made from two tubes. The inner one passes through the center of a strong square brass plate (*ca.* 18 × 18 × 0.7 cm) which carries the worm-drive, counter, clutch, and motor. The outer tube, onto which the toothed disk is fixed, rotates about the inner tube. The disk has a 1.8-cm central hole on each side of which beveled guides are screwed to allow a compensator slide to be slipped into the light beam before it reaches N_2. This unit, being heavy, is best attached directly to its saddle by diverging legs that buttress the square plate; as it is incapable of movement, the other lighter components on the bench must all be positioned in relation to it.

Compensators

The greater part of the work of the Sydney group has involved half-field or whole-field compensators of mica. Both have proved satisfactory for measurements on dilute solutions.

The electrically strained Kerr cell K can be considered as a uniaxial crystal of small double refraction, between limits $\pm(10^{-3}-10^{-5}).2\pi$, with its optic axis lying along the lines of the electric field. Accordingly, monochromatic light, plane polarized at 45° to the field direction, will on leaving the cell behave as two linear waves of the same frequency, polarized in planes which are mutually perpendicular, and traveling along differently polarizable paths.

These two waves recombine on emergence to form an elliptically polarized resultant wave.

Suitable slips for the construction of compensators may be obtained by inserting two sharp needles into opposite ends of a piece of clear mica, starting two cleavages, and tearing apart gently and simultaneously. A fragment with a straight edge is then sealed with Canada balsam halfway across, and between, two microscope cover glasses or polarimeter end-plates (which should be free from strain), taking care not to enclose air. When dry, the sandwich is fixed by means of wax into a beveled brass mount that slides and can be clipped into the guides on the circular scale or disk of IID. The mica then lies halfway across the field of vision afforded by a telescope looking through the prism N_2. Such a compensator is of the Szivessy-Dierkesmann type [8]. For a whole-field compensator, or a *Hilfesplatte* (see Brace compensators), a larger piece of mica is similarly enclosed.

Use of the Szivessy-Dierkesmann Compensator

This compensator is well suited for *visual* measurements on dilute solutions. The mica half-plate contains an optic axis; accordingly, during a complete rotation in the vertical plane, four positions of extinction are found, which lie at 90° to each other and correspond to the optic axis lying parallel to the planes of polarization of the polarizer and analyzer. These positions are called the extinction azimuths. If mica of suitable double refraction is selected, it can be rotated into a position such that, when a voltage is applied across K and causes refractive indices n_p and n_s in the two vectorial components of the light emergent from it, the passage of this light through the mica can result in an inversion of the magnitudes of n_p and n_s, which now become n_s and n_p, respectively. If we assume there is no absorption of light by the mica, then the intensity of illumination in that half of the field of vision not covered by the mica is the same as that which has traversed the mica. Accordingly, we have a match point; there are four such positions. When, therefore, an electric field is acting across the cell, there are eight positions of match throughout a 360° rotation of the half-field plate. The four extinction azimuths observed without the electric field become positions of half-shade when illumination of the field of vision is introduced by the birefringence of the cell. It is found that for a positively doubly refracting dielectric, the azimuthal positions lie as in Fig. 6.2a, and for one that is negatively doubly refracting, they lie as in Fig. 6.2b. The appropriate quadrants for any compensator are determined by observing where the half-shade positions lie for a substance having a B of known sign. The phase difference Δ of the mica can lie between 0 and $\pi/2$. (This value depends on the wavelength of the light used.) The compensator is most accurate when the phase difference δ of the cell is small, as with dilute solutions. The angle β is noted four times during one

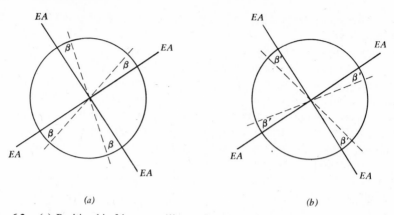

Fig. 6.2. (*a*) Positive birefringence; (*b*) negative birefringence. EA = extinction azimuth, broken lines show half-shade settings.

complete rotation of the mica, the average is found, and then δ is obtained from the expression

$$\tan \delta = (\tan \Delta/2)\sin 2\beta \qquad (6.2)$$

The theory outlined in the original paper [8] shows that Δ must exceed 2δ.

For work with gaseous dielectrics this compensator is of little use because the illumination in the cell usually is insufficient to enable match points to be detected. For solutions of highly polar materials, for example, $PhNO_2$, unless they are quite dilute, the intensity of illumination is too strong to permit accurate matching of half-shade positions. The angle β can be read with good precision up to about $20°$ with a half-shade plate of $\Delta = 4 \times 10^{-2} .2\pi$.

The Brace Compensator

This compensator, as interpreted by Szivessy [9], is the device that was used by Stuart and Volkmann [10] in their measurements of B for vapors and gases. It differs from the Szivessy-Dierkesmann compensator, by incorporating, in addition to the half-shade plate, a *hilfesplatte*—a thin mica plate that covers the whole field of view and that is mounted in a graduated rotatable circular scale fitted with vernier and slow-motion drive.

There are several ways in which the combination of these two plates may be used (see [11]). The great advantage of this compensator lies in its variable sensitivity—for example, in investigations on nonpolar gases, the introduction into the light path of a *hilfesplatte* with its principal axis oriented in a known azimuthal angle, results in illumination of the field. Thus it becomes possible to measure $\pm\delta_H$, the double refraction of this plate, and $\delta_H \pm \delta_c$ (where δ_c is the double refraction of the Kerr cell when the field is applied) by means of a half-shade plate of double refraction Δ. Accordingly, $\pm\delta_c$ may be determined. The *hilfesplatte* acts as a booster in this example. It can also be employed to reduce intensity of illumination with polar materials in solution, the plate being rotated into a fixed azimuth in the "fast" or "slow" direction depending on which reduces the illumination from the birefringence of the Kerr cell. For use in the manner just described, a *hilfesplatte* $\delta_H = 1 \times 10^{-2} . 2\pi$ should be used in a fixed azimuth in conjunction with a rotatable half-shade plate $\Delta = 4 \times 10^{-2} . 2\pi$. Eight half-shade positions are read as with the Szivessy-Dierkesmann compensator; from β, which is calculated and Δ (known), δ_c is determined. For gases, the azimuthal angle of the *hilfesplatte* should be small, 5–10°; the sensitivity is thus increased (see Jerrard [11]). The merit of this compensator—specifically adjustable sensitivity—also involves a disadvantage: its accuracy depends on the careful mounting of *two* plates, whereas the Szivessy-Dierkesmann arrangement requires the mounting of only one plate. The authors recommend the use of the Brace in those fields of measurement in which the half-shade plate alone cannot be used, as with gases; with solutions, the Szivessy-Dierkesmann is the more satisfactory one. The intensity of illumination that results from strongly polar solutes can, of course, also be diminished by either cutting down the light from the monochromator or by operating in the very dilute solution range. Such steps do not entail inaccuracy.

Precision in matching obviously depends on the subjective judgment and photometric discernment of the observer (these vary from person to person and, with one individual, from freshness to fatigue), on the intensity, and on the monochromatism of the illuminated field. On the other hand, the method has the merit that any changes in the light intensity due to scattering in the cell occur generally over the whole area of vision, and thus do not affect the correctness of observed match points.

The utilization of a whole-field compensator in conjunction with photometric detection is described later in this section.

The Babinet-Soleil Compensator

For determination of phase differences greater than $1 \times 10^{-3} \cdot 2\pi$ the Babinet-Soleil quartz wedge commends itself. This device is well known and is adequately described in most textbooks of physics. It is suitable for studies of polar liquids or of solutions of highly polar molecules (e.g., $PhNO_2$ in carbon tetrachloride) when other than low concentrations are to be examined.

The Prism N_2

This is identical with N_1 and is mounted similarly, except that N_2 is crossed with respect to N_1.

Light Sources and Lenses

For routine purposes light from a 100-W quartz-iodine-tungsten lamp, fed from the mains via a transformer with switches to give 6, 8, 10, or the maximum 12 V, is condensed onto a pinhole (diameter *ca.* 0.8 mm) in a blackened aluminum screen. A short-focus lens, using the pinhole as source, transmits a near-parallel beam that strikes an interference filter at 90°, and then enters N_1, K, and so forth. Alternatively a simple monochromator may be installed, in which case its exit slit takes the place of the pinhole. For photometric measurements a voltage stabilizer should be inserted before the transformer. The lamp, the two lenses, the screen, and the filter-holder are separately supported and stand on individual saddles. Initial alignment is helped by placing an aluminum mirror accurately mounted in the plane normal to the optical bench length in the position of the telescope; then, with *either* N_1 *or* N_2 and the filter removed, light should be reflected back onto the pinhole in the screen.

Cells

Cells for Liquids

For the bodies of these cells we recommend thick-walled pyrex glass tubing of internal diameter of about 2.5 cm and length about 30 cm. Details of two useful assemblies are shown in Figs. 6.3 and 6.4. The brass (or stainless steel) electrodes, which constitute the condenser K (of Fig. 6.1), are indicated by P. The specifications that follow are suitable for liquid substances or solutions which have small values of B. For highly polar materials the electrodes are shortened and/or the gap between them widened to give lower voltage gradients. Thus the nature of the dielectric under observation determines the dimensions of the cell to be used.

The ends of the glass tube are ground optically plane, so that they form liquid-tight contacts with the end plates E. A cemented ferrule, cap, and washer holds each E in position; liquid inlet and exit tubes are sealed in as shown.

The emf is introduced by means of mercury in the side arms A_1 and A_2 through platinum-glass seals to the tungsten strips T, the terminations of which are respectively soldered to the platinum and attached to the ends of the electrodes P by means of small brass screws. These electrodes (25 cm long) are made by splitting in two a brass rod, which has an appropriate diameter along its axis, and then running a small "flat" lengthwise at the top and the bottom to assist washing and draining of the cell. This shape was chosen to minimize the volume of liquid required to fill the cell and to simplify the supporting of the lower plate. The opposing faces are milled flat and coated with platinum black to diminish surface reflection from them. Five pairs of

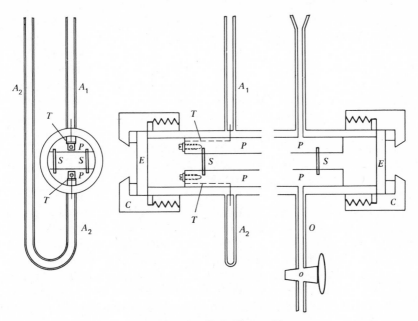

Fig. 6.3. Detail of cell assembly.

hand-ground spacers S, made from thin pyrex glass rod, stand in small holes drilled along the edges of the electrodes, thus giving rigidity and a uniform gap (e.g., 0.320 cm) between the lengths. The end plates E are polarimeter end plates. It is necessary to select only those *truly* free from strain; this may be done by mounting them in a rotatable vertical plane in conjunction with a Brace compensator and testing them; strains causing birefringence of more than approximately $\lambda/20{,}000$ should be thus detectable.

Both the gap between and the lengths of the plates need to be measured as accurately as possible. Since the path through the field exceeds the physical lengths of the plates, the "effective lengths" of the plates must be calculated. Formulas for this purpose were considered by Chaumont [12]. The most suitable appeared to be

$$l_{\text{eff}} = 1 + \left(\frac{d}{\pi}\right)\left\{1 + \left(\frac{e}{d}\right)\log\left[1 + \left(\frac{d}{e}\right)\right]\right\} \tag{6.3}$$

in which e is the thickness of the plates, d is the separation between them, and l is the measured length (for example, with $l = 25$ cm, $e = 0.5$ cm, and $d = 0.32$ cm, the correction amounts to 0.136 cm).

Although theoretically, with the above arrangement, there should be no leakage from the ends of the cell, in practice, there is often a small amount due mainly to the fact that polarimeter end plates are seldom optically planar. Organic solvents thus reach the leather washers and cause the washers to swell inside the caps, C. Pressures are thereby exerted on the optical flats at the ends of the cell, so that the flats become in effect Rayleigh compensators and render uncertain any apparent electric birefringence results obtained. This "background effect" has to be avoided, and it is advisable to seek for birefringence in the cell, without an applied field, before and after each experiment. Of many attempts to overcome these difficulties, the most promising has been the abandonment of ferrules and screw caps and their replacement by turned polythene caps with approximately 1-cm skirts making friction fits over the ends of the cell body. The optical flats can thus be held lightly in position; nonleakage, however, cannot be always achieved.

A more successful design is indicated in Fig. 6.4. It is fabricated by joining two standard Pyrex B 24 sockets coaxially back-to-back in a glass lathe and attaching side tubes as shown for filling and emptying and for carrying electrical leads. The cell is closed by Pyrex B 24 cones, into the smaller tapered ends of which optically flat Pyrex disks are fixed by fusion at 90° to the cone axis. After annealing, the edges of the disk and the first 0.4 cm of the cone are carefully ground so that these parts do not make tight contact when the cones are in their sockets. Cones and sockets are held together by short springs between horns on the respective portions. The filling tubes terminate with B 14 sockets to take corresponding sized stoppers, or thermometers whose bulbs reach to approximately 1 cm above the electrodes, or a sintered glass filter-containing unit through which dust-free liquids are introduced into the cell in amounts sufficient to cover the thermometer bulbs. Rectangular plane-parallel metal plates, separated by pairs of hand-ground pyrex spacers, are maintained in a fixed vertical position by bow-shaped clock springs, two on

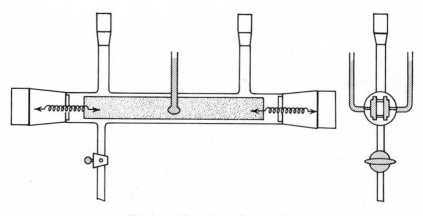

Fig. 6.4. Alternative cell assembly.

each plate, which press against the inside of the glass cell body, and against the inner ends of stout tungsten leads sealed through the cell wall into mercury-holding side arms. (As an example, a cell that we are now using has electrodes 20.25 × 2.0 cm between which a section of dielectric 3.4 mm thick is enclosed vertically.)

The cells are supported on the optical bench by a holder that provides slow-motion screw adjustment in all directions (height and tilt—in both horizontal and vertical planes); a standard saddle allows the cell plus holder assembly to be moved sideways across the bench. For work near room temperatures, experience has shown that it is simplest to enclose all items except the light source and detector in an air thermostat. The construction of water- or oil-jacketed cells is possible but difficult. For higher temperatures we recommend that the cell and its holder be placed in an asbestos-board oven in the ends of which are small holes for the light beam and through the top of which pass the H.T. leads and thermometers. A door gives access to the cell for leveling, filling with liquids, and emptying. Electrical heating tapes are fixed to the inner walls of the oven, lower temperatures being secured by manual control through a "variac," higher ones through a bimetallic thermostat near the floor of the oven.

Cells for Gases

As Kerr effects with gases are generally small, the electrodes can well be several meters long. Most gases must be examined at pressures higher than atmospheric and the cell bodies therefore need to be of thick glass or metal. Supply tubes must be provided for entrance and exit of the dielectric under investigation, and auxiliary equipment is needed for vaporizing liquid materials and for observing temperatures and pressures within the cell. Measurements on gases are practically more difficult to make than those on liquids. Cell closure is a notable problem since spurious birefringences can be caused by heating or pressure strains in the transparent end windows.

Details of one end of a cell used in Sydney [13] are shown in Fig. 6.5. The two plane-parallel brass electrodes (101.0 × 2.2 cm) are held and insulated from each other by carefully machined Teflon rings Fig. (6.5b) spaced at 10-cm intervals. The interelectrode gap is 0.260 cm. The absence of insulators between the plates allows high field strengths to be used without breakdown of the gaseous dielectrics. The cell body is a heavy Pyrex tube (internal diameter 3.8 cm) with tapered ends. Optical windows (diameter 2 cm, thickness 1 cm) of strain-free glass are mounted on brass end plates, B, which are bolted to steel collars, S, behind the tapers (Fig. 6.5a). The window is surrounded by a neatly fitting hard rubber washer, R, which is attached to a threaded brass ring; this unit is screwed into a second brass ring soldered in the center of the end plate. Strain birefringences arising from compression of

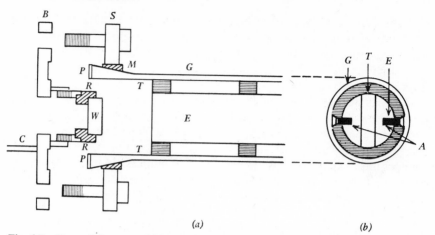

Fig. 6.5. Kerr cell for gases: (*a*) longitudinal section, (*b*) transverse section. E = electrodes, T = Teflon rings, A = countersunk screws, G = glass cell body, W = optically flat end window, B = brass end plate, S = steel collar, M = rubber mounting, R = rubber washer, P = rubber pressure seal, C = copper pressure tubing.

the brass are thus eliminated. The three junctions (glass-rubber, rubber-brass, brass-brass) are sealed with Araldite. A compressible rubber ring, P, acts as a pressure seal between the end plate assembly and the ground ends of the cell body. Evacuation and filling is by copper pressure tubing, C, which is sealed through B. Diaphragm-type line valves serve to isolate the cell. Gas pressures up to 3 atm are measured with an open-column mercury manometer. Vapor manipulation follows in principle that adopted for dielectric constant work [14]. Voltages are applied to the electrodes by thick tungsten leads sealed through the Pyrex jacket, G; an electrostatic voltmeter is connected in parallel to these leads.

Stuart and Volkmann [10] give details of the apparatus they used with success up to temperatures of 230° under pressures not greatly above atmospheric; they also discuss error sources quite fully. Designs of cells for use up to several hundred atmospheres are described in papers by Stevenson and Beams [15], Bruce [16], Bizette [17], Borchert [18], Kuss and Stuart [19], Boyle et al. [20], and Buckingham and Dunmur [21].

Voltage Generation

Voltages variable at choice up to approximately 15 kV are required and are conveniently produced by one of the several circuits now available in which an oscillator supplies power to a radio-frequency transformer having tuned primary and secondary windings, the high voltage ac in the latter being then rectified.

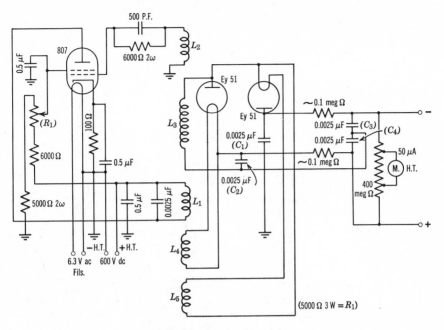

Fig. 6.6. Circuit of voltage generator.

An easily constructed circuit is sketched in Fig. 6.6. Oscillations of approximately 20 kc/sec are generated in L_1 which with L_3 makes up the rf transformer. Feedback is obtained through L_3 to L_2 to maintain oscillations. The high voltage in L_3 is rectified by two EY 51's in a half-wave voltage-doubling circuit, their outputs being connected across the condensers C_1 and C_2, which are charged in alternate halves of the cycle. As they are in series (so that voltages add) this gives, at light loads, an emf approaching twice the crest voltage. Condensers C_3 and C_4 act as a filter. The unit is capable of producing 6–14 kV according to the setting of a potential divider, R_1. The windings L_4 and L_5 supply filament current to the small diodes. Coils L_1 to L_5 are wound on a common former (see Fig. 6.6a). The high voltage dc output is either measured directly on an electrostatic voltmeter or is bled through resistors immersed in wax through a current meter calibrated in kilovolts.

The high voltage and earth connections are made to the cell (Fig. 6.3) by car ignition cables terminating as thin brass rods; each of the latter is completely shrouded by a polythene cylindrical tube, which fits loosely over the mercury-containing side arms A_1 and A_2 and permits handling without shock. It is advisable to use as high a voltage as possible in order to maximize accuracy in reading the volts (since scales on electrostatic voltmeters are more expanded in their upper ranges than in their lower).

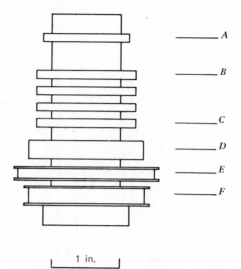

1 in.

Fig. 6.6a. Coil A: 150 turns of 5/41 Litz wire (pi wound). Coils B and C: four coils, each 250 turns of 5/41 Litz wire (pi wound), connected in series. Coil D: 60 turns of 27/41 Litz wire (pi wound). Coil E: 5 turns of 26 B.S. enamel wire. Coil F: 8 turns 26 B.S. enamel wire. Main former: Paxolin tube, outside diameter $1\frac{1}{16}$ inch.

Apparatus involving Photoelectric Photometry

Historical

Beams and his co-workers [15, 16, 22, 23] used vacuum photocells and dc amplifiers to measure Kerr effects in compressed gases. Ellipticities were determined by a comparison of light intensities rather than by compensation. An ingenious feature was the incorporation of both the polarizing and analyzing prisms in the cell, the former being oriented and fixed as usual, the latter being rotatable. Such an arrangement prevented pressure strains on the end plates from affecting the electric birefringences observed, since the plates were outside the prism system. The substances examined were mainly non-polar and, therefore, of exceedingly small Kerr constant. In order to increase the number of molecules per cubic centimeter, pressures up to several hundred atmospheres were used; this had the additional advantage of permitting work at higher electric field strengths (sparking potential increases almost linearly with density).

The theory of the method utilizes the fact that intensity of the light passing N_2 is

$$I = I_o\left[\cos^2(\alpha - \beta) - \sin 2\alpha \cdot \sin 2B \cdot \sin^2\left(\frac{\delta}{2}\right)\right] \tag{6.4}$$

where α and β are the angles made by the planes of vibration of the transmitted light of N_1 and N_2 with the lines of force of the field. Light intensities (of widely differing orders of magnitude) need to be measured with N_1 and N_2, respectively crossed and parallel; I_o may then be eliminated and the required phase difference δ obtained.

Between 1931 and 1936 data were thus obtained for the eight gases: H_2, N_2, O_2, CO_2, CH_4, C_2H_4, C_2H_6, and NH_3 [15, 16, 22, 23]. However, as the parts of polarizing and analyzing prisms are commonly joined together by Canada balsam (or some other soluble and low-melting cement of refractive index $ca.$ 1.55), the method is unsuitable for organic liquids and for gases at elevated temperatures. The advent of photomultipliers of high sensitivity has opened other possibilities, such as the usage of a "simple elliptic compensator" [24, 25], the revival of des Coudres' "relative" techniques [26], and the application of sinusoidal or pulsed voltages to the dielectrics in K (Fig. 6.1). We shall now describe apparatus we are currently using.

Use of a "Simple Elliptic Compensator"

The arrangements described above for visual determinations of δ are modified by replacing the telescope T (of Fig. 6.1) by an RCA 931-A photomultiplier–amplifier combination, using a whole-plate compensator at C and inserting a beam chopper before the polarizer N_1. Figure 6.7 shows the layout now adopted [5–7].

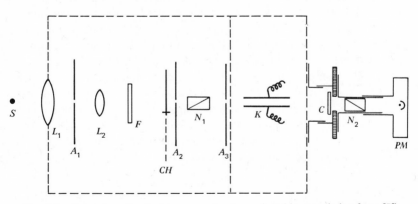

Fig. 6.7. Present routine Sydney apparatus (reproduced by permission from [7]).

The light source S is as noted before but with a voltage stabilizer between the mains and the transformer. CH is a chopper, rotated by a constant-speed motor, to interrupt the light beam 575 times per second. A_1, A_2, and A_3 are circular diaphragms that have openings approximately 0.8, 5, and 5 mm, respectively. PM is the photomultiplier, the output from which is read on a microammeter through a selective amplifier tuned to 575 cps. The azimuths of C corresponding to least illuminations of PM are determined as the means of pairs of angles giving the same currents each side of the minimum. The items between L_1 and C are enclosed within a two-compartment cupboard (see dotted lines in Fig. 6.7) to prevent stray light reaching PM; heating tapes on the back wall maintain the air temperature at 25°. Cells K are usually of the type shown in Fig. 6.4.

In turning the compensator C through 360°, four settings corresponding to minima in the amplifier-output meter are found. A general expression has been written by Jerrard [11] for the light intensity in a beam that has passed successively through a polarizer, two doubly refracting plates, and an analyzer crossed with the polarizer. In our apparatus the Kerr cell is equivalent to one of the plates having a principal vibration direction at 45° to the vibration direction of the polarizer. Jerrard's equation therefore becomes

$$I = \sin^2\left(\frac{\delta}{2}\right) + \sin^2 2\beta \sin^2\left(\frac{\Delta}{2}\right) + 2\sin 2\beta \sin\left(\frac{\Delta}{2}\right)$$

$$\times \sin\left(\frac{\delta}{2}\right)\left[\cos\left(\frac{\delta}{2}\right)\cos\left(\frac{\Delta}{2}\right) - \sin\left(\frac{\delta}{2}\right)\sin\left(\frac{\Delta}{2}\right)\sin 2\beta\right] \quad (6.5)$$

where δ is the retardation produced by the electric birefringence, β is the azimuth of the compensator plate which itself has a retardation of Δ. At the extinction azimuths, I is a minimum, that is, the $d\,I/d\,\beta = 0$; accordingly we have, by differentiation of (6.5), $\tan\delta = -2\sin 2\beta \tan(\Delta/2)$. (The negative sign merely denotes that K and C must act in opposition to one another for compensation to occur.) If before and during application of voltage to the Kerr cell the compensator settings for intensity minima are, respectively, angles a, b, c, d, and a', b', c', d', then β is taken as the mean value of $a - a'$, $b - b'$, and so forth. The compensator factor, $\tan(\Delta/2)$, can be obtained by examining a standard liquid under a definite voltage in a cell of known dimensions, or by direct measurement of C, using either a quarter-wave plate in the Sénarmont arrangement [11] or a quartz-wedge compensator of the Babinet-Soleil type.

Des Coudres Method

As originally introduced [26] this technique involved the compensation of the Kerr effect induced in one cell by that in a second cell placed immediately after the first in the optical alignment. To measure positive or negative birefringence the electrodes in the compensating cell had to be perpendicular or parallel, respectively, to those in the first cell. Then by altering the voltage gradient in the second cell, by varying either the voltage applied to it or the distance between its plates, the birefringence in the first cell could be annulled.

The method was extensively used by Schmidt, Leiser, Lippmann, Hansen, and others up to about 1920, when Szivessy [27] proposed a more convenient compensating device. The majority of the data tabulated in the 1930 edition of the *International Critical Tables* (Vol. VII, 110–112) were obtained with apparatus of the des Coudres type (see Fig. 6.8); Hansen [28] measured the electric birefringence of several gases by its means. The applied voltages could be direct or alternating. Observations were usually expressed relative to the effects induced with a standard liquid (commonly carbon disulphide) contained in the second cell. By the use of high-frequency ac fields, conducting liquids could be studied—for example, Schmidt [29] in this way estimated the Kerr constant of water as 3.7×10^{-7} for red light.

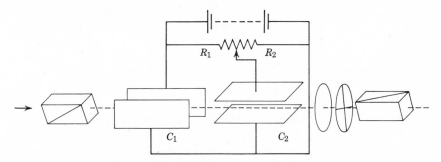

Fig. 6.8. One variant of the des Coudres arrangement, whereby relative voltages may be measured as a ratio of resistances, $V_1/V_2 = (R_1 + R_2)/R_2$.

The des Coudres technique lends itself to photometric detection devices. Cherry et al. [30] have utilized essentially the arrangement of Fig. 6.8 but with the telescope and compensator replaced by an 1P21 photomultiplier; the voltages to the two cells were supplied by separate 115–10,000 V transformers of known step-up ratios under varying load conditions. By noting the voltages needed across the primary circuits to bring the photocurrent to a minimum, those from the secondary circuits could be calculated. The Sydney group have used dc high voltages, independently generated and metered and led to the two cells, in conjunction with a chopped light beam and ac amplification. For routine work, in which liquids of fairly similar birefringence have to be compared, the limited number of significant figures to which voltages can be *read* on electrostatic voltmeters (compared with the precisions by which the same voltage may be *reset* time after time) have caused us to prefer the methods such as the Szivessy-Dierkesmann compensator, the simple elliptical compensator, or the whole-plate compensator. However, the des Coudres principle has lately been revived and modernized by Buckingham and collaborators [20, 21, 31]. Sinusoidal voltages of 3–6 kV rms and about 500 cps frequency are applied to the Kerr cell. The birefringence is nulled by a signal of double frequency acting on a dc-biased compensator cell (containing carbon disulphide). The point of compensation is determined by the output of a phase-sensitive detector whose reference signal is at the doubled frequency. A cw helium-neon laser is the light source, $\lambda = 6328$ Å. By these means electric birefringences have been measured for the inert gases, sulphur hexafluoride, and hydrogen.

Alternating Voltages with a Whole-Plate Compensator

The apparatus involved is as in Fig. 6.7 but without the light chopper. The cell K is connected to a commercial neon-sign transformer producing 50 cps voltages between 0 and 18 kV. Control at the transformer input is by a Variac and monitoring by an ac voltmeter in parallel with the cell terminals. The compensator C is a thin whole plate of mica mounted in a circular scale which with a worm-geared drive permits readings of $0.01°$ to be obtained. The detector is an RCA 931A photomultiplier tube operated at 800 V; its output can either be fed to an oscilloscope or, via a 100-cps tuned amplifier, into a vacuum tube voltmeter.

Operation is as follows: with C set in one of its four extinction azimuths, the 50 cps voltage is applied to K. Since the Kerr effect is proportional to the square of the voltage, the solution in K will become birefringent during each half-cycle, and a 100-cps signal will be seen on the CRO screen. If $I_t = I_o \sin^2(\delta/2)$, where δ is the phase difference in radians between the ordinary and extraordinary rays induced by E_t absolute volts per cm in a material having a Kerr constant B between electrodes of length l; as E_t in the present arrangement is time dependent ($E_t = E_{max} \sin wt$), when δ is not large, I_t is approximately given by $I_o \pi^2 B^2 l^2 E_{max}^4 \sin^4 wt$. If C is now rotated to introduce some birefringence of opposite sign to that occurring in the Kerr cell, the intensity transmitted through the combination, varying with the square of the total birefringence, leads to CRO displays such as those sketched (not to scale) in Fig. 6.9d–e. The initial pattern B is seen when C is in its original extinction azimuth; as C is turned through 360° the trace four times presents the appearance of D. The transition from curve C to curve E through D is sharp and the angle θ by which the compensator has been turned out of its azimuth is easily readable. In practice (because of unsteadiness of the CRO patterns) precision is improved by passing the photomultiplier output through the 100-cps amplifier mentioned above and recording the angles at which voltage (or current) of this frequency becomes minimal; because readings appear to be symmetrical about these minima the position of the last-named may be fixed with accuracy as the averages of pairs of angles giving the same indication before and after traversal of the point sought. A mean value of θ is then struck from the four θ-values obtained during one revolution of C.

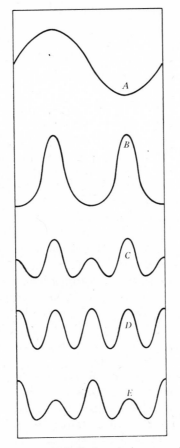

Fig. 6.9.   Patterns B to E: CRO displays during one voltage cycle as in A.

When the objectives sought are the differences between Kerr constants of a solvent and solutions in it, that is, $\Delta B = B_{12} - B_1$, the apparatus is calibrated empirically by finding the θ-values required by a number of mixtures for which the dependence of ΔB upon w_2 (solute concentration expressed as weight fraction) has been established by the half-shade matching technique. The system chlorobenzene/benzene provides easily accessible standards; with $B_1 = 0.410 \times 10^{-7}$ (for Na light and 25°), $10^7 \Delta B = 6.68w_2$ for concentrations of chlorobenzene between 0 and 1 %. Thus for a given cell and compensator an apparatus factor Q is available, for each voltage applied to K, as

$$10^7 Q = \frac{6.68w_2}{(\sin 2\theta_{\text{soln}} - \sin 2\theta_{\text{PhH}})}$$

and incremental Kerr constants for other solutions follow as

$$10^7 \, \Delta B = Q(\sin 2\theta_{\text{soln}} - \sin 2\theta_{\text{solv}}),$$

with the subscripts denoting the dielectric in K to which the angles θ relate.

Pulsed Voltages

The application of voltage to the dielectric in square-wave pulses of short duration is advantageous in cases where such things as conductive media, solutions of macromolecules, or colloids have to be studied. The principles outlined by Kaye and Devaney [33] were brought to practical usefulness by Benoit [34] and O'Konski [35], whose methods, with minor modifications and/or instrumental improvements, have already been utilized in a surprisingly wide-ranging variety of investigations [35–46].

Apparatus [47] constructed in Sydney is indicated in Fig. 6.10. Its development has been greatly helped by the details published by the workers just cited. The components from S to PM are mounted on stands that allow adjustments in three orthogonal planes. The optical bench consists of a length of I-section girder lying within a three-compartment light-tight cupboard. Lettering of items follows that used with Fig. 6.7. The chopper disk and motor CH is operated only during the aligning of the system, a steady light beam being used for actual measurements. A selection of long- and short-path cells for K is advantageous; the smaller cells are conveniently made from tintometer cells in which electrodes are rigidly supported by rods fixed into the glass or micalex cell lids; the larger cells have forms as in Fig. 6.4. The light source S is a quartz-iodine-tungsten lamp fed by a stabilized power supply capable of delivering a maximum of 12 V at 8 amp with a regulation as good as ± 0.01 V; details of this and of the train of units required to generate square-wave high-voltage pulses of predetermined magnitudes and durations are given by Cusiter et al. [47]. The photomultiplier PM (RCA 1P21) has its cathode follower (a 12AT7 tube) housed in a shielding can attached close to the IP21 base. Pulses are triggered by the camera shutter control. The voltages applied to K and the birefringences thus caused are simultaneously displayed, and/or photographed, on the double-beam oscilloscope. Specimens of traces so obtained are shown in Fig. 6.11.

As noted earlier, a phase difference of δ radians induced in the Kerr cell will allow light of intensity I to be transmitted. If I_o is the incident intensity and δ is small, I/I_o is close to $\delta^2/4$, which by (6.1) becomes $\pi^2 B^2 l^2 E^4$. For a given liquid (i.e., for a constant B) a plot of I versus E^4 should be a straight line. During the steady state of an applied pulse, the intensity falling on PM is ksh, where k is a constant unique to the apparatus, s refers to the scale on the calibrated oscilloscope amplifier, and h is the height of the birefringence trace measured in millimeters from the photographic record. The correspond-

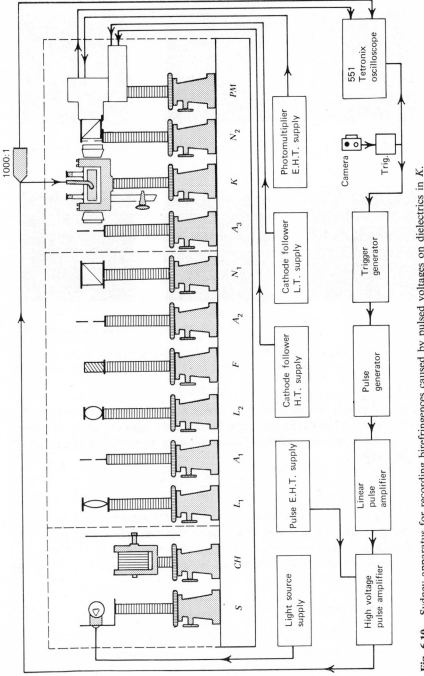

1000:1

S   CH L_1 A_1 L_2 F A_2 N_1 A_3 K N_2 PM

Light source supply

Pulse E.H.T. supply

Cathode follower H.T. supply

Cathode follower L.T. supply

Photomultiplier E.H.T. supply

High voltage pulse amplifier

Linear pulse amplifier

Pulse generator

Trigger generator

Camera

Trig.

551 Tetronix oscilloscope

Fig. 6.10. Sydney apparatus for recording birefringences caused by pulsed voltages on dielectrics in K.

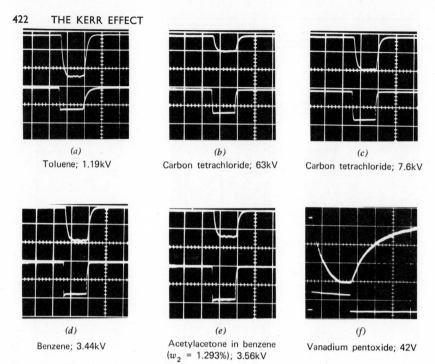

| (a) | (b) | (c) |
| Toluene; 1.19kV | Carbon tetrachloride; 63kV | Carbon tetrachloride; 7.6kV |

| (d) | (e) | (f) |
| Benzene; 3.44kV | Acetylacetone in benzene
 (w_2 = 1.293%); 3.56kV | Vanadium pentoxide; 42V |

Fig. 6.11. Birefringences and voltages versus time. (In each case the upper and lower traces record the birefringence and applied voltage, respectively. In (a) to (e) the pulses are of 70-μ sec duration, in (f) of 10 msec).

ing E is similarly found from the voltage trace. If, successively with two liquids 1 and 2, s is kept the same, and h's are read for progressively increasing E's, then the slopes m_1 and m_2 of the resulting graphs of heights against E^4 will be related as

$$\frac{m_1}{m_2} = \frac{B_1^2}{B_2^2}$$

By making one run on a substance of known Kerr constant, the B of another can be thus deduced. In cases where the rising and falling portions of the birefringence versus time traces are themselves curved, further information on such molecular properties as dimensions, polarities, and relaxation times [34] may also be extracted.

Evaluation of Kerr Constants

Absolute Values of B

The following data must be known accurately: the effective length l' cms of, and the separations d cm between, the rectangular electrodes within a Kerr cell, and the retardation δ radians observable when a voltage difference of V kV is imposed across the electrodes. Then in terms of (6.1), E is 1000 $V/300$ d esu/cm, and the length of the light path through the field is calculable from the physical dimensions of the electrodes by (6.3). With a Szivessy-Dierkesmann half-shade mica compensator, the Δ of which has previously been measured for a range of wavelengths, the mean angle β in (6.2) is determined. With small birefringences δ is commonly reported in degrees; if this quantity is $x°$, $\delta = (x/360)2\pi$ radians and, accordingly, from Kerr's law

$$B = \left(\frac{x}{360l'}\right)\left(\frac{300d}{1000V}\right)^2$$ (6.6)

Two examples follow. The cell contained electrodes 20.25 cm long separated by a gap of 0.34 cm; their thickness above and below the light beam was 0.8 cm. With light of $\lambda = 589$ mμ and an applied voltage of 12 kV, β's were observed as 3.30° and 17.31° with carbon tetrachloride and benzene, respectively, in the cell. The compensator had $\Delta = 14°37'$ at the wavelength used. By (6.3) l' was 20.40 cm. Hence phase differences x appear as 0.845° and 4.167°; these through (6.6) correspond to $10^7 B$ values of 0.084 and 0.413.

Relative Values of B

The methods noted in the section on photoelectric photometry can be recommended when the expression of a Kerr constant as a multiple of that of a standard substance is adequate for the purpose in hand, for example, for the determinations of B's for a series of solutions in the same solvent. As already stated, many earlier workers adopted carbon disulphide as a reference liquid; however its B is large (3.226 × 10^{-7} for Na light at 20°) and values nearer to those under measurement are desirable. During 1953 and 1954 Le Fèvre and Le Fèvre made absolute determinations of B for benzene on 44 separate occasions by the visual method of the preceding paragraph; the mean B and estimated standard deviation emerged (for $\lambda = 589$ mμ and 25°) as (0.410 ± 0.009) × 10^{-7}. Since 1954 all routine work by the Sydney group involving liquid dielectrics has been based on this datum for benzene. Carbon tetrachloride under the same conditions has a B of 0.083 × 10^{-7} [7]. Kerr constants at other wavelengths are given by (6.7) and (6.8):

$$B_\lambda = 0.1726_5 \times 10^{-7} + \frac{82.364 \times 10^{-18}}{\lambda^2}$$ (6.7)

$$B_\lambda = 0.0349 \times 10^{-7} + \frac{16.848 \times 10^{-18}}{\lambda^2}$$ (6.8)

for benzene and carbon tetrachloride, respectively, when λ is inserted as centimeters [7]. The dependences of B_{Na} on temperature for these two solvents appear [48] as

$$10^7 B_{t^\circ} = 0.410 + 0.0022(25° - t°) \text{ for benzene}$$
$$10^7 B_{t^\circ} = 0.083 + 0.0002(25° - t°) \text{ for carbon tetrachloride.}$$

Kerr Constants of Some Liquids and Gases

Kerr constants have been recorded for more than 150 neat liquids; a selection is given in Table 6.1. Asterisked entries are determinations made in the Sydney laboratories; data without references are recalculations from relative values, at 20° and for Na light, listed by Mouton [49].

The Kerr constant of a pure liquid is sensitive to changes in the structure of the molecules making up the dielectric, but not in a way yet satisfactorily treated theoretically. Nevertheless, a few generalizations can be derived from the B's quoted in Table 6.1. For example, among isomeric nonpolar hydro-

Table 6.1.   Kerr Constants of Some Pure Liquids

Liquid		Liquid	
Dioxan[a]	6.8*	Toluene	77
n-Hexane[b]	4.7*	p-Xylene[f]	71.0*
cycloHexane[c]	5.4*	Ethylbenzene[e]	81
Methylcyclohexane[c]	14.4*	Mesitylene[f]	72.6*
cycloHexanol	−923	Chlorobenzene[g]	1242*
Paraldehyde	−2300	o-Dichlorobenzene[e]	4400
Chloroform[d]	−322	m-Dichlorobenzene[e]	893
Chloropicrin	−180	p-Dichlorobenzene[e]	263
Nitromethane	1065	Nitrobenzene[h]	40300
n-Heptane[e]	7.6	Benzyl alcohol	−1538
2-Methylhexane[e]	6.2	Aniline	−163
2,4-Dimethylpentane[e]	5.8	Dimethylaniline	1006
2,3-Dimethylpentane[e]	5.5	Diethylaniline	1042
2,2-Dimethylpentane[e]	5.3	Benzaldehyde	8050
2,2,3-Trimethylbutane[e]	5.1	Phenylhydrazine	ca. 0
n-Octane[e]	8.7	Pyridine	2040
3-Methylheptane[e]	7.7	Quinoline	1503
2,2,4-Trimethylpentane[e]	6.5	Tetralin[e]	140
n-Decane[e]	·10.3	trans-Decalin[i]	10.35*

[a] at 25°, Na light [50].
[b] at 25°, Na light [51].
[c] at 25°, Na light [52].
[d] at 25°, Na light [53].
[e] at 20°, $\lambda = 546$ mμ [54].

[f] at 25°, Na light [48].
[g] at 25°, Na light [55].
[h] at 20°, $\lambda = 546$ mμ [56].
[i] at 20°, Na light [57].

carbons the more branched the chain (i.e., the more nearly spherical the molecule) the smaller is B. With polar substances differences between isomers may be very marked. Generally, B has a range of 40,000 units, which is larger than that covered by any other physical property. Accordingly the measurement of B deserves attention and development: clearly it should have useful analytical applications.

The B quoted for nitrobenzene is the highest of many values in the literature for this substance. Pickara and Konopka [59], using a sawtoothed voltage pulse technique, report B as $(3.74 \pm 0.06) \times 10^{-5}$ at $20°$ for light of 547 mμ. They draw attention to the fact that specimens of nitrobenzene not intensively purified show small conductivities and that such materials display notably reduced B's. As a standard filling for the compensator cell in des Coudres' arrangement, this compound cannot therefore be recommended; carbon disulphide is preferable [31]. In apparently careful work, using dc voltages, Hehlgans [60] found B_{PhNO_2} as 3.86×10^{-5} (at $20°$, 5461 Å), while Friedrich [61] found the lower value of 3.54×10^{-7} (at $23°$, Na light). Dioxan is another liquid whose B appears very dependent on the states of purity and dryness to which the samples have been brought.

Table 6.2 gives a few Kerr constants of gases recorded in Sydney with the apparatus described on pages 411 and 413. Data for some eighty organic

Table 6.2. Kerr Constants of Some Gases at 760 mm Pressure

Gas	$T(°C)$	$10^{11}B$	Gas	$T(°C)$	$10^{11}B$
		A. From [13]; $\lambda = 589$ mμ			
SO_2	20	-17.7	CCl_3F	25	-1.78
CO_2	20	2.35	CHF_3	25	-5.57
NH_3	20	7.01	$CHClF_2$	25	-18.5
CH_3Cl	20	73.4	CCl_2F_2	25	1.43
			$C_2Cl_2F_2$	25	7.28

Gas	$T(°C)$	$10^{15}K$	Gas	$T(°C)$	$10^{15}K$
		B. From [58]; $\lambda = 546$ mμ			
O_2	0	0.45	CH_3Cl	18	36.5
N_2	25	0.14	C_6H_6	113.6	5.5
Cl_2	24	2.3	C_6H_5Cl	153.7	37.2
HCl	18	5.75	$o\text{-}C_6H_4Cl_2$	213.5	56
SO_2	18	-9.4	$m\text{-}C_6H_4Cl_2$	207.5	18.1
NH_3	17.9	3.48	$p\text{-}C_6H_4Cl_2$	213	12.6
$n\text{-}C_{10}H_{22}$	245	3.3	$C_6H_5NO_2$	235.5	146
$n\text{-}C_{12}H_{26}$	275	4.7	Pyridine	146.3	25.7

and inorganic molecules are available from past work. Most of these are tabulated by Stuart in his two monographs [54, 58]; sixteen are quoted in Table 6.2. It should be noted that Stuart and certain other authors have used a wavelength-independent constant K related to B by the expression $K = B\lambda/n$ (where n is the refractive index); this tends to conceal the fact that the degree of agreement between results of independent workers has not always been good. There are severe practical diffculties in measuring double refractions a thousand times smaller than those encountered with liquids. To augment the birefringences, high pressures have sometimes been used [15–23] but the extrapolation of observations so made to 1 atm is clearly a possible source of errors. In several cases the substances examined by Stuart and Volkmann [10] necessitated the use, for the first time, of temperatures well over 100° (e.g., nitrobenzene vapor was examined at 235.5°, n-dodecane at 275°C); troubles due to thermally caused strains in cell end plates had to be overcome, and in all respects the investigation deserves admiration as an outstanding contribution, from experiment, to the development of knowledge concerning the Kerr effect.

3 THEORETICAL INTERPRETATION OF THE KERR EFFECT

Summary

The classical treatment of the Kerr effect is commonly known as the Langevin-Born-Gans theory. It assumes that when a voltage is applied across a dielectric, the molecules therein tend to be oriented into positions of minimum potential energy, but that the full attainment of this orientation is limited by thermal agitations. In other words a system, which was previously isotropic as a whole to the passage of plane polarized light, becomes anisotropic when subjected to an emf. The theory is statistical.

Contributions of Langevin, Born, and Gans

In 1910, Langevin [62] regarded the molecules of a substance as electrically and optically anisotropic, so that, as a result of the orientative action of an electric field upon the dipole induced in each, the material as a whole would become doubly refracting. This picture applied only to nonpolar molecules and implied that all substances should be positively birefringent. Such an idea was inadequate alone, since chloroform, as an example, was known to be negatively doubly refracting.

Eight years later, Born [63] extended Langevin's suggestion to include polar molecules which, because they contain *permanent* resultant electric doublets, tend to orient by means of these, rather than by the weaker induced moments.

In this way the occurrence of negatively as well as positively doubly refracting compounds became explicable.

Finally in 1921 Gans [64] showed by means of a simple relationship how the electrostatic and electro-optical properties of a molecule can be connected.

Outline of the Classical Theory

A full mathematical treatment of the theory is given by Debye [65], and only an indication of the procedure follows here.

Let there be v molecules per cubic centimeter in the dielectric, and let a field E be applied; then, orientation of the molecules occurs with consequent closer packing and there are now $v + dv$ molecules per cubic centimeter. Let a molecule be defined by its electro-optical polarizability tensor components b_1, b_2, and b_3. Let b_1 lie at an angle θ to X, where X, Y, and Z define a set of coordinates with X lying along the direction of the applied electric field, and let ϕ and ψ be the other two angles relating b_2 and b_3 to Y and Z. Let μ_1, μ_2, and μ_3 be the components of $\mu_{\text{resultant}}$ along b_1, b_2, and b_3. Then $dv = e^{-U/kT} \cdot \sin \theta \cdot d\theta \cdot d\phi \cdot d\psi$, this being controlled by the angular displacements of θ, ϕ, and ψ. The potential energy U is next described in terms of the resolved components of the resultant moment and the electrostatic polarizabilities of the molecule, and the result transformed to the X, Y, Z system. Following this, the average moment produced in the model by a superimposed incident light vector at $45°$ to the applied emf has to be considered. The light vector is regarded as being resolved into two equal components perpendicular (s) and parallel (p) to the field. The sum of all such moments in the two directions is found and divided by the number of molecules considered, yielding $\bar{\mu}_p$ and $\bar{\mu}_s$. From the Lorentz-Lorentz relation, E, and γ_p and γ_s (the average polarizabilities parallel and perpendicular to E), n_p and n_s are obtained. Finally, the electrostatic polarizabilities are written in terms of b_1, b_2, and b_3, and the following expression emerges:

$$B = \frac{\pi v(n^2 + 2)^2(\epsilon + 2)^2(\theta_1 + \theta_2)}{27 n \lambda} \tag{6.9}$$

where θ_1 (*the anisotropy term*) is

$$\left[\left(\frac{1}{45kT}\right)\frac{(\epsilon - 1)}{(n^2 - 1)}\right][(b_1 - b_2)^2 + (b_2 - b_3)^2 + (b_3 - b_1)^2] \tag{6.10}$$

and θ_2 (*the dipole term*) is

$$\left(\frac{1}{45k^2T^2}\right)[(\mu_1^2 - \mu_2^2)(b_1 - b_2) + (\mu_2^2 - \mu_3^2)(b_2 - b_3) + (\mu_3^2 - \mu_1^2)(b_3 - b_1)]$$

$$\tag{6.11}$$

where B is the Kerr constant, n the refractive index, and ϵ the dielectric constant of the dielectric, $\mathbf{k}$ is Boltzmann's constant, T the absolute temperature, and λ the wavelength of the light used. The theory excludes those substances which are optically active.

Other Theories

According to (6.9) to (6.11), an isotropic nonpolar molecule should have $B = 0$; in practice *small* values are found for quasi-spherical structures (e.g., SF_6, CCl_4, etc.) and even the inert gases [21]. Voigt [66], supposing that an external electric field would deform the molecules of a dielectric, forecast the occurrence of temperature-independent contributions to B. Le Fèvre and Rao [67] have attempted to relate such deformations with bending-force constants, bond lengths, and so forth, but predicted values are well below those observed.

The Kerr effect has received consideration from the standpoint of the quantum theory, being connected thereby with the Stark effect. Born and Jordan [68] thus find for B:

$$B = \left(\frac{2\pi}{n\lambda}\right)\left[C_o + \frac{(C_1 + D_o)}{\mathbf{k}T} + \frac{D_1}{\mathbf{k}^2 T^2}\right] \tag{6.12}$$

in which C_o and D_o are determined by changes in quantum numbers concerned with relative positions of electrons in the molecule and with total angular momentum. Equation (6.12), it will be noted, contains a temperature invariant term, C_o.

Implicit in the classical treatments is the assumption that an induced moment is rectilinearly proportional to the inducing field. Coulson et al. [69] have demonstrated theoretically that this may not be so, and that atoms and molecules under strong fields could exhibit polarizabilities that exceed those under weak fields. Buckingham and Pople [70], and Buckingham with various collaborators [20, 21, 31], by using the concept of hyperpolarizability whereby moments induced by a field F are given by an expansion of the form $b_o F + aF^2/2 + CF^3/6 + \cdots$ (in which b_o is the "low field" polarizability, and a and c are hyperpolarizability constants) have provided explanations for the non-zero Kerr constants of centro-symmetric molecules. Strictly, a temperature-invariant term should be added to the right-hand side of (6.9) but the correction is hardly significant except when B itself is small [48, 71, 72].

Molecular and Specific Kerr Constants

We here define "molecular and specific Kerr constants," $_mK$ and $_sK$, since these are quantities conveniently computed from observations on solutions. The molecular Kerr constant is related to the constant B by (6.13):

$$_mK = \frac{6\lambda n B M}{(n^2 + 2)^2(\epsilon + 2)^2 d} \tag{6.13}$$

where M is the molecular weight, n the refractive index, ϵ the dielectric constant, and d the density of the medium under examination, while λ is the wavelength of the light used to determine B. The specific Kerr constant is

$$_sK = \frac{6\lambda nB}{(n^2 + 2)^2(\epsilon + 2)^2 d} \tag{6.14}$$

In the anisotropy term θ_1 [(6.10)], the conversion of the electrostatic polarizability terms to b_1, b_2, and b_3 may be made by substituting $_DP/_EP$ for $(\epsilon - 1)/(n^2 - 1)$, where $_DP$ and $_EP$ are the distortion and electronic polarizations, respectively. The $_mK$ of a species may be viewed as the difference, per unit inner field, of molecular refractions R_p and R_s measured in directions parallel and perpendicular to the lines of force in the applied field. Equation (6.13) is the form introduced by Otterbein [73] and used regularly by the Sydney group; some authors (e.g., Briegleb [74] and Stuart [54, 58]) have omitted the 6 from the numerator.

4 DETERMINATION OF MOLAR KERR CONSTANTS

Essential Experimental Data

For Gases

The Kerr constants B must be measured at known temperatures and pressures in order that the related molar volumes ($V = M/d$) can be computed. For pressures below 3 atm, for example, an observed gas pressure P may be expressed (6.13) as an "ideal" pressure P_i by the approximation

$$P_i = P\left(1 + \frac{aP}{R^2T^2}\right)$$

where T is the absolute temperature, R is 0.0821 liter atm deg^{-1} mole^{-1}, and a is the appropriate Van der Waals constant (e.g., 6.71 atm liter^2 mole^{-2} for SO_2). When higher pressures are used gas densities are usually calculated via one of the equations of state [15–23]. By least-squares procedures the B's observed are fitted to rectilinear or quadratic functions of the densities; B's corresponding to low densities are thus obtained. If B is a Kerr constant for a gas under conditions where limiting densities (in moles/cc) are being approached (6.13) becomes simply

$$_mK_{gas} = \frac{2B\lambda}{27d}$$

(since at low pressures both ϵ and n are close to unity).

However, for reasons obvious from an earlier discussion, measurements on gases are practically difficult to make; moreover, many compounds of interest cannot be vaporized without decomposition yet can often be dissolved unchanged in a solvent.

For Solutes

To obtain the molar Kerr constant of a solute the measurement of four physical properties of a solution of known concentration is required. Suitable solvents are those which themselves exhibit small electric birefringences; nonpolar media such as carbon tetrachloride, then benzene, hexane, cyclohexane, or dioxan are recommended in that order. A series of solutions of increasing concentration is made up, and the w_2 of each calculated (w_2 is the weight fraction, i.e., the weight of solute divided by the weight of solvent plus solute). For the individual solutions the following experimental values are then ascertained: (1) the Kerr constant B_{12}; (2) the dielectric constant ϵ_{12}; (3) the density d_{12}; and (4) the refractive index n_{12} (the same λ as for B_{12} being used). In making up the solutions, the nature of the material under examination is kept in mind. For a strongly polar substance of anisotropic nature (such as an aryl nitro-derivative) we should expect a very large Kerr effect, and because the accuracy of measurement using the mica compensators (such as the Szivessy-Dierkesmann compensator) falls with marked increase in the value of B it is advisable to keep the solutions dilute; for the same reason, with nonpolar substances it is helpful to use stronger solutions (say up to $w_2 = 0.05$) when possible.

Le Fèvre and Le Fèvre [4], starting from a conventional alligation formula, have derived the expression (6.15) for the specific Kerr constant of a solute at infinite dilution:

$$_{\infty}(_sK_2) = {_sK_1}(1 - \beta + \gamma + \delta - H\gamma - J\alpha\epsilon_1) \qquad (6.15)$$

where $H = 4n_1^2/(n_1^2 + 2)$, $J = 2/(\epsilon_1 + 2)$, and H, J, and $_sK_1$ are constants for a given solvent at a specified temperature. The following data apply at 25° and $\lambda = 589$ mμ to four commonly used media (CT = carbon tetrachloride, B = benzene, CH = cyclohexane, D = dioxan):

	CT	B	CH	D
ϵ_1	2.2270	2.2725	2.0199	2.2090
d_1	1.58454	0.87378	0.77389	1.0280
n_1	1.45732	1.49795	1.4235	1.4202
$10^7 B_1$	0.083	0.410	0.054	0.068
$10^{14} {_sK_1}$	0.894	7.556	1.340	1.162
H	2.060	2.115	2.013	2.008
J	0.4731	0.4681	0.4975	0.4752

The B quoted for carbon tetrachloride is a revised value, based on many comparisons with benzene and a decision [7] to adopt the Kerr constant of this hydrocarbon [75] as a single standard for all work by the Sydney group.

The values for α, β, γ, and δ are obtained by assuming that the dielectric constants, densities, refractive indexes, and Kerr constants of the solutions depend on the concentrations, as w_2 approaches zero, in a straight-line manner:

$$\epsilon_{12} = \epsilon_1(1 + \alpha w_2) \tag{6.16}$$

$$d_{12} = d_1(1 + \beta w_2) \tag{6.17}$$

$$n_{12} = n_1(1 + \gamma w_2) \tag{6.18}$$

$$B_{12} = B_1(1 + \delta w_2) \tag{6.19}$$

However, in practice B_{12} sometimes shows a curvature when plotted against w_2; when this happens the change in B in passing from solvent to solution (ΔB) is fitted to a power series in w_2.

Let there be n observations of the form $aw_2 + bw_2^2 = \Delta B$. We obtain, by multiplying throughout by w_2 and w_2^2, and adding,

$$a\Sigma w_2^2 + b\Sigma w_2^3 = \Sigma \Delta B \cdot w_2, \quad \text{and} \quad a\Sigma w_2^3 + b\Sigma w_2^4 = \Sigma \Delta B w_2^2$$

therefore,

$$a = \frac{\Sigma \Delta B \cdot w_2 \Sigma w_2^4 - \Sigma \Delta B w_2^2 \Sigma w_2^3}{\Sigma w_2^2 \cdot \Sigma w_2^4 - (\Sigma w_2^3)^2} \tag{6.20}$$

and

$$b = \frac{\Sigma w_2^2 \cdot \Sigma(\Delta B w_2^2) - \Sigma w_2^3 \cdot \Sigma(\Delta B w_2)}{\Sigma w_2^2 \cdot \Sigma w_2^4 - (\Sigma w_2^3)^2} \tag{6.21}$$

hence

$$(\Delta B)_{w_2} = aw_2 + bw_2^2 \tag{6.22}$$

and, when $w_2 = 0$,

$$\frac{d(\Delta B)}{dw_2} = a = B_1 \times \delta$$

On occasions, when a plot of ΔB versus w_2 shows marked curvature, fitting by a three-term power series may be essential. The argument is similar to that giving (6.20), values for a', b', and c' in $\Delta B = a'w_2 + b'w_2^2 + c'w_2^3$ emerging as:

$$a' = \begin{vmatrix} \Sigma w_2 \Delta B & \Sigma w_2^3 & \Sigma w_2^4 \\ \Sigma w_2^2 \Delta B & \Sigma w_2^4 & \Sigma w_2^5 \\ \Sigma w_2^3 \Delta B & \Sigma w_2^5 & \Sigma w_2^6 \end{vmatrix} \div \begin{vmatrix} \Sigma w_2^2 & \Sigma w_2^3 & \Sigma w_2^4 \\ \Sigma w_2^3 & \Sigma w_2^4 & \Sigma w_2^5 \\ \Sigma w_2^4 & \Sigma w_2^5 & \Sigma w_2^6 \end{vmatrix}$$

$$b' = \begin{vmatrix} \Sigma w_2^2 & \Sigma w_2 \Delta B & \Sigma w_2^4 \\ \Sigma w_2^3 & \Sigma w_2^2 \Delta B & \Sigma w_2^5 \\ \Sigma w_2^4 & \Sigma w_2^3 \Delta B & \Sigma w_2^6 \end{vmatrix} \div \text{(same denominator as for } a')$$

$$c' = \begin{vmatrix} \Sigma w_2^2 & \Sigma w_2^3 & \Sigma w_2 \, \Delta B \\ \Sigma w_2^3 & \Sigma w_2^4 & \Sigma w_2^2 \, \Delta B \\ \Sigma w_2^4 & \Sigma w_2^5 & \Sigma w_2^3 \, \Delta B \end{vmatrix} \div \text{(same denominator as for } a')$$

Finally $d(\Delta B)/dw_2$, for $w_2 = 0$, is $a' = B_1 \times \delta$. The two-term equation given above covers the majority of systems that have been investigated.

If the need arises, as it sometimes does, the dielectric constant and/or density increments, $(\epsilon_{12} - \epsilon_1)_{w_2}$ and/or $(d_{12} - d_1)_{w_2}$, should also be fitted to two-term power series, in order to find the values of $\alpha\epsilon_1$ and βd_1 appropriate for $w_2 = 0$.

To illustrate these methods, two sets of actual experimental data are reproduced (Table 6.3) and worked through to the final results in each case. Equations (6.23) and (6.24) are used to compute the specific polarizations and refractions of the solutes at infinite dilution [14]:

$$_\infty(p_2) = p_1(1 - \beta) + C\alpha\Sigma_1 \tag{6.23}$$

$$_\infty(r_2) = r_1(1 - \beta) + D\gamma'n^2 \tag{6.24}$$

where for a given solvent:

$$p_1 = \frac{(\epsilon_1 - 1)}{(\epsilon_1 + 2) \, d_1}$$

$$C = \frac{3}{d_1(\epsilon_1 + 2)^2}$$

$$D = \frac{3}{d_1(n_1^2 + 2)^2},$$

and

$$\gamma'n_1^2 = \frac{(n_{12}^2 - n_1^2)}{w_2}.$$

The constants for the four solvents named in connection with (6.15) are, at 25° and for Na light:

	CT	B	CH	D
p_1 (cc)	0.18319	0.34086	0.32784	0.29528
C	0.10596	0.18809	0.23989	0.16473
r_1 (cc)	0.17198	0.33543	0.32939	0.26026
D	0.11133	0.19063	0.23912	0.18086

Table 6.3. Increments of Dielectric Constant, Density, Refractive Index, and Refractive Index, and Kerr Constant Observed for Solutions Containing Weight Fractions w_2 of Solutes at 25°

$10^5 w_2$	$10^4 \Delta e$	$10^5 \Delta d$	$10^4 \Delta n$	$10^{11} \Delta B$
A. Pentafluorobenzene in Benzene				
1338	149	464	−11	397
3216	353	1140	−28	927
4138	449	1466	−35	1193
5247	573	1870	−45	1544

whence $\Sigma \Delta\epsilon/\Sigma w_2 = 1.09$; $\Sigma \Delta d/\Sigma w_2 = 0.354$; $\Sigma \Delta n/\Sigma w_2 = -0.085$; $\Sigma \Delta n^2/\Sigma w_2 = -0.255$; $10^7 \Sigma \Delta B/\Sigma w_2 = 2.91$

B. Nitrobenzene in Cyclohexane				
163	175	43	1	525
259	293	65	1	829
401	440	93	2	1286
527	587	130	3	1715
692	772	184	5	2276

whence $\Sigma \Delta\epsilon/\Sigma w_2 = 11.10$; $\Sigma \Delta d/\Sigma w_2 = 0.252$; $\Sigma \Delta n/\Sigma w_2 = 0.059$; $\Sigma \Delta n^2/\Sigma w_2 = 0.171$; $10^7 \Delta B = 31.37 w_2 + 217 w_2^2$

Quantities Calculated from the Above Data

	a	b
$\alpha\epsilon_1$	1.09	11.10
β	0.405	0.326
$\gamma' n_1^2$	−0.255	0.171
$_\infty P_2$(cc)	68.6	355.0
$(R_2)_D$(cc)	25.3	32.4
μ(D)[a]	1.43	3.94
γ	−0.057	0.041
δ	7.07	560.2
$10^{12}_\infty(_m K_2)$	91.7	950

[a] Calculated taking distortion polarizations as $1.05 \times R_2$.

At low concentrations $\gamma' n_1^2$ is usually very close to $2n_1(n_{12} - n_1)/w_2$, that is, to $2n_1(\gamma n_1)$. When relationships with w_2 are rectilinear, mean values of $\alpha\epsilon_1$, γn_1, βd, and $B_1\delta$ can be determined as the quotients:

$$\frac{\Sigma \text{ (differences between solutions and solvent)}}{\Sigma w_2}.$$

Standard errors or deviations, if they are desired, may be estimated by using the equations quoted in ref. [4].

About a thousand molar Kerr constants at infinite dilution have been recorded by the Sydney group [76]; other recent data occur in the papers of Cherry et al. [30], Huang with Ng [77] and Chia [78], and so forth. Lists have been given by Stuart [58, 79] and the Le Fèvres [2].

5 ANALYSIS OF MOLAR KERR CONSTANTS

The Anisotropy of Molecular Polarizability

The objective is to extract numerical values for the quantities b_1, b_2, and b_3 that were introduced in the outline of classical theory in Section 3.

When an electric field of unit strength acts upon a molecule the dipole moment induced is, by definition, the polarizability b of the molecule. With spherically symmetrical molecules the directions of action of b and of the field will always be collinear; with a less symmetrical molecule the resultant induced moment will be at an angle to the field depending on the orientation of the molecule in the field. In the case of a completely unsymmetrical molecule there will be only three orientations, out of the infinitude possible, for which this angle is zero; the moments induced parallel to the field for these three particular situations are known as the *principal polarizabilities* b_1, b_2, and b_3 of the molecule under consideration.

The magnitudes and directions of moments induced in any molecule can be calculated whenever the "polarizability ellipsoid," having mutually rectangular half-axes b_1, b_2, and b_3, has been specified [2, 3]. These principal polarizabilities enter the mathematical treatments given by Langevin, Gans, Born, Debye, Raman, Krishnan, Cabannes, de Mallemann, and others to many optical and electrical phenomena, and especially—in the present relevance—to refractivity, light scattering, dielectric polarization, and the Kerr effect.

The total of the three polarizabilities is accessible from the electronic polarization, $_EP$, of a compound:

$$_EP = R_\infty = \frac{4\pi N(b_1 + b_2 + b_3)}{9} \tag{6.25}$$

Electronic polarizations are obtained by relating molar refractions at known wavelengths, λ, to $1/\lambda^2$ and finding the value of R_λ when $1/\lambda^2 = 0$ (equations for this purpose are cited in ref. [14], pp. 17, 18). A table for the approximate calculation of $_EP$'s has been given by Le Fèvre and Steel [80]. From (6.25) we have:

$$b_1 + b_2 + b_3 = 1.1891 \times 10^{-24} \times R_\infty \text{ cc}$$

The squares of the differences between the principal polarizabilities are connected with depolarization factors, Δ, by (6.26)

$$\frac{10\,\Delta}{6-7\,\Delta} = \frac{(b_1-b_2)^2+(b_2-b_3)^2+(b_3-b_1)^2}{(b_1+b_2+b_3)^2} \tag{6.26}$$

(where Δ is the ratio of the intensity of horizontally to vertically polarized light in Rayleigh scattering examined at $90°$ to the incident beam which is itself proceeding in the horizontal plane). The right-hand side of (6.26) can be evaluated for solutes at infinite dilution in carbon tetrachloride [81].

A sum $(\theta_1 + \theta_2)$ is available from the measured molar Kerr constant. By (6.9) and (6.13) we have:

$$_mK = \frac{2\pi N(\theta_1 + \theta_2)}{9} \tag{6.27}$$

so that

$$\theta_1 + \theta_2 = 2.3782 \times 10^{-24} \times {_mK}$$

The molar Kerr constant strictly should be the $_mK_{gas}$ of Section 4, but $_\infty(_mK_2)$ values—via (6.15)—must be used for solutes. From experiment it appears that $_\infty(_mK_2)$ is usually close to $_mK_{gas}$. Expansions of θ_1 and θ_2 have already been given as (6.10) and (6.11); in (6.10) the quotient (distortion polarization) divided by (electronic polarization) may be substituted for $(\epsilon - 1)/(n^2 - 1)$ since the latter ratio cannot be estimated with certainty for a solute at infinite dilution [2, 3].

Separation of θ_1 and θ_2 is possible through (6.25) and (6.26). Then if

$$A = b_1 + b_2 + b_3$$

$$B = 2b_1 - b_2 - b_3 = \frac{45k^2T^2\theta_2}{\mu^2}$$

$$C = (b_1-b_2)^2+(b_2-b_3)^2+(b_3-b_1)^2$$

and the resultant molecular moment is taken as collinear with b_1, we have

$$b_1 = \frac{(A+B)}{3},$$

and

$$b_2(\text{and } b_3) = \left(\frac{A}{3}\right) = (B/6) \pm \frac{(6C-3B^2)^{0.5}}{6}$$

In the general case, when μ acts at known finite angles with the 1, 2, and 3 directions, the components μ_1, μ_2, and μ_3 must be inserted into (6.11), and computation is more tedious. With nonpolar molecules, however, for which $b_1 \neq b_2 \neq b_3$, there are difficulties; these are mentioned later (pp. 439–441).

Some examples of polarizability ellipsoid specifications, drawn from $_\infty(_mK_2)$ values, are given in Table 6.4. Le Fèvre and Le Fèvre [2] and Stuart [54, 58] have listed b_i's derivable from electric birefringence measurements on gases. Comparison of data from the two sources suggests that molecular polarizability ellipsoids can be determined in nonpolar media not less dependably than in the vapor phase; however, solvent effects are detectable and sometimes predictable [2, 3, 82, 88].

Table 6.4. Some Molecular Principal Polarizabilities

Molecule	$10^{24}b_1$	$10^{24}b_2$	$10^{24}b_3$	Reference
CS_2	13.08	5.58	5.58	82
CH_3Cl	5.09	4.11	4.11	75
CH_3Br	6.56	4.99	4.99	75
CH_3I	8.62	6.69	6.69	83
$(CH_3)_3CCl$	10.82	9.41	9.41	85
$(CH_3)_3CBr$	12.94	10.26	10.26	75
$(CH_3)_3CI$	16.00	11.91	11.91	83
CH_3CN	5.56	3.67	3.67	84
$(CH_3)_3CCN$	10.84	9.16	9.16	84
$CHCl_3$	6.73	9.01	9.01	81
$CHBr_3$	9.11	12.58	12.58	5
CHI_3	15.90	17.85	17.85	5
CH_3CCl_3	9.62	10.40	10.40	5
C_6H_6	11.15	11.15	7.44	86
C_6H_5Cl	14.78	12.55	8.21	87
C_6H_5Br	16.83	13.01	8.92	87
C_6H_5I	19.71	15.88	9.96	87
C_6H_5CN	16.27	11.34	8.30	84

Anisotropic Bond Polarizabilities

Assuming that a bond $X-Y$ is anisotropically polarizable, with a longitudinal polarizability b_L^{XY} and two transverse polarizabilities b_T^{XY} and b_V^{XY}, *and that such polarizabilities are tensorially additive*, the principal molecular polarizabilities of a molecule XY_2 with an angle YXY of $2\theta°$ may be written

$$b_1^{XY_2} = 2[b_L^{XY} \cos^2 \theta + b_T^{XY} \sin^2 \theta]$$
$$b_2^{XY_2} = 2[b_T^{XY} \cos^2 \theta + b_L^{XY} \sin^2 \theta]$$

and

$$b_3^{XY_2} = 2b_V^{XY}$$

Further, from R_∞ for XY_2 we have, via (6.25) twice the sum $(b_L^{XY} + b_T^{XY} + b_V^{XY})$. By application of this type of argument to data from molecules with definite stereo-structures, the principal polarizabilities of bonds can be extracted. In fact, however, because—when X is a carbon atom—X must be linked by two single bonds or one double bond to other atoms, the number of unknowns exceeds the number of equations. For example, the chlorinated alkanes with tetrahedral bond angles will all yield numerical values corresponding to $(b_L - b_T)^{CCl}$ minus $(b_L - b_T)^{CH}$. Originally Le Fèvre and Le Fèvre ([2] and references therein) assumed the C—H bond to be isotropically polarizable; subsequent work [52] has shown this to be approximately correct. The polarizabilities of some four dozen bonds have been estimated on such a basis, by the Sydney group, from source information gained from experiments on solutions. A selection of these b_i's is in Table 6.5.

Table 6.5. Some Anisotropic Bond Polarizabilities in Cubic Å Units

Bond XY	b_L^{XY}	b_T^{XY}	b_V^{XY}	Bond Environment
C—H	0.65	0.65	0.65⎫	Cyclohexane and related
C—C	0.97	0.26	0.26⎭	hydrocarbons
C=C	2.80	0.73	0.77	$CH_2 : CCl_2$
C≡C	3.79	1.26	1.26	Acetylene
C—F	1.25	0.4	0.4	MeF
C—Cl	3.18	2.20	2.20	MeCl
C—Cl	3.59	1.99	1.99	EtCl
C—Cl	3.94	1.81	1.81	Me_3CCl
C—Cl	4.2	1.95	1.5	C_6H_5Cl
C—Br	4.65	3.1	3.1	MeBr
C—Br	5.3	2.7	2.7	EtBr
C—Br	5.98	2.58	2.58	Me_3CBr
C—Br	6.2	2.4	2.2	C_6H_5Br
C—I	6.70	4.77	4.77	MeI
C—I	7.59	4.49	4.49	EtI
C—I	9.19	3.68	3.68	Me_3CI
C—I	9.1	5.3	3.3	C_6H_5I
C—O	0.89	0.46	0.46	1,3,5-Trioxan
C=O	2.30	1.40	0.46	Acetone
N—H	0.50	0.83	0.83	NH_3
N—C	0.57	0.69	0.69	Me_3N
C—S	1.88	1.69	1.69	Me_2S
C—(CN)	3.64	1.75	1.75	MeCN
C—(CN)	4.03	1.54	1.54	$Me_3C\cdot CN$

As would be expected, the anisotropy of X—Y is not constant throughout all molecular environments; it changes markedly when X—Y is in conjugation with double bonds (e.g., compare the carbon-halogen bonds in alkyl halides with those in aryl halides). Tentative polarizabilities for the C—P, C—As, C—Sb, C—Bi, O—P, O—As, and N=N linkages are given in ref. [3], p. 50; the O—O bond in di-t-butyl peroxide has $b_L = 0.62$, and $b_T = b_V = 1.04 \times 10^{-24}$ cc [89].

Bond Polarizabilities and Bond Properties

The anisotropic polarizabilities of bonds buried within some structures are not directly measurable by experiment. Recourse must be had to relationships between bond polarizability components and other accessible bond physical properties. By simple electrostatic theory the polarizability of a conducting sphere of radius r is r^3. Correlations of bond lengths with one of the tensor components of bond polarizabilities have proved useful. The b_L^{XY} of a bond X—Y in different structural situations often conforms to

$$10^{24} b_L^{XY} = a + b[r^3 + c(r_{single}^3 - r^3)] \tag{6.28}$$

where a, b, and c are numerical constants, r is the intercenter separation of X and Y, and r_{single} is the corresponding distance for a single bonded union [90]. With bonds having polarizability ellipsoids of rotation there is no further problem since the means $(b_L + 2b_{T=V})/3$ or $(2\theta_{L=T} + b_V)/3$ are available from tables [80]. Equation (6.28) has been used in the study of the conformations of the triphenyl derivatives of N, P, As, Sb, and Bi [91].

Longitudinal and "vertical" polarizabilities of C⋯C bonds can also be easily computed a priori as functions of r_{cc} by (6.29):

$$\left. \begin{array}{l} 10^{24} b_L = 5.077 + 4.683r - 4.786r^2 \\ 10^{24} b_V = 11.183 - 12.384r + 3.435r^2 \end{array} \right\} \tag{6.29}$$

Using these relations it is possible to deduce the principal polarizabilities of *nonpolar* aromatic hydrocarbons [86] by calculating, say b_3, thereby reducing three unknowns to two, and thus making b_1 and b_2 accessible through (6.25), (6.27), (6.15), and (6.10). Various analogous expressions for carbon-oxygen bonds, for example,

$$10^{24} b_L^{CO} = 39.106r - 17.646r^2 - 19.059$$

are currently under test [92]. Intercenter distances in most bonds are now known accurately [93].

Longitudinal polarizabilities of bonds XY have been related to the infrared and Raman stretching frequencies (in cm^{-1}) by the rectilinear dependences of $Q = (1/r_{XY}^2)(b_L^{XY}/\overline{M}_{XY})^{1/3}$ upon v_{XY} in different series of compounds [94, 95]. Provided that values for a and b in $v_{XY} = aQ + b$ are known ($\overline{M}$ is the reduced mass of XY), b_L^{XY} can be ascertained from recorded r_{XY} and the easily observed v_{XY}. An examination of *s-trans/s-cis* equilibria in alkyl nitrites [96] is cited to illustrate this procedure.

Allusion was made in the first part of this section to difficulties in evaluating b_1, b_2, and b_3 for nonpolar molecules. These arise because when $\mu = 0$ so does θ_2. Equations (6.25) to (6.27) then become *two* equations containing *three* unknowns—hence the value of empirical and other routes whereby one of the three b's may be calculated a priori. Refractive indexes of crystals, magnetic birefringence (Cotton-Mouton effect) measurements, Verdet constants, and even the dimensions of scale models, have sometimes been invoked to break the impasse (ref. [2], pp. 287/8).

Of course, when from symmetry it is obvious that $b_1 = b_2$, or $b_2 = b_3$, the molecular ellipsoid sought should be a body of revolution; a mathematical solution is then possible from (6.27) and (6.25), irrespective of whether the substance is polar or nonpolar. Equation (6.27) supplies a value for $(b_i - b_j)^2$ while (6.25) provides $(b_i + 2b_j)$.

Another device worth mentioning depends on the observation that with markedly anisotropic and strongly polar solutes θ_2 is usually large and of the order 10^{-34}, whereas θ_1 is commonly small and of the order 10^{-35}; in such cases the latter can often be neglected without seriously affecting the values ultimately calculated for b_1. This fact may be helpful when no reliable light-scattering determinations exist to provide θ_1 in the way described above. As before (p. 435), $b_1 = (A + B)/3$, but B is now taken from the *whole* quantity $0.2378 \times 10^{-23} {}_\infty({}_mK_2)$. It is not feasible to substitute b_1 into A and B because each of these equations thereby gives the sum $(b_2 + b_3)$.

The case of nitrobenzene serves as an illustration: here ${}_\infty({}_mK_2) = 950 \times 10^{-12}$, ${}_EP = 30.9$ cc, ${}_DP = 36.2$ cc, $\mu = 3.94$ D, and $\Delta = 0.056$; from these $b_1 = 15.8 \times 10^{-24}$, $b_2 = 14.1 \times 10^{-24}$, and $b_3 = 6.9 \times 10^{-24}$. When θ_1 is put as zero, b_1 becomes 15.9×10^{-24} and $(b_2 + b_3)$ is 20.8×10^{-24}.

Unfortunately for many solutes only refractivity, polarity, and Kerr effect data are easily accessible. Camphor (with ${}_EP = 44.0$ cc, $\mu = 3.09$ D, and ${}_\infty({}_mK_2) = 115 \times 10^{-12}$) is an example. When first examined no light-scattering data were available. Putting θ_1 at 1×10^{-35} (by analogy with *cyclo*hexanone) gave $b_1 = 18.0 \times 10^{-24}$ and $(b_2 + b_3) = 34.0 \times 10^{-24}$. Later measurements showed that, in fact, θ_1 is 6.4×10^{-35}; this leads to $b_1 = 17.9 \times 10^{-12}$ and $(b_2 + b_3) = 34.2 \times 10^{-12}$. The last quantity, when separated, yields $b_2 = 21.3 \times 10^{-12}$ and $b_3 = 12.9 \times 10^{-24}$.

6 TENSORIAL ADDITIVITY OF BOND POLARIZABILITIES

General

The argument used in Section 5, page 436, assumed the validity of such additivity. However, as is noted in comments on Table 6.5, the anisotropy so deduced for a bond between a given atom pair is not constant throughout all molecules containing such bonds. Part of this inconstancy is probably due to the incorporation into the values in Table 6.5 of various "secondary polarizabilities." Some of these, theoretically, should be predictable by methods outlined by Silberstein [97], others are due to chemical causes, for example, conjugation ("mesomeric" and "inductomeric" effects) and/or to the inclusion, by unknown amounts, of polarizabilities of nearby lone-pair orbitals. The Silberstein calculations are often tediously long and are inevitably unreliable through ignorance of the dielectric constant of interatomic space. Papers by Pitzer [98], Smith and Mortensen [99], Davies [100], Bolton [101], Bunn and Daubeny [102], Murrell and Musgrave [103], Vuks [104], Philippoff [105], Clément and Bothorel [106], and Sachsse [107] are relevant to certain aspects of the present problem. It is clear that, except for a gaseous diatomic molecule, *absolute* values of the principal polarizabilities of a bond cannot be precisely learned from experiment.

Nevertheless, *apparent* bond polarizabilities, such as those in Table 6.5, drawn from measurements on solutes, have considerable stereochemical usefulness, as accumulated data on a wide range of molecules have shown [76]. A safe viewpoint is that these polarizabilities are empirical and should be applied in molecular situations analogous to those from which they have come. The assumed tensorial additivity of bond polarizability ellipsoids has, by now, often been tested practically by comparing molar Kerr constants determined via (6.15) with those calculated a priori by methods that involve a reversal of the steps described in Section 5, pages 434 and 435.

A priori Computations of Molar Kerr Constants

Within a three-dimensional model of the molecule under consideration a set of rectangular axes, X, Y, and Z, is arbitrarily placed. Bond polarizabilities appropriate to the kind of molecule under consideration are selected. One first imagines a unit field acting, for example, along X and computes the moments b_{xx}, b_{yx}, and b_{zx} which it induces in the X, Y, and Z directions; then the process is repeated for the field along the Y, and finally along the Z-axis. Solution of the cubic equation

$$\begin{vmatrix} (b_{xx} - \lambda) & b_{xz} & b_{xz} \\ b_{yx} & (b_{yy} - \lambda) & b_{yz} \\ b_{zx} & b_{zy} & (b_{zz} - \lambda) \end{vmatrix} = 0 \qquad (6.30)$$

yields the required principal polarizabilities, b_1, b_2, and b_3, for the structure in question and permits the calculation of the direction cosines which locate these b's within the X, Y, Z coordinates. Extraction of the three roots of (6.30) is lengthy by "longhand" methods but can be quickly performed on any computer for which a standard eigenvalue-eigenvector program is available. The moment components μ_1, μ_2, and μ_3 need also to be estimated, either from $\mu_{\text{resultant}}$ (if its direction of action within the molecular framework is known) and the direction cosines found from (6.30), or by vectorial summation of all the bond moments involved. The molar Kerr constant can be evaluated by means of (6.10), (6.11) and (6.27). In (6.10) $(\epsilon - 1)/(n^2 - 1)$ may be taken as $_DP/_EP$ (see p. 429). A detailed explanation of this type of calculation is given by Eckert and Le Fèvre [108].

For many purposes (e.g., if one desires only to compare calculated and found $_mK$'s) it is unnecessary to diagonalize the polarizability tensor, since θ_1 and θ_2 can be obtained directly from the coefficients $b_{xx}, b_{yy}, b_{zz}, b_{xy} = b_{yx}, b_{yz} = b_{zy}$, and $b_{zx} = b_{xz}$ (referred to X, Y, Z axes) by (6.31) and (6.32)

$$45\mathbf{k}T\theta_1 = (b_{xx} - b_{yy})^2 + (b_{yy} - b_{zz})^2 + (b_{zz} - b_{xx})^2 + 6b_{xy}^2 + 6b_{xz}^2 + 6b_{yz}^2 \tag{6.31}$$

$$45\mathbf{k}^2T^2\theta_2 = \mu_x^2(2b_{xx} - b_{yy} - b_{zz}) + \mu_y^2(2b_{yy} - b_{xx} - b_{zz})$$
$$+ \mu_z^2(2b_{zz} - b_{xx} - b_{yy}) + 6b_{xy}\mu_x\mu_y + 6b_{xz}\mu_x\mu_z + 6b_{yz}\mu_y\mu_z \tag{6.32}$$

If $_DP/_EP = 1.1$, $\theta_1 = 5.942 \times$ [right side of (6.31)] $\times 10^{11}$, $\theta_2 = 13.13 \times$ [right side of (6.32)] $\times 10^{24}$, and $_mK$ (calc) $= 4.2052(\theta_1 + \theta_2) \times 10^{23}$ for $T = 298°$ (absolute).

7 STEREOCHEMICAL APPLICATIONS

Significance of Algebraic Sign of $_mK$

When the resultant dipole moment of a molecule acts in a direction that is perpendicular (or near-perpendicular) to the axis of maximum polarizability, the sign of θ_2 will, by (6.11), be negative. This is understandable since the orientative effects of a field on the permanent dipole (of order, 10^{-18} esu) will outweigh those on the induced moment (of order, 10^{-24} esu) along b_{max}. Accordingly, in the Kerr cell, the less polarizable path will be that traversed by the resolved component of the light which is parallel to the applied field; this light vector will travel more quickly than the component which is perpendicular to the lines of electric force and therefore in expression (6.1), n_s will be greater than n_p, so that B becomes negative. As $_mK$ contains B [see (6.13)], $_mK$ must also be negative. Certain molecules, for example, bipyridyl [109] may show small positive $_mK$'s yet possess θ_2's. Nevertheless a substance exhibiting negativity of $_mK$ must certainly be polar. Thianthren and phenothiazine have negative $_\infty(_mK_2)$'s in benzene solutions [110, 111]; this is interpreted as due to these molecules being folded about lines joining the hetero atoms. Thus the algebraic sign of the molar Kerr constant may often have a qualitative structural significance.

Some Quantitative Examples

Bis-(N-alkylsalicylaldiminato) Beryllium Complexes

These contain a Be atom surrounded by two oxygen and two nitrogen atoms. If planar, the complex would be either *cis* or *trans*, for which the $_mK$'s predicted (when R in

$$[C_6H_4(O-)\cdot CH:NR]_2Be$$

is CH_3) are, respectively, $+720 \times 10^{-12}$ and $+60 \times 10^{-12}$. The observed [112] value is -1705×10^{-12}, indicating a near-tetrahedral environment for the central Be atom. When R is larger (e.g., a CMe_3 group) the $_mK$ measured is less negative, consistently with steric distortions in the direction of planarity.

Conformations of 1,2-Diketones

In a molecule of formula $R\cdot CO\cdot CO\cdot R$, when R is an anisotropically polarizable group, there are two conformational problems: the degree θ to which the R's are twisted about the bonds $(R)-(CO)$, and the degree ϕ to which each half of the structure is twisted about the $(CO)-(CO)$ bond. Benzil [113], furil [113], and α-pyridil [114] are relevant instances. If θ and ϕ denote rotations away from the flat *cis* extreme forms, then in all three cases θ is within $\pm 5°$, while ϕ is approximately $97°$, $118-119°$, and $81-82°$, respectively. The calculated molar Kerr constants move rapidly with ϕ (e.g., with α-pyridil the $_mK$'s for $81°$ and $82°$ are -250×10^{-12} and -288×10^{-12}, by measurement the value is -268×10^{-12}).

Diphenyl Disulphide

Polarity considerations show the planes containing the two $C-S$ bonds to be twisted $85°$ about the $S-S$ line; they cannot give information concerning the phenyl groups. The observed $_\infty(_mK_2)$ is -11.5×10^{-12}. Expected $_mK$'s are -14×10^{-12} or $+167 \times 10^{-12}$ when both phenyl rings are either coplanar with or perpendicular to their adjoining $C-S-S$ planes, respectively [115].

Various Aryl Ketones

The $_\infty(_mK_2)$ for acetophenone is $+387 \times 10^{-12}$ by experiment versus -112×10^{-12} and $+398 \times 10^{-12}$ when calculated for forms with the acetyl group plane at $90°$ and $0°$, respectively, to the C_6H_5 plane. As expected, therefore, acetophenone appears to be near flat. Acetylmesitylene [116] and acetyldurene [117], however, have $_\infty(_mK_2)$'s of -106×10^{-12} and -196×10^{-12}, which correspond to ϕ's of $73°$ and nearly $90°$. For the fully planar forms the $_mK$'s would be $+338 \times 10^{-12}$ and $+591 \times 10^{-12}$, and for the orthogonal forms -150×10^{-12} and -191×10^{-12}. The steric effect of 2,6-disubstitution by methyl is thus much more plainly demonstrated by $_mK$'s than by dipole moments (which run, for the three ketones mentioned, as 2.96 D, 2.81 D, and 2.81 D); nevertheless it was through polarity measurements that "steric inhibition of resonance" was first recognized in this series [118].

Further examples include diphenyl- and dimestitylketones [119], the $_\infty(_mK_2)$'s of which are $+126 \times 10^{-12}$ and -202×10^{-12}; the values calculated when the aryl groups are twisted by ϕ's of 42° and 43° from the planar configuration of the former molecule are $+131 \times 10^{-12}$ or $+116 \times 10^{-12}$, and by ϕ's of 60° or 61° in the latter are -192×10^{-12} or -208×10^{-12}.

Steric interactions between 9-substituents and *peri*-hydrogens are evident in cases [120] such as 9-acetylanthracene, for which the $_mK$ observed is -643×10^{-12}, while $_mK$'s calculated are $+1189 \times 10^{-12}$, -648×10^{-12}, or -705×10^{-12} for out-of-plane angles of 0°, 80°, or 90°. With 9-benzoylanthracene the $_\infty(_mK_2)$ is -1000×10^{-12}, the predicted $_mK$ for a configuration having the C=O group in the plane perpendicular to the anthryl radical, and with the C_6H_5 group twisted 45° out of this plane, is -993×10^{-12}.

In these and other flexibility problems, the angles quoted merely specify the "effective" conformations (see ref. [3], p. 63); it is not suggested that single forms alone occur. Properly, the measured $_\infty(_mK_2)$'s are mean values for populations of conformations executing rotational oscillations with temperature-dependent amplitudes:

$$\langle _mK \rangle = \frac{\int_{-\pi}^{+\pi} {}_mK(\phi) \cdot \exp[-U(\phi)/kT] \, d\phi}{\int_{-\pi}^{+\pi} \exp[-U(\phi)/kT] \, d\phi} \tag{6.33}$$

While usually the function $_mK(\phi)$ is easy to express, that for the potential energy $U(\phi)$—arising from various hindrances to free internal rotation—is more complicated. Attempts to apply (6.33) to 1,2-dichloroethane, and to a number of benzyl and alkyl halides, are described in refs. [121–123].

Conformational Equilibria in Cyclic Molecules

Two examples are cited. The first concerns morpholine and its *N*-methyl-derivative, both of which have "chair" structures in which the N—H or N—Me bonds can be disposed equatorially or axially. Molar Kerr constants indicate equatorial/axial ratios for morpholine as $(37 \pm 4)/(63 \pm 4)$ in cyclohexane and $(13 \pm 6)/(87 \pm 6)$ in benzene, and for *N*-methylmorpholine as $(57 \pm 4)/(43 \pm 4)$ in cyclohexane and $(98 \pm 5)/(2 \pm 5)$ in benzene. Solute-solvent interactions are inferred which stabilize an axial disposition of N—H in morpholine and an equatorial one for N—Me in the N-methyl derivative, when these molecules are dissolved in benzene. For morpholine, a type of (H)···(benzene π-electrons) bonding is presumed to occur more readily when N—H is axial than when it is equatorial; for *N*-methylmorpholine a (N)···(benzene π-electrons) inductive association is supposed [124, 125].

The second example concerns the equatorial/axial ratios k deduced [126] for cyclohexyl chloride, bromide, and iodide as solutes in carbon tetrachloride at 25°. In order, these were 2.2_4, 1.5_6, and 1.6_7. By the equation $\Delta G_x^0 = RT \ln k$ estimates of conformational free energies (in kcal/mole) follow as 0.48, 0.27, 0.30; by various other methods (listed in ref. [127]) values have been found within the ranges 0.3–0.5, 0.24–0.94, and 0.3–0.43. The results via $_mK$'s seem reasonable in showing the equatorial isomer to be more stable than the axial by some 250–500 cal/mole in each case. In all such enquiries the effects of concentration of solute, and nature of solvent should be considered —points demonstrated by work on 2-halogenocyclohexanones [51], 1,2-dicyanoethane [128], and so forth—and when ratios in different media are to be compared the k's should desirably always be extrapolated to infinite dilution.

Configurations of Solvent-Solute Complexes

Solvents, besides altering the conformations of flexible molecules, may affect the apparent $_\infty(_mK_s)$'s of rigid ones. Attempts have been made to correlate observed $_\infty(_mK_2)$'s with factors such as solute polarity ([3], p. 46), solute shapes or dimensions [82], and/or known physical properties of the medium [53, 55]. In addition, possible H-bondings between protons of a solute and the π-electrons of aromatic solvents must be envisaged as a fairly common occurrence. Although such associations are loose, they have stereo-specificity. The last point is illustrated [88] by a comparison of the $_mK$'s of fiuoroform and chloroform as gases with those observed at infinite dilution in benzene:

	CHF$_3$	CHCL$_3$
$10^{12}{}_\infty(_mK_2)_{benzene}$	-33	-41
$10^{12}{}_mK_{gas}$	-5.3	-25.8

Two extreme modes of attachment can be foreseen: (a) in which the H$-$C bond of CHX$_3$ is coplanar with the C$_6$H$_6$ ring, or (b) in which this bond is perpendicular to the benzene plane. The resultant moment of the complex will act in (a) along, and in (b) at 90° to, the direction of maximum polariza-bility. By (6.11), therefore, $_\infty(_mK_2)_{benzene}$ should be more positive than $_mK_{gas}$ in situation (a), and more negative in situation (b). Experiment favors the alternative (b). Generalized arguments on the same lines can be applied to a wide range of solute types; further evidence from ^{1}H nmr spectra and mole-cular relaxation times now exists ([88], 1968 ref.).

Optically Active Solutes

The electric birefringences of solutions of such solutes can be measured by the standard methods of Section 2. Eckert and Le Fèvre [129] compared a number of optically active solutes with their inactive forms. It was noted that plots of the birefringences versus concentrations diverged increasingly as solutions were made stronger, but that near-identical readings could be secured by extrapolations to infinite dilution.

Eckert and Le Févre [108] subsequently investigated cholesteryl chloride, bromide, and iodide, and the related molecules Δ^5-cholestene and Δ^5-cholestenone. The equatorial attachment of the halogens in the 3-position was clear from the observed and calculated positive ${}_mK$'s of the halides. By contrast, epi-cholesteryl chloride had the negative ${}_\infty({}_mK_2)$ predictable for an axial disposition of C—Cl in this isomer. Apart from being the first physical determination of the conformations of these two chlorides, the work incidentally showed that the replacement of OH by halogens in the reactions of cholesterol with thionyl chloride, aluminium iodide, and so forth, occurs with retention of configuration. Another point of interest is that decision between axial or equatorial conformations for the isomeric chlorides, m.p. 96° and 107°, could not be made on the basis of polarity since both forms had, within experimental error, the same dipole moment.

Conclusion

Many other examples of the application of the Kerr effect to structural, conformational, and mechanistical problems in chemistry are summarized up to 1965 on pages 56–62 of ref. [3]. Their range and variety demonstrates that the Sydney group has developed a relatively inexpensive simple procedure, involving a minimum of arguments by analogy, which deserves inclusion among the standard methods of modern stereochemistry.

8 ELECTRIC DOUBLE REFRACTION IN DISPERSED AND CONDUCTIVE SYSTEMS

Original Observations

Kerr [1] first noted that dispersed systems of microscopic particles showed electric birefringence; the phenomenon was later studied by Chaudier [130]. An effect in colloids was reported by Cotton and Mouton [131] and the subject was taken up by Freundlich [132], Bergholm and Björnstahl [133], Heller [134], and others.

Size and Cause of the Effect

The Kerr effects observable in such dispersions can be enormous and as much as 10^6 times that observed in nitro-benzene; qualitatively they may be explained by a macroscopic version of the molecular model. Individual particles, having complete randomness of rotation and translation, will, if anisotropically polarized by an external field, be subjected to a torque conducive to bringing the axis of maximum polarization parallel to the lines of force. If the particle should have the character of a permanent dipole, then its orientation in the field will be controlled by the direction of action of this doublet. In either case, the torque increases in proportion to the volume of the particle and with the square of the field strength; thus the Kerr constant B [see (6.1)] may be taken as:

$$B = \frac{(n_p - n_s)}{\lambda E^2} = \frac{\delta v_{12}}{2\pi E^2 l v_2} \tag{6.34}$$

where v_2 is the volume of the dispersed phase, v_{12} the volume of the dispersed system, and l the cell thickness. With molecular mixtures, thermal agitation is the factor controlling the degree of orientation of the molecules; on a macroscopic scale the Brownian movement takes its place. As the particle size increases, the disordering influence decreases, and the validity of (6.34) diminishes.

Relationship between Kerr's Law and Particle Size

For molecules of diameter less than 10 Å Kerr's law holds almost up to sparking potential, for particles of approximately 1000 Å in diameter a diminution of the exponent of E begins to display itself at very low voltage gradients [135, 136].

Experimental Techniques

Early Methods

Several workers used the polarizing microscope to study colloidal solutions and suspensions of solids in liquids subjected to sinusoidal alternating fields of about 50-cycle frequency. This technique has a disadvantage. A frequency should exist above which birefringence due to orientation becomes negligible. According to Heller [134] this $v_{critical} = \mathbf{k}T/8\pi\eta r$ (where η is the viscosity and r the radius). The critical frequency is not necessarily at all high, for example, the negative double refraction of benzopurpurin falls off rapidly at about 10^6 cps; with certain other substances confused results have been reported, colloidal clays being an instance where the birefringence is apparently negative below 1000 cps and positive above [137].

There is also the difficulty that each particle may be made from intrinsically birefringent molecules, which can build up into rods or plates, in such a way as to yield differently anisotropic wholes.

Methods Now in Use

The introduction of the cathode-ray oscilloscope (CRO) and the photo-multiplier tube brought new opportunities for investigation, especially when these instruments were combined with pulsed electric fields. Pauthenier [138] had used such fields in studies of electrostriction, producing them from condenser discharges. The first apparatus of a more convenient kind was described by Kaye and Devaney [33]. Pulses of a few microseconds duration, generated every hundredth of a second, were raised to appropriate voltages by amplification, and then applied to a Kerr cell situated within the usual optical arrangement of crossed prisms (as in Fig. 6.1) except that a photo-multiplier had replaced the telescope-observer combination. The signals from the multiplier passed, via a video amplifier, to the Y-plates of a CRO. By these means the shape of the response curve (in essence, a function of B^2E^4 plotted against time) was directly displayed. When orientation did not lag behind the field, the wave shape of the optical pulse was very nearly the same as that of the electrical pulse, but when relaxation phenomena occurred, distortion to forms such as those indicated in Fig. 6.11(f) were seen.

The use of pulsed fields diminishes heating in the material under examination and appears to make accessible more information than could formerly be gained with sinusoidal fields.

Mathematical treatment of the oscilloscope records has been supplied by Benoit [34], who applied his calculations to studies of tobacco mosaic virus, thymonucleic acid, and other systems, showing that rotary diffusion constants can be deduced quite simply.

In practice today a double-beam oscilloscope is used [34–47] and the signals from the birefringence and the applied voltage photographed simultaneously (as in Fig. 6.11). With macromolecular solutes such records commonly show three regions in which the birefringence (a) is increasing, (b) is steady, and (c) is decaying with time. Respectively, these regions reveal (a) whether the solute has a permanent dipole moment, (b) the magnitude of its Kerr constant, and (c) its rotary diffusion constant. To extract such information, equations relating the intensity of birefringence to time must be fitted to the traces (cf. Benoit [34]).

The decay of birefringence in region (c) occurs exponentially, and—with a monodisperse phase or a single macromolecule—can be written

$$y = \exp(-12Dt) \qquad (6.35)$$

where y represents the intensity at time t, and D is the rotary diffusion constant of the solute under examination. For an ellipsoidal particle of length $2a$ and breadth $2b$, D is

$$D = \frac{3kT[\ln(2a/b) - 0.5]}{8\pi\eta a^3} \qquad (6.36)$$

(or, if the particle is regarded as cylindrical, the 0.5 is replaced by 0.8). The relaxation time τ for the species involved is $1/6D$; thus τ depends strongly on the length and much less on the axial ratio—which enters as $\ln(2a/b)$.

Apart from Benoit's work already cited, notable applications of this "dynamic birefringence" technique have been Tinoco's study of the fibrin-fibrinogen reaction [36], Itzhaki's investigations on rat thymus deoxyribonucleoprotein preparations [43], Norman and Field's assessment [38] of the breakage of DNA rods by x-rays, and observations by Shah et al. [44] on bentonite suspensions. Consultation of refs. [34] to [47] will show the scope of the technique.

The region (b) of the birefringence-time curves has been used to estimate Kerr constants of weakly conducting liquids. Thus Orttung and Meyers [46] have recorded $10^7 B$ for water at 25° as 3.72, 2.89, and 2.72, relatively to carbon bisulphide, for $\lambda\lambda$ of 436, 546, and 578 mμ respectively, using 2–7 kV pulses of 70 μsec length; further, they have successfully examined glycine and certain simple amino acids in aqueous solution. The Sydney group has similarly avoided conduction troubles that previously prevented measurements on such things as aliphatic acids, ethyl acetoacetate, acetylacetone, dimedone, and phenols [47].

References

1. J. Kerr, *Phil. Mag.*, **50**, 337, 446 (1875).
2. C. G. Le Fèvre and R. J. W. Le Fèvre, *Rev. Pure Appl. Chem.*, **5**, 261 (1955).
3. R. J. W. Le Fèvre, "Molecular Refractivity and Polarizability," in *Advances in Physical Organic Chemistry*, V. Gold, Ed., Vol. 3, Academic, London-New York, 1965, p. 1.
4. C. G. Le Fèvre and R. J. W. Le Fèvre, *J. Chem. Soc.*, **1953**, 4041.
5. R. J. W. Le Fèvre and G. L. D. Ritchie, *J. Chem. Soc.*, **1963**, 4933.
6. *Idem.*, *ibid.*, **1965**, 3520.
7. R. J. W. Le Fèvre and S. C. Solomons, *Australian J. Chem.*, **21**, 1703 (1968).
8. G. Szivessy and A. Dierkesmann, *Ann. Physik*, **11**, 949 (1931); *Z. Instrumentenk.*, **52**, 337 (1932).
9. G. Szivessy, *Z. Instrumentenk.*, **57**, 49, 89 (1937).
10. H. A. Stuart and H. Volkmann, *Ann. Physik*, **18**, 121 (1933).
11. H. G. Jerrard, *J. Opt. Soc. Am.*, **38**, 35 (1948).
12. L. Chaumont, *Ann. Physik*, **5**, 31 (1916).
13. R. J. W. Le Fèvre and G. L. D. Ritchie, *J. Chem. Soc.*, **1965**, 3520.
14. R. J. W. Le Fèvre, *Dipole Moments*, 3rd ed., Methuen, London, 1953, Chap. 2.
15. E. C. Stevenson and J. W. Beams, *Phys. Rev.*, **38**, 133 (1931).
16. C. W. Bruce, *Phys. Rev.*, **44**, 682 (1933).
17. H. Bizette, *Compt. Rend.*, **202**, 304 (1936).
18. L. H. Borchert, *Physik. Z.*, **39**, 156 (1938).
19. E. Kuss and H. A. Stuart, *Physik. Z.*, **42**, 95 (1941).

20. L. L. Boyle, A. D. Buckingham, R. L. Disch, and D. A. Dunmur, *J. Chem. Physics*, **45**, 1318 (1966).

21. A. D. Buckingham and D. A. Dunmur, *Trans. Faraday Soc.*, **64**, 1776 (1968).

22. G. G. Quarles, *Phys. Rev.*, **46**, 692 (1933).

23. W. M. Breazeale, *Phys. Rev.*, **48**, 237 (1935); **49**, 625 (1936).

24. *cf.*, F. A. Jenkins and H. E. White, *Fundamentals of Optics*, McGraw-Hill, New York, 1957.

25. *cf.*, N. H. Hartshorne and A. Stuart, *Crystals and the Polarising Microscope*, 3rd ed., Arnold, London, 1960.

26. T. des Coudres, *Verhandl. Ges. deutsch Naturforsch. Ärtz*, **65**, 67 (1893).

27. G. Szivessy, *Z. Physik*, **2**, 30 (1920).

28. D. E. Hansen, *Dissertation*, Karlsruhe (1912).

29. W. Schmidt, "Electrische Doppelbrechung in gut und schlecht isolirenden Flüssigkeiten," *Dissertation*, Gottingen, 1901.

30. L. V. Cherry, M. E. Hobbs, and H. A. Strobel, *J. Phys. Chem.*, **61**, 465 (1957).

31. A. D. Buckingham and B. J. Orr, *Proc. Roy. Soc.*, *A*, **305**, 259 (1968).

32. R. J. W. Le Fèvre, R. K. Pierens, and K. D. Steels, *Australian J. Chem.*, **19**, 1769 (1966).

33. W. Kaye and R. Devaney, *J. Appl. Phys.*, **18**, 912 (1947).

34. H. Benoit, *Compt. Rend.*, **228**, 1716 (1949); Contribution à l'étude de l'effect Kerr présenté par les solutions diluées de macromolecules rigides, Thesis, Strasbourg, 1950; *Ann. Physik*, **6**, 561 (1951); *J. Chim. Phys.*, **47**, 719 (1950); **49**, 517 (1952).

35. C. T. O'Konski and B. H. Zimm, *Science*, **111**, 113 (1950).

36. I. Tinoco, *J. Am. Chem. Soc.*, **77**, 3476, 4486 (1955).

37. C. T. O'Konski and J. B. Applequist, *Nature*, **178**, 1464 (1956).

38. A. Norman and J. A. Field, Jr., *Arch. Biochem. Biophys.*, **70**, 257; **71**, 170 (1957).

39. C. T. O'Konski, K. Yoshioka, and W. H. Orttung, *J. Phys. Chem.*, **63**, 1558 (1959).

40. I. Tinoco and K. Yamaaska, *J. Phys. Chem.*, **63**, 423 (1959).

41. S. Krause and C. T. O'Konski, *J. Am. Chem. Soc.*, **81**, 5082 (1959).

42. P. Ingram and R. G. Jerrard, *Nature*, **196**, 57 (1962).

43. R. F. Itzhaki, *Nature*, **194**, 1241 (1962); *Biochem. J.*, **100**, 211 (1966); *Proc. Roy. Soc. B*, **164**, 75, 411 (1966).

44. M. J. Shah and C. M. Hart, *I.B.M. J. Res. Devt.*, **7**, 44 (1963).

45. M. J. Shah, D. C. Thompson, and C. M. Hart, *J. Phys. Chem.*, **67**, 1170 (1963).

46. W. H. Orttung and J. A. Meyers, *J. Phys. Chem.*, **67**, 1905, 1911 (1963).

47. M. A. Cusiter, R. J. W. Le Fèvre, D. Millar, R. K. Pierens, and K. D. Steel, *Australian J. Chem.*, in press.

48. C. G. Le Fèvre and R. J. W. Le Fèvre, *J. Chem. Soc.*, **1959**, 2670.

49. H. Mouton, *International Critical Tables*, Vol. 7, 1930 ed., p. 110.

50. M. J. Aroney, R. J. W. Le Fèvre, and J. Saxby, *J. Chem. Soc.*, **1963**, 4938.

51. C.-Y. Chen and R. J. W. Le Fèvre, *J. Chem. Soc.*, **1965**, 3700.

52. R. J. W. Le Fèvre, B. J. Orr, and G. L. D. Ritchie, *J. Chem. Soc. B*, **1966**, 273.

53. R. J. W. Le Fèvre and A. J. Williams, *J. Chem. Soc.*, **1961**, 1671.

54. H. A. Stuart, *Die Struktur des Freien Moleküls*, Springer, Berlin-Gottingen-Heidelberg, 1952, pp. 460–463.

55. R. J. W. Le Fèvre and A. J. Williams, *J. Chem. Soc.*, **1964**, 562.

56. F. Gabler and P. Sokub, *Naturwiss.*, **24**, 570 (1936).

57. C. G. Le Fèvre and R. J. W. Le Fèvre, *J. Chem. Soc.*, **1957**, 3458.

58. H. A. Stuart, *Molekülstruktur*, Springer, Berlin-Heidelberg-New York, 1967.

59. A. Piekara and R. Konopka, *Brit. J. Appl. Phys.*, **12**, 50 (1961).

60. F. Hehlgans, *Z. Physik*, **33**, 378 (1932).

61. H. Friedrich, *Physik. Z.*, **38**, 318 (1937).

62. P. Langevin, *Radium*, **7**, 249 (1910), *Compt. Rend.*, **151**, 475 (1910).

63. M. Born, *Ann. Physik*, **55**, 177 (1918).

64. R. Gans, *Ann. Physik*, **64**, 481 (1921); **65**, 97 (1921).

65. P. Debye and H. Sack, *Handbuch der Radiologie*, Marx, Leipsig, 1934, Vol. 6, Ch. 2.

66. W. Voigt, *Ann. Physik*, **4**, 197 (1901); *Wiedem. Ann.*, **69**, 297 (1899).

67. R. J. W. Le Fèvre and D. A. A. S. N. Rao, *J. Chem. Soc.*, **1956**, 708.

68. M. Born and P. Jordan, *Elementare Quantenmechanik*, Springer, Berlin, 1930, p. 259.

69. C. A. Coulson, A. Maccoll, and L. E. Sutton, *Trans. Faraday Soc.*, **48**, 106 (1952).

70. A. D. Buckingham and J. A. Pople, *Proc. Phys. Soc. (London)*, **A68**, 905 (1955).

71. Ref. [2], pp. 307–310; ref. [3], pp. 65–73.

72. A. D. Buckingham and B. J. Orr, *Quart. Rev.* (Chem. Soc. London), **21**, 195 (1967).

73. G. Otterbein, *Physik. Z.*, **35**, 249 (1934).

74. G. Briegleb, *Z. Physik. Chem.*, **B14**, 97 (1931).

75. C. G. Le Fèvre and R. J. W. Le Fèvre, *J. Chem. Soc.*, **1954**, 1577.

76. In about 200 papers, mostly prefixed by the series title "Molecular Polarisability," mainly in the *J. Chem. Soc.* and the *Australian J. Chem.*, since 1953.

77. H. H. Huang and S. C. Ng, *J. Chem. Soc.* (*B*), **1968**, 582.

78. L. H. L. Chia and H. H. Huang, *J. Chem. Soc.* (*B*), **1968**, 1369.

79. H. A. Stuart, in *Landolt-Börnstein, Zahlenwerte und Funktionen*, 6th ed., Springer, Berlin-Göttingen-Heidelberg, 1962, Vol. 2, No. 8, p. 886.

80. R. J. W. Le Fèvre and K. D. Steel, *Chem. Ind.*, **1961**, 670.

81. R. J. W. Le Fèvre and B. P. Rao, *J. Chem. Soc.*, **1957**, 3644.

82. R. S. Armstrong, M. J. Aroney, C. G. Le Fèvre, R. J. W. Le Fèvre, and M. R. Smith, *J. Chem. Soc.*, **1958**, 1474.

83. R. J. W. Le Fèvre and B. J. Orr, *J. Chem. Soc.*, **1965**, 5349.

84. R. J. W. Le Fèvre, B. J. Orr, and G. L. D. Ritchie, *J. Chem. Soc.*, **1965**, 2499.

85. R. J. W. Le Fèvre and B. J. Orr, *J. Chem. Soc.* (*B*), **1966**, 37.

86. R. J. W. Le Fèvre and L. Radom, *J. Chem. Soc.* (*B*), **1967**, 1295.

87. R. J. W. Le Fèvre and B. P. Rao, *J. Chem. Soc.*, **1958**, 1465.

88. R. J. W. Le Fèvre, G. L. D. Ritchie, and P. J. Stiles, *Chem. Comm.*, **1966**, 326; *J. Chem. Soc.* (B), **1968**, 148.

89. M. J. Aroney, R. J. W. Le Fèvre, and R. K. Pierens, *Australian J. Chem.*, **20**, 2251 (1967).

90. R. J. W. Le Fèvre, *Proc. Chem. Soc.*, **1958**, 283.

91. M. J. Aroney, R. J. W. Le Fèvre, and J. D. Saxby, *J. Chem. Soc.*, **1963**, 1739.

92. R. J. W. Le Fèvre, Rev. Pure and Appl. Chem. **20**, 67 (1970).

93. L. E. Sutton, *Tables of Interatomic Distances and Configuration in Molecules and Ions*, Chem. Soc. Special Publs., No. 11 (1958), No. 18 (1965).

94. R. J. W. Le Fèvre, *Proc. Chem. Soc.*, **1959**, 365.

95. C.-Y. Chen and R. J. W. Le Fèvre, *J. Chem. Soc.* (*B*), **1966**, 40.

96. R. J. W. Le Fèvre, R. K. Pierens, D. V. Radford, and K. D. Steel, *Australian J. Chem.*, **21**, 1965 (1968).

97. L. Silberstein, *Phil. Mag.*, **33**, 92, 215, 521 (1917).

98. K. S. Pitzer, *Advan. Chem. Phys.*, **2**, 79 (1959).

99. R. P. Smith and E. M. Mortensen, *J. Chem. Phys.*, **32**, 502, 508 (1960).

100. P. L. Davies, *Trans. Faraday Soc.*, **48**, 789 (1952).

101. H. C. Bolton, *Trans. Faraday Soc.*, **50**, 1261, 1265 (1954).

102. C. W. Bunn and R. Daubeny, *Trans. Faraday Soc.*, **50**, 1173 (1954).

103. J. N. Murrell and M. J. P. Musgrave, *Trans. Faraday Soc.*, **63**, 2849 (1967).

104. M. F. Vuks, *Opt. i Spektroskopiya*, **2**, 494 (1957).

105. W. Philippoff, *J. Appl. Phys.*, **31**, 1899 (1960).

106. C. Clément and P. Bothorel, *J. Chim. Phys.*, **61**, 1262 (1964).

107. G. Sachsse, *Physik. Z.*, **36**, 357 (1935).

108. J. M. Eckert and R. J. W. Le Fèvre, *J. Chem. Soc.*, **1962**, 1081.

109. P. H. Cureton, C. G. Le Fèvre, and R. J. W. Le Fèvre, *J. Chem. Soc.*, **1963**, 1736.

110. M. J. Aroney, R. J. W. Le Fèvre, and J. D. Saxby, *J. Chem. Soc.*, **1965**, 571.

111. M. J. Aroney, G. M. Hoskins, and R. J. W. Le Fèvre, *J. Chem. Soc.* (*B*), **1968**, 1206.

112. R. W. Green, R. J. W. Le Fèvre, and J. D. Saxby, *Australian J. Chem.*, **19**, 2007 (1966).

113. P. H. Cureton, C. G. Le Fèvre, and R. J. W. Le Fèvre, *J. Chem. Soc.*, **1961**, 4447.

114. R. J. W. Le Fèvre and P. J. Stiles, *J. Chem. Soc.* (*B*), **1966**, 420.

115. M. J. Aroney, R. J. W. Le Fèvre, R. K. Pierens, and H. L. K. The, *Australian J. Chem.*, **21**, 281 (1968).

116. R. J. W. Le Fèvre, G. L. D. Ritchie, P. H. Gore, J. A. Hoskins and L. Radom, *J. Chem. Soc.* (*B*), **1969**, 485.

117. M. J. Aroney, R. J. W. Le Fèvre, and M. G. Corfield, *J. Chem. Soc.*, **1964**, 648.

118. R. G. Kadesch and S. W. Weller, *J. Am. Chem. Soc.*, **63**, 1310 (1941).

119. P. H. Gore, J. A. Hoskins, R. J. W. Le Fèvre, L. Radom, and G. L. D. Ritchie, *J. Chem. Soc.* (*B*), **1967**, 741.

120. R. J. W. Le Fèvre, L. Radom, and G. L. D. Ritchie, *J. Chem. Soc.* (*B*), **1968**, 775.

121. R. J. W. Le Fèvre and B. J. Orr, *Australian J. Chem.*, **17**, 1098 (1964).

122. R. J. W. Le Fèvre, B. J. Orr, and G. L. D. Ritchie, *J. Chem. Soc.*, **1965**, 3619.

123. C.-Y. Chen, R. J. W. Le Fèvre, and S. C. Paul, *J. Chem. Soc. (B)*, **1967**, 503.

124. M. J. Aroney, C.-Y. Chen, R. J. W. Le Fèvre, and J. D. Saxby, *J. Chem. Soc.*, **1964**, 4269.

125. F. G. Riddell, *Quart. Rev.*, **21**, 364 (1967).

126. C. G. Le Fèvre, R. J. W. Le Fèvre, R. K. Pierens, and R. Roper, *Proc. Chem. Soc.*, **1960**, 117.

127. E. L. Eliel, N. L. Allinger, S. J. Angyal, and G. A. Morrison, *Conformational Analysis*, Interscience, New York, 1965, Chap. 7.

128. R. J. W. Le Fèvre, G. L. D. Ritchie, and P. J. Stiles, *Chem. Comm.*, **1966**, 846; *J. Chem. Soc. (B)*, **1967**, 819.

129. J. M. Eckert and R. J. W. Le Fèvre, *J. Chem. Soc.*, **1961**, 2356.

130. J. Chaudier, *Compt. Rend.*, **137**, 248 (1903); **142**, 201 (1906); **149**, 2021 (1909).

131. A. Cotton and H. Mouton, *Ann. Chim. Phys.*, **145**, 289 (1907).

132. K. Diesselhorst, H. Freundlich, and W. Leonhardt, *Elster-Geitel Festschrift*, Braunschweig, 1915, p. 462; H. Freundlich, *Z. Elektrochem.*, **22**, 27 (1916).

133. C. Bergholm and Y. Björnstahl, *Physik. Z.*, **21**, 137 (1920).

134. W. Heller, *Rev. Mod. Phys.*, **14**, 390 (1942).

135. H. Muller, *Phys. Rev.*, **55**, 508, 792 (1939).

136. J. Errera, J.-Th.-G. Overbeek, and H. Sack, *J. Chim. Phys.*, **32**, 681 (1935).

137. F. J. Norton, *Phys. Rev.*, **55**, 668 (1939).

138. M. J. Pauthenier, *J. Phys. Radium*, **2**, 183, 384 (1921); **5**, 312 (1924); **6**, 1 (1925).

General References

Beams, J. W., "Electric and Magnetic Double Refraction," *Rev. Mod. Phys.*, **4**, 133 (1932).

Debye, P., and H. Sack, *Handbuch der Radiologie*, Marx, Leipzig, 1934, Vol. 6, Chap. 2.

Gray, A., "Obituary of John Kerr (1824–1907)," *Proc. Roy. Soc.*, **A82**, 629 (1909).

Le Fèvre, C. G., and R. J. W. Le Fèvre, "The Kerr Effect, Its Measurement and Applications in Chemistry," *Rev. Pure Appl. Chem.* (Australia), **5**, 261–318 (1955).

Le Fèvre, R. J. W., "Molecular Refractivity and Polarizability," in *Advances in Physical Organic Chemistry*, V. Gold, Ed., Vol. 3, Academic, London-New York, 1965, pp. 1–90.

Stuart, H. A., *Molekülstruktur*, Springer, Berlin-Heidelberg-New York, 1967.

Chapter **VII**

ELLIPSOMETRY

N. M. Bashara, A. C. Hall, and A. B. Buckman

I INTRODUCTION

Two types of measurements can be made in conventional ellipsometry: (1) determination of the properties of a surface and (2) determination of the properties of a partly transparent film on a known substrate. In the first even slight contamination cannot be tolerated if the optical or dielectric constants, real and imaginary, are to be determined to a high accuracy. This extreme sensitivity of ellipsometry to even slight contamination in itself has resulted in the use of the technique to determine when a surface is clean in comparative LEED-ellipsometry measurements. In the many applications where the error in surface properties due to surface contamination is acceptable, the ellipsometer gives both components of the refractive index (the usual index and the extinction coefficient) in a single measurement. Studies in ultra-high vacuum should give accurate surface properties on many materials which when cleaned remain free of contamination for long periods in high vacuum. Although this technique is under study in many laboratories, obtaining strain-free windows is still a major problem. The extreme sensitivity of the ellipsometer to thin films on a surface makes the technique particularly attractive for the study of such films. Furthermore, the studies can be carried out *in situ* so long as the environment is at least partly transparent to the light. Oxidation kinetics in gaseous and liquid environments, adsorption from solution, and things of this kind are amenable to study by the method. The digital computer has opened new areas of study. Large numbers of trial solutions can be generated and compared to experimental data.

It is well known in ellipsometry that even the very thinnest films—monolayers—can produce observable effects in light reflection. This is confirmed by theory, which shows that thickness changes of less than 1 Å are experimentally accessible. Although surface layers are usually assumed to be plane-parallel, homogeneous and isotropic, the experimental sensitivity is such that films which are emphatically anisotropic, in fact discontinuous, can be studied. Equations applying to this case have not yet been derived from first principles, but a number of modifications have been made of existing formulas in seeking to account for results on partially covered surfaces. Aside from difficulties of theoretical interpretation, there are many problems in preparing clean surfaces, introducing known fractional surfaces coverages, or measuring coverages by independent techniques. Nevertheless, it appears certain that ellipsometry can be used in favorable cases to detect films at coverages below 10%.

Ellipsometry also offers additional analytical power in the study of perturbations in optical or dielectric constants caused by time-varying electric fields, sample temperatures, or mechanical stresses. It will be shown how spectra of the perturbation-induced changes in the dielectric constants can be obtained directly from modulated ellipsometry measurements, and how the available experimental parameters can be used to determine the extent of the modulation into the sample surface. The photon-energy spectra of the changes in the dielectric constants, obtained directly by modulated ellipsometry, are useful in determining the topology and energy of critical points corresponding to interband electronic transitions in the sample being studied. The depth into the sample to which the modulation extends can in some cases be critical to the determination of the changes in dielectric constants, as well as giving information on sample surface conditions.

Abstract

The essentials of the analysis of measurements on clean, isotropic surfaces and filmed surfaces are presented at an introductory level. The published literature is referenced and discussed, particularly for studies of very thin surface coverage where the great power of the technique is most evident. The advantages of ellipsometry over normal incidence measurements to study perturbations in the optical or dielectric properties are also discussed.

2 ANALYSIS OF THE CLEAN ISOTROPIC SURFACE

Introduction

Two configurations are most commonly used in ellipsometric measurements to determine the two optical constants of a clean surface. In the earlier method [1] the order of the components is monochromatic light source, polarizer, specimen, compensator, analyzer, and detector. Holmes [2] has discussed the general theory of this arrangement. The method [3, 4] discussed in this section places the compensator before the specimen and is shown schematically in Fig. 7.1. In this configuration linearly polarized light of a particular wavelength from the polarizer [5] becomes, in general, elliptically polarized passing through the compensator [6–8]. Reflection from the specimen introduces changes in the magnitudes and phases of the electric field components of the light. The measurement involves adjustment of the polarizer and analyzer for minimum light to the detector (the null condition).

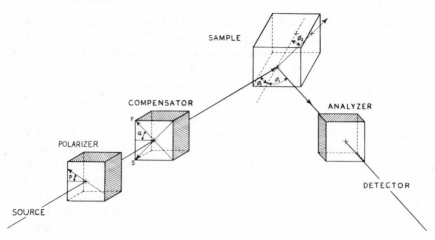

Fig. 7.1. The assumed geometry. The angle of incidence is ϕ_1 and the angle of refraction is ϕ_2. The plane of incidence is defined by the direction of the light ray and the normal to the test surface. The azimuthal angles (with respect to the plane incidence) P and Q are for the polarizer plane of transmission and the fast axis of the compensator, respectively. In the analysis in Section 2, Q is assumed to be $\pm 45°$.

The parameters, ψ (related to magnitude) and Δ (phase), are determined from azimuthal angle measurements of polarizer, analyzer, and compensator and other knowns of the system. The two optical or dielectric constants of the specimen are then determined from the parameters ψ and Δ through the Fresnel reflection coefficients [9].

A simple and direct presentation on an introductory level will be made and involves analysis of the effect of each of the components in the optical path, successively. Much of the notation and general approach used by Holmes [2] and Oldham [4] will be followed.

Hall [9a] has summarized the early developments in ellipsometry and Smith [9b] has presented a general analysis for both ellipsometric arrangements, that is, compensator before and after the specimen.*

Preliminary Analysis

Figure 7.2 defines the changes that linearly polarized light undergoes as it passes through the compensator and is incident on the specimen. In Fig. 7.2 the azimuthal relations of Fig. 7.1 are retained if the reader imagines himself looking into the light beam with positive angles measured counterclockwise (ccw). The electric component (throughout, only the electric vector is considered) of the light leaving the polarizer is perpendicular to the direction of transmission (the direction of the light ray) and is at an azimuthal angle P to the plane of incidence (defined by the light ray and the normal to the test surface), as shown in Figs. 7.1 and 7.2. It is also assumed to be of unit amplitude. Its plane of transmission is defined by the light vector at azimuth P and the direction of the light ray. Since the light source is monochromatic, the light vector varies sinusoidally in length. This time dependence will enter implicitly into the convention that will be used in characterizing the compensator. We will assume a time dependence $E \cos \omega t$, where E is the amplitude, ω is the angular frequency, and t is time. It will be convenient to use polar notation, so that $E \cos \omega t$ will be written $E e^{i\omega t}$, it being understood that what is meant is $\mathrm{Re}(E e^{i\omega t})$. It will also be convenient (necessary when considering the compensator) to resolve the linearly polarized light exiting the polarizer into orthogonal components, which are in phase and perpendicular to the direction of the light ray. The amplitude of each component is a function of the azimuthal angle P. The ensuing analysis is concerned with the variation of the relative amplitudes and phase differences between the resolved components due to the elements in the path of the light ray. The time dependence of the light vectors is irrelevant to the calculations; only the azimuthal angles are of concern.

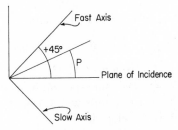

Fig. 7.2. The azimuthal relations of the polarizer and the fast axis of the compensator looking into the light beam and with the assumed geometry of Fig. 7.1.

The compensator is either a quarter-wave plate or more conveniently a Soleil-Babinet compensator if the wavelength dependence of the specimen is being studied. The compensator transmits along two axes, fast and slow. The components of the linear light from the polarizer incident on these axes are, first, that incident on the fast axis

$$E_i^f = \cos\left(\frac{\pi}{4} - P\right) \qquad (7.1)$$

and second, that incident on the slow axis

$$E_i^s = \sin\left(\frac{\pi}{4} - P\right) \qquad (7.2)$$

for the case where the fast axis is at an azimuthal angle of $+45°$ with respect to the plane of incidence. The components transmitted by the compensator are (1) for the fast axis

$$E_t^f = E_i^f, \tag{7.3}$$

and (2) for the slow axis

$$E_t^s = TE_i^s e^{-i\beta}, \tag{7.4}$$

due to the attenuation and phase difference between the fast and slow axes. Equation (7.4) implies that the wave component transmitted by the slow axis lags that of the fast axis by the angle β. Also, implicit in the notation is the convention that the time dependence between the transmitted and incident wave for either axis is given by $E_t^{s,f} = E_i^{s,f} T^{s,f} e^{i\omega(t-x/v_{s,f})}$ where $T^{s,f}$ is the ratio of the transmitted to incident amplitudes for either axis, x is the distance traversed by each wave component, and $v_{s,f}$ is the wave velocity for the axes. With this notation, then,

$$\frac{E_t^s}{E_t^f} = \frac{(E_i^s/E_i^f)(T^s/T^f)e^{+i\omega(t-x/v_s)}}{e^{+i\omega(t-x/v_f)}}.$$

This reduces to the ratio of (7.4) to (7.3) where

$$T = T^s/T^f \quad \text{and} \quad \beta = (2\pi/\lambda_{\text{vac}})(n_s - n_f)$$

for a quarter-wave plate, for example.

Following Holmes [2] and Oldham [4] it will be assumed that the compensator is not ideal, that is, T is not unity and β is not $90°$. Expressions for calculation of T and β will be developed.

Incident on the specimen in the plane of incidence are components from both the fast and slow axes of the compensator. By using (7.1) to (7.4), the p-light (parallel to the plane of incidence) incident on the specimen is found to be

$$E_{pi} = \cos\left(\frac{\pi}{4} - P\right)\cos\left(\frac{\pi}{4}\right) + \sin\left(\frac{\pi}{4} - P\right)\sin\left(\frac{\pi}{4}\right)Te^{-i\beta}, \tag{7.5}$$

$$= A + Be^{-i\beta} \tag{7.5a}$$

and similarly the s-light (perpendicular to the plane of incidence) incident on the specimen is

$$E_{si} = \cos\left(\frac{\pi}{4} - P\right)\sin\left(\frac{\pi}{4}\right) - \sin\left(\frac{\pi}{4} - P\right)\cos\left(\frac{\pi}{4}\right)Te^{-i\beta}, \tag{7.6}$$

$$= C + De^{-i\beta}. \tag{7.6a}$$

A planar test surface with optical properties $n_2^* = n_2 - ik_2$ is immersed in a material whose properties are $n_1^* = n_1 - ik_1$, where n is the refractive index and k is the extinction coefficient. Light from the compensator passes through medium 1 and is incident on the test surface at an angle of incidence ϕ_1. Part of this light is reflected at the angle ϕ_1 and part is refracted at the angle ϕ_2. The test specimen is assumed to be isotropic so that Snell's law is obeyed, that is, $n_1^* \sin \phi_1 = n_2^* \sin \phi_2$.

The reflected and transmitted electric field components are proportional to the incident field components.

$$E_{pr} = pE_{pi} \tag{7.7}$$

$$E_{sr} = sE_{si} \tag{7.8}$$

$$E_{pt} = t_p E_{pi} \tag{7.9}$$

$$E_{st} = t_s E_{si}, \tag{7.10}$$

where in (7.7) E_{pr} and E_{pi} are the reflected and incident electric field components, respectively, and p is the Fresnel reflection coefficient for p-light. A similar expression for s-light is given in (7.8). The relations between the transmitted and incident p- and s-components are given in (7.9) and (7.10) where t_p and t_s are the transmission coefficients for the p- and s-components, respectively.

The Fresnel reflection and transmission coefficients can be written in terms of the material constants. The form of the coefficients is based on the conventions used in Fig. 7.3 [9c].

$$p = \frac{n_2^* \cos \phi_1 - n_1^* \cos \phi_2}{n_2^* \cos \phi_1 + n_1^* \cos \phi_2} \tag{7.11}$$

$$s = \frac{n_1^* \cos \phi_1 - n_2^* \cos \phi_2}{n_1^* \cos \phi_1 + n_2^* \cos \phi_2} \tag{7.12}$$

$$t_p = \frac{2n_1^* \cos \phi_1}{n_2^* \cos \phi_1 + n_1^* \cos \phi_2} \tag{7.13}$$

$$t_s = \frac{2n_1^* \cos \phi_1}{n_1^* \cos \phi_1 + n_2^* \cos \phi_2}. \tag{7.14}$$

The ratio of the Fresnel reflection coefficients, p/s, in terms of the materials constants is generally complex and is commonly written

$$\frac{p}{s} = \tan \psi e^{i\Delta}, \tag{7.15}$$

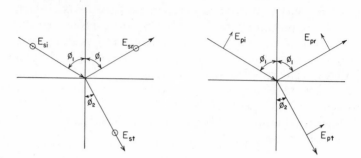

Fig. 7.3. The conventions used in defining the Fresnel coefficients. The horizontal axis is the plane of the specimen. The s-light vectors are out of the page. E_{si}, E_{sr}, E_{st} and E_{pi}, E_{pr}, E_{pt} are the incident, reflected, and transmitted components of the s- and p-light, respectively; ϕ_1 and ϕ_2 are the angle of incidence and refraction, respectively.

(tan ψ and $e^{i\Delta}$ are a magnitude and phase factor, respectively) and in terms of the electric field vectors is, from (7.7) and (7.8),

$$\frac{p}{s} = \frac{E_{pr}/E_{pi}}{E_{sr}/E_{si}}. \tag{7.16}$$

When (7.15) and (7.16) are combined the result is

$$\frac{E_{pr}}{E_{sr}} = \frac{E_{pi}}{E_{si}} \tan \psi e^{i\Delta}. \tag{7.17}$$

Equation (7.17) is the general equation of ellipsometry which through (7.11), (7.12), (7.15), and (7.16) relates the ellipsometer parameters ψ and Δ to the material constants of the clean surface. A similar relation will be obtained later for a substrate with an overlay film.

Polarizer Zone Relations-Ideal Compensator

Two pairs of polarizer–analyzer readings called zones [4, 10, 11] and arbitrarily labeled (P_2, A_2), (P_4, A_4) will give the null condition for a compensator azimuth of $+45°$. The conditions on polarizer readings for null are selected first. Assume an ideal compensator, that is, $T = 1$, $\beta = 90°$. The ratio of (7.5) and (7.6) written in polar form is

$$\frac{E_{pi}}{E_{si}} = e^{i(2P-\pi/2)}. \tag{7.18}$$

For this case (7.17), also in polar form, is

$$\frac{E_{pr}}{E_{sr}} = \tan \psi e^{i(2P-\pi/2+\Delta)}. \tag{7.19}$$

The null condition requires that the light to the analyzer be linearly polarized, that is, E_{pr} and E_{sr} are in phase, so that the phase of the reflected light at null is either π or 0. The polarizer setting is labeled P_2 for the first condition and P_4 for the latter. To obtain a phase of π in (7.19) requires that

$$P_2 = \frac{3\pi}{4} - \frac{\Delta}{2}, \qquad \text{for zone 2,} \tag{7.20}$$

or rotating P by π, which is also acceptable, gives

$$P_2 = -\frac{\pi}{4} - \frac{\Delta}{2}. \tag{7.21}$$

Similarly, to obtain a phase factor of zero requires

$$P_4 = \frac{\pi}{4} - \frac{\Delta}{2}, \qquad \text{for zone 4,} \tag{7.22}$$

or rotation of P by π,

$$P_4 = \frac{5\pi}{4} - \frac{\Delta}{2}. \tag{7.23}$$

When the ideal compensator azimuth is $-45°$ to the plane of incidence

$$P_1 = \frac{\Delta}{2} - \frac{\pi}{4}, \qquad \text{for zone 1} \tag{7.24}$$

and

$$P_3 = \frac{\Delta}{2} + \frac{\pi}{4}, \qquad \text{for zone 3.} \tag{7.25}$$

Determination of Δ

Equation (7.17) can be written

$$\frac{E_{pr}/E_{sr}}{E_{pi}/E_{si}} = \frac{|E_{pr}|/|E_{sr}|}{|E_{pi}|/|E_{si}|} e^{i(\Delta_{pr} - \Delta_{sr} - \Delta_{pi} + \Delta_{si})} \tag{7.26}$$

where the magnitudes and phases of the field components are written explicitly. The null measurement involves adjustment of polarizer and analyzer so that linearly polarized light is incident on the analyzer. This means that $\Delta_{pr} = \Delta_{sr}$ and the phase difference introduced by the sample is $\Delta = \Delta_{si} - \Delta_{pi}$. Then, $\tan(\Delta_{pi}) = \tan(\Delta_{si} - \Delta)$, which when rearranged is

$$\tan \Delta = \frac{\tan \Delta_{si} - \tan \Delta_{pi}}{\tan \Delta_{si} \tan \Delta_{pi} + 1}. \tag{7.27}$$

From (7.5a) and (7.6a)

$$\tan \Delta_{pi} = \frac{-B \sin \beta}{A + B \cos \beta} \tag{7.28}$$

and

$$\tan \Delta_{si} = \frac{-D \sin \beta}{C + D \cos \beta}. \tag{7.29}$$

Substituting (7.5) and (7.6) explicitly into (7.27), then

$$\tan \Delta = \frac{2T \sin \beta \sin(\pi/4 - P)\cos(\pi/4 - P)}{\cos^2(\pi/4 - P) - T^2 \sin^2(\pi/4 - P)}. \tag{7.30}$$

The calculation of Δ in the general form of (7.30) requires that T and β be known and that the polarizer setting for zone 2 or zone 4 is also known. Equation (7.30) applies to compensator azimuth $-45°$ except that the argument is $(\pi/4 + P)$ for the sine and cosine functions [4].

Determination of T

Divide (7.30) by $\cos^2 U$, where $U = (\pi/4 - P)$, to obtain

$$\tan \Delta = \frac{2T \sin \beta \tan U}{1 - T^2 \tan^2 U}, \tag{7.31}$$

and U can take on two values $(\pi/4 - P_2)$ or $(\pi/4 - P_4)$, depending on whether the null condition has been obtained in zone 2 or 4. When the two values of $\tan \Delta$ in (7.31) obtained for the polarizer readings in zones 2 and 4 are equated, the resultant can be simplified to

$$T^2 = -\frac{1}{\tan U_2 \tan U_4}, \tag{7.32}$$

where

$$U_2 = \frac{\pi}{4} - P_2 \quad \text{for zone 2,}$$

$$U_4 = \frac{\pi}{4} - P_4 \quad \text{for zone 4.}$$

For zones 1 and 3 with compensator azimuth $-45°$

$$T^2 = \frac{-1}{\tan U_1 \tan U_3} \tag{7.33}$$

where

$$U_1 = \frac{\pi}{4} + P_1 \qquad \text{for zone 1,}$$

$$U_3 = \frac{\pi}{4} + P_3 \qquad \text{for zone 3.}$$

Determination of ψ

Substituting either (7.20) or (7.21) into (7.17), then $E_{pr}/E_{sr} = \tan \psi e^{i\pi} = -\tan \psi$ and the azimuth of the linearly polarized light incident on the analyzer is in the second or fourth quadrants for the zone 2 condition. However, the null condition requires that the analyzer be crossed with the incident linearly polarized light, so that in the crossed position (which is what is measured) the analyzer azimuth is in the first or third quadrants. Figure 7.4 illustrates the

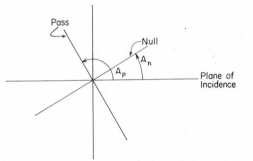

Fig. 7.4. Analyzer azimuthal relations for the pass and null conditions for compensator azimuth $\pm 45°$. The pass azimuth is in the second and fourth quadrants, the null azimuth is in the first and third quadrants. The figure shows one pair of pass (A_p) and null (A_n) azimuth angles.

conditions. From Fig. 7.4 $A_p = A_n \pm \pi/2$. Then $E_{sr}/E_{pr} = \tan(A_n \pm \pi/2) = -1/\tan A_n$. Since the analyzer at null is the quantity that is measured, the subscript n is dropped. Equation (7.17) can then be written

$$\frac{E_{pr}}{E_{sr}} = -\tan A = \frac{E_{pi}}{E_{si}} \tan \psi e^{i\Delta}. \tag{7.34}$$

Since E_{pr} and E_{sr} are in phase (linear polarization from the specimen) the right side of (7.34) must be real, which requires that the magnitude of $E_{pi}/E_{si} \, e^{i\Delta}$ be found, that is, that the quantity be multiplied by its complex conjugate. When this operation is carried out, substituting from (7.5) and (7.6) for E_{pi} and E_{si} and setting $U = (\pi/4 - P)$ for zones 2 and 4,

$$\tan \psi = \pm \tan A \left[\frac{\cos^2 U + T^2 \sin^2 U - 2T \cos U \sin U \cos \beta}{\cos^2 U + T^2 \sin^2 U + 2T \cos U \sin U \cos \beta} \right]^{1/2}. \tag{7.35}$$

In this general form a determination of ψ at a compensator setting $+45°$, polarizer azimuth at P, and analyzer azimuth at A, both T and β of the compensator are needed. When $T = 1$ and $\beta = 90°$ then (7.35) reduces to $\tan \psi = \pm \tan A$, or $\psi = \pm A$. Label the analyzer readings A_2 and A_4 for zones 2 and 4, respectively. For the ideal compensator P_2 is given by (7.20) and (7.21). Substituting into (7.34) $E_{pr}/E_{sr} = -\tan \psi$. Also, since $E_{pr}/E_{sr} = -\tan A$, then, for zone 2

$$\psi_2 = +A_2.$$

Similarly, substituting the null values for an ideal compensator for P_4 from (7.22) and (7.23) into (7.34) $E_{pr}/E_{sr} = \tan \psi$ and $\psi_4 = -A_4$.

Equation (7.35) holds for compensator azimuth $-45°$ when $U = (\pi/4 + P)$ and two values are possible for P in either zones 1 or 3. In the ideal compensator case where $\tan \psi = \pm \tan A$, $\psi_1 = +A_1$ for zone 1, and $\psi_3 = -A_3$ for zone 3.

Determination of β

Oldham [4] obtains an expression for β (Oldham's δ) of the following form when compensator azimuth is at $+45°$ to the plane of incidence.

$$\cos \beta = \frac{\tan A_2 \sin U_2 \cos U_4 - \tan A_4 \sin U_4 \cos U_2}{T \sin U_2 \sin U_4 (\tan A_4 - \tan A_2)}$$
$$+ \frac{\cos^2 U - T^2 \sin^2 U}{2T \sin U \cos U}, \quad (7.36)$$

where the subscripts denote either zone 2 or 4 and $U = (\pi/4 - P)$ can take on two values, U_2 or U_4. If the first term on the right of the equality is divided through by $\cos U_2 \cos U_4$ and the second term is divided through by $\cos^2 U$, then (7.36) can be written

$$\cos \beta = \frac{\tan A_2 \tan U_2 - \tan A_4 \tan U_4}{T \tan U_2 \tan U_4 (\tan A_4 - \tan A_2)} + \frac{1 - T^2 \tan^2 U}{2T \tan U}. \quad (7.37)$$

In the second term on the right in (7.37), let $U = U_2$. Multiply the denominator by T/T and substitute for T^2 from (7.32) to obtain

$$\frac{1 - T^2 \tan^2 U}{2T \tan U_2} = \frac{1 + \tan U_2/\tan U_4}{-2/T \tan U_4}$$
$$= -T/2(\tan U_4 + \tan U_2).$$

Then (7.37) can be written

$$\cos \beta = \left(\frac{T}{2}\right) \frac{(\tan A_2 + \tan A_4)(\tan U_2 - \tan U_4)}{\tan A_2 - \tan A_4}. \quad (7.38)$$

Similarly with compensator at $-45°$ to the plane of incidence, $\cos \beta$ is given by an expression like (7.38) except that the subscripts 1 and 3 replace 2 and 4, respectively.

Determination of Surface Properties

The ratio of p/s can be written

$$\frac{p}{s} = \frac{\tan(\phi_1 - \phi_2)/\tan(\phi_1 + \phi_2)}{-\sin(\phi_1 - \phi_2)/\sin(\phi_1 + \phi_2)}$$

$$= -\cos(\phi_1 + \phi_2)/\cos(\phi_1 - \phi_2)$$

$$= \tan \psi e^{i\Delta}.$$

Also,

$$\frac{1 - \tan \psi e^{i\Delta}}{1 + \tan \psi e^{i\Delta}} = \frac{\cos \phi_1 \cos \phi_2}{\sin \phi_1 \sin \phi_2}$$

$$= \frac{(n_2^{*2} - n_1^{*2} \sin^2 \phi_1)^{1/2}}{n_1 \sin \phi_1 \tan \phi_1}.$$

Then

$$\frac{n_2^2 - k_2^2 - n_1^2 \sin^2 \phi_1 - i2n_2 k_2}{n_1^2 \sin^2 \phi_1 \tan^2 \phi_1} = \frac{\cos^2 2\psi - \sin^2 2\psi \sin^2 \Delta - i \sin 4\psi \sin \Delta}{(1 + \sin 2\psi \cos \Delta)^2},$$

If n_1^* is assumed real and $n_1 = 1$ and equating real and imaginary parts of the previous expression, then the real (ϵ_2') and imaginary (ϵ_2'') parts of the dielectric constant can be written

$$\epsilon_2' = n_2^2 - k_2^2$$

$$= \sin^2 \phi_1 \left\{ 1 + \tan^2 \phi_1 \frac{(\cos^2 2\psi - \sin^2 2\psi \sin^2 \Delta)}{(1 + \sin 2\psi \cos \Delta)^2} \right\},$$

$$\epsilon_2'' = 2n_2 k_2$$

$$= \sin^2 \phi_1 \tan^2 \phi_1 \frac{\sin 4\psi \sin \Delta}{(1 + \sin 2\psi \cos \Delta)^2},$$

and

$$n_2 = \{\tfrac{1}{2}[\epsilon_2' + (\epsilon_2'^2 + \epsilon_2''^2)^{1/2}]\}^{1/2},$$

$$k_2 = \{\tfrac{1}{2}[-\epsilon_2' + (\epsilon_2'^2 + \epsilon_2''^2)^{1/2}]\}^{1/2}.$$

Experimental Considerations

Among the most demanding of the experimental factors that must be observed is careful alignment of the instrument particularly for determination of the optical or dielectric constants of surfaces. McCracken, et al. [10] have outlined a suitable alignment procedure. Winterbottom [11] has discussed the reflections from the various optical components that may be encountered. Surface roughness can affect the accuracy of the measurement, and generally the most desirable surface is one that is mirrorlike. The problem of roughness and its effect on ellipsometer measurements is not well defined [11a,b].*

Considerable attention is being given to the automation of the nulling procedure for applications where changes occur too rapidly for manual operation. Winterbottom [12], Takasaki [13], and Ord [14] have discussed various approaches. Other work is under way in a number of laboratories [15a–e].

The importance of cleanliness in characterizing the true properties of a surface cannot be overemphasized [16, 17]. While the ideal procedure for such studies would be either in situ fabrication and/or cleaning by heating or other means with the specimen under ultra-high vacuum, the birefringence of optical windows can be a severe problem.*

Disagreement in results between zone pairs can exist, for example, between zones 2, 4 and 1, 3 [4, 10]. Oldham suggests this may be due to compensator misalignment [4].

Hall [18] has found that the optical constants of Au films are independent of angle of incidence. This should be generally true for homogeneous, isotropic films.

Error analysis has received considerable attention recently [18a,b]. Complementary studies using ellipsometry and low-energy electron diffraction have been reported [18c–e]. Considerable effort of this type is under way in a number of laboratories.

3   ISOTROPIC FILM ON ISOTROPIC SUBSTRATE

Introduction

The overlay film is assumed to be partly transparent, otherwise the problem reverts to that of an opaque film discussed in Section 2. The model for this problem is illustrated in Fig. 7.5. To treat a specific example, the filmed substrate is assumed to be immersed in air. The incoming light at an angle of incidence, ϕ_o, undergoes multiple reflections in the partly transparent film. At point A, part of the light is reflected and part is refracted at the first interface according to Snell's law. The refracted light is transmitted through the film and incurs a phase lag $\delta_o = (2\pi/\lambda_o)n_1^* d_1 \cos \phi_1$, from A to C, where λ_o is

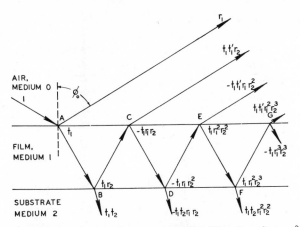

Fig. 7.5. Multiple reflections in a partly transparent film. The phase factor, δ_o, associated with each component is not shown on the figure (from [20a] with permission).

the free space wavelength, n_1^* and d_1 are the complex refractive index and thickness, respectively, of the film and ϕ_1 is the angle of refraction at the first interface.

The total reflected light for each of the p- and s-components is then,

$$R = r_1 + t_1 t_1' r_2 e^{-i2\delta_o} - t_1 t_1' r_1 r_2^2 e^{-i4\delta_o} + t_1 t_1' r_1^2 r_2^3 e^{-i6\delta_o} - \cdots,$$

$$= r_1 + t_1 t_1' r_2 e^{-i2\delta_o}[1 - r_1 r_2 e^{-i2\delta_o} + r_1^2 r_2^2 e^{-i4\delta_o} - \cdots],$$

$$= \frac{r_1 + (t_1 t_1' r_2 e^{-i2\delta_o})}{1 + r_1 r_2 e^{-i2\delta_o}},$$

$$= \frac{r_1 + r_2 e^{-i2\delta_o}}{1 + r_1 r_2 e^{-i2\delta_o}} \tag{7.39}$$

where use has been made of the relations

$$r' = -r \quad \text{and} \quad tt' = 1 - r^2,$$

the primed factors denoting the given factor for reverse propagation.

Both the p- and s-components of the light will have a total reflection coefficient of the form of (7.39).

Then,

$$\frac{R_p}{R_s} = \tan \psi e^{i\Delta} \tag{7.40}$$

which is the ellipsometer equation for the filmed substrate, or

$$\tan \psi e^{i\Delta} = \left(\frac{p_1 + p_2 e^{-i2\delta_o}}{1 + p_1 p_2 e^{-i2\delta_o}} \right) \left(\frac{1 + s_1 s_2 e^{-i2\delta_o}}{s_1 + s_2 e^{-i2\delta_o}} \right) \tag{7.40a}$$

where p_1, p_2 are the Fresnel reflection coefficients for p-light at the first and second interface, respectively, and s_1, s_2 are the Fresnel coefficients for s-light at the first and second interface, respectively. The individual coefficients are of the form given in (7.11) and (7.12), except that in Fig. 7.5 the angle of incidence is ϕ_o; the constituents of the first interface are n_o and n_1^* and of the second interface are n_1^* and n_2^*, where n_o is the refractive index of air (real) n_1^* and n_2^* are the complex indices of the film and substrate, respectively.

The Film Properties

The problem is undetermined, two quantities are measured, Δ and ψ; however, in the most general case three unknowns are involved, the complex refractive index and the thickness of the film. The substrate properties are assumed to be known, although the demands on knowledge of substrate characteristics are not as severe as for determination of the clean substrate properties.

McCrackin and Colson [19] have investigated the possible use of a number of techniques to secure additional information to make up for the indeterminacy of the ellipsometer measurement, which yields only a given pair of values of Δ and ψ. In practice one of the most commonly used methods is to measure Δ and ψ for films of different thickness and then curve-fit the experimental values to Δ and ψ plots, which are generated (by computer), for various assumed optical constants and thicknesses. For the conditions assumed by McCrackin and Colson films in the thickness range of 100 to 1000 Å could be studied in this way, but films less than 100 Å thickness could not be studied.* The conference, " Recent Developments in Ellipsometry," documents a number of papers dealing with filmed surfaces [19a–g].

When one of the three properties of the film is determined, the other unknowns can be calculated from ellipsometer measurements using the general

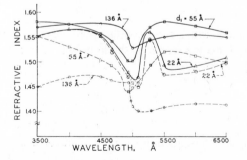

Fig. 7.6. A comparison of refractive index obtained by curve fitting and from the linear approximation where higher-order terms in d/λ are neglected (from [20a] with permission).

relationship (7.40*a*). A simplified (linear) form of this equation is sometimes used in which higher-order terms in d_1/λ in the expansion of the exponentials are neglected. Bashara and Peterson [20*a*] found considerable disparity (Fig. 7.6) between film properties using the linear approximation of (7.40*a*) and values obtained by the curve-fitting method for film thickness greater than 25° Å.

Where film absorption is so small as to be assumed negligible, the problem becomes determinate, that is, only the real part of the refractive index and film thickness are unknown. Then a simplified form of (7.40*a*) is used [20*b*]. However, considerable error can occur (the results are often meaningless) if film absorption is not negligible even if it is very small. Peterson and Bashara [20*b*] found that for film absorption so small that only 1% of the light is absorbed (e.g., 100 Å-thick film, $n = 1.4$, $k = 0.01$, $\phi_o = 70°$) highly erroneous results are obtained if the film absorption is neglected. Figure 7.7 is an example of this for thickness calculations.

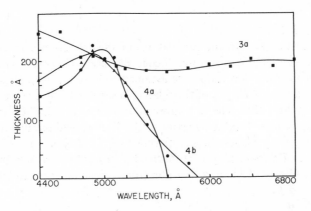

Fig. 7.7. The thickness of a thin film as a function of wavelength assuming film absorption can be neglected (taken from [20*b*] with permission). The figure shows that two of the equations that result from the analysis give meaningless results. The reasons for this discrepancy are explained in the reference.

The error in the determination of film properties which incorrectly neglect film absorption is dependent on substrate and film properties. The substrate dependence is illustrated in Fig. 7.8.

Anomalous Absorption

Bashara and Peterson [20*a*] found an increase in film absorption with decrease in film thickness for polybutadiene films on Au and Cr substrates (Figs. 7.9 and 7.10). These observations are believed to be anomalous and may be due to misapplication of isotropic theory to what is essentially an

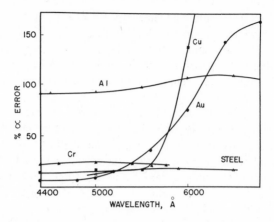

Fig. 7.8. The figure shows the effect of the substrate on the determination of the properties of a film ($n = 1.4$, $k = 0.01$) but where k is assumed zero (taken from [20b] with permission).

anisotropic situation. In the calculations of the extinction coefficient shown in Figs. 7.9 and 7.10 the isotropic theory was used. However, in the vicinity of the interface the film cannot be assumed to have the properties of a bulk, isotropic material. Analysis of the relative electric fields in the vicinity of the interface confirm this. The film absorption behavior exhibited in Figs. 7.9 and 7.10 correlates with the behavior of the electric field parallel to the plane of the film as can be seen in comparing Fig. 7.11 with Figs. 7.9 and 7.10. Johnson and Peterson [21] have outlined a procedure for determination of the relative electric fields in a thin film from ellipsometer measurements.

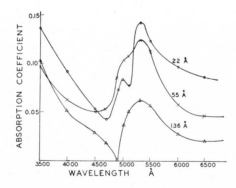

Fig. 7.9. The anomalous absorption of a thin polybutadriene film on Au showing increase in the absorption (extinction coefficient) with decrease in film thickness (taken from [20a] with permission).

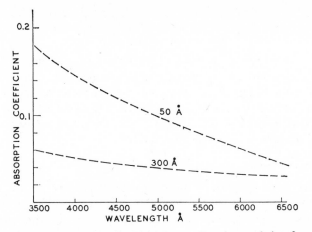

Fig. 7.10. Results similar to those for Fig. 7.9 on a Cr substrate (taken from [20a] with permission).

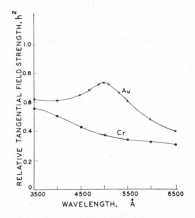

Fig. 7.11. The relative electric field at the interface between a polybutadiene film and the substrates shown. The wavelength dependence of the field correlates with the wavelength dependence of the absorption shown in Figs. 7.9 and 7.10 (taken from [20a] with permission).

4 PARTIAL SURFACE COVERAGE

Introduction

The study of reflecting surfaces that are partly covered by thin films is by adaptation of the methods, such as are given in Sections 2 and 3, applied to clean or film-covered surfaces. Very often inferences must be drawn from results obtained near the limits of experimental capability, and many conclusions are as yet qualitative or intuitively appealing rather than rigorously founded. In this outline illustrative examples are given from experiments on surfaces that are thought to be clean, film-covered, or in some intermediate state.

In treating reflection of polarized light from clean surfaces it is useful to categorize dielectrics and absorbing media separately. Dielectrics enjoy considerable historical priority in the field. Furthermore, although there are formal similarities in the equations for reflection from dielectrics on one hand, and from conductors and semiconductors on the other, the phenomena are physically different in a number of ways.

Transparent Media

The great preponderance of work on transparent systems has concerned external reflection: the case in which the incident and reflected beams propagate in the medium (1) of lower dielectric constant, ϵ_1, while the refracted beam is in the medium (2) of higher dielectric constant, ϵ_2. Although ellipsometry can be done via internal reflection [22, 22a], which indeed has certain advantages, it will not be considered in the following account. Also, because it is usually so, ϵ_1 will be taken as unity unless specified.

If the transition between transparent media is abrupt; that is, there is a discontinuous change between bulk dielectric constant at the interface, then an incident plane polarized beam will give rise to a reflected beam which is also plane polarized. The amplitudes E_{pr} and E_{sr} of the principal components in the reflected beam depend upon the incident amplitudes, E_{pi} and E_{si}, the angle of incidence, φ_1, and on the refractive indices, $n_1 = \sqrt{\epsilon_1}$ and $n_2 = \sqrt{\epsilon_2}$. The equations relating these quantities are Fresnel's and, via Snell's law, can be put in a convenient form containing the angle of refraction:

$$E_{pr} = E_{pi} \cdot \frac{\tan(\varphi_1 - \varphi_2)}{\tan(\varphi_1 + \varphi_2)}, \qquad E_{sr} = -E_{si} \cdot \frac{\sin(\varphi_1 - \varphi_2)}{\sin(\varphi_1 + \varphi_2)}.$$

It can be shown from these equations that at Brewster's angle,

$$\varphi_1 = \varphi_B = \tan^{-1}\left(\frac{n_2}{n_1}\right),$$

the p-component vanishes, and also that the phase difference, Δ, between the reflected components is either 0 or π, depending on whether the angle of incidence is greater or less than Brewster's angle.

Actually, real interfaces do not conform to the postulates of Fresnel's theory, for the optical transition between contiguous media is more or less gradual. This may be due to a continuous variation of density, as with the liquid-vapor interface, to the existence of a worked or leached layer as on metals and glasses, to adventitious contamination, or to sundry other causes [23]. The effect of this continuous transition is that the p-component reaches a minimum in the vicinity of Brewster's angle but does not vanish. Also, the phase difference, Δ, varies continuously between 0 and π, being $\pi/2$ in the vicinity of Brewster's angle. Consequently, for $\varphi_1 \approx \varphi_B$ the reflected light is elliptically polarized. Since E_{pr} is relatively very small, the resultant vibration can be represented as a highly eccentric ellipse with its minor axis near the plane of incidence.

The degree of ellipticity, defined as a coefficient, ρ, the ratio of minor to major axis, will depend upon the angle of incidence, φ_1, wavelength, λ, refractive index, n_2, of the substrate, and the nature of the transition zone. Equations relating these parameters were given by van Ryn van Alkemade and by Drude [1]. For an incident beam with $E_{pi} = E_{si}$, that is the plane of polarization in 45° azimuth,

$$\frac{E_{pr}}{E_{sr}} = -\frac{\cos(\varphi_1 + \varphi_2)}{\cos(\varphi_1 - \varphi_2)} \cdot \left[1 + \frac{i4\pi}{\lambda} \cdot \frac{\epsilon_2}{1 - \epsilon_2} \cdot \frac{\cos \varphi_1 \sin^2 \varphi_1}{\sin^2 \varphi_1 - \epsilon_2 \cos^2 \varphi_1} \cdot \eta \right]. \quad (7.41)$$

The transition zone is characterized by a dielectric constant, ϵ, which depends upon the structure of the transition so that,

$$\eta = \int_1^2 \frac{(\epsilon - 1)(\epsilon - \epsilon_2)}{\epsilon} \cdot dz. \quad (7.42)$$

At Brewster's angle,

$$\frac{E_{pr}}{E_{sr}} = i\frac{\pi}{\lambda} \cdot \frac{(1 + \epsilon_2)^{1/2}}{1 - \epsilon_2} \cdot \eta, \quad (7.43)$$

or, in terms of phase difference and ratio of real amplitudes:

$$\rho = \pm\frac{\pi}{\lambda} \cdot \frac{(1 + \epsilon_2)^{1/2}}{1 - \epsilon_2} \cdot \eta, \quad \text{and} \quad \Delta = \pm\frac{\pi}{2}.$$

If the phase difference is 90°, these imply that the ellipse is oriented with minor axis in the plane of incidence and major axis perpendicular to it.

The ellipticity coefficient, the ratio of axes, is identical to E_{pr}/E_{sr}. Thus, if ρ is measurable, it can be used to derive the value of η, and by assigning a mean dielectric constant $\bar{\epsilon}$ to the transition layer,

$$\eta = \frac{(\bar{\epsilon} - 1)(\bar{\epsilon} - \epsilon_2)}{\bar{\epsilon}} \cdot d, \quad (7.44)$$

and it is possible to compute either $\bar{\epsilon}$ or the layer thickness d, provided one of the two is known in advance. It is readily seen that the ellipticity coefficient, ρ, is positive if $\bar{\epsilon} < \epsilon_2$, and negative if $\bar{\epsilon} > \epsilon_2$. This is consistent with the observation that most materials exhibit a positive coefficient, and negative values are found only with substances of very low refractive index [24].

For given film thickness, d, the maximum positive ellipticity is obtained for $\bar{n} = \sqrt{n_2}$, which provides a means of calculating the minimum film thickness corresponding to any observable positive coefficient. Vasicek [25] has pointed out that for $\bar{n} < n_2$, the ellipticity coefficient is not very sensitive to small variations in $\bar{n}$, so that in practice it is unnecessary to have a highly accurate value of $\bar{n}$ in order to obtain a reasonably good estimate of d. Conversely, values of $\bar{n}$ calculated from known d tend to be somewhat approximate. For negative ellipticity coefficients, $|\rho|$ increases with increasing $\bar{n}$. There is, of course, no way of estimating a minimum film thickness in this case. On the other hand, since $\bar{n}$ is not bounded by values for the adjacent media it is possible, at least in principle, to attain ellipticities of great magnitude, which would make experimental observations relatively easy [26].

The usual form of the van Ryn–Drude approximate equation implies that the dielectric tensor is isotropic throughout the transition zone. Buff [27] has pointed out the error of this assumption and, gives the following equation:

$$\rho = \frac{\pi}{\lambda} \cdot \frac{(1 + \epsilon_2)^{1/2}}{1 - \epsilon_2} \left[\int_1^2 (\epsilon_T - \epsilon_N) dz + \int_1^2 \frac{(\epsilon_N - 1)(\epsilon_N - \epsilon_2)}{\epsilon_N} dz \right], \quad (7.45)$$

where ϵ_N, ϵ_T are the normal and tangential components of the dielectric tensor. Strictly speaking, the interface is intrinsically anisotropic, and ϵ_T, ϵ_N must be known if thickness is to be computed. Sivukhin [28] derived equations which take into account effects caused by orientation of anisotropic molecules, for instance at a liquid surface. At Brewster's angle,

$$\rho = \frac{\pi}{\lambda} (1 + \epsilon_2)^{1/2} \cdot (\gamma_z - \gamma_x), \quad (7.46)$$

where $\gamma_{z,x}$—parameters derived from molecular anisotropy constants— characterize the transition zone and are effectively independent of field intensity and frequency. Similar results were obtained by Strachan [29] in analyzing the optical effects of adsorbed monolayers. In both cases the optical properties of the transition zone are defined in terms of molecular quantities, without reference to a bulk refractive index as used by Drude.

The foregoing formulas do not apply to film thicknesses much in excess of $0.02\ \lambda$, for which equations involving higher orders of d/λ must be used [30]. For such systems it is observed that Brewster's angle ($\varphi_1 = \tan^{-1} n_2$), the principal angle (for which $\Delta = \pm \pi/2$), and the quasi-polarizing angle (minimum E_{pr}) do not coincide, and the differences can be calculated in higher-order theory. Both theory and experiment show that for very thin films these differences are scarcely accessible to measurement. In considering partial surface coverage the first-order equations are adequate.

Early studies of reflection, often made with the hope of characterizing reflecting substrates, were commonly vitiated by failure to use surfaces free of contamination or gross structural defects. This is not particularly surprising since the ellipticity of the reflected beam was thought to be related to some intrinsic bulk property, and the role of surface films was not at all understood. Jamin's results on many solids and liquids are typical. Although his apparatus was capable of moderate accuracy ($\rho \approx 1.10^{-3}$), the value of ρ for water was given as $-577 \cdot 10^{-5}$, which is not only numerically much too great, but even of the wrong sign. Before considering more recent data, and their compatibility with theory, some mention will be made of the technical side.

Experimental Requirements

The experimental procedures for ellipsometry with transparent systems are the same, in principle, as those used on absorbing substrates. In practice the requirements are somewhat different. The reflectance, $I_r = (I_{sr} + I_{pr})/2$ of a typical transparent medium at Brewster's angle is small; for example, 0.096 for $n_2 = 1.6$, compared to that of a metal near the principal angle. In general, too, a very thin dielectric film produces a smaller absolute change in ellipticity when deposited on a transparent substrate than it does on a metallic mirror. For instance, in the case of oxide films on aluminum, Vasicek [32] finds a change in ellipticity angle ($\gamma = \tan^{-1} \rho$) of $0.1°$ per angstrom change in thickness. In a relatively favorable case a very thin film on glass might have about one-tenth as much effect. Because the reflected intensity is relatively low, and since the accuracy of determining the plane of polarization in the reflected compensated beam varies as the square root of its brightness, it is clear that a very intense light is needed. Useful sources have included the sun, high-pressure mercury arcs, and lasers.

As ρ is defined for equal incident amplitudes, E_{pi} and E_{si}, it is obvious that measurements could be made using incident monochromatic natural light. Frazer [33] devised a technique, later modified by Hofmeister [34], that avoids use of compensators. A polarizing prism is inserted either before or after reflection, and the intensities of the p- and s-components are measured photometrically. The ellipticity coefficient is the square root of the intensity ratio. In this method the sign of ρ is not immediately evident from extinction settings alone.

Using conventional apparatus it is customary to have the incident azimuth not at $45°$ but much closer to the plane of incidence. Thus the measured ellipticity is much larger than ρ, which is calculated from the known azimuth. Bouhet [35] and Kizel' [36] used azimuths of $4°$ and $2°$, respectively, in studies of liquid surfaces. Kizel' judged extinction visually, while Bouhet made use of an elaborate half-shade system to achieve visual balance. Rayleigh [31] observed extinction visually by setting the analyzer at a fairly large, convenient angle β ($30-45°$), extinguishing with the polarizer at some small angle, α, then reversing the analyzer to $-\beta$ and finding the new polarizer extinction angle, α'. Then, $\rho = \tan \beta \cdot \tan(\alpha - \alpha')/2$. By reversing the initial setting—either α or β could be used—the difference of the other is measured on a doubled scale, and the scale zero settings need not be established with great accuracy. Incidentally, it is worth noting that by setting $\beta = 30°$, the polarizer azimuth was brought to about 0.04°, compared to Bouhet's 4°. Furthermore, Rayleigh's apparatus was much simpler than Bouhet's, but their values of ρ for a clean water surface differ by only 5%, and the still more elaborate experiments of Bacon [37, 37a] give a result, $\rho = 42 \cdot 10^{-5}$, identical to Rayleigh's.

This illustrates a point that deserves emphasis: visual observation of extincinto is still a very useful technique, particularly when angular readings cannot be made to better than a large fraction of a minute. Although it has long been known that half-shade measurements are superior in principle to extinction measurements, the superiority is often not evident in practice. The dark adapted eye, though easily fatigued, is an extremely sensitive detector in the proper spectral region. As Rayleigh [38] pointed out, in the hands of an experienced observer, an extinction becomes, in effect, a kind of half-shade procedure. Also, for a number of reasons, most practical half-shades fall far short of optimum performance.

Wholly objective measurements are possible using photomultiplier and microphotometer [39]. These techniques are mandatory in spectral regions outside the range of visual sensitivity, or when automatic operation is needed. Most ellipsometers lack the mechanical accuracy to take full advantage of the maximum capabilities of photoelectric detection. However, even where automatic operation is intended, it cannot be too strongly urged that provision be made for visual inspection of the whole optical train. It should be possible to focus on each element; aperture, polarizer, cell windows (if any), specimen, compensator, and analyzer. Where this cannot be done it is often difficult to isolate and correct optical problems. In addition, it is useful to be able to examine the reflecting surface for localized effects.

In brief, for conventional measurements involving very thin films on transparent substrates one requires:

1. Very bright light.
2. A highly collimated [35] monochromatic beam.
3. That analyzer and polarizer should be Glan-Thompson, or equivalent prisms of the highest quality (nicol prisms or dichroic sheet are not ideal for these purposes). Both elements should be set in accurately divided circles readable to at least 0.01°.
4. A compensator; quarter-wave plates are standard for measurements at a single wavelength, otherwise Babinet-Soleil [6] compensators are used.

If, in addition to measuring ellipticity, it is desired to carry out principal angle measurements, the graduated spectrometer circle must be capable of settings accurate to 0.001°.

Liquid Surfaces

It was understood early [23] that the study of liquid surfaces, which presents great experimental difficulties, is nevertheless likelier to yield fundamental results than the study of solids. Liquids require no processing to insure specular reflection, and there is also an independent means of detecting superficial contamination—change of surface tension—that is not applicable to solids. In doing ellipsometry on clean surfaces there are two main objectives: to learn something about the structure of the transition zone, or to check the relevant optical theory. Thus it would be desirable to measure ρ, and Δ, or φ_p on clean surfaces as a function of wavelength and temperature for many homologous series of liquids. In fact, mostly because of their practical difficulty, studies have been restricted to measurement of ρ at a single wavelength over restricted temperature ranges.

Following Rayleigh a considerable amount of work was done by Raman and Ramdas [42], Bouhet [35], and more recently by Kizel' [36, 43] and Kinosita [44] and Yokota [44, 45]. As might be expected, the results of various workers can differ substantially: Raman reported $75 \cdot 10^{-5}$ for the water surface. But it is certain that for reflection from clean surfaces of pure liquids ρ is always positive, varying from $221 \cdot 10^{-5}$ for methylene iodide to $34 \cdot 10^{-5}$ for formic acid. For most liquids ρ lies well within these extreme values. From Drude's equations it is possible to infer that ellipticity increases with increasing bulk refractive index. Indeed, it is true that the extreme values just cited are for liquids of greatly different index. However, isoamyl alcohol and isoamyl ether, with very similar indices, have $\rho = 100 \cdot 10^{-5}$, and $\rho = 140 \cdot 10^{-5}$, respectively. By contrast, carbon tetrachloride ($n_F = 1.462$) and isobutyl alcohol ($n_F = 1.397$) have ellipticity coefficients of $84 \cdot 10^{-5}$ and $87 \cdot 10^{-5}$, respectively. Liquids with high surface tension tend to have small ρ; for example, water, or glycerine, $\rho = 64 \cdot 10^{-5}$.

A transition layer can properly be characterized by refractive index only if its thickness is large in the scale of molecular dimensions [28]. But, because of the lack of sensitivity noted by Vasicek, it may be that in so far as the Drude equations are sound, an approximate or "effective" refractive index is sufficient for calculation. The fact that minimum thicknesses are of molecular dimensions suggests that this is so. The Drude transition layer on water is between 3 and 10 Å, for instance. In the great majority of cases [35] there is close agreement between the minimum thickness, molecular dimensions from x-ray data, and the quantity $(m/\delta)^{1/3}$, where m is the molecular mass and δ the liquid density.

In comparing experimental and theoretical values for ellipticity coefficients of clean liquid surfaces, Kizel' made use of Sivukhin's microscopic theory. On this basis the optical contrast between surface and bulk is attributed to orientation of surface molecules rather than to a density gradient. Naturally, the experimental values of ρ are lower than the calculated ones for highly anisotroptic molecules, and greater for isotroptic molecules. The parameters $\gamma_{x,z}$ are calculated and applied on the assumption of a fully oriented monolayer. In reality the transition layer should be thicker and partly disordered. As with refractive index, the relationship of ρ to chain length, though evident, is not emphatic. There is no obvious correlation with molecular anisotropy, dipole moment, or the details of molecular structure. This is to be expected, since the net effect is a result of several interacting causes.

Kizel' and Stepanov [46] have confirmed that, as predicted by the theories of Drude and Sivukhin, ρ is inversely proportional to the wavelength. The change in ρ at various wavelengths is not due only to this relationship, of course, but also to differences in dispersion of film and substrate. This factor has not been much investigated and as yet very few results are available at wavelengths other than 5461 Å.

Kinosita and Yokota, and Kizel' have made measurements on a number of liquids over limited ranges of temperature. Insofar as they can be compared, the results are somewhat discrepant. For carbon tetrachloride Kizel' finds $\rho = 1 \cdot 10^{-3}$, almost independent of temperature from 0 to 140°C, while Yokota finds a change of ρ from $8 \cdot 10^{-4}$ at -20°C to more than $25 \cdot 10^{-4}$ at 30°C. Similar increases are noted for α-methylnaphthalene and water. Using the observed ellipticities and Drude's equation, the minimum transition layers are calculated to increase from 7 Å at 20°C to 18 Å at 60°C. Such changes conflict with the predictions of molecular theory and with evidence from the variation of surface tension with temperature [47]. It is not clear whether this reflects the inapplicability of Drude's equation or experimental difficulty.

Contrary to these results, Kizel' finds that ρ decreases, in general, with increasing temperature, and that there are three modes of behavior: (1) imperceptible decrease, (2) slight decrease; (3) sharp drop. Those substances showing no apparent change are weakly anisotropic, while the second and third categories are more so, and differ from each other in the temperature dependence of the effective anisotropy. In several cases measurements were made in the vicinity of the freezing point. A notable increase in ρ was observed on supercooling and was attributed to structural changes in the interface before freezing.

It has been mentioned that work on liquids is exacting, and it is particularly difficult to assess the role of impurities which, if surface active, are preferentially adsorbed at the interface. It seems possible that the effects of surface impurities may be used, in particular cases, to decide whether Drude's or Sivukhin's interpretation of the cause of ellipticity is more plausible. The results obtained so far, though important, emphasize the need for further broad studies.

Solid Surfaces

When the dielectric surface is solid, experimental procedure is simplified, but theoretical problems prevail. Grinding, polishing, etching, freezing, and even cleavage may work subtle and irreproducible changes in solid surfaces. Rayleigh, for instance, was able to produce positive, zero, and negative ellipticity coefficients by polishing a given glass surface [40]. Lummer and Sorge reported changes in reflection induced by exerting pressure upon the lateral surfaces of glass prisms [41].

It is true that the vanishing of E_{pr} and, thereby, of ρ is a necessary condition for a clean surface, but it is by no means sufficient. It was noticed by Rayleigh [40] that reflection from glass can be quite insensitive to surface coverage, probably because of a close match in refractive indices of film and substrate. The state of solid surfaces depends crucially upon many details of their preparation, which is the subject of a large literature [48]. Yokota, Kinosita, and Sakata [49] have shown that high index glasses tend to develop layers by leaching of surface components, while low index glasses are liable to densification of the surface material on polishing. Vasicek [26] found surface films on all glasses, the least effect being $\rho = 6 \cdot 10^{-4}$. Fused quartz can be largely freed of its surface layer by HF etching [50] or, as recently disclosed, by unconventional polishing techniques [19c].

Absorbing Media

Ellipsometric studies on absorbing media are far more numerous than corresponding work on dielectrics. There are several reasons for this, including the following:

1. Ellipsométry provides one of the simplest and best methods of obtaining the relevant optical constants [51]: refractive index and extinction coefficient.

2. It can yield information crucial to the understanding of phenomena important in current technology: corrosion, oxidation, catalysis, lubrication, and so forth.

3. Results can be correlated with predictions of metal theory [52].

4. As far as increased reflectance is beneficial, the systems are much easier to work with than dielectrics.

Calculation of the complex index, $\tilde{n} = n - ik$ is fairly straightforward in principle:

$$n^2 - k^2 = \frac{\sin^2 \varphi_1 \tan^2 \varphi_1 (\cos^2 2\psi - \sin^2 2\psi \sin^2 \Delta)}{(1 + \sin 2\psi \cos \Delta)^2} + \sin^2 \varphi_1$$

$$2nk = \frac{\sin^2 \varphi_1 \tan^2 \varphi_1 \sin 4\psi \sin \Delta}{(1 + \sin 2\psi \cos \Delta)^2}.$$

Unfortunately, both Δ and ψ depend not only upon the bulk constants, but also upon the nature of the surface film and, for solids, the microtopography of the surface. As already mentioned, these effects are greater with absorbing media. Most commonly, a surface film causes a decrease in Δ and an increase in ψ, as compared to the values for a clean surface. Vasicek [53], therefore, took the cleanest surface to be the one exhibiting maximum Δ and minimum ψ. The extraction of true optical constants in the presence of surface films has been of concern to experimentalists [16, 54] and to theoreticians [54a] for many years. Unfortunately, very little is known, as yet, as to how to allow for the effects on apparent optical constants of surface structural features such as facets, scratches, and pits.

Although there is much literature on theoretical and experimental studies of the optical properties of metals and semiconductors, it will be possible to briefly illustrate only a single, rather favorable, example here. As with dielectrics, it is preferable to consider a liquid, for which clean, reproducible surfaces are more easily managed. Table 7.1 shows the results obtained over a

Table 7.1. Optical Constants of Hg at 5461 Å

Author	n	k
Meier [55]	1.46	2.86
Reeser [56]	1.538	3.05
Ellerbroek [57]	1.33	3.63
Tronstad and Feachem [58]	1.61	3.1
Emberson [59]	1.458	2.93
Schulz [60]	1.20	3.33
Leyluk, Shklyarevskii, and Yarovaya [61]	1.68	2.98
Smith and Stromberg [62]	1.485	3.061
Smith [63]	1.602	2.954

long period, by workers using different systems and techniques. The temperature dependence between 23 and 115°C is given as [63]

$$n = 0.00036\,T + 1.593, \qquad k = -0.00165\,T + 2.995$$

These values show that substantial variations can be expected even when great care and ingenuity have been exercised.

The results on solid conductors are generally less concordant, for the reasons already mentioned. It is obvious that when studying adsorption, or other surface phenomena, on conductors by the techniques of ellipsometry, it is very advisable to measure the substrate optical constants, rather than to rely on known values.

Partial Coverage

It is well known that even the thinnest surface layers can affect the way polarized light is reflected from a transparent or absorbing mirror. This suggests that ellipsometry can be used not only to analyze such films quantitatively (thickness) or qualitatively (refractive index), but to obtain information on surface coverage when the film is discontinuous. This, of course, is what is done when ellipsometry is used to measure adsorption at low relative pressure.

In the case of a dielectric substrate that is initially perfectly clean, a transparent film of molecular thickness has an optical effect only at angles of incidence near Brewster's angle. At this angle it is possible to determine ρ, and Δ is known to be $\pm \pi/2$. This is the situation described by Drude's equations (7.43) and [64],

$$\tan \Delta = \frac{4\rho\epsilon_2}{(1 + \epsilon_2)^{1/2}} \cdot \frac{\sin \varphi_1 \tan \varphi_1}{\tan^2 \varphi_1 - \epsilon_2}. \tag{7.47}$$

The effect of the film is to increase ρ from 0 to some finite value, and to change Δ by 90°. It is seen at once that the ellipticity coefficient varies linearly with thickness, but Δ does not because, although there is a small change in ρ, the factor $\tan^2 \varphi_1 - \epsilon_2$ is zero.

On conducting substrates the effect of a very thin film is different. There are relatively small changes in the values $\bar{\Delta}$ and $\bar{\psi}$ that characterize the clean surface. Near the principal angle of incidence for a gold substrate covered by a film with $d = 10$ Å and $n = 1.45$; $\bar{\Delta} = 88.86°$, $\bar{\psi} = 38.10°$, $\Delta = 88.12°$, and $\psi = 38.19°$. The sign and magnitude of the changes produced by the layer depend upon $\tilde{n}, n, \varphi_1$ ρ, and d, but in general,

$$\Delta - \bar{\Delta} = -\alpha d, \quad \text{and} \quad \psi - \bar{\psi} = \beta d.$$

That is, the changes are linear, small, and calculable from known values [63, 63a]. Drude's equations are clear in showing that as film thickness diminishes, the absolute values of $\rho, \Delta - \bar{\Delta}$, and $\psi - \bar{\psi}$ decrease linearly. The equations say nothing as to the nature of the variation as film thinning ceases, and fractional coverage, θ, starts to decline. The real situation is probably still more complicated because of changes in orientation and aggregation of the remaining molecules. An exact solution of the problem will be quite difficult, but some more or less intuitive arguments have been put forth. In one of them it is supposed that as the light begins to "see" the bare surface, a change in the course of ρ as a function of the amount adsorbed should occur. Conversely, the first increment of adsorption on a bare surface should have a greater effect than the last on a surface that is nearly covered

[65]. For the all dielectric case it has been suggested that the ellipticity coefficient at unit coverage, $\bar{\rho}$, is related to that at fractional coverage θ, by $\rho = \bar{\rho}\sqrt{\theta}$. Here it is assumed that I_p is proportional to coverage, θ, and coherence, which can change only the distribution of intensity, is neglected.

An alternative formulation assumes that fractional coverage, θ, by a film of thickness d is equivalent to full coverage by a film of thickness θd [66]. In other words, it is immaterial whether or not bare substrate is exposed, and changes which were linear in thickness continue to be linear as coverage approaches zero. Archer [67] has derived equations, formally similar to Drude's linear equations, for this case:

$$\Delta - \bar{\Delta} = -\theta\Sigma\alpha_i'\sigma_i', \qquad \text{and} \qquad \psi - \bar{\psi} = \theta\Sigma\,\beta_i'\sigma_i',$$

Where σ_i' are scattering indices characteristic of the film [29, 68], and α', β' are analogous to the Drude constants for bulk adsorbate [63a].

The most direct experimental evidence on these matters comes from studies of films spread on liquid surfaces, for in these experiments it is usually possible to have independent knowledge of adsorbate concentration. Considerable work has also been done on solids, for which auxiliary information must come from conventional adsorption measurements. In all cases the effects of contamination can be very serious, since the changes sought are small.

Hofmeister [34] calculated the refractive indices of fatty acids by application of the Drude equation to monolayers spread on aqueous surfaces. A film balance made possible the determination of ρ as a function of pressure. However, the mean value of ρ for the water surface alone was $-4.3 \cdot 10^{-2}$, indicating the presence of an initial layer, probably organic. Furthermore, the magnitude of the initial ellipticity is a substantial fraction of the total value. Thus, it is not quite certain that changes in ρ were due solely to changes in the state of the film initially spread.

On expansion from the close-packed layer the decrease in ρ is in general considerably larger than predicted by either a square root or a direct proportionality with coverage. At higher areas, between 50 Å^2 per molecule and 80 Å^2 per molecule, the value of $\rho = -5.3 \cdot 10^{-3}$ remained practically constant for stearic acid. Bouhet [35] found similar results for palmitic acid, with $\rho = -98 \cdot 10^{-5}$. Clearly, Bouhet's surface was not seriously contaminated, and the reason for the constancy of ρ is unknown. Both authors found that indices calculated from Drude's formula are lower than bulk values, apparently because of water trapped in the monolayer. Ellipsometry was also able to demonstrate the coexistence of surface phases independently of force—area curves derived from film balance data.

Studies of insoluble organic films on mercury have also been reported. It is probable that fatty acids are oriented with their long axes parallel to the liquid surface, until increasing pressure causes the organic chains either to pile up or to tilt at an angle to the surface. For stearic acid the molecular area is about 120 Å² and corresponds to a pressure near 40 erg/cm². Partial coverage must be studied at pressures below this, and the film thickness is equivalent to the molecular width, or slightly over 4 Å. Smith [63] has shown that $\Delta - \bar{\Delta}$ is linear against surface excess, Γ, in mol/cm² both above and below unit coverage. In the former instance increasing surface excess implies increasing thickness, but below $\theta = 1$, diminishing Γ probably implies change of coverage only. In this connection there is an interesting discontinuity in the curve of $\Delta - \bar{\Delta}$ against area per molecule near 120 Å² and a second discontinuity near 80Å².

Feachem and Tronstad [68] also made ellipsometric studies of fatty acid films on mercury; a curve of $\Delta - \bar{\Delta}$ against area per molecule for lauric acid shows a discontinuity near 40 Å² per molecule. Their other results are broadly compatible with known film behavior, but thickness values calculated from Drude's equation are systematically too great.

Early investigations of the adsorption of water vapor and methanol on glass and NaCl were reported by Frazer [33, 69] and Silverman, but the data at low relative pressure are too few to be more than suggestive. In the case of methanol on NaCl [70] it is found that adsorption at 200° C occurs in two ranges, between $2 \cdot 10^{-5}$ torr and $2 \cdot 10^{-4}$ torr. At each step a monolayer is formed, the first being characterized by a nonlinear change of ρ with pressure, and the second by a linear change.

In a recent study [71] of water vapor adsorption on lithium fluoride, both ellipticity and changes of φ_p were measured. The apparent film index attained the bulk value at a relative pressure of 0.6, and film thickness of nearly 80 Å. Still more recently [72] the adsorption of water vapor on single crystal faces of lithium fluoride, potassium chloride, and rubidium chloride was examined ellipsometrically. Different faces of crystal polyhedra were observed to adsorb different thicknesses of water, and adsorption was seen to be stepwise from $3 \cdot 10^{-8}$ to $1 \cdot 10^{-5}$ relative pressure. In all cases the substrates carried an initial adsorbed layer, and there is no evidence as to anomalous refractive index or the course of ρ at coverage below unity. Further investigations of this kind, involving adsorption of polar and non-polar molecules on surfaces of high energy and low energy are needed.

Miller and Berger [73] compared adsorption measurements obtained by ellipsometric and radiochemical techniques using polar organic molecules on metals. Data obtained by simultaneous application of both methods to vapor adsorption indicate that the optical thickness is slightly greater than the radiotracer result; 7 and 16% for tagged capric acid on chromium and nickel, respectively. This may, of course, be due to adsorption of inactive material. At all coverages the agreement between the two methods is generally good.

Meyer [18c] has correlated ellipsometric measurements of chemisorption on silicon single crystals with data from volumetric adsorption studies involving the same gases on high-area silicon powders. In addition to the expected increase in $\bar{\Delta} - \Delta$, there is an increase in $\bar{\psi} - \psi$ also. This is considered to be caused by changes induced in a thin surface layer that is present initially. Values of $\bar{\Delta} - \Delta$ are calculated from both Drude's equation and two-dimensional theory using refractive indices derived from the Lorentz-Lorenz equation and atomic parameters. It is shown that information on surface specificity and the nature of the chemisorption bond are accessible.

Finally, Chiu and Genshaw [74] have used ellipsometry to follow anion adsorption from aqueous solution as a function of potential on mercury electrodes and have compared the results with data obtained by electrocapillary methods. This is a more complex case than any of those previously cited because the film index includes contributions from both ions and water molecules. The ionic part is calculated from the Lorentz-Lorenz equation and is linearly combined, as a function of coverage, with the value of bulk water. Also, in addition to the layer of ions adsorbed at the interface, which affects the surface-tension, there is a more or less diffuse layer of counter ions which, presumably, contributes to the optical measurement. Nevertheless, results of the two methods are in good agreement even below one-tenth monolayer.

Ellipsometry can be applied to adsorption measurements both above monolayer coverage and well below it. For some systems it may well be the best technique presently available. Final disposition of these questions awaits comparison of detailed results obtained simultaneously by ellipsometry and by low-energy electron diffraction [18d, e] or microgravimetric methods.*

5 MODULATED ELLIPSOMETRY

Introduction

"Modulated ellipsometry" denotes a group of possible experiments where very small changes are detected in the optical properties of the sample under investigation. The optical properties are changed by an externally applied perturbation such as electric or magnetic field, mechanical stress, or sample temperature variation. Extremely small changes can be determined that would be lost in experimental errors if measured in the usual way. The primary application has been in the study of electronic structure in solids, but it should also be applicable to the study of effects of various perturbations on such things as adsorbed surface films and layer depositions since ellipsometry is uniquely suited to the study of surface phenomena. It may also find use in the study of chemical reactions at surfaces.

General Theory

Variations in the optical properties of part or all of the sample will appear as variation in the angles ψ and Δ measured by the ellipsometer. Since the variations are generally too small to detect by simply comparing ellipsometer readings taken with and without the external perturbation, the following experiment [75], shown schematically in Fig. 7.12, is useful. The arrangement of apparatus is the same as for conventional ellipsometry with the exception of the external modulation of the sample and the phase-sensitive detection of the photomultiplier output. The photomultiplier signal can be written as

$$S = S_n + I_o \, \mathcal{R}\{\sin^2(P - P_n) + \sin^2(A - A_n)\} \qquad (7.48)$$

where S_n is the signal when the ellipsometer is set for minimum transmission through the instrument, P is the polarizer azimuth (all azimuths are measured counterclockwise from the plane of incidence looking into the light beam), A is the analyzer azimuth, I_o the incident light intensity, P_n and A_n are the null settings for polarizer and analyzer, respectively, and $\mathcal{R}$ is the oblique-incidence reflectance of the sample at the angle of incidence ϕ. For small offsets of analyzer and polarizer from null (7.48) reduces to

$$S = S_n + I_o \, \mathcal{R}\{(P - P_n)^2 + (A - A_n)^2\}. \qquad (7.49)$$

Let the applied modulation of the optical properties be proportional to some periodic function $f(t)$, and let a reference signal proportional to $f(t)$ be supplied to the lock-in amplifier. The effect of the modulation of the optical properties of the sample is to make time functions of $\mathcal{R}, P_n$, and A_n as follows

$$\mathcal{R} = \mathcal{R}_0 + \delta \mathcal{R} f(t)$$
$$P_n = P_{n0} + \delta P_n \, (t) \qquad (7.50)$$
$$A_n = A_{n0} + \delta A_n \, f(t)$$

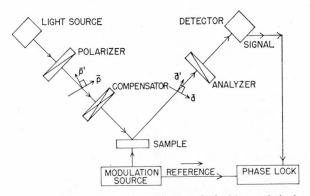

Fig. 7.12.   Experiment schematic (from [75] with permission).

where the zero subscripts denote values in the absence of modulation and the δ-quantities the magnitudes of the changes caused by external modulation. Substituting (7.50) into (7.49) and neglecting terms of second or higher order in $\delta\mathscr{R}$, δP_n, or δA_n yields

$$S = S_n + I_0\mathscr{R}_0\{(P - P_{n0})^2 + (A - A_{n0})^2\} + I_0\{\delta\mathscr{R}[(P - P_{n0})^2 + (A - A_{n0})^2]$$
$$- 2\mathscr{R}_0(P - P_{n0})\delta P_n - 2\mathscr{R}_0(A - A_{n0})\delta A_n\}f(t). \tag{7.51}$$

Equation (7.51) consists of a dc term that is the same as (7.49), the signal in the absence of modulation, plus a time varying term which follows the modulating function $f(t)$. Additional terms, all of which are of second or higher order in the δ-quantities have been neglected. Since $f(t)$ is periodic, the second term of (7.51) can be detected with a lock-in amplifier tuned to the fundamental frequency of $f(t)$. Setting $P = P_{n0}$ in (7.51) and taking the ratio of the ac to dc signal gives

$$\left.\frac{S}{S_{dc} - S_n}\right|_{P = P_{n0}} = \frac{\delta\mathscr{R}}{\mathscr{R}} - \frac{2\delta A_n}{(A - A_{n0})}, \tag{7.52}$$

where S_{dc} is the dc part of (7.51) and has been corrected for residual null signal S_n. Similarly, with $A = A_{n0}$

$$\left.\frac{S}{S_{dc} - S_n}\right|_{A = A_{n0}} = \frac{\delta\mathscr{R}}{\mathscr{R}} - \frac{2\,\delta P_n}{(P - P_{n0})}. \tag{7.53}$$

Taking the ratios in (7.52) and (7.53) can be done either by an analog computer during the experiment, or by direct calculation from measured ac and dc readings. Two readings each of the ratios of (7.51) and (7.52) with A and P set on opposite sides of A_{n0} and P_{n0}, respectively, will give maximum difference in the right-hand terms in these equations, and will allow solution for $\delta\mathscr{R}/\mathscr{R}$, δA_n and δP_n with maximum precision for a given set of experimental conditions.

Equations relating ψ and Δ to P_n and A_n have been given earlier. The modulation-induced changes $\delta\psi$ and $\delta\Delta$ are found by differentiating these equations, taking proper care that the equations governing the correct zone are employed in the process.

The oblique-incidence modulated reflectance signal $\delta\mathscr{R}/\mathscr{R}$ constitutes an additional experimental parameter and, therefore, may be employed in calculation of quantities that would otherwise need to be assumed or independently measured, such as film thickness.

It is now necessary to compute $\delta\mathscr{R}/\mathscr{R}$ in terms of the Fresnel reflection coefficients of the sample R_p and R_s and their ratio $\tan\psi e^{i\Delta}$. Assume the ellipsometer to be in the configuration of Fig. 7.12 with the compensator set for quarter-wave retardation and its fast axis at an angle $\pi/4$ to the plane of

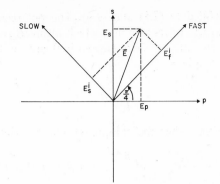

Fig. 7.13. Polarization directions through the compensator set at $\pi/4$, with plane-polarized field $\bar{E}$ incident.

incidence. This situation is shown schematically in Fig. 7.13, where the light beam direction is out of the paper. If E_p and E_s are the complex amplitudes of the parallel- and perpendicular-polarized components, respectively, of the electric fields incident on the compensator, a coordinate transformation yields

$$E_i^f = \frac{1}{\sqrt{2}}(E_p + E_s) = \frac{1}{\sqrt{2}}\left(\frac{E_p}{E_s} + 1\right)E_s$$
$$E_i^s = \frac{1}{\sqrt{2}}(E_s - E_p) = \frac{1}{\sqrt{2}}\left(1 - \frac{E_p}{E_s}\right)E_s$$

(7.54)

where E_i^f and E_i^s are defined as for (7.1) and (7.2). The transmitted complex amplitude ratio of a general waveplate is given by [2, 4]

$$\frac{E_t^f}{E_t^s} = Te^{i\beta}\left(\frac{E_i^f}{E_i^s}\right)$$

(7.55)

where $0 \le T \le 1$, and β is the angular retardation of the slow axis with respect to the fast. We will assume the compensator to be ideal, that is, $T = 1$ and $\beta = \pi/2$. Then, inverting the coordinate transformation (7.54) gives

$$\frac{E_{pi}}{E_{si}} = -\frac{1 + i\gamma}{1 - i\gamma},$$

(7.56)

where

$$\gamma = \frac{1 + E_p/E_s}{1 - E_p/E_s},$$

(7.57)

and E_{pi} and E_{si} are the complex electric field amplitude in the p and s directions, respectively, which are transmitted by the compensator and incident on the specimen, as defined in (7.5) and (7.6). The polarizer produces plane polarized light regardless of its azimuth, which means E_p/E_s and γ are purely real numbers. Thus, the ratio of the intensities in the two polarizations, I_p and I_s, incident on the sample is

$$\frac{I_p}{I_s} = 1 \tag{7.58}$$

for all polarizer settings. Then the oblique reflectance is just

$$\mathscr{R} = \frac{R_s R_s^* I_s + R_p R_p^* I_p}{I_s + I_p} = \tfrac{1}{2}(|R_p|^2 + |R_s|^2). \tag{7.59}$$

Equation (7.59) can be written in terms of $\tan^2 \psi$ and either R_p or R_s. We choose to retain R_s, whereby we obtain

$$\mathscr{R} = \tfrac{1}{2}|R_s|^2 (\tan^2 \psi + 1),$$

and

$$\frac{\delta\mathscr{R}}{\mathscr{R}} = \frac{\delta(|R_s|^2)}{|R_s|^2} + 2 \tan \psi \, \delta\psi. \tag{7.60}$$

To summarize, the experiment described above provides the usual parameters of ellipsometry, ψ and Δ. In addition three experimental parameters, $\delta\psi$, $\delta\Delta$, and either $\delta(|R_p|^2)/|R_p|^2$ or $\delta(|R_r|^2)/|R_s|^2$, are obtained from measurements on the observed ac signal at different polarizer and analyzer settings. The calculation of these sample parameters is independent of how the sample optical properties are perturbed and requires only that this perturbation be time periodic.

Optical Theory of Modulation Mechanisms

This section deals with the transformation of the experimental data ψ, Δ, $\delta\psi$, $\delta\Delta$, and $\delta\mathscr{R}/\mathscr{R}$, all of which are measured as functions of photon energy $\hbar\omega$, into the dielectric constant $\tilde{\epsilon} \equiv \epsilon_1(\hbar\omega) - i\epsilon_2(\hbar\omega)$ and the changes induced by modulation $\delta\epsilon_1(\hbar\omega)$ and $\delta\epsilon_2(\hbar\omega)$. The quantities ϵ_1, ϵ_2, $\delta\epsilon_1$, and $\delta\epsilon_2$ will be here assumed to be isotropic. Although this condition is not necessarily satisfied in all the experiments to be discussed, it greatly simplifies the analysis. The technique can be generalized to the anisotropic case in certain cases of relatively high symmetry using data from measurements at more than one angle of incidence and/or crystal orientation.

The distinctions between the isotropic cases considered here have to do primarily with homogeneity, variation of ϵ_1, ϵ_2, $\delta\epsilon_1$, and $\delta\epsilon_2$ with distance into the sample. In ascending order of complexity, the various possible isotropic configurations are the following.

Case I: Bulk Sample, Homogeneously Modulated.   This case occurs with a bulk sample whose temperature is modulated. When the analysis is generalized to the anisotropic case, experiments with uniaxial stress [76, 77] also fall into this category. The unmodulated dielectric constant in this case, is given by

$$\tilde{\epsilon} = \epsilon_a \sin^2 \phi \left\{ \tan^2 \phi \left[\frac{1 - \tan \psi e^{i\Delta}}{1 + \tan \psi e^{i\Delta}} \right]^2 + 1 \right\}, \qquad (7.61)$$

where ϵ_a is the dielectric constant of the ambient medium surrounding the sample. Forming the total differential of (7.61) with respect to ψ and Δ yields $\delta\tilde{\epsilon} = \delta\epsilon_1 - i\delta\epsilon_2$, which is solved for $\delta\epsilon_1$ and $\delta\epsilon_2$ by equating real and imaginary parts.

Case II: Bulk Sample with Modulation Only in a Thin Surface Layer.   The theory of modulated ellipsometry measurements for the case where $\delta\epsilon_1$ and $\delta\epsilon_2$ are nonzero only over a short distance from the sample surface has been treated by Buckman and Bashara [78], assuming that $\delta\epsilon_1$ and $\delta\epsilon_2$ are constant over the volume in which they are nonzero. The approximation implicit in this is the replacement of a complicated spatial variation in optical properties by a constant average value. Such an approximation is valid if the total distance over which the average is taken, that is, the depth into the sample to which $\delta\epsilon_1$, $\delta\epsilon_2$ are not negligible, is considerably less than the photon wavelength λ [79–81]. This modulation of a surface layer only occurs where the modulation is an electric field normal to the sample surface and the sample is a highly doped semiconductor or a metal. Complicating the analysis is the fact that electric field modulation, because of its vector nature, is not isotropic even in an isotropic crystal. The analysis below may, however, be adequate for polycrystalline samples, where all orientations of crystal with respect to surface and crystal with respect to electric field are present simultaneously.

The model to be analyzed is shown in Fig. 7.14 and is very similar to the model for a filmed surface discussed earlier. In addition, the film dielectric constant is the same as that of the substrate in the absence of modulation, and is just $\tilde{\epsilon} + \delta\epsilon_1 - i\delta\epsilon_2$ in the presence of modulation. The typical very small values of $\delta\epsilon_1$ and $\delta\epsilon_2$ observed experimentally [75, 78] suggest a Taylor series

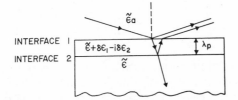

Fig. 7.14.   Optical model for a bulk sample with modulation of a surface layer only.

approach to solution of the problem. Starting with the equation for a filmed surface

$$\tan \psi e^{i\Delta} = \frac{p_1 + (p_2 + p_1 s_2 s_1)e^D + p_2 s_1 s_2 e^{2D}}{s_1 + (s_2 + s_1 p_1 p_2)e^D + s_2 p_1 p_2 e^{2D}} \equiv G_o \qquad (7.62)$$

where

$$D = -4\pi i \frac{\lambda_p}{\lambda} (\tilde{\epsilon}' - \epsilon_a \sin^2 \phi)^{1/2}, \qquad (7.63)$$

λ_p is the effective penetration depth to which $\delta\epsilon_1$ and $\delta\epsilon_2$ are nonnegligible, and $\tilde{\epsilon}'$ is the complex dielectric constant of the surface layer, $\tilde{\epsilon} + \delta\epsilon_1 - i\delta\epsilon_2$. To make the appropriate Taylor series expansion, we form the total differential of $\tan \psi e^{i\Delta}$ for the filmed surface about the point $\tilde{\epsilon}' = \tilde{\epsilon}$, or equivalently $s_2 = p_2 = 0$. These values of s_2, p_2 and $\tilde{\epsilon}'$ are those which obtain in the absence of modulation. The resulting complex equation can be solved directly to yield

$$\delta\tilde{\epsilon} \equiv \delta\epsilon_1 - i\,\delta\epsilon_2$$

$$= \frac{(\delta\psi \sec^2 \psi + i\,\delta\Delta \tan \psi)e^{i\Delta}}{\dfrac{\partial G_0}{\partial D}\dfrac{\partial D}{\partial \tilde{\epsilon}'} + \displaystyle\sum_{i=1}^{2}\left(\dfrac{\partial G_0}{\partial p_i}\dfrac{\partial p_i}{\partial \tilde{\epsilon}'} = \dfrac{\partial G_0}{\partial s_i}\dfrac{\partial s_i}{\partial \tilde{\epsilon}'}\right)}. \qquad (7.64)$$

The required partial derivatives, evaluated at conditions in the absence of modulation are

$$\frac{\partial G_0}{\partial D} = 0$$

$$\frac{\partial D}{\partial \tilde{\epsilon}'} = -\frac{2\pi i d}{\lambda(\tilde{\epsilon}' - \epsilon_a \sin^2 \phi)^{1/2}}$$

$$\frac{\partial G_0}{\partial p_1} = \frac{1}{s_1}, \frac{\partial G_0}{\partial s_1} = -\frac{p_1}{s_1^2}$$

$$\frac{\partial G_0}{\partial p_2} = \frac{(1 - p_1^2)e^D}{s_1}$$

$$\frac{\partial G_0}{\partial s_2} = \frac{-p_1(1 - s_1)^2 e^D}{s_1^2}. \qquad (7.65)$$

For a unique determination of $\delta\tilde{\epsilon}$ from (7.64), it is apparent that D and hence λ_p must be known. Usually a fit for λ_p can be obtained by calculating $\delta\mathcal{R}/\mathcal{R}$ or the normal incidence modulated reflectance $\delta R/R$ (if this has been measured) for a range of λ_p using the $\delta\tilde{\epsilon}$ obtained from (4.64) for each assumed

λ_p, and comparing with experiment [78, 82]. Thus, a set of measurements gives ϵ_1, ϵ_2, $\delta\epsilon_1$, $\delta\epsilon_2$, and λ_p.

The validity of the above optical models depends on the relationship of λ_p to the penetration depth of the photons. If the photon penetration depth is much less than λ_p, the second interface in Fig. 7.14 has no appreciable effect on ψ and Δ, and homogeneous modulation as in Case I can be assumed. If the opposite is true, the model discussed in Case II is valid providing $\lambda_p \ll \lambda$. In the intermediate case where $\lambda \approx \lambda_p$ and the photon penetration depth is comparable to or larger than both of these, a more complicated model would be required which takes account of the explicit variation of $\delta\tilde{\epsilon}$ with distance into the sample. The effect of such a variation on normal-incidence $\delta R/R$ has been calculated by Aspnes and Frova [83]. No similar calculation of the effect on $\delta\psi$ and $\delta\Delta$ is available at this writing.

Case III: Thin Film Samples.   Another parameter, the film thickness d, enters into the calculation and must be either measured independently or eliminated by multiple ellipsometer measurements at several angles of incidence or with several ambient media with widely varying ϵ_a. It will be seen later that measurements at multiple angles of incidence are generally needed to resolve sample anisotropies. Further, measurements with several different ambient media can complicate interpretation by altering surface conditions in unknown ways. Hence, an independent thickness determination by interferometry or other means is recommended, if this is feasible.

We again assume that the modulation is inhomogeneous and nonzero only to a depth λ_p into the film-substrate system. Subject to the limitation discussed in Case II, we can replace the space dependence of $\delta\tilde{\epsilon}'$ by an average value across λ_p. The resulting optical model is shown in Fig. 7.15 for the case $\lambda_p < d$. Equation (7.64) is still valid; however, the partial derivatives, evaluated under conditions holding in the absence of modulation, are different. Specifically, p_2 and s_2 are not zero, due to the presence of interface 3. The quantities p_2 and s_2 must be calculated by assuming a film of dielectric constant $\tilde{\epsilon}$, of thickness $d - \lambda_p$ on a substrate of infinite thickness and with dielectric constant $\tilde{\epsilon}_s$, using the equation for a filmed surface. It should be noted that for this calculation, the "ambient medium" has the same dielectric constant as the film, $\tilde{\epsilon}$.

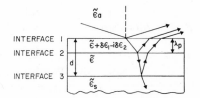

Fig. 7.15.   Optical model for a thin-film sample with $\lambda_p < d$.

Where $\lambda_p > d$, possible modulation of $\tilde{\varepsilon}_s$ by the perturbation must also be taken into account. Analogy with conventional ellipsometer measurements on thin films suggests that a reference measurement of the effect of the modulation on a bare substrate be used to estimate $\delta\tilde{\varepsilon}_s$, the change in substrate parameters. The denominator of the right-hand side of (7.64) is then modified to include partials of p_2 and s_2 with respect to $\tilde{\varepsilon}_s$.

Another possible complication in modulated ellipsometry for studies of films is modulation of d. In electric field modulation of insulating films the two oppositely charged surfaces of the film will be forced toward each other. Thermal expansion can introduce thickness modulation in thermal-modulation experiments. Holden and Ullman [84] and Frova and Migliorato [85] have observed normal incidence electroreflectance signals from tantalum oxide films due to thickness modulation. The contribution of thickness modulation should be separable from that of $\delta\tilde{\varepsilon}'$ because the wavelength dependence is simply that of an interferometer. The contribution to the signal from thickness modulation should be maximum at wavelengths near interference peaks, and hence measurements on several films of various thicknesses should make possible separation of the response due to thickness modulation.

A summary of the influence of finite λ_p, modulation-induced anisotropy, thickness modulation, and substrate modulation on the various types of experiment is given in Table 7.2. Consideration of the properties of the system to be studied and the properties of each kind of experiment can aid in the selection of the type of experiment to be done.

Under certain conditions where the crystal is highly symmetric with respect to the reflecting surface, sample anisotropies can be included in the optical models of any of the above cases with only a slight increase in the mathemati-

Table 7.2. Characteristics of Modulation Types

Type	λ_p	Additional Anisotropy?	Modulation of d?	Modulation of Substrate
Electric field				
Bulk	Finite	Along $\vec{E}$	No	—
Film	Finite	Along $\bar{E}$	Possible[a]	Depends on λ_p
Piezo-stress				
Bulk	∞	Along stress	No	—
Film	d	Along stress	No	Always
Thermal				
Bulk	∞[b]	No	No	—
Film	∞	No	Possible	Always

[a] For insulating films.
[b] For samples of reasonably high thermal conductivity.

cal complexity of the analysis. The conditions for high symmetry with respect to the reflecting surface are illustrated in Fig. 7.16. The principal axes of the crystal (or both crystals in the case of an interface between two anisotropic media) are aligned with the axes of a coordinate system whose axes are:

(*a*) the surface normal;

(*b*) the normal to the plane of incidence;

(*c*) the intersection of the plane of incidence with the surface.

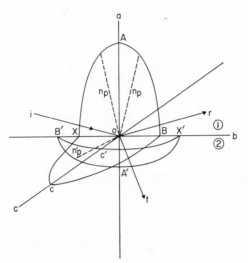

Fig. 7.16. Reflection at a surface between two anisotropic media, under highly symmetric surface conditions. The *b-c* plane is the surface plane and the *a-b* plane is the plane of incidence.

In this case [86] the incident ray *i*, the reflected ray *r*, and the transmitted ray *t* all lie in the plane of incidence. For treatment of this problem using a complex dielectric constant tensor, both dielectric constant and electrical conductivity must have the same principal axes. Inspection of Fig. 7.16 shows that the *p*-polarized component of *i*, *r*, or *t* is equivalent to the extraordinary ray and the *s*-polarized component to the ordinary ray. Therefore, if XAB and XCB are taken to be the intersections of the index ellipsoid [87, 88] of medium 1 with the plane of incidence and surface plane, respectively, and $X'A'B'$ and $X'C'B'$ the same intersections for medium 2 the following observations can be made.

1. The index of refraction seen by the *p*-polarized components of *i* and *r* is given by the length of the line n_p, drawn perpendicular to *i* and *r*.

2. The index of refraction seen by the *p*-polarized component of *t* is given by the length of the line n_p'.

3. The index of refraction seen by the *s*-polarized components of *i* and *r* is given by the length of the line *OC*, and that seen by the *s*-polarized component of *t* is given by the length of the line *OC'*.

4. Finally, for any interface, Snell's law is obeyed for the ordinary ray, while a different refractive law holds for the extraordinary ray [86].

For anisotropic media, the concept of polarization should refer to the direction of $\bar{D}$, not $\bar{E}$. However, since the incident and reflected rays see the same refractive index for this symmetric case, the reflection coefficients for $\bar{D}$ will be the same as those for $\bar{E}$. Hence, all that is required is to rewrite the Fresnel coefficients *p* and *s* for each interface in the system, taking account of anisotropy, and these will combine to form the effective R_p and R_s in the same way as for isotropic media. The refracted angles ϕ will be different for *p*- and *s*-polarized components, as will the refractive indices. The determination of the additional parameters requires more experimental measurement. For an ellipse, if two points in any one quadrant are known, the major and minor axes can in principle be determined. Hence for the determination of the lengths *OA* and *OB*, which are $\epsilon_a^{1/2}$ and $\epsilon_b^{1/2}$, respectively, measurement at two angles of incidence are required. For determination of *OB* and *OC*, hence $\epsilon_b^{1/2}$ and $\epsilon_c^{1/2}$, a rotation of the sample through 90° about the surface normal interchanges ϵ_b and ϵ_c in the equations and again allows determination. For uniaxial crystals, one of the operations of sample rotation or angle of incidence alteration may produce a degenerate dielectric tensor,

$$[\tilde{\epsilon}] = \begin{bmatrix} \tilde{\epsilon}_a & 0 & 0 \\ 0 & \tilde{\epsilon}_b & 0 \\ 0 & 0 & \tilde{\epsilon}_c \end{bmatrix},$$

and hence would not contribute to resolution of the anisotropy.

In principle, computation of $\delta\tilde{\epsilon}_a$, $\delta\tilde{\epsilon}_b$, and $\delta\tilde{\epsilon}_c$ proceeds as before, by forming the total differential of $\tan \psi e^{i\Delta}$. Three complex equations of the type

$$e^{i\Delta_j}(\delta\psi_j \sec^2 \psi_j + i \, \delta\Delta_j \tan \psi_j) = \left(\frac{\partial G_0}{\partial \tilde{\epsilon}_a}\right)_j \delta\tilde{\epsilon}_a + \left(\frac{\partial G_0}{\partial \tilde{\epsilon}_b}\right)_j \delta\tilde{\epsilon}_b + \left(\frac{\partial G_0}{\partial \tilde{\epsilon}_c}\right)_j \delta\tilde{\epsilon}_c,$$

$$(7.66)$$

where $j = 1, 2, 3$ represents three nondegenerate experimental geometries. Nondegeneracy simply means that the partial derivatives and the left side of (7.66) are sufficiently different from each other for each different *j* to yield independent equations. In practice, however, the equations may still be degenerate within experimental errors, rendering the above method invalid.

Experimental

The steps in a modulated ellipsometry experiment have been outlined in the preceding sections. Here we wish to discuss in more detail aspects of sample preparation and application of the modulation. The considerations governing these facets of the experiment are not very different from those of normal incidence modulated-reflectance experiments. They have recently been reviewed thoroughly by Seraphin [89] and only the major points will be considered here.

Electric Field Modulation

Three experimental methods have been used. The most popular, largely because of its experimental convenience is the electrolytic method [75, 78, 90–92]. The sample and a counterelectrode are immersed in an electrolytic solution and an ac voltage between the electrodes sets up a very strong (up to 10^7 V/cm) electric field at the sample surface. This field drops to zero over a distance λ_p into the sample, where λ_p depends on the sample resistivity. A dc bias voltage is usually also applied between the sample and counterelectrode in order to set the value of the sample surface potential about which modulation due to the ac voltage takes place. Modulating voltages of the order of 1 V or less are found to produce phototube signals sufficient to be recovered from the noise.

If we want to study the clean sample surface, as is usually the case, the electrolyte, bias voltage, and modulating voltage magnitude must be carefully chosen to avoid chemical reactions and adsorption at the sample surface [93]. A measurement of test cell capacitance versus bias voltage [93] is a possible check on film formation at the sample surface, and on the actual value of the surface potential of the sample. Standard ellipsometer measurements over a period of time also serve as a check on film formation. A measurement of the dc bias at which the experimental signal goes through zero is an optical check on the zero of surface potential [92], since $\delta\tilde{\varepsilon}$ does not depend on the sign of the electric field.

The field-effect [94–96] method of electric field modulation has the advantages inherent in all all-solid-state system: absence of chemical reactions, workability at low temperatures, and absence of adsorption. A thin dielectric layer (Mylar or an evaporated film of dielectric material) is placed over the sample surface, and the field is applied between the sample and a transparent conducting electrode (i.e., SnO_2) on the opposite side of the insulating layer. Although these overlayers do not seem to appreciably affect the normal-incidence modulated reflectance signal, their effect on ellipsometric measurements, as seen from the preceding discussion, can be quite complicated. For low-temperature work and for samples that readily react or adsorb, however, it may be the only applicable method. Care must be taken to eliminate or take account of modulation of the thickness of the insulating layer by capacitive forces. Use of thin-film dielectrics seems to alleviate this problem [89]. Applied voltages necessary to produce a signal increase with increasing dielectric thickness, as would be expected.

Transverse electric field modulation is applicable only to samples of very high resistivity in the plane of the surface. Here the field is applied between two conducting electrodes typically a few tenths of a millimeter apart on the sample surface [97–98].

Piezo-Modulation

For these experiments, the sample is rigidly mounted on a piezoelectric transducer which is made to periodically strain the sample in one direction [77]. The analysis is simplified by the fact that the sample surface remains relatively clean and free of overlayers. However, Garfinkel et al. [77] report several difficult technical problems associated with this experiment. The vibrating transducer can introduce synchronous vibrations in the dynode structure of the phototube, giving rise to a spurious electric signal at the modulation frequency. Also, lateral motion of the light beam after reflection, caused by sample vibration, will give rise to a spurious signal if there is any nonuniformity in transmission through optical elements or in sensitivity over the photocathode surface. Fortunately, as with thickness modulation in the electric field experiments, these errors are wavelength independent, and possibly can be separated from the data.

Thermal Modulation

Thermal modulation of the sample optical properties is usually accomplished by passing a current pulse through either the sample itself or a strip heater behind the sample [99, 100]. Although this method is free from mechanical vibrations, the need for overlayers in front of the sample, or modulation-induced anisotropy, the results are somewhat more difficult to interpret [89]. Also, the need to rapidly dissipate large amounts of heat for every modulation cycle can pose experimental problems.

In addition to these characteristics, which are true for normal incidence as well as ellipsometric measurements, modulated ellipsometry generally requires more care in sample preparation. Surface roughness on a large scale, such as that caused by etching of semiconductors not properly cleaned beforehand, must be avoided. Such roughness, in addition to scattering the incident light and degrading experimental sensitivity, causes variation in the polarization state of the reflected light from point to point in a cross section of the beam. The effect of lateral variation in polarization state of the light beam is to make the minima in the $\sin^2$ terms in Eq. (7.40) shallower which, in turn, degrades, sometimes very drastically, the sensitivity in the modulated measurements by making $\partial S/\partial P_n$ and $\partial S/\partial A_n$ much smaller.

Further, polarization measurements at oblique incidence are much more sensitive than normal incidence reflectance measurements to surface conditions such as contaminating films. This can be either advantageous or disadvantageous depending on whether or not surface conditions are included in the analysis of the data. It may not be necessary that the surface be ideal, but at least it must be reproducibly controlled.

Results

As an illustration of the kind of results to expect, we shall consider modulated ellipsometry measurements on Au and Ag polycrystalline films using electric field modulation [75, 78, 82]. Since the films used were thick enough to be opaque, and since Thomas-Fermi screening in noble metals leads us to expect electric field penetration depths of the order of tenths of angstroms, the results are analyzed according to the model of Case II.

The electrolyte used was 1 N KCl. The modulating voltage was a 21-Hz sine wave with a dc bias such that the sample went sinusoidally from 0 to -1.5 eV with respect to the Pt foil counterelectrode. Several ellipsometer measurements at the same photon energy during a typical run showed no significant changes in ψ or Δ over the time it took to make the run. This suggests that if a surface film was in the process of forming, by chemical reaction or adsorption, during the run, its optical properties were not such that it could be detected by ellipsometry. Therefore, a surface contamination layer was not included in the model for analysis.

The modulated ellipsometric data for Au taken at an angle of incidence of 70° are shown in Fig. 7.17. Even in the raw data, the spectral structure is quite pronounced.

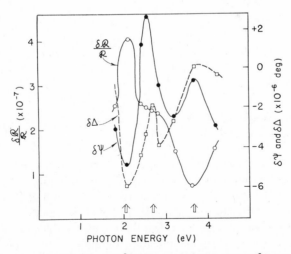

Fig. 7.17. The raw data $\delta\psi$, $\delta\Delta$, and $\delta\mathcal{R}/\mathcal{R}$ versus photon energy for an opaque, polycrystalline Au film. Electric field modulation with $\phi = 70°$ (from [75] with permission).

The reduced data for both Au and Ag are shown in Figs. 7.18 and 7.19. In both cases, a fit for λ_p was obtained by comparison of our calculated normal incidence $\delta R/R$ for an assumed λ_p with the experimental $\delta R/R$ of Feinleib [91]. Requiring only order of magnitude agreement between experimental and calculated $\delta R/R$ restricts possible values of λ_p to $0.1\ \text{Å} \le \lambda \le 0.9\ \text{Å}$ for both metals.

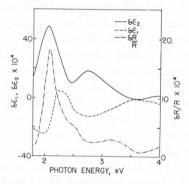

Fig. 7.18. Electric field induced $\delta\epsilon_1$, $\delta\epsilon_2$ and computed $\delta R/R$ versus photon energy for polycrystalline, opaque Au (from [78] with permission).

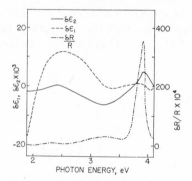

Fig. 7.19. Electric field induced $\delta\epsilon_1$, $\delta\epsilon_2$ and computed $\delta R/R$ versus photon energy for polycrystalline, opaque Ag (from [78] with permission).

Because of the presence of plasma as well as interband transitions in the photon energy range studied, and because of the great inhomogeneous broadening of the spectral peaks due to spatial variation of the electric field, only tentative assignments of the origins of the peaks can be made. However, neglecting line shape and considering only the signs and locations in energy of the peaks strongly suggests identification of some of them on the basis of Aspnes' theory [101–103] as interband transitions. The identification of other peaks is even less clear.

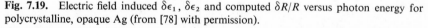

The accuracy with which these small signals can be measured is approximately $\pm 20\%$ for the lower level signals. Greater accuracy is obtained at the strong peaks.

Discussion

We are concerned here with exposition of a general experimental method, rather than interpretation of specific results. Therefore the points to be discussed concern general characteristics of the experiment. As with studies of modulated reflectance at normal incidence, the purpose of modulated ellipsometry is the measurement of small perturbations in $\tilde{\epsilon} = \epsilon_1 - i\epsilon_2$ caused by an external modulation, as functions of photon energy $\hbar\omega$.

The intrinsic advantage of modulated ellipsometry is that measurement of ψ, Δ, $\delta\psi$, $\delta\Delta$, and $\delta\mathcal{R}/\mathcal{R}$ at a single photon energy and angle of incidence provides four more parameters than does a similar measurement of $\delta R/R$ at normal incidence. The use of these additional parameters has been illustrated in the preceding sections for determination of λ_p and for determination of $\delta\epsilon_1$ and $\delta\epsilon_2$ independently of each other, that is, without recourse to the Kramers'-Kronig analysis. These quantities must either be determined by another Kramers-Kronig analysis of the measured absolute reflectance, with all the stability and other experimental problems inherent in such a measurement, or by ellipsometer measurements of ψ and Δ. If ellipsometry is to be used to obtain ϵ_1 and ϵ_2 anyway, the incorporation of modulated ellipsometry is easily accomplished, since it only requires reading ac signals accompanying the dc phototube output.

The availability of more information per measurement can be exploited to an even greater extent in studies of more complicated systems such as thin films, ordered overlayers, surfaces with contaminating films, and anisotropic samples or modulation.

Balancing these advantages is the fact that a measurement of ψ, Δ, $\delta\psi$, $\delta\Delta$, and $\delta\mathcal{R}/\mathcal{R}$ at one wavelength and one angle of incidence actually involves taking readings at four different combinations of polarizer and analyzer settings. Thus, modulated ellipsometry data must be taken point by point, while modulated reflectance experiments are amenable to automatic recording as λ is swept through the spectrum.

Conclusion

Modulated ellipsometry has been discussed here from the point of view of applications in the study of energy band structure where modulated reflectance measurements have found a great deal of use. However, potential future applications for the method exist in areas of perturbation effects on surface reactions, diffusion of impurities into surfaces, and studies of surface states. The only requirement is that the small changes of ψ and Δ due to perturbations can be made periodic in time.

Acknowledgments

NMB is indebted to John A. Johnson for many helpful discussions on Section 2. ABB is happy to recognize helpful suggestions from Donald G. Schueler on points of interpretation of reflectance measurements in Section 5.

NMB and ABB express their appreciation to the U.S. Office of Naval Research and the National Science Foundation for support of their research in ellipsometry in the Electrical Materials Laboratory, University of Nebraska. The work of the Laboratory is also aided by unrestricted grants-in-aid from the 3M Company and the Mobil Research and Development Corporation.

ACH thanks the Mobil Research and Development Corporation for permission to publish part of this manuscript.

* Added in Proof. The following papers, which either have been published or are in press, provide added information on topics covered in the chapter.

1. D. W. Berreman, " Reflectance and Ellipsometry when Submicroscopic Particles Bestrew a Surface," *J. Opt. Soc. Am.*, **60**, 499 (1970). An important addition to the literature on partial surface coverage and surface roughness.

2. R. M. A. Azzam and N. M. Bashara, " Unified Analysis of Ellipsometry Errors Due to Imperfect Components, Cell-Window Birefringence and Incorrect Azmuth Angles," and " General Treatment of the Effect of Cell Windows in Ellipsometry," *J. Opt. Soc. Am.*, **61**, 600, 773 (1971).
    In these two papers, sources of error are analyzed and procedures for their correction are presented.

3. J. A. Johnson and N. M. Bashara, " Multiple Angle of Incidence Ellipsometry of Very Thin Films," *J. Opt. Soc. Am.*, **61**, 457 (1971). Multiple angle of incidence measurements are utilized to reduce random errors in measurements on very thin films where experimental sensitivity is relatively low.

4. J. A. Johnson and N. M. Bashara, " General Equations of Symmetrical Ellipsometer Arrangements," *J. Opt. Soc. Am.*, **60**, 221 (1970). Advantage is taken of the symmetry inherent in the arrangement where the compensator appears before or after the specimen to present the general equations for determining the ellipsometer parameters Δ and ψ from the experimental measurements.

5. L. G. Holcomb and N. M. Bashara, " Measurement of Elasto-optic Effect in Absorbing Materials by Ellipsometry," *J. Opt. Soc. Am.*, **61**, 608 (1971). The theory of piezoreflectance by ellipsometry is presented and advantages of the technique are discussed in experiments on silver.

References

1. P. Drude, *The Theory of Optics*, C. R. Mann and R. A. Millikan, Trans., Dover, New York, p. 258; P. Drude, *Wiedem. Ann.*, **64**, 159 (1898).
2. D. A. Holmes, *J. Opt. Soc. Am.*, **57**, 466 (1967).
3. R. J. Archer, *J. Opt. Soc. Am.*, **52**, 970 (1962).
4. W. G. Oldham, *J. Opt. Soc. Am.*, **57**, 617 (1967).
5. F. A. Jenkins and H. E. White, *Fundamentals of Optics*, McGraw-Hill, New York, 1957, p. 488.
6. H. G. Jerrard, *J. Opt. Soc. Am.*, **38**, 35 (1948).
7. D. A. Holmes, *J. Opt. Soc. Am.*, **54**, 1115 (1964).
8. D. A. Holmes, *J. Opt. Soc. Am.*, **54**, 1340 (1964).
9. A. Vasicek, *Optics of Thin Films*, North Holland, Amsterdam, 1960, p. 31.
9a. A. C. Hall, "Recent Developments in Ellipsometry," N. M. Bashara, A. B. Buckman, and A. C. Hall, Eds., *Surface Sci.*, **16**, (1969).
9b. P. H. Smith, in [9a].
9c. R. H. Muller, in [9a].
10. F. L. McCrackin, E. Passaglia, R. R. Stromberg, and H. L. Steinberg, *J. Res. Nat'l. Bur. Stds.*, **67A**, 363 (1963).
11. A. B. Winterbottom, "Optical Studies of Metal Surfaces," *The Royal Norwegian Sci. Soc. Rpt. No. 1*, F. Bruns, Trondheim, 1955, p. 149.
11a. C. A. Fenstermaker and F. L. McCrackin, in [9a].
11b. D. K. Burge, J. M. Bennett, R. L. Peck, and H. E. Bennett, in [9a].
12. A. B. Winterbottom, "Ellipsometry in the Measurement of Surfaces and Thin Films," E. Passaglia, R. R. Stromberg, and J. Kruger, Eds., *Nat'l. Bur. Stds. Misc. Publ. 256*, U.S. Gov't. Printing Office, Washington, 1964, p. 97.
13. H. Takasaki, *J. Opt. Soc. Am.*, **51**, 463 (1961).
14. J. L. Ord and B. L. Wills, *Appl. Opt.*, **6**, 1673 (1967).
15a. H. G. Jerrard, "A High Precision Photoelectric Ellipsometer," in [9a].
15b. I. Wilmanns, in [9a].
15c. J. L. Ord, in [9a].
15d. B. D. Cahan and R. F. Spanier, in [9a].
15e. H. P. Layer, in [9a].
16. D. K. Burge and H. E. Bennett, *J. Opt. Soc. Am.*, **54**, 1428 (1964).
17. H. E. Bennett and Jean M. Bennett, *Physics of Thin Films*, G. Hass and R. E. Thun, Eds., Academic, New York, 1967, Vol. IV, p. 1.
18. A. C. Hall, *J. Opt. Soc. Am.*, **55**, 911 (1965).
18a. H. G. Jerrard, "Sources of Error in Ellipsometry," in [9a].
18b. F. Lukes, in [9a].
18c. F. Meyer and G. A. Bootsma, in [9a].
18d. R. H. Muller, R. E. Steiger, G. A. Somorjai, and J. M. Morabito in [9a].
18e. J. J. Carroll and A. J. Melmed, in [9a].
19. F. L. McCrackin and J. P. Colson, in [12, p. 61].

19*a*. H. Yokota, H. Sakata, M. Nishibori, and K. Kinosita, in [9*a*].

19*b*. H. Yokota, M. Nishibori and K. Kinosita, in [9*a*].

19*c*. R. J. King and M. J. Downs, in [9*a*].

19*d*. B. J. Bornong, in [9*a*].

19*e*. C. J. Dell'Oca and L. Young, in [9*a*].

19*f*. C. L. McBee and J. Kruger, in [9*a*].

19*g*. L. A. Weitzenkamp, in [9*a*].

20*a*. N. M. Bashara and D. W. Peterson, *J. Opt. Soc. Am.*, **56**, 1320 (1966).

20*b*. D. W. Peterson and N. M. Bashara, *J. Opt. Soc. Am.*, **55**, 845 (1965).

21. J. A. Johnson and D. W. Peterson, in [9*a*].

22. R. W. Ditchburn and G. A. Orchard, *Proc. Phys. Soc. (London)*, **67B**, 608 (1954).

22*a*. T. R. Young and J. M. Fath, in [12, p. 349].

23. V. A. Kizel', *Soviet Phys. Usp.*, **10**, 485 (1968).

24. R. W. Wood, *Physical Optics*, Macmillan, New York, 1934, p. 415.

25. A. Vasicek, *J. Opt. Soc. Am.*, **47**, 565 (1957).

26. A. Vasicek, *J. Opt. Soc. Am.*, **37**, 979 (1947).

27. F. P. Buff, *1966 Saline Water Conversion Report*, U.S. Government Printing Office, Washington, D.C., p. 26.

28. D. V. Sivukhin, *Soviet Phys. JETP*, **3**, 269 (1956).

29. C. Strachan, *Proc. Cambridge Phil. Soc.*, **29**, 116 (1933).

30. S. Nomura and K. Kinosita, *J. Phys. Soc. Japan*, **14**, 297 (1959).

31. Lord Rayleigh, *Phil. Mag.*, **33**, 1 (1892).

32. A. Vasicek, *Czech. J. Phys.*, **4**, 204 (1954).

33. J. H. Frazer, *Phys. Rev.*, **33**, 97 (1929).

34. E. Hofmeister, *Z. Physik.*, **36**, 137 (1953).

35. C. Bouhet, *Ann. Phys.*, **15**, 5 (1931).

36. V. A. Kizel', *Soviet Phys. JETP*, **26**, 228 (1954).

37. R. C. Bacon, Dissertation, Stanford University, 1933.

37*a*. J. W. McBain, R. C. Bacon, and H. D. Bruce, *J. Chem. Phys.*, **7**, 818 (1939).

38. Lord Rayleigh, *Phil. Trans. Roy. Soc. (London)*, **176**, 343 (1885).

39. R. J. Archer, *Ellipsometry*, Gaertner Scientific Corporation, 1968, p. 7.

40. Lord Rayleigh, *Phil. Mag.*, **16**, 444 (1908).

41. O. Lummer and K. Sorge, *Ann. Phys.*, **31**, 325 (1910).

42. C. V. Raman and L. A. Ramdas, *Phil. Mag.*, **3**, 220 (1927).

43. V. A. Kizel', *Soviet Phys. JETP*, **2**, 520 (1956).

44. K. Kinosita and H. Yokota, *J. Phys. Soc. Japan*, **20**, 1086 (1965).

45. H. Yokota, *J. Phys. Soc. Japan*, **21**, 200 (1966).

46. V. A. Kizel' and A. F. Stepanov, *Soviet Phys. JETP*, **4**, 458 (1957).

47. F. P. Buff and R. A. Lovett, in *Simple Dense Fluids*, H. L. Frisch and Z. W. Salsburg, Eds., Academic, New York, 1968, p. 17.

48. L. Holland, *The Properties of Glass Surfaces*, Wiley, New York, 1964.

49. H. Yokota, K. Kinosita, and H. Sakata, *Japan. J. App. Phys.*, **3**, 805 (1964).

50. Lord Rayleigh, *Proc. Roy. Soc. (London)*, **156A**, 507 (1937).

51. R. W. Ditchburn, *J. Opt. Soc. Am.*, **45**, 743 (1955).

52. F. Abeles, Ed., *Optical Properties and Electronic Structure of Metals and Alloys*, Wiley, New York, 1966.

53. A. Vasicek, in [12, p. 25].

54. I. N. Shklyarevskii, V. P. Kostyuk, and V. R. Karas, *Opt. Spectr.*, **23**, 76 (1966).

54a. R. C. MacLaurin, *Proc. Roy. Soc.* (*London*), **77A**, 211 (1906).

55. W. Meier, *Ann. Phys.*, **31**, 1017 (1910).

56. C. A. Reeser, *Arch. Neerland*, **6**, 225 (1923).

57. J. Ellerbroek, *Arch. Neerland*, **10**, 42 (1927).

58. L. Tronstad and C. G. P. Feachem, *Proc. Roy. Soc.* (*London*), **A145**, 115 (1934).

59. R. M. Emberson, *J. Opt. Soc. Am.*, **26**, 443 (1936).

60. L. G. Schulz, *J. Opt. Soc. Am.*, **47**, 64 (1957).

61. L. G. Leyluk, I. N. Shklyarevskii, and R. G. Yarovaya, *Opt. Spectr.*, **16**, 484 (1964).

62. L. E. Smith and R. R. Stromberg, *J. Opt. Soc. Am.*, **56**, 1539 (1966).

63. T. Smith, *J. Opt. Soc. Am.*, **57**, 1207 (1967).

63a. R. C. Smith and M. Hacskylo, in [12, p. 83].

64. See [1, p. 293].

65. A. C. Hall, *J. Phys. Chem.*, **70**, 1702 (1966).

66. T. Smith, *J. Opt. Soc. Am.* **58** 1069 (1968).

67. R. J. Archer in [12 p. 259].

68. C. G. P. Feachem and L. Tronstad *Proc. Roy. Soc.*, **A145**, 127 (1934).

69. J. H. Frazer, *Phys. Rev.*, **34**, 644 (1929).

70. S. Silverman, *Phys. Rev.*, **34**, 644 (1930).

71. K. Kinosita, K. H. Kojima, and H. Yokota, *Japan. J. App. Phys.*, **33**, 2089 (1962).

72. W. Bayh and H. Pflug, *Z. Angew. Phys.*, **25**, 358 (1968).

73. J. R. Miller and J. E. Berger, *J. Phys. Chem.*, **70**, 3070 (1966).

74. Y.-C. Chiu and M. A. Genshaw, *J. Phys. Chem.*, **72**, 4325 (1968).

75. A. B. Buckman and N. M. Bashara, *J. Opt. Soc. Am.*, **58**, 700 (1968).

76. U. Gerhardt, *Phys. Rev.*, **172**, 651 (1968).

77. M. Garfinkel, J. J. Tiemann, and W. E. Engeler, *Phys. Rev.*, **148**, 695 (1966).

78. A. B. Buckman and N. M. Bashara, *Phys. Rev.*, **174**, 719 (1968).

79. R. Jacobsson, *Progress in Optics*, E. Wolf, Ed., North-Holland, Amsterdam, 1965, Vol. V, p. 247.

80. P. Drude, *Wiedem. Ann.*, **43**, 26 (1891).

81. L. Rayleigh, *Proc. Roy. Soc.*, **A86**, 207 (1912).

82. A. B. Buckman, in [9a].

83. D. E. Aspnes and A. Frova, *Solid State Comm.*, to be published.

84. B. J. Holden and F. G. Ullman, *J. Electrochem. Soc.*, **116**, 280 (1969).

85. A. Frova and P. Migliorato, *Appl. Phys. Letters*, **13**, 328 (1968).

86. L. Landau and E. Lifshitz, *Electrodynamics of Continuous Media*, Pergamon, New York, 1960, pp. 323–324.

87. G. N. Ramachandran and S. Ramaseshan, *Handbuch der Physik*, S. Flügge, Ed., Springer-Verlag, Berlin, 1961, Vol. XXV and references cited therein.

88. M. Born and E. Wolf, *Principles of Optics*, 3rd rev. ed., Pergamon, New York, 1964, pp. 665 ff.

89. B. O. Seraphin, "Modulated Reflectance" in *Optical Properties of Solids*, F. Abeles, Ed., North-Holland, Amsterdam 1969.

90. K. Shaklee, J. Pollack, and M. Cardona, *Phys. Rev. Letters*, **14**, 1069 (1965).

91. J. Feinleib, *Phys. Rev. Letters*, **16**, 1200 (1966).

92. Y. Hamakawa, P. Handler, and F. Germano, *Phys. Rev.*, **167**, 709 (1968).

93. B. O. Seraphin, *Surface Sci.*, **8**, 399 (1967).

94. B. O. Seraphin, R. B. Hess, and N. Bottka, *J. Appl. Phys.*, **36**, 2242 (1965).

95. B. O. Seraphin, *Phys. Rev.*, **140**, A1716 (1965).

96. R. Ludeke and W. Paul, *II–VI Semiconducting Compounds*, 1967 Intern. Conference, Providence, D. G. Thomas, Ed., Benjamin, New York, pp. 123–135.

97. V. Rehn and D. Kyser, *Phys. Rev. Letters*, **18**, 848 (1967).

98. E. W. Williams and V. Rehn, *Phys. Rev.*, **172**, 798 (1968).

99. B. Batz, *Solid State Commun.*, **4**, 241 (1966).

100. C. N. Berglund, *J. Appl. Phys.*, **37**, 3019 (1966).

101. D. E. Aspnes, *Phys. Rev.*, **147**, 544 (1966).

102. D. E. Aspnes, *Phys. Rev.*, **153**, 972 (1967).

103. D. E. Aspnes, P. Handler, and D. J. Blossey, *Phys. Rev.*, **166**, 921 (1968).

INDEX